AF440110

The Science of
Weather and Environment

The Science of
Weather and Environment

Navale Pandharinath

M.Sc. (Maths), M.Sc. (Statistics)

Director (Retd.),
Indian Meteorological Deparment,
Hyderabad.

BSP BS Publications

A unit of **BSP Books Pvt., Ltd.**

4-4-309/316, Giriraj Lane, Sultan Bazar,
Hyderabad - 500 095
Phone : 040 - 23445605, 23445688

Published by :

BS Publications

A unit of **BSP Books Pvt., Ltd.**

4-4-309/316, Giriraj Lane, Sultan Bazar,
Hyderabad - 500 095
Phone : 040 - 23445605, 23445688
e-mail : info@bspbooks.net

ISBN: 978-93-52300-38-9 (HB)

Dedicated to My Parents

Late Smt. Narasubai R. Navale
Late Shri. Rukmaji Rao Narsoji Rao Navale

Preface

The book is primarily written as Weather Science (Meteorology), however in view of its wide application in environmental studies, the basic relevant subject matter of environment has been included. It is a specialised subject and is taught at Undergraduate, Postgraduate levels and it is now contemplated to be introduced at junior college level also.

The introduction to meteorology traces its history and development and provides the wide applicability in other science subjects like Hydrology, Agriculture, Aviation, transportation, trade, human health etc. Non-meteorological natural disasters, in general, are abetted in their fury by the meteorological parameters. Similarly in case of air pollution, water pollution, weather parameters play vital role in promoting or dissipating the events. The first few Chapters deal with the solar energy input, atmospheric constituents, radiation laws, water in the atmosphere, surface wind, visibility, atmospheric pressure, clouds, aerosols, precipitation processes, thunderstorms, hailstorms, lightning, tornadoes and water spouts. Basic principle of Radar and its application in detection of clouds, thunderstorms, tornadoes and tropical disturbances have been presented. Subsequent chapters dealt with important aspects of tropical meteorology–easterly waves, westerly waves, jet-streams, Bienneal oscillations, ENSO, Monsoons, Tropical cyclones, together with natural meteorological hazards. Application of artificial Satellite based imagery data in detection of tropical cyclones, its movement, storm surges and Disaster Warning System through INSAT have been discussed. Climate, climate classification by Koppen, Thronthwait, Budykov, geological time scale and Indian climatology discussed.

Giving a break in weather events, non-meteorological natural disasters–Earthquakes, Avalanches, Landslides, Tsunamis are briefly presented. Turning to man made hazards–Air pollution, Water pollution, Sound pollution, Soil contamination, Sewage, Acid rains, Antarctic Ozone hole, hardness, softness of water, water purification, water famine are briefly discussed. Towards the end, defining Environment, Ecosystem, movement of Energy in Ecosystem, Food Chain, Productivity, Evolution, Forest wealth, Wildlife, Domestic animals,

Conservation of natural resources are briefly dealt. Keeping in view of the ever increasing demand of energy, renewable natural energy source of Wind Energy is presented briefly. Finally a Chapter on Aeronautical Meteorology has been introduced, which is the real source for about a century for the development of meteorology as a science. At present, there is an explosive development in Commercial Aviation activity in India, which is facing acute shortage of pilots. It is, in this context, the last chapter has been added to serve the needs of trainee pilots.

I am sure, this book will serve as Text/Reference to Undergraduates and Postgraduates, Colleges/Universities and scientific community.

Author

Acknowledgements

In writing this book, which spread over a period of decade after retirement, I collected the relevant material from various sources, published papers, books, particularly WMO publications and IMD forecasting manuals, Met-monographs, IMD Observational Organisation standard brief, Cyclone tracks, Climatological Atlas etc. Really speaking I lost many of the source references from which material collected. Though I have given a few references which are incomplete. I express my sincere thanks to the Director General of Meteorology, India Meteorological Department. Also express my indebtedness to S/Shri. Dr. Y.P. Rao, K. Rama murthy, G.S. Mandal, G.R. Gupta, K. Veeraraghavan, U.S. Desai, V.S. De, G.C. Asnani, M. Krishnamurthy, G. Appa Rao for using their published material.

I am greatly indebted to Dr. R.R. Kelkar, Director General of Meteorology (Retd). India Meteorological Department for his valuable suggestions and encouragement in writing books on meteorology. Lastly, I express my indebtedness to Shri Nikhil Shah, Proprietor, B.S Publications for patronage and to the staff specially K.S. Raju, Laxminarayana, Naresh, and Sandhya for their untiring help in bringing this book.

Any suggestions for the improvement of the book and corrections shall be thankfully acknowledged.

Author

Contents

Introduction to Meteorology

Meteor meaning a shining body temporarily appearing in the sky or a shooting star. The word meteorology stands for the science of weather. The phenomena of weather and climate have universal jurisdiction and they do not owe fealty to any man or man made institution. Weather is a subject which is of interest to every-one whether rich or poor, learned or illiterate and all through the year. From the dawn of civilization man has been dependent on weather. The very basis of flourishment of ancient civilizations was favourable weather and their downfall was adverse weather. Ancient Egyptian and Indus valley civilizations are examples of this kind. The tasks of prescientist, the priests and sooth sayers of ancient civilizations of Mesopotamia, Chaldaea, India and China were much devoted to weather. In some of the early religions, specific deities presided over the weather phenomena and claimed control of weather and its modification. In ancient India, Indra was the god of the rain. In Vedas and in later mythology, Indra was the regent of the atmosphere and of the east quarter. His world was called Swarga. Thunder, lighting and rain was attributed to his work. Indradhanush (rainbow) was his weapon. Varuna was the regent of the oceans and of the western quarter. He was represented with a noose in hand. In ancient China (fourth century BC) the Taoist genie Liu Thien Chun was the Comptroller-General of Crops and Weather. Long before the origins of this rain-maker and weather forecaster the Chinese had produced agrometeorological texts. In India, one thousand years before Christian era, the Rigveda dealt on the meteorology of the Punjab and north-west India. In the pantheon of the pre-Colombian deities (weather god) the rain-maker occupied an important place.

In the middle of the fourth century BC the study of atmospheric environment began in the western world. The subject of weather as a science began in Greece with Aristotle's Meteorologica. Based on weather observations and reasons his pupil and successor, Theophrastus compiled a small treatise on wind and signs of weather. This work remained conclusive (definitive) for about two thousand years. Meteorologica came to medieval European knowledge through Arabs, who preserved that Greek work. During this period, based on the work of the two Greeks, Dante described about weather in La Divina Comedia. An early raingauge was described in India by a Rishi date back to

400 BC was a simple bowl (cylinder) about 45 cm diameter and the Rishi suggested that sowing seeds should be regulated according to the amount of rain water registered in the bowl.

Brief History Development of Modern Science of Weather

The air thermometer was invented about 1600 AD by Santorio (however Galileo claimed to be its originator). Although rainguages were in use for many centuries earlier in China, India, Korea and Palestine but the modern rainguage was invented by Castelli in 1639. In about 1644 AD, Evangelista Torricelli invented the barometer. At about this time various forms of hygrometer and anemometer appeared. In 1644 an English physicist Robert Hooke invented pressure tube anemometer to measure wind pressure. In 1659, Robert Boyle discovered the relationship between pressure (p) and volume (V). In 1735, Hadely established a relationship between trade winds and the rotation of the earth. In 1752, Benjamin Frankline studied the atmospheric electricity. In 1783, Lavoisier and in 1800 Dalton laid physical basis of meteorology.

In 1780, the Societas Meteorologica Palatina (SMP) started (in Mannhiem) with 39 weather observatories (14 in Germany, 4 in USA and the rest in other countries). These observatories had calibrated instruments of barometer, thermometer, hygrometer and some with windwanes and rainguages. The first international compilation was made by Lamarck and the first weather map based on the data from SMP was made by H.W. Brandes in Leipzig in 1820. At the same time W.C. Redfield in NewYork prepared a series of charts of hurricanes. Between 1820-50 it was established the existence of characteristic patterns of pressure, wind associated with depression and Anti-cyclones and weather patterns, together with empirical rules for their development, movement and the sequence of weather changes. All these investigations were done only after the events but no weather forecast was made.

The invention of electric telegraph by Samuel Morse in 1843 revolutionised the possibilities of weather forecasting particularly storm warning. In 1850, based on telegraphic reception of weather data first weather map was prepared, publicly displayed in Washington D.C. and in France in 1855. Industrial Revolution in western Europe caused rapid strides in science and technology and promoted international trade through ships. Maritime transportation security demanded precise, reliable, regular weather information. To meet this demand of maritime meteorological problems, the First International Meteorological Conference was held in Brussels in August 1853. This conference standardised the instruments, Marine Met Observations and exchange cooperation. Meteorological institutes came into being and felt the need of terrestrial exchange of weather observations. As meteorology began to take universal character, standard symbols, standardisation of observations become a necessity. To meet this necessity a preparatory conference of meteorologist held at Leipzig on 14 August 1872. World's foremost meteorologist met at the conference and reached an agreement on standardisation of methods of observation and analysis and use of single set of symbols.

The First International Meteorological Congress was held in Vienna from 2-16 September 1873. It was attended by 32 representatives of 20 Governments. The Vienna Congress was a milestone in the history of international cooperation in meteorology and reflected its universal nature. The congress created a permanent seven member committee. It laid down standardised instructions and procedures for weather observations on land and universal telegraphic weather code. On 14 April 1879, the Second International Meteorological Congress met in Rome and it was attended by 40 meteorologists representing 18 countries. It formed an International Meteorological Committee (IMC) with nine members which replaced the 7 member permanent committee. The members were non-governmental weather experts. IMC continued the international cooperation in meteorology for a period of about 70 years. The first International Polar Year 1882-83 entrusted the IMC certain tasks which is continuing. IMC published or encouraged the publication of meteorological studies and reports. IMC brought out International Meteorological Tables and instructions on the observations cloud movements and arranged conference on Agricultural and Forestry Meteorology in Austria (in September 1880). IMC which mainly comprised of European meteorologists gradually became defunct. In its place First Conference of Directors Meteorological Services (DMS) was held in Munich on 26 August 1891 and was attended by 31 DMS which include 4 from USA, two from Australia, and one from Brazil. The conference elected an International Meteorological Committee (IMC) comprising 14 members, who created first permanent Technical Commission. The organisation had no funds and IMC members surved in a personal capacity- out on voluntary basis. IMC worked till the break of first world war 1914. IMC prepared and published International Cloud Atlas in 1896. Under the commission of aeronautics, great advances were made in collecting data in the free atmosphere. During 1914 -18 international meteorological cooperation practically ceased, but rapidly revived in 1919 with the Conference of Directors held at Paris (in September 1919).

Before the First World War meteorology was in infancy stage, during 1914 -19 it was in adolescence state. However great developments were made in Radio and Aviation. Application of meteorology to air navigation came into being. The status of International Meteorological Organisation (IMO) and its role in international affairs was recognised until the outbreak of Second World War. Again IMO activities were subdued for five years. Permanent secretariat came into being in Netherlands, which was under the control of IMC president. Conference of Directors stressed the need to convert the organisation into inter-governmental. In 1935 at Warsaw Conference of Directors, Commission for the Application of Meteorology to Air navigation was replaced by International Commission for Aeronautical Meteorology (ICAeM) an inter-governmental organisation. This was done due to rapid development of Civil Aviation.

During inter war period, Technical commission system was introduced for international cooperation in meteorology. The commission covered the following subjects or activities. Terrestrial Magnetism, Atmospheric Electricity, Solar Radiation, Exploration of the Upper atmosphere, Synoptic Weather Information, Maritime Meteorology, Agricultural Meteorology, Application of Meteorology to Aerial Navigation, Investigation of Waves of Explosion, Study of Clouds, the Polar Year, the Reseau Mondial and Polar Meteorology and Climatology.

In Norway, the Bergen school developed methods of weather study pertaining to air mass and frontal analysis. Initially, in the early 1900s, upper-air sounding was carried using Aeroplane but Aerometeorographs came into existence in 1920-30. Subsequently these were replaced by Radiosonde (in France - 1927) and a practical radio transmitter was developed in USSR in 1930.

WMO and its Activities

IMO, the non-governmental organisation which guided international meteorological cooperation since 1873 (Vienna Congress) ceased to exist in 1951. It was replaced by WMO (World Meteorological Organisation) an inter-governmental organisation. It is a specialised agency of the United Nations. The transformation of IMO into WMO amply shows the practical importance of meteorology. The aims of WMO were to facilitate world-wide cooperation in establishing network of weather observatories, to promote the development of centres capable of providing the services of meteorology, to promote rapid exchange of weather information, standardisation of meteorological observations and their publication, to help forward the application of weather science to human activities and to encourage research and training in meteorology. The conference in October 1946, signed the WMO convention by 31 governments but it came into force on 23rd March 1950. This date is being celebrated annually as World Meteorological Day. On 17 March 1951 a conference of Directors of IMO held its last meeting and formally dissolved the Organisation. On 19 March 1951 the first Congress of WMO opened in Paris.

The science of meteorology is one of the important sciences and the meteorologists have an important role to play in nation building or human affairs. WMO had 30 members when the WMO convention came into force (23 March 1950). The membership rose to 83 countries in 1955 at the Second Congress held at Geneva. It steadily rose to 111 members at the Fourth Congress held in Geneva by realising its importance.

Uniform meteorological practices achieved with WMO Technical Regulations. A World Climatic Atlas was prepared, for radio-teletypewriter and facsimile method used for communication purpose. The scheme of Northern Hemispheric Meteorological Telecommunication provided for RTT transmission stations in Frankfurt, New Delhi, Tokyo, Moscow and New York. This helped in preparing the hemispherical weather maps for weather forecasting. WMO participated with more than 100 meteorologists in the International Geophysical Year (IGY) which was one of the largest global science research programmes ever undertaken. Under the IGY the studies of the earth's interior, its crust and oceans, the atmosphere and the sun were undertaken. By (1963) the time of Fourth Congress of WMO, more than 3000 voluntary weather observing ships were providing meteorological information from the world's oceans and seas. A number of WMO technical notes on various meteorological topics were published. These technical notes are of great help not only to meteorologists but to agriculturists (Agrometeorology) aviators (Aeronautical meteorology), health organisation (Atmospheric pollution) and IMCO (maritime meteorology). WMO provided technical assistance service with education and training to many developing countries.

The advent of weather satellites (global observation), electronic computers (data exchange and processing), World Weather Watch (WWW) came into existence which was recongised by United Nations General Assembly (Resolution 1721 C (XVI) in 1963.

At the beginning of the centenary year the membership of the organisation rose to 136. The WMO decided to set up three basic commissions: (i) Commission for Basic System (CBS), (ii) Commission for Instruments and Methods of Observation (CIMO) and (iii) Commission for Atmospheric sciences (CAS). Further decided to have five application commissions. They are (i) Commission for Aeronautical Meteorology (CAeM), (ii) Commission for Agricultural Meteorology (CAgM), (iii) Commission for Marine Meteorology (CMM), (iv) Commission for Special Applications of Meteorology and Climatology (CoSAMC) and (v) Commission for Hydrology (CHy). WMO continued to produce a flow of Technical Notes which include, Guide to meteorological instruments and observing practices. The WMO official languages are: English, French, Russian and Spanish. As a specialised agency of UN, WMO established collaboration with other members of UN family, in particular FAO, IMCO, IAEA, ICAO, ITU, UNESCO and WHO. WMO provided the essential input of meteorological information to increase world food production, to improve shipping and marine safety, to benefit aviation and increase air safety, to improve communication networks for the dissemination of meteorological information. WMO collaborated with UN Regional Economic Commission on water resources development and management.

Fifth WMO Congress (held at Geneva in 1963) approved World Weather Watch (WWW) plan and Global Atmospheric Research Programme (GARP) in celebration with International Council of Scientific Unions (ICSU).

The objectives of WWW was: (i) A Global Observing System (GOS) which includes observational networks and other observational facilities (weather satellite data etc.,) (ii) A Global Data Processing System (GDPS), which contains meteorological centres (for processing of data and archival), (iii) Global Telecommunication System (GTS), which comprises of telecommunication facilities and arrangements for rapid exchange of observations and processed data, (iv) A research programme (GARP) and (v) a programme in education and training. By 1972, there were 8500 surface observatories, 5500 merchant ships, plus ocean weather ships, commercial aircraft and weather satellites for global coverage of data. GDPS works through World Meteorological (Met) Centres (WMC), regional Met Centres and National Met centres. The WMC were established in Melbourne, Moscow and Washington, which provide analysis and prognosis on global basis. The regional Met centres (there were more than 20 by 1972 end) prepare detailed analyses and prognosis for their regions. This includes hurricanes and typhoon warnings.

In 1960 and 1970 much of meteorological aspects of agriculture was concerned to the problem of food production, that is in support of world campaign against hunger. In 1971, WMO brought out Guide to agricultural meteorological practices. During the introduction of supersonic flight, WMO collaborated with ICAO and contributed for the increasing safety and efficiency of civil aviation. At the end of 1972, WMO contributed for the collection of basic hydrological network, such as standardisation of instruments and installations, methods of observation and processing of data. UN General Assembly requested WMO in its Resolution 2733 (XXV) to take up Tropical Cyclone Project to mitigate its disastrous effects to save human lives and to reduce the ravages caused by tropical cyclones in different parts of the world. The interaction of man with environment led to the study of environmental pollution which WMO closely worked with WHO, IAEA and other organisations.

The Role of Meteorology in Economic Development

Weather and climate knows no political boundaries and have all pervading influence on man and his economic development. Climate determines the natural vegetation suitability of a locality for human habitation, and also provides the information of abundance or lack of water resources. On a shorter scale weather determines the most suitable period of various farming operations, dispersal of pollutants in atmosphere from factories, economic operation of a dam, use of gas/coal in our domestic life.

The basic needs of a man are food, clothing, housing and personal health and happiness. These can be considered as human factors, while material factors are agriculture, industry, trade, water resources management, insurance. Man's greatest challenge in present time is to feed the growing population. This can only be achieved by climate based agriculture.

Food

Man must eat to live. Food comes from plants or animals, the later are dependent on plants. Both are governed by the complex influence of weather and climate. The photosynthesis process in plant is dependent on solar energy. Ever since man began to cultivate the land and domesticate his animals he took into consideration of weather. However, recently during green revolution and freedom from hunger campaign, meteorology began to play its important role in food production, transportation,storage and consumption. The use of sound knowledge of meteorology leads to prosperity and avoids disasters.

Clothing

The essential aim of clothing is protection from the extremes of heat and cold and civilised living. Clothing must provide protection to maintain core body temperature of about 37 °C or else it may be a contributory factor for ill health. Traditional clothing is built up on weather and availability of materials.

Housing

A house creates its own internal climate which should provide comfort to the inhabitants. Like clothing traditional styles, sites and methods of housing have evolved based on climatic conditions but modern housing of multistoried complexes do not meet the comforts without air conditioners. Again these complexes must bear the heat and wind loads. Actually Indian Vastu Shastra is nothing but application of climate in house building.

Health

Although bad food, poor clothing, unsuitable houses are all contributory factors for bad health but weather itself is a factor in many ways for the spread of diseases. Epidemics were always weather based. Though practically there is no rapport between meteorologist and doctors still we know that many doctors advise some patients for climate based health resorts.

Recreation

Man cannot work the entire day, he needs some form of leisure or relaxation. This is particularly required in those parts of the world where the climate is highly variable. There are very few areas where the weather is reliable in its behaviour. In most of the places man requires a good sense of weather.

Education

Weather is a subject of universal interest and effects every man and nation, a knowledge of the subject is essential to every individual from all walks of life. Fore warned is fore armed, consequently knowledge of weather is essential and it must begin from school. Meteorology as a subject must be introduced in all classes in schools.

The Material Factor

The correct application of the science of weather is an investment in both personal and national fortunes. An old saying is ' the year that bears fruit and not the field' and in India "a good monsoon is a good harvest". These reflect that the nature of crop yield depends more on rainfall rather than on soil. The interface between agriculture and weather is very complex. Weather effects are different one crop to another and one variety to another variety. Some stages in plant life are more weather sensitive than others. The successive stages in life cycle of crop – namely preparation of ground, sowing, germination, cultivation, growth, ripening, harvesting, drying, storage, transportation - are all radically effected by weather conditions. All phases of a agricultural work depends on meteorological factors and these must be correctly assessed for profitable yields. Land use for flourishing activities has to be studied in relation to suitability of climate, soil and the needs of the nation. Land use must be planned in consultation with agrometeorologist to overcome the world of hunger, shelter from wind and drifting sand, irrigation in places of deficient rainfall or drought, and protection against frost etc. should be done. It may be noted that irrigation helps to increase production even in places where rainfall is reliable and high. With advanced technology the ability of man to resist climate change has increased. He is manipulating to overcome adverse weather conditions for habitation and food production, however he cannot control weather beyond a limited extent.

Great revolution has been made in building materials for construction and succeeded in changing indoor climate. Losses of food in storage on an average is about 20% of the total production and these losses mostly related to weather conditions during growth, harvest, transport to storage, pests and diseases. Though pests and diseases are not of meteorological nature yet their existence are practically controlled by weather and climate. Pests are most serious in tropics while diseases are more in temperate conditions.

Trades and Industry

It looks trade has little to do with meteorology except with respect to indoor climate. But certain commodities demand increases with weather. Food and clothing sales are influenced by weather. Large firms are conscious of weather based demands. True assessment of weather factors greatly helps in reducing critical losses. Meteorological assistance is of great value and help in industrial planning and operation stages.

Importance of Water and its Resource Management

World's fresh water resources constitute about 5.8% of the global water, while seas and oceans contain 94.2%. The amount of fresh water that man actually uses is about 0.3% of the hydrosphere or 5.2% of fresh water, but this plays an important role in human life and economy. Because of this insignificant proportion of fresh water the hydrosphere economists profess that a water famine threatens mankind. However the danger of water famine may arise only from inefficient use of water resources than from lack of water. Completely on use, water remains as water but may change its state (vapour). All water consumed by human beings and animals is evaporated. The water shortage will be created by pollution of rivers, lakes by animal wastes, municipal and industrial sewage. The source for all fresh water on land is rain or snow that falls from clouds and the main loss of water from the earth is evaporation.

Water economy of a country has become one of the most important sectors of its economic activities. A country's economic development largely depends on the amount of geoghraphical distribution of water resources. The global demand for water has increased dramatically over the last century. Water withdrawal from the existing sources increased more than sixfold which was more than double the population growth during the same period. This rise was the result of increased irrigation, industrial growth and rising water consumption in domestic use. Man's average water requirement is about 2.5 liters per day but to improve his living condition he is using 250 to 500 liters for his household needs and 1000 times for industrial consumption.

Water is vital to both plants and animals. Animals contain very high percentage of water in their blood. Man is nearly 70% water by weight. Any loss of water in the body by more than 10% is fetal.

Water resources may be considered in three groups of purpose. (i) Use of water from streams, lakes: power generation, navigation, recreation, preservation of aqualife (fish), wild life and pollution abatement. (ii) Diversion of water for consumptive use: Irrigation, municipal and rural supplies and industrial use. (iii) Excess water damage control: Agriculture drainage, flood control, urban storm drainage, erosion and sedimentation.

Insurance

In modern life of a man insurance is becoming a necessity. Besides life insurance there are insurances for many other things such as goods, house, car, motor cycle etc., marine, and crop insurance. People insure these things against risk, mainly from weather, fire, natural hazards and theft. The rates of insurance generally determined by climate statistics. Insurance companies have to fix the premium which is not too high to discourage and yet reasonably high to present themselves with reasonable profit. Hence insurance companies should consult meteorologist before determining insurance rate. Even in clearing insurance claims the possible contributory cause of loss, accident has to be determined from professional meteorologist.

Weather Disasters

Extreme natural events become disasters when they affect human settlements, economic and social activities. Weather either causes or influences non-weather disasters. According to one study, the world over disasters statistics during 1947-80 indicates that the number of deaths in Tropical cyclones, Hurricanes, Typhoons rank first (about 41%), followed by earthquakes (37%), Floods (16%). Thunderstorms, Tornadoes (2.4%), Snow/storms (1%), Heat waves (0.6%) and Avalanches Landslides, Tidal waves account about 2%. Thus weather hazards account 62%. Non-atmospheric hazard forecasting is very difficult and not yet reached the stage of practical use, while the meteorological hazards may be predicted nearly precise and more practicable. Natural disasters will remain the same in future as at present but a lot more can be done to reduce the deaths and destruction they cause.

Education and Publicity

Education begins at schools and it should continue throughout life. The subject of meteorology taught at schools is practically very little. There are very few text books on the subject and they have very severe limitations. The spread of knowledge or awareness among people is mainly achieved through education and publicity. A lot of publicity takes place only during adverse weather not about the subject but about its adverse effect. Weather is of universal interest yet there are insignificantly few colleges and universities where the weather science is taught. This is a great drawback or injustice to the subject and the consequences are reflected in weather disasters. Weather is a news and must be treated in that perspective. Like news weather is a perishable commodity and it waits for none. In this age of technology the contributory value of meteorology is really great for mankind.

The Sun and the Atmosphere of Earth

Introduction

The naked eye stars in the sky is called celestial sphere. At the dawn of civilization man thought that the earth was the centre of the universe and it was fixed, while the sun, the moon and the planets were moving across the sky. From the astronomical studies it is now understood that the sun is one of the normal sized star in the milky way galaxy. The milky way galaxy itself is rotating around its nucleus in accordance with the law of gravitation. The distance of the sun from the galactic centre is about 30,000 light years (1 light year = distance travelled by the light in one year, speed of light = 3×10^8 m/s). The galaxies are the building blocks of the universe. There are about 10 billion (10^9) galaxies in the universe. The link between the distant stars and the earth is the electromagnetic radiation emitted by them. The light (radiation) emitted by the stars enable us to detect its source of direction, composition, temperature and its velocity towards the earth or away from it (Doppler effect).

According to the Bigbang theory, the universe is expanding. All matter in the universe initially was packed in a compact superdense agglomeration. It was hurled in all direction with a cataclysmic explosion, called BigBang. Since then the mass is moving away from that nuclear centre. The recession of galactic mass depended on the initial velocity. Higher the velocity of matter farther is its recession.

According to Hubbles principle, all galaxies are receding from the earth. The speed of recession is proportional to its distance. No light reaches us (the earth) from the distant galaxies which are away from 20 billion light years. This distance is called the boundary of the observable universe. However the latest astronomical findings that the universe (that is observable universe) is expanding.

The early man believed that the earth was the centre of the universe. The sun, stars and the planets were moving, around it. In 1545 AD, Poland's astronomer Nicholas Copernicus propounded the modern heliocentric theory of the solar system. According to which the sun is at the centre and at one of the foci of the ellipse. All planets, including the earth, are revolving in elliptical orbits around the sun. Keplers laws of planetary motion supported this heliocentric theory. The solar system as a whole is rotating and moving in the milkyway galaxy with a velocity of about 400 km/sec, which is the absolute velocity of the sun. The origin of the solar system is very complex and still obscure.

The Sun

The mass of the sun is about 1.99×10^{30} kg, radius 6.97×10^{8} m (or diameter $\simeq 14 \times 10^{5}$ km), while the mass of the earth is 6×10^{24} kg, radius 6.4×10^{6}m. The gravitational attraction of the sun is about 274 m/sec^2, which is about 25 times that of the earth (gravitation of the earth is 9.81m/s^2). About 99.87% of the mass of the solar system is contained in the sun.

Photosphere

The visible surface of the sun is called photosphere, its temperature is about $6000°$ K, thickness about 161000 km and consists hot gases under high pressure. Photosphere produces radiations of the visible part of the solar spectrum, but the energy is produced in the deeper central region of the sun's sphere. The surface of the photosphere changes constantly without any rhythm. It appears as mottled or granular due to boiling motion of the gases, however the granules are not visible to the nacked eye. They are about a million in number with about 10 minutes life. Besides granules the photosphere shows bright (hot) regions called faculae and dark (cold) regions called sunspots. Sunspots are the regions of strong magnetic field.

Chromosphere

During total solar eclipse a reddish ring of light observed around the sun's photosphere, which is called chromosphere or colour sphere of the sun. The depth of the chromosphere varies 3000 to 5000 km above the photosphere. It consists of hydrogen and helium gases at low pressure but has temperature about 20000 K Fig. 1.1. Fibrilles and spicules are observed in chromosphere. Fibrilles are extended structures approaching granules, located around the centre of the solar disc, while spicules are located at the edge of the disc with life period of about 5 minutes, diameter 750 Km. In chromosphere temperature rises slowly from 4400 K to about 1.5 to 2.0 million K. Between photosphere and chromosphere there is a cooler gas layer with temperature about 4000 K, thickness about 500 km. This is called reversing layer and is responsible for Fraunhofer absorption lines. Above this layer temperature rises.

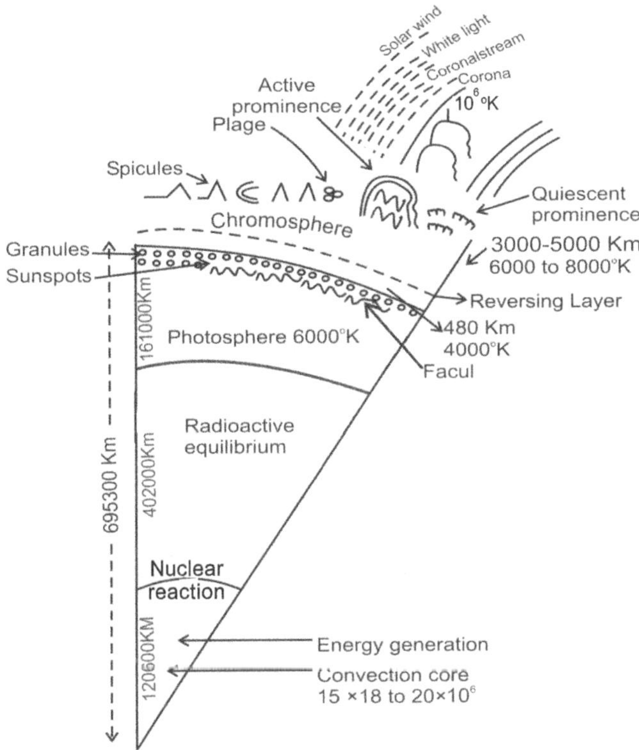

Fig. 1.1 Schematic representation of strcuture and activities of the sun with the exaggerated photosphere, chromosphere.

Corona

Above chromosphere lies corona, which spreads far out into space with temperature 1.5 to 2.0 million K. During total solar eclipse corona shows spectacular light of halo. Solar winds originate in Corona of speed exceeding 100 km/s and the plasma particle of corona exceed the escape velocity of the sun's atmosphere. This is called solar wind, depends on the solar activity. Associated with solar wind, solar mass of about 4×10^8 kg/s escapes from the sun.

Solar Activity

When the frequency of observable features of the sun(such as sunspots, solar wind, solar flares) are more than normal, the sun is said to be "active" or "disturbed" , but when they are abscent the sun is said to be "quiet". Based on the observations of sunspots, the solar activity has a cycle of 11 years. Sunspots begin to appear at latitude 35° on both hemispheres of the sun. This is sunspot minima phase. As the cycle advances the number of spots increase in size and number and spread to the solar equator, which is sunspot maxima phase. Subsequently these spots decrease gradually. Again a new cycle begins with sunspots reappearing at Lat 35°.

Sunspots are regions of cooler temperature and lower energy. The other observable features where energy emission is more than normal are called active regions. A Facula appears before sunspot and persists even after the visible appearance of the sunspot. Faculae are also observed at high latitudes of the sun in the absence of sunspots.

Associated with the sunspot groups in photosphere, increased solar activity is observed in chromosphere. Large bright regions are called plages. Sometimes clouds of bright gas observed projecting outwards from chromosphere which are called Prominences. Rapid changes are noticed in the neighbourhood of prominences which are clearly not known but shows chromospheric bursts, coronal rays and holes.

The magnetic field in sunspots observed to exceed thousand times as compared to the magnetic field outside the sunspots. The polarity of the magnetic field in sun spot area changes in 22 years, which is twice the period of sunspot cycle. There is no clearcut theory of solar activity. No sunspots observed during 1645 to 1715 (70 years) and there does not appear 11 year sunspot cycle before 1645. These facts were brought to the light by E. Maunder, the superintendent of Greenwich observatory in 1890. The 70 year (1645 – 1715) period of sunspot minima or complete absence of it is called Maunder Minima.

Cosmic Rays

The high energy particles of the galactic origin (outside the solar system) are called cosmic rays. The solar activity influences the intensity of cosmic rays that are falling on the earth. Cosmic rays propagate along the galactic magnetic filed and are deviated by the earth's magnetic field (Moon has no magnetic field and has no radiation belts around it). During quiet sun, the solar wind is weak, and as a result solar magnetosphere shrinks and allows the cosmic rays to penetrate to the earth.

The sun's thermal radiation remains constant and is not effected by the solar activity but it effects short wave radiation (λ - less than 0.1 μm).

The effect of solar gravity on the earth's biosphere is insignificant compared to the gravity of the earth itself.

Central Core

The central core of the sun has thickness of about 120600 km and has convection temperature 15 to 20 million K. Above this convective core there is a radiative equilibrium layer of depth 402000 km, temperature ranging 10 million K to 6000 K. Together with convective core and radiative equilibrium region is characterized by nuclear reactions. This is the source region of solar radiative energy.

Solar Flares

The solar flares are the result of gigantic nuclear explosions that occur in the central core. Solar flares are the most spectacular activity of the chromosphere and has a short period of life. It consists of intense eruption and brightness in its neighbourhood for a few minutes. The brightness gradually diminishes and subsides within a few hours and returns to original state. There may be hundred odd flares daily but a few large flares may occur in a year. The large flares emit electromagnetic radiation along with UV, visible, IR, x-rays, radio-waves and high-energy charged particles of plasma. These particles may have velocities as much as 10^5 km/s. The plasma particles consists of 91% ionised helium, which on reaching the earth interacts with atmosphere and earth's magnetic field. It causes ionospheric storm (high frequency radio black out) and magnetic disturbances. The solar wind associated with these flares have velocities 300-800 km/s.

The Heterogenous Rotation of the Sun

The sun itself rotating in forward direction (coinciding with planets revolution) about an axis which makes an angle of about 7° 15' with the normal to the plane of ecliptic and the whole of solar system itself moving in space. The north pole of the sun is seen on earth between 7 June to 7 December and the south pole of the sun is seen during 8 December to 6 June. Unlike the earth, (which has all its particles same angular velocity) the gaseous sphere of the sun rotates a full turn at its equator in 25 days and at its poles in 35 days. This heterogenous rotation of the sun is called differential. The angular velocity of the sun's rotation also changes with depth, however the convection of the sun, its magnetic field and differential rotation are all related and interact with one another.

Characteristics of the Sun's Radiation

In very hot materials electrons (negatively charged particles) are in continuous motion at high speeds, collide with one another and cause vibrations. This gives rise to electromagnetic waves. Very high speed particles vibrate at high speeds produce short waves. Slower particles collision produce long waves. In the sun, the gases which are at high temperature (> 6000 K) produce electrically charged particles with different speeds. As a result of their collision, electromagnetic waves of different wave lengths are emitted by the sun. These waves, constitute the electromagnetic spectrum of the sun (Fig. 1.2), which extends from gamma rays ($\lambda = 10^{-13}$ m) to radiowaves ($\lambda = 10^3$ m). This range is 10^{16} m. However about 99% of the solar radiation lies between (λ = 0.15 – 4.0 μm). Of this about :

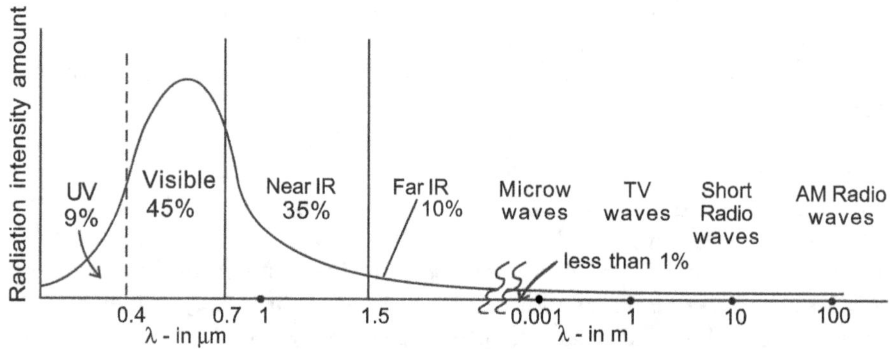

Fig. 1.2 Sun's electromagnetic spectrum not to scale.

(i)   9% lies in UV-radiation (0.15 μm – 0.38 μm): This causes photochemical effects, bleaching and sunburn etc.

(ii)   44% lies in visible light (0.38μm – 0.7 μm i.e., violet to red).

(iii) 46% lies in IR (0.7 to 2.3 μm): This causes radiant heat with some chemical effects.

All these radiations travel with speed of 3×10^8 m/s in vacuum.

According to Wien's law :

$$\lambda_{max} = \frac{Constant}{T} = \frac{2897 \text{ μm K}}{T}$$

or

$$\lambda_{max} \simeq \frac{3000 \text{ μm K}}{T}$$

Sun emits energy at average surface temperature of 6000 K and the earth at its average surface temperature of 288 K

For the sun : $\lambda_{max} = \dfrac{3000 \text{ μ m K}}{6000 \text{ K}} = 0.5$ μm.

That is, the sun emits maximum amount of radiation energy at wave length λ max = 0.5 μm

For the earth : $\lambda_{max} = \dfrac{3000\ \mu m\ K}{\cdots\cdots\cdots} \simeq 10\ \mu m$ [taking 288 K $\simeq$ 300 K]

That is, the earth emits maximum amount of radiation energy at wavelength $\lambda_{max} = 10\ \mu m$.

Because of these facts the terrestrial radiation is called long wave radiation and solar radiation is called short wave radiation.

To Find the Radius of the Sun

Sun's diameter = earth sun distance × angular diameter of the sun

(Arc = Radius × Radian)

Earth-sun distance $\simeq 1.5 \times 10^{11}$ m

Angular diameter of the sun observed from the earth = 9.3×10^{-3} radians

Sun's diameter = $1.5 \times 10^{11} \times 9.3 \times 10^{-3} \simeq 13.95 \times 10^{8}$ m

$\therefore$ the radius of the sun = 6.97×10^{8} m

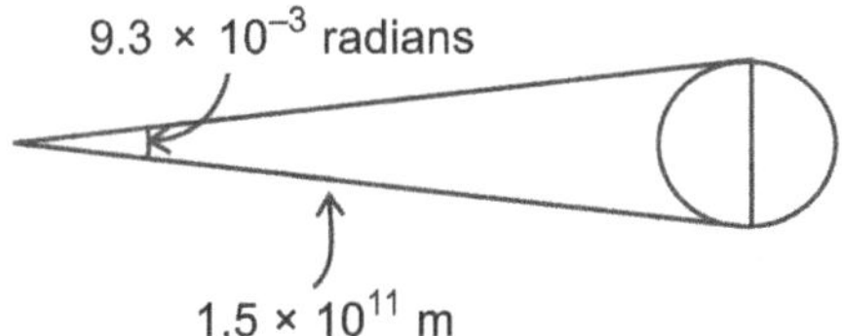

To find the Gravitational Attraction of the Sun and its Density

The mass of the sun = 1.99×10^{30} Kg

Radius of the sun $\simeq 6.97 \times 10^{8}$ m

Universal gravitational constant G = 6.67×10^{-11} N-m^2/kg^2

$$g = \frac{G \cdot m_1 \cdot m_2}{R^2} = \frac{6.67 \times 10^{-11} \times 1 \times 1.99 \times 10^{30}}{[6.97 \times 10^{8}]^2}$$

$$= 274\ \text{m/s}^2$$

density of the sun $= \dfrac{m}{V} = \dfrac{1.99 \times 10^{30}}{\dfrac{4}{3}\pi(6.97 \times 10^{8})^3}$ since $V = \dfrac{4}{3}\pi R^3$

where R = radius of the sun

$$= 1.41 \times 10^{3}\ \text{kg/m}^3$$

Electromagnetic Spectrum

All objects in the universe emit radiant energy as long as their temperature is more than 0 K, but ceases when it is less than 0 K. As the temperature of the object increases it sends more and more energy into space and obeys Stefan Boltzman law viz

$$R \propto T^4$$

Where

R = maximum rate of radiation emitted per unit surface area of the object

T = Temperature of the object in degrees Kelvin

When an object emits and receives energy at the same rate, it is said to be in radiative balance with its surroundings. If it is not in radiative balance either it receives or loses energy depending on its temperature less than or more than its surroundings. By this the object get warms up or cools down.

Radiant energy travels in space/vacuum in the form of electromagnetic waves. Radiant energy waves possess both electric and magnetic properties, consequently they are called electromagnetic waves. Electromagnetic waves travel in all directions away from the source with the speed of light ($C = 3 \times 10^8$ m/s). The wave length λ and frequency f are related by the formula $C = f\lambda$.

Human eyes are sensitive to waves of certain wave length, which are called light waves. Radiant energy waves which travel through space/vacuum as light waves is called visible radiation. Visible radiation spectrum consists of VIBGYOR (Violet, Indigo, Blue, Green, Yellow, Orange and Red) light. The average wave lengths of these are:

Violet light	0.400 μm
Indigo light	0.424 μm
Blue light	0.49 μm
Green light	0.575 μm
Yellow light	0.585 μm
Orange light	0.647 μm
Red light	0.71 μm

The range of all possible radiation wave lengths is called electro magnetic spectrum. The electro-magnetic spectrum consists of Gamma rays, X-rays, ultraviolet (UV) radiation, Visible light, Infrared (IR) radiation, microwaves, short and long radio waves. The wave length (λ) and frequency (f) of these waves are given Table 1.1.

Table 1.1

	Radiation	Wave length λ in meters	Frequency f in Hertzs /second (Hz/s)
1.	Gamma rays emanates in nuclear explosions	$<10^{-10}$	$\geq 10^{18}$
2.	X -rays (Medical use)	10^{-7}	$5 \times 10^{15} - 10^{19}$
3.	UV -light	$0.3 \times 10^{-6} - 10^{-8}$	$10^{15} - 2 \times 10^{16}$
4.	Visible light /violet Red (Rainbow part)	0.4×10^{-6} 0.8×10^{-6}	3.7×10^{14} 10×10^{14}
5.	Principal solar radiation (UV, visible and IR)	$10^{-7}-10^{-5}$	$10^{15} - 10^{13}$
6.	IR (thermal radiation) Max. radiation at room temperature	$0.8 \times 10^{-6} - 10^{-4}$ 8×10^{-6}	$3.7 \times 10^{14} - 10^{12}$ 0.3×10^{14}
7.	Microwaves or very short Radiowaves	10^{-4} - 1	$10^{12} - 10^{8}$
8.	UHF (Ultra High Frequency) TV channels 14 -83	0.1 - 1	$3 \times 10^{9} - 3 \times 10^{8}$
9.	VHF (TV channels 2-13) FE - Radio	1-10	$3 \times 10^{8} - 3 \times 10^{7}$
10.	Short Radio Waves	10 - 100	$3 \times 10^{7} - 3 \times 10^{6}$
11.	Long Radio Waves	200 - 600	$5 \times 10^{6} - 1.5 \times 10^{5}$

Solar Energy

The fusion furnace of the sun converts about 4 million tons (4×10^{9} kg) of hydrogen into helium and radiates an amount of energy 3.8×10^{23} Kw/s as a by product. The solar energy incident on $1m^2$ area held normal to the sun's rays at the outer boundary of the atmosphere in one second is 1.38×10^{3} W/m^2. It is virtually constant and called solar constant (accepted by IGY).

A house with peak load requirement of 2.5 KW would require a solar collector of 3.6 m^2 with 100% efficiency. The total solar energy received at the earth in one day is equivalent to 35 lakh nuclear explosions or 10000 hurricanes or 10^{8} thunderstorms or 10^{11} Tornadoes. If this energy were stored it will satisfy all the world needs of domestic and industrial use for about 100 years.

According to one estimate, the solar radiation received at the outer boundary of the atmosphere is about 1.74×10^{14} KW. Of this about (30%) 5.2×10^{13} KW is reflected back to the space as shortwave radiation. 8.2×10^{13} KW (about 47%) is directly absorbed by the land, oceans and atmospheric system and the remaining 4.0×10^{13} KW (about 23%) is utilised in driving the hydrologic cycle through evaporation, convection and precipitation. About 3.7×10^{11} KW ($\simeq 0.21\%$) of the incident solar energy is utilised (as input into Biosphere) for driving winds, waves, convection and sea currents. 4×10^9 KW (0.023% of the incident solar energy) is used in photosynthesis process on earth. The green canopy of plant kingdom consumes solar energy about (0.01 ly/min) 7 watt m^2. The total solar energy intercepted by the earth = 2.55×10^{18} Cal/min or 3.67×10^{21} Cal/day.

Solar energy per unit area expressed in langley (ly) or Kilo-langley (kly).

$$1 \text{ ly } = 1 \text{ Cal/cm}^2,$$

It is estimated that sun radiates each minute about 56×10^{26} Cal of energy in all.

Definition

Solar Constant (S)

The amount of solar radiation (energy) incident on unit area in unit time on a surface held at right angles to the solar beam at the outer boundary of the atmosphere.

$$\text{The solar constant S} = \frac{56 \times 10^{26} \text{ Cal / min}}{4\pi \, (1.5 \times 10^{13} \text{ cm})^2}$$

$$= 2.0 \text{ ly/min or } 1359 \text{ w/m}^2$$

The total solar energy intercepted by the earth in unit time

$$= \pi \, Re^2 \, S$$

$$= \frac{22}{7} \, (6.40 \times 10^8 \text{ cm})^2 \, S \text{ cal/cm}^2$$

$$= 2.55 \times 10^8 \text{ cal/min}$$

$$= 3.67 \times 10^{21} \text{ cal/day}$$

(where Re = Radius of the earth)

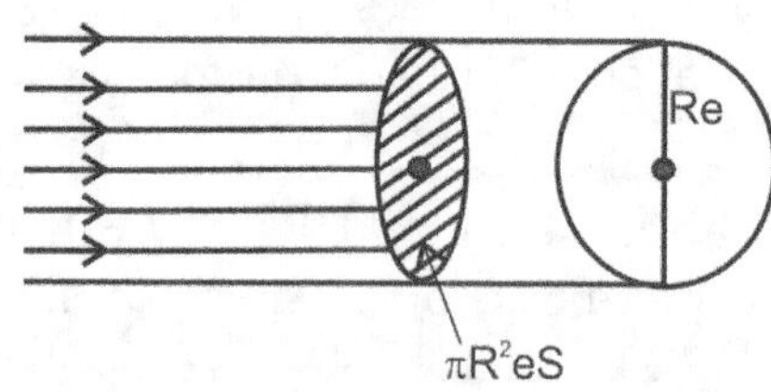

Note :   1 cal/cm^2/min = 0.6975 KW/m^2

2S = 1.3950 KW/m^2

This intercepted energy if spread uniformly over the whole of the earth, the amount of solar energy received on unit area in unit item at the top of the atmosphere (Q_s) is

$$Qs = \frac{\pi Re^2 S}{4\pi Re^2} = \frac{S}{4} = 0.5 \text{ ly/min}$$

$$= 263 \text{ k ly/year}$$

$$= 339.75 \text{ W/m}^2$$

On an average the mean cloud coverage of the earth is slightly more than 50%. Under this condition the distribution of insolation (incoming solar radiation) is given Fig. 1.3

Reflection and back scattering 35% ($\simeq$ 92 kly/yr)

Absorption by ozone 2% ($\simeq$ 5 kly/Yr)

Absorption by clouds, water vapour, dust etc. 20% (53 kly/year)

Absorption by the earth's surface 43% (113 kly/year)

The earth atmosphere together absorbes 65% (171 kly/year)

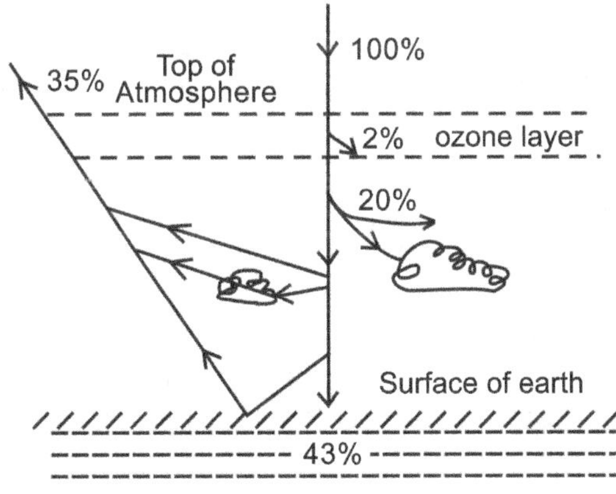

Fig. 1.3 Disposition of insolation in earth atmosphere.

Measurement of Solar Radiation

A simple sunshine recorder (glass ball) burns a card during sunshine and thus gives the number of hours of sunshine in a day. From which monthly, seasonal and annual averages are worked out.

$$\text{Albedo of a surface} = \frac{\text{Global radiation from the sun and sky reflected by the surface}}{\text{Global radiation from the sun and sky incident on the surface}}$$

A number of sophisticated instruments are used for measuring radiation are given below as classified by WMO.

Radiometer (or Actinometer) is a generic term used for any instrument that is used for measuring radiation energy. The basic types of radiometers are 1. Pyrheliometer 2. Pyranometer 3. Pyregeometer 4. Pyrradiometer 5. Net Pyrradiometer.

Pyrheliometer is used for measuring intensity of direct insolation.

Pyranometer is used for measuring global radiation (ie. combined intensity of direct insolation + diffused sky radiation)

Pyregeometer is used for measuring hemispherical long wave radiation of the earth together with reflected atmospheric radiation.

Pyrradiameter (or total hemispherical radiometer) is used for measuring hemispherical long wave radiation together with global radiation.

Net pyrradiometer is used for measuring net all-wave radiation flux

Definition of Some Terms

Terrestrial radiation includes longwave radiation emitted by the earth and its atmosphere.

Total radiation is the sum of solar and terrestrial radiation

$$\text{Albedo of the earth} = \frac{\text{Radiation reflected by the earths surface}}{\text{Radiation incident on the earth's surface}}$$

Insolation

Intensity of direct solar radiation on a horizontal surface or simply incoming solar radiation.

Attenuation of Solar Radiation

It is the loss of solar energy caused by scattering by air molecules or by selective absorption by certain molecules (eg. water-vapour, ozone, CO_2 etc)

Diffuse or Sky Radiation

The solar radiation which is indirectly received at the surface of the earth after being scattered or diffusely reflected by the atmosphere is called sky radiation.

Direct Solar Radiation

Incoming solar radiation from the solid angle of the sun's disc on a surface perpendicular to the axis of the cone, including scattered and unreflected solar radiation in that cone.

Global Solar Radiation

It is the downward direct and diffused solar radiation incident on a horizontal surface.

Insolation at the surface of the earth, in the absence of atmosphere, is a function of (i) solar output (ii) distance from the sun. (iii) altitude of the sun and (iv) the length of the day.

Radiation

All objects irrespective of their size, emit radiation when the object's temperature is more than 0 K. Thus radiation depends on the temperature of the object and it ceases when the object acquires a temperature of zero degree Kelvin (0 K) or less.

A dull or black surface emits more radiation than a silver polished surface. Radiation emitted by a surface is proportional to area of its surface and its temperature. A body at higher temperature emits more radiation than at lower temperature.

A black body (or surface) absorbs more radiation falling on it as compared to a bright polished surface which reflects more. Different surfaces absorb radiant energy falling on them differently (wave lengths).

Abosorptive power (a_λ) of a surface for a certain temperature T, and wave length (λ) is given by

$$a_\lambda = \frac{\text{The amount of radiation absorbed by the surface}}{\text{The amount of radiation incident on its surface (at the same temperature)}}$$

Black Body

A hypothetical body (or surface) which absorbs all the radiation falling on it at any temperature without reflecting or transmitng any part of it is called a black body.

Gray Body

A body which at a certain given temperature emits a fixed proportion of black body radiation at that temperature in all wave lenths is called a graybody.

Note : A White body which scatters the visible radiation falling on it may act as a black body for different wave length of radiation. For example, snow acts as a black body for wave length greater than 1.5 µm. A good absorber is a poor reflector and poor absorber is a good reflector.

Emissive Power or Spectral Density

Radiant energy distributed over a range of wave lengths, when radiation at a particular wave length λ is considered (i.e., in the infinitesimal wave length interval λ to $\lambda + d\lambda$) it is called spectral density or monochromatic.

The emissive properties of a black body is given by a number of laws. A few important laws are given below.

Planck's Law

The spectral density of the radiance of a black body is a function of temperature (T) and wavelength (λ)

Lambert's Cosine Law

The flux per unit solid angle emitted in any direction from a unit radiating surface is proportional to the Cosine of the angle between the direction of radiation and normal to the surface.

Stefan-Boltzman Law

It states that the total emittance of a black-body is proportional to the fourth power of its absolute temperature. This law generally used for estimating temperatures of hot bodies.

Surfaces which absorb more heat than they reflect are called good absorbers. Surfaces which reflect more heat than they absorb are called reflectots.

Wiens Displacement Law

For a given temperature the wavelength of maximum radiance (λ_m) is given by

$$\lambda_{max} = \frac{constant}{T} \text{ or } \frac{0.2898 \times 10^{-12}}{T} \text{ m/K}$$

Where T is temperature in $^{\circ}K$

Kirchhoff's Law

The ratio of the total radiant emittance (e_λ) of a gray body to its absorptance (a_λ) is equal to the total radiant emittance of black body (e_b) at the same temperature.

i.e.,
$$\frac{e_\lambda}{a_\lambda} = e_b$$

Emissivity (ε_λ)

For a particular temperature, wavelength

$$\varepsilon_\lambda = \frac{L_\lambda}{L_{\lambda B}} = \frac{\text{Radiance of a radiating object}}{\text{Black body radiance}}$$

In case of black body $\varepsilon_\lambda = 1$, for gray body $\varepsilon_\lambda = $ constant, which is less than 1.

Atmosphere is a Selective Absorber

We have learnt that incoming solar radiation mostly consists in short waves while terrestrial radiation in IR (infrared) region of long waves. Terrestrial radiation is also called nocturnal radiation. The global IR value is about 258 kly/yr, which is slightly less than insolation value of 263 kly/yr. The surface of the earth with its average temperatutre of 288 K ($\sim$300 K) emits and absorbs radiation energy as a gray body in IR region.

Atmosphere is virtually transparent to insolation (short wave radiation) but it readily absorbs terrestrial (IR long wave) radiation. Atmosphere is thus selectively absorbs and emits radiation, consequently called selective absorber. The selective absorbers are water-vapour (absorbing range 1 to 8 μm and greater than 12 μm), Carbondioxide (CO_2: absorbs at 4 μm and between 13 to 17 μm), Ozone (O_3 : absorbs UV-radiation between 0.2 to 3 μm and at 9.6 μm). Molecular oxygen (O_2 : absorbs UV-radiation below wavelength of 0.2 μm). However, practically no radiation is absorbed by these gases and water vapour in the range of 8 - 11μm. This wavelength range 8 - 11μm is called atmospheric window.

Atmospheric gases constituents CO_2, O_3 and water vapour in the (lower atmosphere) lower troposphere allow the incoming solar radiation to pass through it but prevent the terrestrial radiation to escape to the space as in a glass house (green house). The gases CO_2, O_3, atomic oxygen, methane, N_2O, chlorofluro carbon (CFC's) which prevent terrestrial radiation to escape to the space are called green house gases. In the absence of green house gases effect, the surface of the earth would have been cooler by 30 to 40 °C (average global temperature would have been –20 °C instead of the present 15 °C)

Radiative Balance and Horizontal Transport of Heat

Past available records show that the average temperature of the earth as a whole is about 15 °C, and is virtually constant. This shows that the amount of heat energy that is received by the earth (mainly as short wave radiation) is equal to the amount of heat energy that is lost by the earth (by way of long wave IR - radiation). On the whole the earth-atmosphere system radiates nearly the same amount of energy as it is receiving. However the latitudinal distribution of incoming radiation to the earth and the outgoing radiation from the earth are different (not equal, see Fig. 1.4). The figure shows that near the equator, more energy is absorbed than emitted by the earth. The earth atmosphere system shows surplus energy between latitudes 0 to 35° (area hatched with cross lines), while more energy emitted to space than received between latitudes 35° to 90° indicating deficient energy in these latitudes (area hatched with dots). Since the earth as a whole is neither becoming warmer nor cooler, shows that the excess heat energy absorbed in low latitides is transported to higher latitudes otherwise low latitude regions would have much higher temperatures and regions in mid and higher latitudes would have been much cooler. This transfer of heat from low to higher latitudes is carried out by the action of wind and ocean currents. It has been calculated that ocean currents carry about 30% required heat transport equator to pole wards.

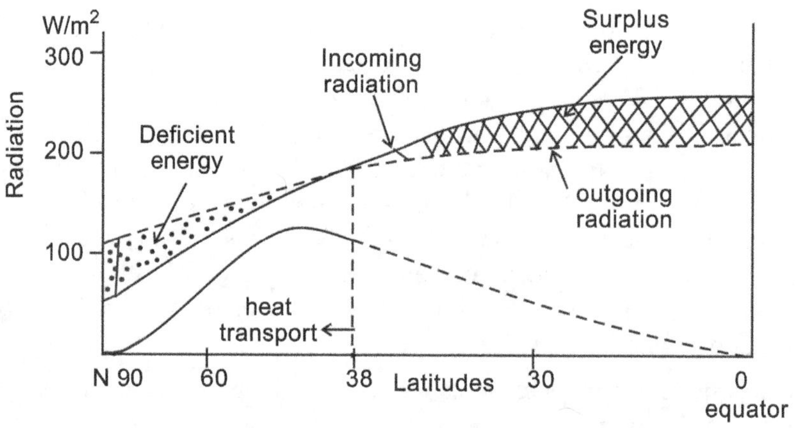

Fig. 1.4 Mean radiation balance northern hemisphere, solid line incoming solar radiation, dotted line outgoing earth radiation.

Solar Battery

Photoelectric cell or solar battery is a device for generating electric power from sunlight Fig. 1.5. It generates power at the rate of 100 watts / m² of illuminated

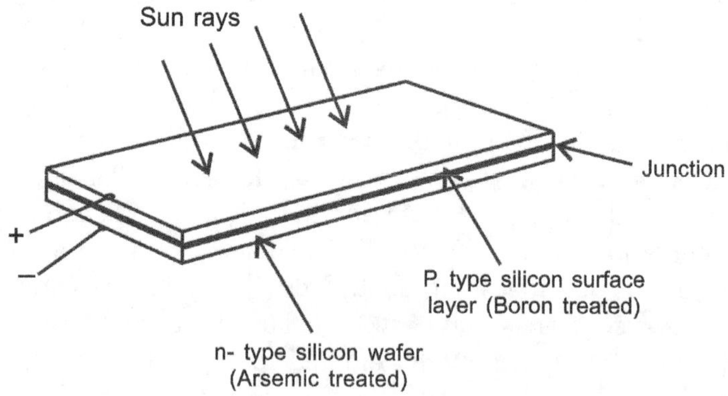

Fig. 1.5 Solar battery cell.

surface area. The basic unit of a typical solar battery is a thin wafer of extra pure silicon, containing less than 1 ppm of inpurity. For solar battery a tiny amount of arsenic is added. This silicon-arsenic wafer is called n-type silicon. Similarly if the silicon wafer is doped with tiny amount of boran then it is called p-type silicon. The combination of these two wafers is called p-n junction.

In the solar battery, one electrical lead is attached to the surface and the other to the body of wafer. The device is a battery with postitive terminal at the p-contact and the negative terminal at the n-contact. A series of such wafers (or so called solar battery cells) are used side-by-side.

Atmosphere of the Earth

Air, which we call, is a mechanical mixture of gases. The various gases, solid and liquid particles in the air, that envelope the earth are bound to it by the gravitational attraction. This envelope of gases, solid and liquid particles is called atmosphere. The atmosphere extends above the surface of the earth to great heights. There is no sharp boundary between the atmosphere and the outer space. However, conventionally, the height of the atmosphere is taken as 1000 km. More than 50% of the mass of the atmosphere lies below an altitude of 5.5 km and 98% of the mass lies below an altitude of 30 km (Above 700 km of altitude a vary thin atmosphere of Hydrogen and Helium atoms exists). The lowest one kilometer, which is called planetary boundary layer, contains 10% of the mass of the atmosphere. All biological and human activities are confined to this planetary boundary level baring aircraft flights. Infact the vertical dimension of the atmosphere is very shallow compared to the size of the earth, it is as thin as the skin on an apple. The mass of the atmosphere is about 5.6×1018 kg (mass of the earth is about 6×1024 Kg), while the mass of oceans (Hydrosphere) water is about 1.4×1021 kg. This shows that the mass of the oceans is more than [2]50 times the mass of the atmosphere. The density of the dry air at the surface is 1.225 Kg/m3 and at an altitude of 5.5 km this drop^s to 0.66 Kg/m3 and at 30 km altitude it is about 0.013 Kg/m3.

Geological evidence suggests that the age of the earth is about 4.5×10^9 years (4.5 billion years) and life began on earth between 0.6 to 1.0 billion years after earth has formed (i.e., life began on earth some where 4 to 3.5 billions years ago). It is estimated that in each second about one Kg of hydrogen from the upper atmosphere escapes into outer space. The mass of oxygen in the atmosphere is 1018 kg, that is one fifth of the mass of the atmosphere. The plant kingdom over the globe produces oxygen about 3×10^6 kg (3 million kg) per second which is consumed by the living beings on the whole earth. About 12 kg of cosmic dust per second falls from space on the earth.

Weather and Climate

The physical state of atmosphere at any given location rarely exhibits steady state even during short intervals of time. This physical state of the atmosphere constitutes weather at a location when considered over a short period of time. Weather is broadly described by the parameters temperature, pressure, relative humidity, wind direction and speed, cloud, precipitation. Climate on the other hand is the average weather over a place including variability and extremities. For this purpose the average has to be taken for a period of not less than 30 years. Both weather and climate are inseparable parts of meteorology. Settlements and civilizations have risen and fallen with the favourable and unfavourable climatic conditions. Weather knows no political boundaries and it favours or frowns equally on all nations. Weather and climate have all pervading influence on man.

Composition of the Atmosphere

The composition of the atmospheric gases, are practically in the same proportion (that is homogeneous) up to an altitude of 80 to 90 Km with the exception of water vapour, ozone and carbon-dioxide which are variable. The composition of dry air at sea level is given in Table 1.2.

Table 1.2 Composition of the atmospheric gases.

Constituent	Symbol	% by volume	Total weight $\times 10^{17}$ Kg
Nitrogen	N_2	78	38.65
Oxygen	O_2	21	11.84
Argon	Ar	0.93	0.66
Carbon-dioxide	CO_2 variable	~ 0.03	~ 0.023
Neon	Ne	180×10^{-5}	0.00064
Helium	He	52×10^{-5}	0.00004
Methane	CH_4	15×10^{-5}	0.00004
Krepton	Kr	10×10^{-5}	–
Hydrogen	H	5×10^{-5}	–
Xenon	Xe	0.8×10^{-5}	–
Ozone	O_3	variable	–

Non-gaseous constituents, (called aerosols) such as dust, smoke, salt particles from sea spray, water particles are all variable.

All living organisms are composed of Carbon, Nitrogen, Oxygen, Hydrogen which are also the basic chemical elements of water and air shells of earth. A large part of living matter contains in green plants which entrap solar energy and construct complex compounds by photosynthesis. The main sources of plant feeding are CO_2 and Water. Plants use about 2% of the incident solar radiation for photosynthesis process. If we assume incident solar energy as 0.5 langleys per min near the earth, then 0.01 ly/min or 7 w/m^2 is consumed by plants.

The plant kingdom provides about 10^{17} kg of biomass annually and an equal quantity of oxygen. An average size of tree supplies about 3500 kg of oxygen per year which is sufficient for three people (because of this the plant kingdom is called the green lungs of the earth). A man requires about 3.13 kg of oxygen or 15 kg of air daily.

At any location water vapour varies from nearly absent to about 4 % by volume (or about 3% by weight) of the atmosphere. However by continuous interchange in the hydrologic cycle, in a year the water vapour produces precipitation about 40 times greater in volume. It is practically absent above altitude of 10 to 12 km. Water vapour is fed to the atmosphere by evaporation of water from surface water bodies or by transpiration of plant kingdom and transported upwards by turbulence or connective currents. This will be discussed in detail in hydrologic cycle. The volume of the water vapour in the atmosphere is about 14×10^{12} m^3 or 0.001 percent of the volume of the earth's hydrosphere. Changes in the moisture concentration below an altitude of 6 km creates all weathers.

The total mass of the CO_2 in the atmosphere is about 0.023×10^{17} Kg while the mass of CO_2 in ocean water is about 1.4×10^{17} Kg which is more than 60 times the amount of CO_2 in the atmosphere. Sea water plays a peculiar role in respect of dissolved natural gases of N_2, O_2, CO_2 and H_2S. These gases as said earlier are closely related to living matter on the land and Sea. CO_2 enters the atmosphere by human and animal breathing, decay and burning of material containing carbon, and volcanic activity. More than 90% of the earth's CO_2 is dissolved in the sea waters. The solubility of CO_2 in sea water varies with temperature, so it enters or leaves the ocean waters with changes in temperature. Because of this the concentration of CO_2 in the atmosphere, particularly near the surface, changes. A large part of the atmospheric CO_2 is removed by the plant kingdom as mentioned above.

Ozone concentration in the atmosphere varies with altitude, latitude and time of the day. In general ozone is found in the atmosphere in variable small quantities up to an altitude of 50 Km. It is mainly found in the stratosphere with maximum concentration between altitudes 18–20 Km in a globe encircling ozone layer. The total ozone column above the earth shields the earth's surface from the lethal UVB (Ultra Violet Radiation).

UVB induces skin cancers, damages eye and suppress immune systems in human beings. UVB affects the productivity of aquatic and terrestrial ecosystem. One percent decline of ozone in atmosphere results in an estimated 3% rise in the potential incidence of skin cancer in humans.

Ozone molecule (O_3) consists of three atoms of oxygen. In the upper stratosphere ozone is formed by the absorption of UV-radiation by oxygen at wavelength 0.18 μm. This splits oxygen molecule (O_2) into two atomic oxygen atoms.

$$O_2 + UVB \rightarrow O + O$$

The atomic oxygen atom (O) reacts with an oxygen molecule (O_2) to form an ozone molecule.

$$O \text{ (atomic)} + O_2 \rightarrow O_3 + \text{energy (exothermic reaction)}$$

Ozone molecules then tend to sink in the atmosphere to altitudes between 15–25 km. Ozone thus formed absorbs all wave lengths of UV-radiation below 0.3 μm. Even in these altitudes the concentration of ozone is still very small compared to oxygen (one ozone molecule to 10^6 normal oxygen molecules). The ozone molecules from the lower stratosphere entering into troposphere gradually decompose into oxygen and thus keep maintain the balance. Ozone is present near the surface of the earth in highly varying concentrations as part of photochemical smogs. Small amounts of ozone is formed near the ground by electrical discharges.

The composition of the atmosphere above 80 Km altitude is not homogeneous and the gases are less mixed. In these altitudes most of the gases are found in ion-atomic state which is caused by solar UV and X-radiation (If a neutral atom loses an electron it becomes positive ion, but if it gains an electron it becomes negative ion. This process is called ionization). Ionization also occurs below 80 Km altitude but in a lesser degree. Here the free electrons and ions mix together and neutralize to form molecules. In higher altitudes (above 100 Km) protons (ionized hydrogen atoms) and free electrons dominate but the atmosphere is very thin which gradually merges with the interplanetary gas.

Interplanetary Gas

The space between the planets is not empty but contains very low density material and hot gases mainly in the form of electrons and protons. This is called interplanetary gas. The planets and the earth in their orbital motions around the sun pass through this inter-planetary gas. The upper most earth's atmosphere at very high altitudes (above 1000 km) merges gradually with this interplanetary gas.

Vertical Divisions of the Atmosphere
Based on Temperature

In 1962 World Meteorological Organization (WMO) decided to divide the atmosphere into four regions based on temperature change with altitude. These are :

(i)   The Troposphere   (ii)   The Stratosphere

(iii)   The Mesosphere   (iv)   The Thermosphere

These are shown in the Fig. 1.6

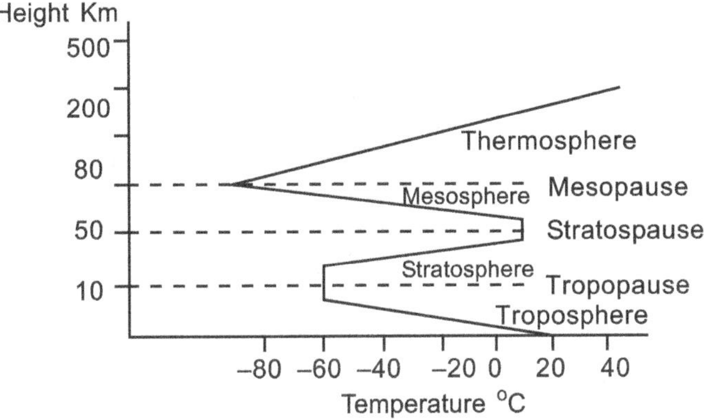

Fig. 1.6 Vertical divisions of atmosphere based on temperature.

Note : Altitude refers to height above mean sea level (amsl)

Troposphere

The lowest layer of the atmosphere adjacent to the earth's surface is called troposphere. Tropos in Greek meaning 'turn'. The temperature turns (increases) above the troposphere and the sun turns above tropics during solstices. The altitude of troposphere in the equatorial region is about 18 Km, which drops to 8 Km in the polar regions [Decrease of temperature with increasing altitude is called lapse rate. Layers where temperature increases with increasing altitudes are called inversion layers].

In troposphere, in general, temperature falls with increasing height. On some occasions, particularly in winter, inversion layers are found near the surface of the earth. There may be found some shallow inversion layers in troposphere. The average lapse rate in troposphere is 6.5 °C/km (3.6° F/1000 ft). Troposphere contains about 75% of the mass of the atmosphere and virtually all the water vapour and aerosols. All weather systems and associated cloud systems are practically confined to troposphere only.

The top of the troposphere is called tropopause where temperature begins to rise or isothermal. Pause meaning break, indicating break in two regions of the vertical atmosphere. Tropopause is defied as follows. The lowest level at which the lapse rate decreases to less than or equal to 2 °C/km at least for a layer of 2 Km layer and above does not exceed 2 °C/km. The altitude of tropopause varies with latitude. The height of the tropical tropopause is about 18 Km (lower temperature about –80 °C) while that of polar tropopause is about 8km (constant temperature about –50 °C). Tropopause is not a continuous layer. In middle latitudes two tropopause are found, one with tropical characteristics and another with extra tropical characteristics, in between these two there is sub-tropical jet (around lat 30°). Double tropopause with less marked characteristics are observed with polar jet in between them (around lat 60°).

The altitudes of tropopause change from day to day and also in association with the movement of synoptic systems. The characteristics of tropopause change with time and place. There is sharp rise in temperature above the tropical tropopause, but in middle latitude tropopause there is slight fall of temperature. Troposphere is mainly warmed up by the underlying earth surface, consequently temperature falls with increasing altitude. The colder, denser air in the upper part of the troposphere sinks by gravity along with traces of ozone and forces the warm air near the surface upward. Thus sets in convection currents, clouding and heat transfer.

Stratosphere

The second stratum that begins, above the tropopause is called stratosphere. It extends above tropopause to an altitude 50 Km. There is an isothermal layer above tropopause to about 20 Km altitude, above which temperature rises gradually to about 32 Km altitude and then rises rapidly. The upper layer of the stratosphere has temperature equal to that of earth's surface temperature. The importance of this region lies in the presence of ozone and sudden warming in polar area, quasi biennial wind oscillation in equatorial area. As mentioned above the concentration of ozone lies at about 25 to 50 altitude, which absorbs the harmful solar radiation between wavelengths 220 nm to 290 nm. There is no ozone above stratosphere. Ozone absorbs about 2% of the insolation and as a result of which stratospheric temperature rises to more than 0 °C (but less than thermospheric temperature). There is not much convection in the stratosphere because it has cold temperature at its bottom and warm temperature at the top.

The small layer above the stratosphere, where temperature ceases to rise (isothermal) and that it separates the stratosphere with third region (of mesosphere) where temperature again falls, is called *stratopause*.

The Mesosphere

The region above the stratopause is called Mesosphere and it extends up to an altitude of 80 Km. In Greek 'Meso' means "medium". This region is characterized by fall of temperature above stratopause (> 0 °C) to about –95 °C near 80 Km altitude. This region presents noctilucent clouds in higher latitudes during summer. The top of the mesosphere is called mesopause. The three regions troposphere, stratosphere and mesosphere together is called homosphere. As stated earlier the composition of the atmosphere is nearly constant, except ozone, water vapour and carbon-dioxide in homosphere. The mean structure of temperature in the mesosphere, the temperature upto 100 Km altitude is shown in the Fig. 1.7. Both in the troposphere and the mesosphere the temperature in general decreases with altitude but in the stratosphere the temperature increases with altitude baring in winter in high latitudes. The stratospheric latitudinal temperature distribution is accounted by the radiative heating of ozone. The temperature is maximum in summer polar region and minimum in winter polar region due to solar radiative flux at the solstices. At an altitude above 60 km the temperature increases from summer pole to winter pole.

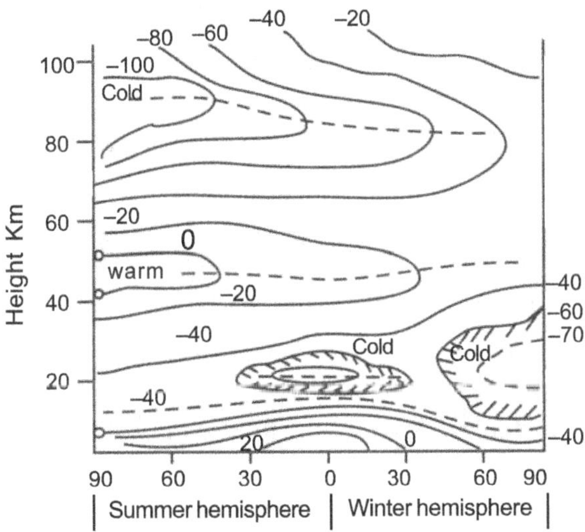

Fig. 1.7 Mean meridional cross section of temperature in °C. Dotted lines indicate tropopause, stratopause and mesopause.

The mean zonal (East-west direction) wind profiles indicate approximately thermal wind balance with temperature profile. Easterly Jet in summer hemisphere and westerly Jet in winter hemisphere with wind maxima occurring at about 60 Km altitude. In winter there is high latitude westerly Jet in lower stratosphere, which is called polar night Jet. This is related to stratospheric warming. In the equatorial stratosphere there exists quasi-biennial oscillations which will be the discussed separately.

The Thermosphere

The region above mesopause is called thermosphere or hetrosphere. Here the composition of the atmosphere is not homogeneous. The main gases stratify by their molecular weights. The lowest layer predominates with nitrogen and oxygen molecules [90 to 150 km], then oxygen (150–500 km) and at the top (500–1000 Km) hydrogen and then inter-planetary gas. The temperature of the thermosphere rises to 1000 to 1200 °K at 400 km but it

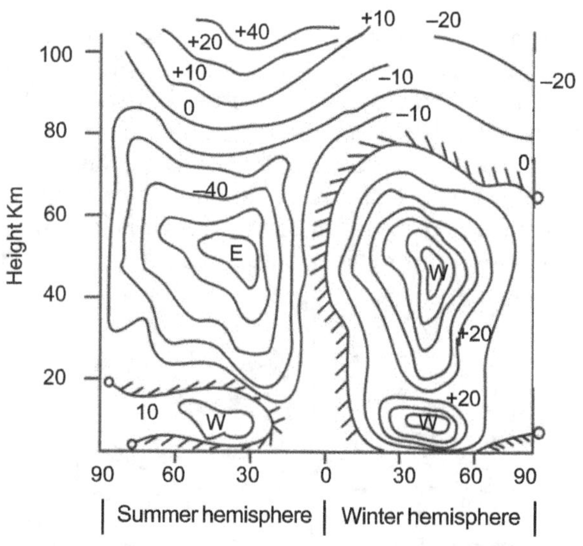

Fig. 1.8 Mean zonal wind in mps at the time of solstices.

has gas of very thin density 3×10^{-12} Kg/m^3 and pressure 10^{-8} mm of mercury. In this region gases are found mostly in atomic state due to photo-dissociation by the short wave radiation of the sun and they are not in the same proportion as in homosphere.

There is temperature rise above mesopause. Oxygen atoms are seen in the lower thermosphere which gradually increases with altitude. At about 130–150 Km altitude about 70% of the oxygen is found in atomic state due to the action of UV-radiation of the sun. At further up there are only hydrogen, helium atoms.

The Ionosphere

The upper atmosphere which is characterized by the presence of dense ions and free electrons is called ionosphere. These ions cause reflection of radio waves. The ionosphere is sub divided into D-region (50–90 km), E-region (90–140 km) and F-region (F_1 and F_2) (140–250 km, 250–500 km).

The D-Region

This occurs between 50 to 90 km (lies mainly in mesosphere). It reflects low-frequency radio waves, but absorbs medium and high frequency waves. This region is entirely dependent on solar radiation and disappears during nights. The D-region is well-developed during solar flares. Because of this, during solar flares complete break down in medium and high frequency radio communication takes place. This is called sudden ionospheric disturbances (SID).

The E-Region

This occurs between altitudes 90 to 140 km. This region strongly reflects medium and high frequency radio waves. E-region begins to weaken after sunset, but does not disappear completely. However during polar nights it disappears completely. The lower part of the E-region is marked by the recombination of ions and electrons. Under special conditions and irregular times a "sporadic E-layer" occurs. On such occasions the motion of ions and electrons can be detected by their effects on radio transmission.

The F-Region

This is further sub-divided into F_1 and F_2 regions extends above 140 km.

(a) **F_1-Region :** It is found only during day time when the sun is fairly high. When the sun is low and at night it merges with the higher F_2-region. F_1-region is important in the propagation of medium and high frequency radio waves.

(b) **F_2-Region :** This region is important in long-distance radio communication. Depending on location and conditions the ionization density reaches a peak at altitudes of 250–500 km. It gradually diminishes above 500 km but it has no upper limit. The upper F_2-region contains protons (ionized atomic hydrogen) and electrons. In fact this region is similar to inter planetary gas.

The Dynamo Region

In the region of 80–130 km altitude ions, atoms and molecules of neutral gases move together. The mean velocity of electrons are different from that of ions. Atmospheric tidal oscillations occur in this region with the creation of ionospheric currents. These ionospheric currents are called "dynamo currents", which are caused by the motion of the conducting atmosphere across the earth's magnetic field. This region is called the "dynamo region" because these currents are similar to the currents generated by a moving conductors in a dynamo. These electric currents are responsible for the regular variations of earth's magnetic field.

The regular variation of earth's magnetic field follows a 24 hr daily pattern. They are related to the solar or lunar day, and over a period of 24 hrs they pass through a regular cycle of increases and decreases. These are believed to be the effect of electric currents of tidal oscillations at altitudes 80 to 130 km referred above. Irregular variations may also occur suddenly in short intervals of time few seconds to days. These irregular variations are caused by the particles ejected from the sun. During solar winds magnetic storms occur particularly when these winds approach the outer atmosphere. As stated in solar flares, during intense solar activity charged solar particles enter the upper atmosphere and cause electrical phenomena.

This electrical phenomena in turn cause fluctuations in direction and strength of the earth's magnetic field. When the irregular variations are prominent magnetic storm results. During the period of magnetic storms communication system will be jammed or blackout over the entire world. Frequent magnetic storms would occur during the period of maximum sunspot activity, which has about eleven years cycle. The solar particles would be trapped and controlled by the earth's magnetic field. These particles become concentrated in two zones of radiation which surround the earth above the magnetic equator. These bands are called Van Allen radiation belts.

Exosphere

As stated earlier, the ionosphere consists of ions, electrons and neutral particles, extends outward till it merges with the inter planetary gas. It may be noted that even at about 1200 km altitude the number of neutral particles are more or less same as that of electrons. However at about 500-600 km altitude the density of neutral particles is so small that collision between them is very rare. The mean free path is very large, as a result some particles escape from the earth's gravitational attraction. Because of this the region above 500–600 km is named exosphere. It is estimated that each second 1 kg of hydrogen escapes from the earths atmosphere to outer space.

In contrast to the neutral particles, the electrically charged particle movements are controlled by the earths magnetic field. The collision between them still occurs at levels above 500–600 km. As a result the electrically charged particles do not escape from the earth's atmosphere.

Cosmic rays coming from the space undergo deviation in the earth's magnetic field and hence intensity depends on latitude. In the vicinity of the equator the deviation is manifested strongly, and the particles subjected to large deviation do not enter the atmosphere (this is latitude effect). The positively charged particles of the cosmic rays deviate towards east and the negatively charged particles towards west (this is called east-west effect). Intensity also depends on the longitude (this is called longitude effect).

Van Allen Radiation Belts

Radiation belts have been found at the outer most atmosphere. There are two delimited (assigning of boundaries) regions with high intensity of cosmic ionizing radiation. These belts being formed due to trapping of charged particles by the earths magnetic field. The inner belt extends from about 600 km to 6000 km while the outer belt 20000 to 60000 km from the earth. Inner radiation belt consists of high energy protons while the outer radiation belt formed by electrons of solar origin. Radiation belts are characteristic of all celestial bodies which have magnetic field. The moon has no magnetic field of its own and hence it has no radiation belts around it.

Questions

1. What are the building blocks of Universe Big Bang Theory Hobble's principal, observable universe and the link between the distant starts and the earth.

2. What is the mass of the sun and the solar system? Describe the principal features of photosphere chromosphere and corona of the sun.

3. Write briefly on (i) Solaractivity (ii) Maunder minima, (iii) Solarflares, (iv) Heterogenous rotation of the sun and (v) Cosmic rays.

4. Write briefly the principal characteristics of the sun and electromagnetic spectrum.

5. Define solar constant and the estimates of solar energy received on earth in one day. Give the distribution of incoming solar radiation on earth.

6. Define albedo of a surface and explain sky radiation, global radiation direct solar radiation and attenuation of solar radiation.

7. Define Black body, Gray body and Spectral density.

8. State Planck's law, Stefan-Boltzmann law, Kirchhoff's law and Emissivity.

9. Atmosphere in a selective absorber why? What are its consequences.

10. Describe briefly the radiative balance of the earth and horizontal transport of heat.

11. What is atmosphere? How is the mass distributed in the atmosphere with respect to height (vertical)?

12. Distinguish between weather and climate. What are its effects on civilization?

13. What is homogenous atmosphere? Write its composition.

14. What are the green lungs of the earth? What is the man's daily requirement of air? What is photosynthesis and the daily output of oxygen by a normal tree?

15. What do you know about the variation of water vapour at any location? How is that water vapour produces precipitation about 40 times greater in volume in a year?

16. Write briefly on the ozone formation in the atmosphere and ozone hole. What would happen to the life on earth with the depletion of ozone in the atmosphere?

17. Write briefly on the WMO-vertical divisions of the atmosphere based on temperature.

18. Write the principal features of troposphere, stratosphere and mesosphere.

19. Write briefly on (i) The thermosphere (ii) The ionosphere, (iii) The dynamo region, (iv) Exosphere (v) Van Allen radiation belts.

CHAPTER 2

Physical Variables

The state of atmosphere is determined by a number of parameters (or state variables). The atmospheric state is said to be steady or static if it doesnot change with time or in equilibrium state. However, in general atmosphere is dynamic, that is in motion. The principal state variables of atmosphere are pressure (p), temperature (T), specific (or molar) volume, and moisture (or humidity). The most of the other parameters of atmosphere are the complex combination of these and a few of them are considered in thermodynamic variables.

Mathematically, there exists a functional relationship among these parameters such as

$$V = f\ (P,\ T,\ M)$$

where V = volume of gas, P = Pressure, T = temperature and M = mass

$$\text{Pressure} = \frac{\text{normal force}}{\text{area}} = \frac{\Delta F}{\Delta A}$$

where   F = force, A = area

$$\text{Density} = \rho = \frac{\text{mass}}{\text{volume}} = \frac{M}{V}$$

$$\alpha \qquad = \text{specific volume} \ \frac{V}{M} = \frac{1}{\rho}$$

Definition

An ideal (or perfect) gas is one in which there are no forces of molecular interaction.

Ideal gas obeys the following four laws.

1. *Boyle's Law* : Temperature remaining constant (isothermal process) the volume (V) of a given mass of gas is inversely proportional to the pressure (p)

$$V \propto \frac{1}{p} , \text{ when T is constant}$$

2. *Charle's Law* : Pressure remaining constant (isobaric process) the volume of a given mass of gas is directly proportional to the absolute temperature.

$$V \propto T, \text{ when p is const.}$$

3. *Gay-lussac's Law* : Volume remaining constant (isochoric process) the pressure of a given mass of gas is directly proportional to its absolute (Kelvins) temperature.

$$P \propto T, \text{ when V is constant.}$$

4. *Avogadro's Law* : Equal volumes of different ideal gases at the same temperature and pressure contain equal number of molecules. Or Moles of different gases occupy the same volume at the same temperature and pressure.

Moles of all ideal gases at STP (T = 0 °C, or 273 °K; P = 1 atm or 760 mm Hg or 1.013×10^5 pa)occupy 22.414 liters of volume. Or one cubic cm (cc) of any ideal gas at STP contains 2.69×10^{19} molecules. This number is called Loschmidt's's number.

At STP the mass of 22.414 liters of any ideal gas is equal to its molecular weight expressed in grams.

Thus the mass of 22.414 liters of oxygen and hydrogen at STP weighs 32 gm and 2 gm respectively.

Mole

One mole of any substance is a quantity of matter expressed in grams (or kg) and numerically equal to its molecular weight.

OR

A gram molecule or mole (kilogram molecule or k-mole) is the amount of substance expressed in grams (kilograms) is equal to its molecular weight.

Molecular Weight

The molecular weight of a substance may be defined as the number of times its molecule is heavier than $\frac{1}{12}$ th the weight of one C^{12}-atom.

Gram atomic weight (molecular weight) of an element is defined as the quantity of the element (substance) in grams numerically equal to the atomic weight (molecular weight) of the substance.

Ideal Gas Equation

Combining the first three gas laws we get

$$\frac{PV}{T} = \text{Constant}$$

The equation of state for an ideal gas is given by

$$PV = nRT, \qquad\qquad .....(2.1)$$

where $\quad$ n = number of moles $= \dfrac{M}{\mu}$

μ = Molecular mass,

M = Mass

R = Universal gas constant

$\quad = 8.314 \times 10^3$ J k-mol^{-1} K^{-1}

$\quad = 8.314 \times 10^7$ erg mol^{-1} K^{-1}

$\quad = 1.987$ cal mol^{-1} K^{-1}

$\quad = 0.0821$ liter atm mol^{-1} K^{-1}

R is numerically equal to the work done by one mole of ideal gas when it is heated isobarically by one degree.

eq. (2.1) may be written as

$$PV = \frac{M}{\mu} R T$$

or $\qquad P\dfrac{V}{M} = \dfrac{R}{\mu} T$

or $\qquad P = \rho\, R'\, T$

or $\qquad P\,\alpha = R'\, T \qquad\qquad .....(2.2)$

where $\quad \alpha = \dfrac{V}{M} =$ specific volume, $R' = \dfrac{R}{\mu}$

eq. (2.2) is a second form of ideal gas equation.

Example 2.1

Find the volume of one mole of ideal gas at STP (standard temperature and pressure)

Solution :

At S T P, $P = 1.01325 \times 10^5$ pa

$T = 273$ K

$n = 1$

$R = 8.314$ J mol^{-1} K^{-1}

Ideal gas law $PV = nRT$

$$V = \frac{nRT}{p} = \frac{1 \times 8.314 \times 273}{1.01325 \times 10^5}$$

$$V = 22.414 \times 10^{-3} \text{ m}^3$$

$$= 22.414 \text{ liters}$$

Avogadro Number (N_A)

N_A = number of molecules in a gram molecule

= number of atoms in a gram atom.

$= 6.023 \times 10^{23}$ mol^{-1}

$= 6.023 \times 10^{26}$ kmol^{-1}

The mass of 22.414 liters of any ideal gas at STP is equal to its molecular weight expressed in grams.

Thus the mass of 22.414 liters of oxygen at STP weighs 32 gm.

Unit

1. In SI (mks) units, where unit of pressure (p) = 1Pa = 1N m^{-2}

Volume 1m^3

$P = 1.013 \times 10^5$Pa

$V = 0.022414$ m^3

$n = 1$mol

$T = 273$K

$$R = \frac{pV}{nT}$$

$$R = \frac{1.013 \times 10^5 \text{pa} \times 0.022414}{1 \text{ mol} \times 273\text{K}}$$

$$= 8.314 \text{ J mol}^{-1} \text{ K}^{-1}$$

2. In CGS units where pressure unit is dyne/cm^2

   Unit of volume cm^3

$$p = 76 \times 13.6 \times 980 \text{ dynes / cm}^2$$

$$V = 22414 \text{ cm}^3$$

$$T = 273\ ^\circ K$$

$$n = 1 \text{ mol}$$

$$1J = 10^7 \text{ ergs.}$$

$$R = \frac{pV}{nT} = \frac{76 \times 136 \times 980 \times 22414}{1\,\text{mol} \times 273K} \text{ dynes cm}^{-2}\,\text{cm}^3$$

$$= 8.314 \times 10^7 \text{ ergs mol}^{-1}\,K^{-1}$$

For a fixed mass of an ideal gas the product nR is constant.

Hence $\qquad \dfrac{P\,V}{T} = $ constant

For different temperatures $\qquad T_1, T_2, T_3....$etc.

with corresponding pressures $\quad P_1, P_2, P_3....$etc.

$\qquad\qquad$ and volumes $\qquad V_1, V_2, V_3....$etc.

we have $\dfrac{P_1 V_1}{T_1} = \dfrac{P_2 V_2}{T_2} = \dfrac{P_3 V_3}{T_3} ...... = $ Const.

For dry air $\qquad R_d = 287$ J Kg^{-1} K^{-1},

$\qquad\qquad\qquad C_p = 1004$ J Kg^{-1} K^{-1}

and $\qquad\qquad C_v = 717$ J Kg^{-1} K^{-1}

No. of moles of dry air $n_d = 28.97$

Model Variation of Pressure with Altitude in the Atmosphere

The hydrostatic equation (which will be proved later) is given by

$$dp = -g\,\rho dz \qquad\qquad .....(2.3)$$

Where g = gravity, z = altitude.

Ideal gas equation is PV = nRT $\qquad\qquad .....(2.4)$

where $\qquad\qquad n = \dfrac{M}{\mu} \qquad\qquad .....(2.5)$

$\qquad$ M = mass, μ = molecular mass, n = number of moles

From eq. (2.4) and eq. (2.5) we have $PV = \dfrac{M}{\mu} RT$

$$\text{or} \qquad p = \frac{M}{V} \frac{RT}{\mu}$$

$$\text{or} \qquad p = \rho \frac{RT}{\mu}$$

$$\therefore \qquad \rho = \frac{p\mu}{RT} \qquad\qquad .....(2.6)$$

Since ρ varies with pressure, substituting eq. (2.6) in eq. (2.3)

$$\text{we have} \qquad dp = -g\frac{p\mu}{RT}\, dz$$

$$\text{or} \qquad \frac{dp}{p} = -\frac{g\mu}{RT}\, dz \qquad\qquad .....(2.7)$$

Integrating eq. (2.7) in the limits p_1 to p_2 with corresponding altitudes z_1, z_2 we have

$$\int_{p_1}^{p_2} \frac{dp}{p} = \frac{-g\mu}{RT} \int_{z_1}^{z_2} dz$$

$$\ln \frac{p_2}{p_1} = -\frac{g\mu}{RT}(z_2 - z_1) \qquad\qquad .....(2.8)$$

Note : Here T and g assumed as constant.

Let p_0 be the pressure at sea level where altitude $z = 0$ and p be the pressure at altitude z

From eq. (2.8) we have

$$\ln \frac{p}{p_0} = \frac{-g\mu}{RT}(z - 0)$$

$$\text{or} \qquad p = p_o\, e^{-\frac{g\mu}{RT}z} \qquad\qquad .....(2.9)$$

This is the required model equation of pressure variation with height in isothermal atmosphere

Example 2.2

Assuming isothermal atmosphere, find the pressure at the top of Mount Everst (P_{ME}) given its altitude as 8880 m and $\mu = 28.8 \times 10^{-3}$ kg/mol, $g = 9.8$ m/s^2.

Solution :

The value of
$$\frac{g\mu}{RT}z = \frac{9.8 \times 28.8 \times 10^{-3} \times 8880}{8.314 \times 273}$$

$$= \frac{2506.3}{2269.7} = 1.10$$

Using eq. (2.9) we have

$$P_{ME} = (1.01325 \times 10^5\ e^{-1.10})$$
$$P_{ME} = 0.336 \times 10^5 \text{ Pa}$$
$$P_{ME} = 0.333 \text{ atm.}$$

$P_0 = 1.01325 \times 10^5$ Pa at sea level

1 atm $= 1.01325 \times 10^5$ Pa

Example 2.3

Assuming temperature varies linearly (in troposphere) as $T = T_o - \lambda Z$, find an expression of pressure variation with altitude.

Given
$$T = T_o - \lambda Z \qquad\qquad .....(2.10)$$

$$\therefore \qquad dT = -\lambda dZ$$

Substituting this value in eq. (2.7) Viz $\dfrac{dp}{p} = -\dfrac{g\mu}{RT}\,dz$

we have
$$\frac{dp}{p} = \frac{g\mu}{RT}\frac{dT}{\lambda}$$

Integrating
$$\int_{P_o}^{P} \frac{dp}{p} = \frac{g\mu}{R\lambda} \int_{T_o}^{T} \frac{dT}{T}$$

where

at msl T_o = Temperature P_o = pressure, $Z = 0$ and at altitude Z pressure p and temperature T

$$\ln \frac{p}{P_o} = \frac{g\mu}{R\lambda} \ln \frac{T}{T_o} = \ln \left(\frac{T}{T_o}\right)^{\frac{g\mu}{\lambda R}}$$

$$\therefore \qquad p = P_o \ln \left(\frac{T}{T_o}\right)^{\frac{g\mu}{\lambda R}} \qquad\qquad .....(2.11)$$

This is the required model expression of pressure variation with altitude under the assumption that temperature changes linearly with altitude.

Meteorological aspects of Thermodynamics

Some Definitions

Specific Heat of a substance is the quantity of heat required to raise the temperature of unit mass (1gm, 1mol, 1kg) through one degree celsius.

Thermal Capacity of a body is the quantity of heat required to raise the temperature of the body through one degree celsius.

> Note : Specific heat of a body gives the thermal capacity of the body per unit.

Water Equivalent of a body is numerically equal to its thermal capacity.

Specific Heat of a Gas at Constant Volume c_v.

The amount of heat required to raise the temperature of unit mass (1gm) of gas through one degree celsius keeping its volume constant (in isochoric process).

Specific Heat of a Gas at Constant Pressure c_p.

The amount of heat required to raise the temperature of unit mass (1gm) of gas through one degree celsius, keeping its pressure constant (in isobaric process)

The specific heat corresponding to one gram molecule of a gas instead of unit mass are called gram molecular or molar specific heats and these are denoted by C_p, C_v in isobaric and isochoric precess respectively and they are related to c_p, c_v as follows.

$$C_p > C_v, \quad C_p = \mu\, c_p, \quad C_v = \mu\, c_v$$

$$C_p - C_v = \frac{R}{J} = \frac{M}{\mu} R$$

$$= AR$$

where $\quad A = \dfrac{1}{J}$

$\quad J$ = mechanical equivalent of heat

$\quad R$ = gas constant.

$$\frac{C_p}{C_v} = \frac{c_p}{c_v} = \gamma = \text{adiabatic exponent}$$

$$W = JH$$

where

$$H = \text{heat supplied}$$

$$W = \text{work done.}$$

$$\text{Molar heat} = \text{Specific heat} \times \text{molecular weight of the gas.}$$

Law of Conservation of Energy

Energy is neither created nor destroyed, it only changes from one state to another. This is first law of thermodynamics.

OR

The total energy of an isolated system remains constant, irrespective of the process occurring in the system. This implies that the motion of matter (in atmosphere) can neither be created nor destroyed, it can change from one form to another.

$$\Delta i = Q + \omega'$$

where $\quad \Delta i$ = change in internal energy of the system

$\quad\quad Q$ = Quantity of heat transferred to the system

$\quad\quad \omega'$ = work done on the system by external bodies

Heat capacity of a body (C) is defined by the relation

$$C = \frac{\delta Q}{dT}$$

Where $\quad \delta Q$ = infinitesimal heat transferred to the body

$\quad\quad dT$ = small change in temperature of the body T to T + dT.

Specific Heat (c)

If c is the heat capacity of a unit mass of homogeneous substance, then

$$c = \frac{C}{M}$$

where $\quad M$ = mass of the substance.

$\quad\quad M = n\mu$

$\quad\quad n$ = number of moles

$\quad\quad \mu$ = molecular mass

$\quad\quad C_\mu = c.\mu$ molecular heat capacity.

$$cM = \frac{\delta Q}{dT}$$

$$\text{or} \qquad \delta Q = cMdT$$

$$= c\, n\, \mu\, dT$$

$$\delta Q = nC_\mu dT$$

$$C\mu = \frac{1}{n}\frac{\delta Q}{dT} \qquad \qquad .....(2.12)$$

The molar heat capacity of water $\simeq 75.3$ J/mole Kelvin

First law of Thermodynamics

Let heat be supplied to a system which is capable of doing work.

The quantity of heat absorbed by the system is equal to the sum of the external work performed by the system and increase in its internal energy.

$$\delta Q = di + \delta w \qquad \qquad .....(2.13)$$

$$\text{or} \qquad dH = di + dw \qquad \qquad .....(2.14)$$

Where

$p\alpha = R' T$ ideal gas equation

δQ = added energy

di = increase in internal energy

δw = work done (by the gas against the surrounding)

dH = heat quantity added.

For an ideal gas

$$di = C_\upsilon\, dT$$

$$\delta w = pd\alpha$$

Substituting these values in eq. (2.13)

we have

$$\frac{dQ}{dt} = C\upsilon\,\frac{dT}{dt} + p\,\frac{d\alpha}{dt}$$

$$\text{or} \qquad \delta Q = C\upsilon\, dT + pd\alpha \qquad \qquad .....(2.15)$$

This is the thermodynamic equation.

For adiabatic process (in which heat quantity remains constant)

$$dh = C_v\, dT + pd\alpha \qquad \qquad .....(2.16)$$

$$\text{where} \qquad \frac{dh}{dt} = H \text{ and } C_p\text{-}C_v = R'$$

$$pd\alpha = R'dt - \alpha dp$$

and $\qquad dh = C_p\, dT - \alpha dp$

or $\qquad \dfrac{dQ}{dt} = Cp\, \dfrac{dT}{dt} - \alpha \dfrac{dp}{dt}$ ......(2.17)

Equation eq. (2.16) and eq. (2.17) represent the first law of thermodynamics used in meteorology

Note : $\qquad C_v = \left(\dfrac{dh}{dT}\right) \alpha$ constant, $C_p = \left(\dfrac{dh}{dT}\right) p$ constant

For adiabatic process we get.

$$\frac{T}{T_1} = \left(\frac{\alpha}{\alpha_1}\right)^{\frac{-R'}{C_v}}$$

$$\frac{T}{T_1} = \left(\frac{p}{p_1}\right)^{\frac{R'}{C_p}} = \left(\frac{p}{p_1}\right)^{K}$$

$$\frac{\alpha}{\alpha_1} = \left(\frac{p}{p_1}\right)^{\frac{-C_v}{C_p}} \quad \text{or} \quad \frac{p}{p_1} = \left(\frac{\alpha}{\alpha_1}\right)^{\frac{-C_p}{C_v}}$$

$$\theta = T \left(\frac{p_1}{p}\right)^{\frac{\gamma-1}{\gamma}}$$

(where $\qquad k = \dfrac{R'}{Cp} = \dfrac{\gamma-1}{\gamma}$)

or $\qquad \theta = T \left(\dfrac{p}{p_0}\right)^{-K}$

This is Poisson's equation, where $P_0 = 1000$ hPa and $\theta =$ potential temperature

Definition of Potential Temperature (θ)

It is the temperature attained by a sample of dry air when the sample is compressed or expanded adiabatically from a given state to the pressure level of 1000 h P_a.

θ is constant during adiabatic process.

The another form of thermodynamic equation is given by

$$\frac{d}{dt} (\ln \theta) = \frac{1}{C_p} \frac{H}{T}$$

Some Properties of Ideal Gases : In meteorology atmospheric gases are assumed to obey ideal gas law.

1. In isobaric process (p = const)

   (i) $\dfrac{V}{T}$ = const   (obeys Charle's law)

   (ii)   Work done in the process $\delta w = pdV$

       or $w = p(V_2 - V_1)$

   (iii)   Amount of heat transferred

$$\delta Q = C_p \, dT$$

       or $H = C_p (T_2 - T_1)$

   (iv)   Change in internal energy

$$di = C_v \, dT$$

       or $\Delta i = C_v (T_2 - T_1)$

   (v)   Heat capacity $C_p = \dfrac{M}{\mu} \cdot \dfrac{R'}{K}$

2. In isothermal process (T = const)

   (i)   PV = const   (Obeys Boyle's law)

   (ii)   Work done $\delta W = pdV$

$$W = \frac{M}{\mu} RT \ln \left(\frac{V_2}{V_1} \right) = \frac{M}{\mu} RT \ln \left(\frac{p_1}{p_2} \right)$$

   (iii)   Amount heat transferred

$$\delta Q = \delta w$$

       or $Q = W$

   (iv)   Change in internal energy $di = 0$,

   (v)   Heat capacity $C_T = \pm \infty$

3. In adiabatic process ($\delta Q = 0$)

   (i) $p\,\alpha^\gamma$ = const or PV^γ = const, $PT^{\frac{\gamma}{1-\gamma}}$ = const and

$$VT^{\frac{1}{\gamma-1}} = \text{const}$$

$$\gamma = \frac{C_p}{C_v}$$

(ii) Work done $\delta w = pdV = -di$

$$\text{or} \quad w = C_v\,(T_1 - T_2) = \frac{1}{\gamma - 1}\,(p_1 v_1 - p_2 v_2) = -\Delta i$$

(iii) Heat transfered amount $\delta Q = 0$ or $Q = 0$

(iv) Change in internal energy

$$di = -pdv = -\delta w$$
$$= C_v\,dT$$
$$\Delta i = -w = C_v\,(T_2 - T_1)$$

(v) Heat capacity $C_{ad} = 0$.

4. In isochoric process, $V = \text{const}$

(i) $\dfrac{P}{T} = \text{const}$ \qquad (Gay-Lussac's law)

(ii) Work done $\delta w = 0$ or $W = 0$

(iii) Heat transferred amount $\delta Q = C_v\,dT$ or $Q = C_v\,(T_2 - T_1)$

(iv) Change in internal energy

$$di = C_v\,dT$$
$$\text{or} \quad Q = C_v\,(T_2 - T_1)$$

(v) Heat capacity $C_v = \dfrac{M}{\mu}\,\dfrac{R}{\gamma - 1}$

Entropy and Second Law of Thermodynamics

A thermodynamic process which proceeds with one of its state variable being kept constant is called Iso-process.

Entropy

The physical quantity which describes the ability of a system to do work is called entropy of the system.

Entropy may be taken as randomness of a process or disorder. Addition or removal of heat to a process causes increase or decrease in disorder or randomness.

$$\text{Let} \qquad \oint \frac{dQ}{T} = 0 \text{ or } \oint d\phi = 0$$

$$\text{where} \qquad d\phi = \frac{dQ}{T}$$

Then ϕ is called entropy

Change in entropy

$$\Delta\phi = \int_a^b \frac{dQ}{T} = \int_a^b d\phi$$

$$\Delta\phi = \phi_b - \phi_a = \frac{Q}{T}$$

Isentropic Process

A thermodynamic process in which entropy of the system remains constant.

Entropy of an ideal gas ($pV = RT$) is given by

$$\oint \frac{dh}{T} = \oint d\phi = 0 \quad \text{where } \phi = c_p \, d \, (\ln T) - R \, d \, (\ln p)$$

Relation between entropy (ϕ) and potential temperature θ is given by the relation $\phi = C_p \ln \theta + const$

That is,

Entropy is proportional to the logarithmic of potential temperature.

Geopotential (ψ)

The potential energy imparted to a unit mass when it is lifted from mean sea level to a height (z) is called geopotential. It is defined as

$$d\psi = gdz$$

$$\psi = 0 \text{ when } z = 0$$

The value of g at msl is 9.8 m/s^2

The geopotential of unit mass, one metre above msl is about 9.8 m/s^2

A unit of geopotential meter (gpm) is defined as

$$\psi = \frac{1}{9.8} \int_o^z gdz, \text{ where } \psi \text{ (height) is in geopotential meters.}$$

Example 2.4

Convert 2998 m into geopotential meters where g = 9.806 m/s^2

$$\psi = \frac{1}{9.8} \int_0^{2998} gdz = \frac{1}{9.8} \int_0^{2998} 9.806 \, dz = 3000 \text{ gpm}$$

Thermodynamic Diagrams

Thermodynamic diagrams provide graphical display of atmospheric process such as isobaric, isothermal, dry adiabatic, pseudo adiabatic processes that the atmosphere is subjected. These diagrams also contain constant values of saturation mixing ratio lines. The desirable characteristics are (i) Area enclosed by the lines is proportional to the change in energy or work done during the process. (ii) The fundamental lines as many as possible should be straight. (iii) The angle between the isotherms and dry adiabats may be as large as possible, say about 90°.

Equal Area Transformation

Let the variables A and B represent two thermodynamic variables, $\propto$, - p as axes (as shown in the Fig. 2.1). Let there be a one-to-one correspondence between α, -p and B, A diagrams such that: area enclosed on $\propto$, - p diagram (A_1) = area enclosed on B, A diagram (A_2). From equal area - $\oint p\,d\propto$ =

$$\oint A\,dB, \text{ or } \oint (p\,d\propto + A\,dB) = 0 \qquad \qquad .....(2.18)$$

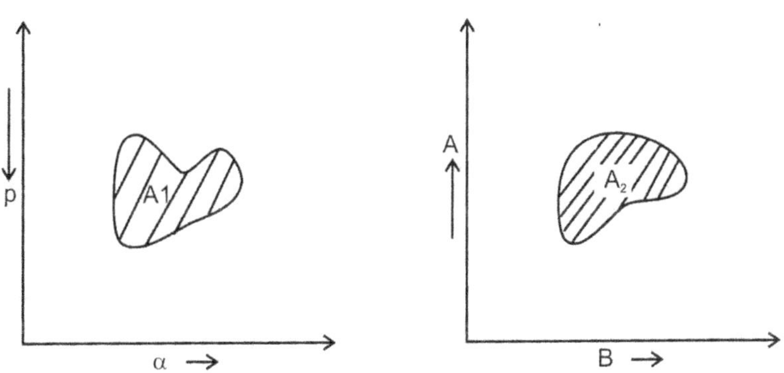

Fig. 2.1 Thermodynamic diagrams.

Equation (2.18) is true only when $p\,d\propto + A\,dB$ is an exact differential, say d S $(\propto,B)$, which is a function of $\propto$ and B. From the condition of as exact differential we have

$$dS (\propto, B) = \frac{\partial S}{\partial \propto}\, d\propto + \frac{\partial S}{\partial B}\, dB = p\,d\propto + A\,dB \qquad .....(2.19)$$

Comparing the like terms in (eq. 2.19) we have

$$p = \frac{\partial S}{\partial \propto},\ A = \frac{\partial S}{\partial B}$$

$$\text{or} \quad \frac{\partial p}{\partial B} = \frac{\partial^2 S}{\partial \alpha\, \partial B}, \quad \frac{\partial A}{\partial \alpha} = \frac{\partial^2 S}{\partial B\, \partial \alpha}$$

$$\therefore \quad \frac{\partial p}{\partial B} = \frac{\partial A}{\partial \alpha} \quad \left(= \frac{\partial^2 S}{\partial \alpha\, \partial B}\right) \qquad\qquad \ldots\ldots (2.20)$$

If condition eq. (2.20) satisfied, then the areas in (α, p) and (B, A) diagrams are equal.

The Emagram

Let $B = T$, in above equal area transformation analysis

$$\text{Then} \quad \left(\frac{\partial A}{\partial \alpha}\right)_T = \left(\frac{\partial p}{\partial T}\right)_\alpha \quad \text{from eq. (2.20)} \qquad \ldots\ldots(2.21)$$

We know equation of state is given by $p\,\alpha = RT$ or $p = \dfrac{RT}{\alpha}$

$$\left(\frac{\partial p}{\partial T}\right)_\alpha = \frac{R}{\alpha} \qquad\qquad \ldots\ldots(2.22)$$

Substituting eq. (2.22) in eq. (2.21) we have

$$\left(\frac{\partial A}{\partial \alpha}\right)_T = \frac{R}{\alpha} \qquad\qquad \ldots\ldots\ldots(2.23)$$

Integrating eq. (2.23) w.r.t α we have

$$\int \left(\frac{\partial A}{\partial \alpha}\right)_T d\alpha = \int \frac{R}{\alpha} d\alpha$$

$$A = R \ln \alpha + F(T) \qquad\qquad \ldots\ldots(2.24)$$

where integral constant $F(T)$, is a function of T.

$$p\alpha = RT \text{ equation of state.}$$

$$\ln\, p + \ln\, \alpha = \ln\, R + \ln\, T \text{ (taking logarithms)}$$

$$\text{or} \quad \ln \alpha = \ln R + \ln T - \ln p \qquad\qquad \ldots\ldots(2.25)$$

Substituting eq. (2.25) in eq. (2.24) we have

$$A = R\, (\ln R + \ln T - \ln p) + F(T)$$

$$A = -R \ln p + [R \ln R + R \ln T + F(T)] \qquad\qquad \ldots\ldots(2.26)$$

Now select $F(T) = - [R \ln R + R \ln T] \qquad\qquad \ldots\ldots(2.27)$

From eq. (2.26) and eq. (2.27) we have $A = -R \ln p$

With this selection we have A = – R ln p, B = T   .....(2.28)

Thus the new coordinates of thermodynamic diagram are T, –R ln p. The plot of this is called Emagram.

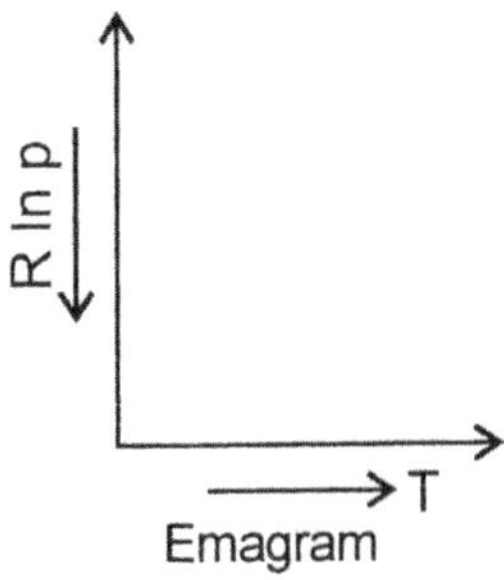

Properties of Emagram

(i)   Area is proportional to energy

(ii)   Abscissa is temperature, which is in linear scale, while ordinate (R ln p) is in logarithmic scale of pressure. Negative sign indicate pressure fall with increasing height.

(iii) Isobars and isotherms are straight and mutually perpendicular.

(iv)   The angle between adiabats and isotherms is about 45°.

(v)   Dryadiabats are logarithmic curves

(vi)   Psuedo-adiabats are markedly curved.

(vii) Saturation mixing ratio lines are gently curved, θ_e lines are curved, w_s lines are slightly curved.

The T - φ gram

As in Emagram

let   B = T, then we get

A = R ln ∝ + F(T)   .....(2.29)

We know $$\frac{T}{\theta} = \left(\frac{p}{1000}\right)^k = \left(\frac{RT}{\propto 1000}\right)^K$$ .....(2.30)

(from Poisson's equation and gas equation)

Where $$k = \frac{R}{C_p}$$

          R = gas constant

For dry air R = 2.87 × 10 ergs/gm°C

          C_p = Specific heat of dry air in isobaric process = 0.24 cal/gm

Taking logarithms

$$\ln T - \ln \theta = k \, [\ln R + \ln T - \ln \alpha - \ln 1000]$$

$$\ln \alpha = -\frac{1}{k} \, [\ln T - \ln \theta] + \ln T + \text{Constant} \qquad \dots(2.31)$$

$$(\text{Constant} = \ln R - \ln 1000)$$

$$\left(\frac{T}{\theta}\right)^{\frac{1}{k}} = \left(\frac{RT}{\alpha 1000}\right) \quad \text{from eq. (2.30)}$$

$$\text{or} \qquad \alpha = \left(\frac{RT}{1000}\right)\left(\frac{\theta}{T}\right)^{\frac{1}{k}}$$

Taking logarithms

$$\ln \alpha = \ln R + \ln T - \ln 1000 + \frac{C_p}{R} \, (\ln \theta - \ln T)$$

$$R \ln \alpha = R \, (\ln R + \ln T) - R \ln 1000 + C_p \, (\ln \theta - \ln T)$$

$$= C_p \ln \theta + R \ln R + R \ln T - C_p \ln T - R \ln 1000$$

$$R \ln \alpha = C_p \ln \theta + (R - C_p) \ln T + \text{constant}$$

$(\because R \ln R \text{ and } R \ln 1000 \text{ are constants})$

$$\text{or} \qquad R \ln \alpha = C_p \ln \theta + G(T) \qquad \dots(2.32)$$

$$[\text{where } G(T) = (R - C_p) \ln T + \text{Constant})$$

From eq. (2.29) and eq. (2.32) we have

$$A = C_p \ln \theta - G(T) + F(T)$$

Now select $G(T) = - F(T)$, which gives

$$A = C_p \ln \theta.$$

We know Entropy $\phi = C_p \ln \theta + \text{Constant}$

$$\therefore \qquad A = C_p \ln \theta \simeq \phi$$

Thus the new axes of coordinates are T (= B) and ϕ (= A = C_p ln θ)

With this coordinates the thermodynamic diagram is called T – ϕ gram (Te-phi gram)

The equations of isobars on T - ϕ gram are obtained from Poisson's

equation $\dfrac{T}{\theta} = \left(\dfrac{p}{1000}\right)^{K} = \text{Constant (for isobars)}$

$$\ln T - \ln \theta = \text{const}$$

$$\text{or} \qquad \ln \theta = \ln T + \text{const}$$

Properties of T - φ gram

(i)   The coordinates of T - φ gram are T and ln θ. That is one axis (T) is linear (or isotherms are straight)

(ii) Isobars are logarithmic curves, which slope upward to the right. As T - increases the slope of the isobar decreases.

(iii) The Psuedo-adiabats are appreciably curved.

(iv) The saturation mixing ratio lines are nearly straight.

(v)   The angle between isotherms and adiabats is 90°.

(vi) Area is proportional to energy.

Phase Change

Phase Change is a transition of a substance from one phase to another. Water can exist in three states solid (ice), liquid (water) and gas (water vapour). As temperature decreases and pressure increases an ideal gas can change from gas phase to the liquid phase, or the solid phase. A substance can exist in either solid, liquid or gas phase or in two phases simultaneously or in all the three phases along the triple line.

At any point on the solid-liquid, solid-vapour or liquid-vapour surfaces, two phases can exist in equilibrium and along the triple line all three phases can coexist. A vapour, at the pressure and temperature, at which it can exist in equilibrium with liquid is termed saturated vapour and the liquid is called saturated liquid.

A typical phase diagram of P-T projection is given in Fig. 2.2. Keeping pressure constant, if a substance is heated, then the points on the dotted line L_1 shows different phases as it passes. Melting point and boiling point are shown as p_1, p_2 respectively.

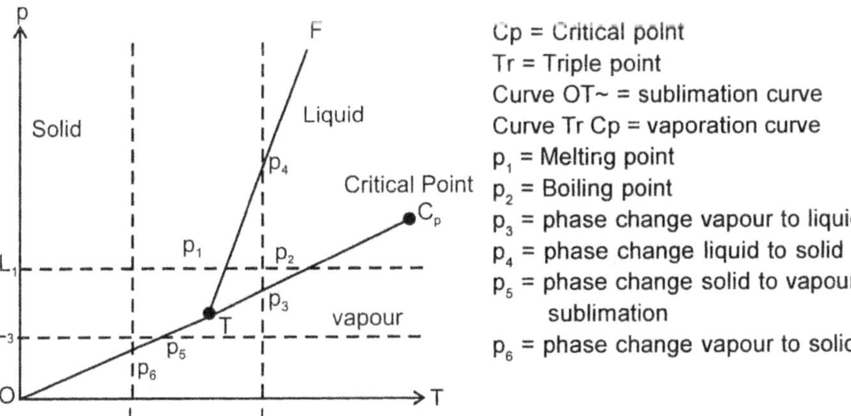

Fig. 2.2 A Typical p-T phase diagram showing sublimation curve vaporisation curve and fusion curve.

Keeping temperature constant, if a substance is compressed, then the points on the dotted line L_2 shows phase changes vapour to liquid and liquid to solid which are shown by the points p_3 and p_4 respectively.

At low pressure and by constant pressure heating material transforms solid to vapour state p_5. The dotted line L_3 represents this process which is called sublimation. Example, solid CO_2 undergoes sublimation (there is no liquid phase). At low temperature, below triple point Tr, if a substance is compressed by increasing pressure vapour directly changes to solid phase shown by point p_6. This process is shown by the dotted line L_4.

Triple point (Tr)

For any substance there is one pressure and temperature at which all three phases namely solid, liquid and gas can coexist. This single point is called triple point.

Critical Point

Liquid and vapour phases can exist together at the critical point and the corresponding values of pressure (p), temperature (T) and specific volume (α) are called the critical pressure, critical temperature and critical specific volume.

Equation of Continuity

Equation of continuity expresses the conservation of mass, that is mass is neither created nor destroyed. It expresses that the rate of change of generation of mass within a given volume is balanced by an equal outflow of mass from the volume. It implies that there are no sources or sinks of mass anywhere in the atmosphere or a given air parcel will retain its mass.

The mathematical equation of continuity is given by

$$\frac{d\rho}{dt} + \rho \nabla \cdot \vec{V} = 0$$

or
$$\frac{\partial \rho}{\partial t} + \vec{V} \cdot \nabla \rho + \rho \nabla \cdot \vec{V} = 0 \qquad \qquad .....(2.33)$$

or
$$\frac{d\rho}{dt} = - \rho \left[\frac{\partial u}{\partial x} + \frac{\delta v}{\partial y} + \frac{\delta w}{\partial z} \right]$$

$$= - \rho \left[\text{div } \vec{V}_H + \frac{\partial w}{\partial z} \right]$$

or
$$\frac{1}{\rho} \frac{d\rho}{dt} + \nabla \vec{V}_H + \frac{\partial w}{\partial z} = 0 \qquad \qquad .....(2.34)$$

where
$$\overrightarrow{V_H} = iu + jv$$

$$\rho = \text{density}$$

$$\vec{V} = \text{wind velocity vector}$$

$$= iu + jv + kw$$

u, v, w are components of velocity

$$\nabla = i\,\frac{\partial}{\partial x} + j\,\frac{\partial}{\partial y} + k\,\frac{\partial}{\partial z}$$

$$\nabla_H = i\,\frac{\partial}{\partial x} + j\,\frac{\partial}{\partial y}$$

Equation of continuity may be interpreted as, if a mass is compressed vertically, it will be expanded horizontal or increase in density or both.

Scale Analysis

In the atmosphere the term $\dfrac{1}{\rho}\dfrac{d\rho}{dt}$ averages about 10% of other two terms

and hence the last above eq. (2.34) can be approximated as

$$\nabla_H \cdot \overrightarrow{V_H} = -\frac{\delta W}{\delta z} \qquad\qquad .....(2.35)$$

Eq. (2.35) implies that surface level convergence will be associated with upward motion at some height indicating some weather (rain etc). On the contrary surface level divergence will be associated with downward motion, indicating fair weather or fog.

For incompressible fluid $\dfrac{d\rho}{dt} = 0$. Equation of continuity eq. (2.33) reduces

to
$$\rho\,\nabla \cdot \vec{v} = 0$$

or
$$\nabla \cdot \vec{v} = 0$$

or
$$\frac{\partial u}{\partial x} + \frac{\partial v}{\partial y} + \frac{\partial w}{\partial z} = 0$$

Axes of reference fixed to the surface of earth

Note : Cartesian coordinate axes are used (unless otherwise stated), in which x-axis points eastwards, y-axis northwards and z-axis upward (zeinth)

Hydrostatic Equation

If a fluid is in equilibrium, then every portion of the fluid is in equilibrium.

Consider a small elemental area $A = \delta x \cdot \delta y$, volume $\delta v = \delta x \cdot \delta y \cdot \delta z$ as shown in Fig. 2.3.

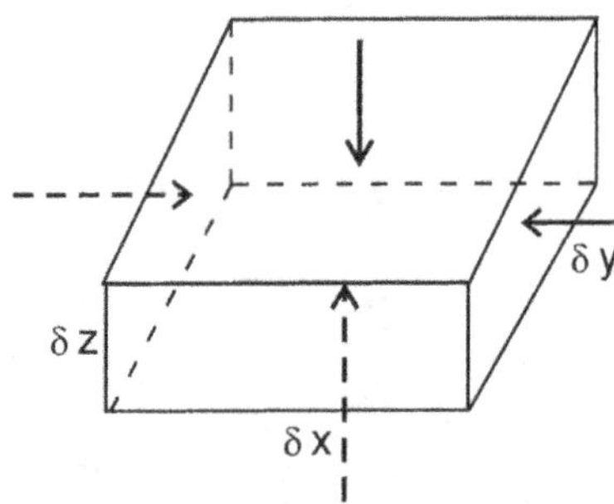

Fig. 2.3 An elemental fluid volume in equilibrim.

Let ρ be the density of the fluid

The mass of the element $= \rho\, Adz$ and weight $dw = g\, \rho\, Adz$

At each of the surface, the forces exerted on the element by the fluid are perpendicular to it. The resultant horizontal forces due to fluid pressure is zero.

Since the fluid is at rest (static) the resultant of vertical forces is also zero. In the vertical direction, pressure force and weight of the fluid acts.

Let p be the pressure on the lower face and be the pressure on the upper face. Then the upward force is pA, while the downward force is (p + dp) A plus weight of the fluid element dw.

$\therefore$ the vertical equilibrium is given by

$$p\,A = (p + dp)\,A + dw$$

$$p\,A = (p + dp)\,A + g\rho\, Adz$$

Simplifying

$$0 = dp + g\,\rho\, dz$$

i.e., $$dp = -\,g\rho dz \quad \text{or} \quad \delta p = -\,g\,\rho\,\delta z \qquad \text{.....(2.36)}$$

This is the fundamental Hydrostatic equation of meteorology.

Integrating eq. (2.36) between p_1, p_2 with corresponding z_1, z_2 we have

$$\int_{p_1}^{p_2} dp = -g\rho \int_{z_1}^{z_2} dz$$

$$p_2 - p_1 = -g\rho(z_2 - z_1) \qquad \qquad \dots(2.37)$$

In hydrostatic equilibrium, pressure at any point is equal to the weight of the column of air on unit cross-section area over that point.

For all practical purposes the hydrostatic equation may be regarded as the vertical component of equation of motion. Further hydrostatic equation tells us that pressure fall with height is steep or rapid where the density of air is largest and virtual temperature is least.

Vertical pressure gradient force $\left(\dfrac{dp}{dz} = -g\rho \right)$ nearly balances gravity force.

Equation for Variation of Density with Height

We know $\qquad dp = -g\,\rho\,dz \qquad$ (Hydrostatic equation) $\qquad \dots(2.38)$

$$p = \rho\, R'\, T \qquad \text{(Ideal gas equation)} \qquad \dots(2.39)$$

Taking logarithm of eq. (2.39)

$$\ln p = \ln \rho + \ln R' + \ln T$$

Differentiating w.r.t z we have

$$\frac{1}{p}\frac{dp}{dz} = \frac{1}{\rho}\frac{d\rho}{dz} + \frac{1}{T}\frac{dT}{dz} \qquad (R' = \text{constants})$$

Using eq. (2.38) this becomes

$$\frac{-g\rho}{p} = \frac{1}{\rho}\frac{d\rho}{dz} + \frac{1}{T}\frac{dT}{dz}$$

$$\text{or} \qquad \frac{1}{\rho}\frac{d\rho}{dz} = -\frac{g\rho}{p} - \frac{1}{T}\frac{dT}{dz}$$

$$= -\frac{g}{RT} - \frac{1}{T}\frac{dT}{dz}$$

$$\frac{1}{\rho}\frac{d\rho}{dz} = -\frac{1}{T}\left(\frac{g}{R'} + \frac{dT}{dz} \right) \qquad \dots(2.40)$$

eq. (2.40) is the required equation of variation of density with height.

Let us consider some hypothetical atmospheres which may be useful as model in atmospheric studies.

1. Homogeneous Atmosphere

If we assume the atmosphere as homogeneous then ρ = const. The above eq. (2.40) reduces to

$$\frac{g}{R'} + \frac{dT}{dz} = 0$$

or $\qquad \dfrac{dT}{dz} = -\dfrac{g}{R'} \qquad$ for dry air $R' = 287 \ m^2/s^2$

$$= \frac{-9.8}{287} = -0.0341 \ C^\circ/m$$

$$= -34.1 \ ^\circ C/km.$$

$\therefore$ the lapse rate of homogeneous atmosphere is $-34.1 \ C^\circ/Km$

To find the height of homogeneous atmosphere; we have

when $\qquad p = p_0 \quad z = 0$

and $\ $ Let $z = H \qquad$ when $p = 0$

$dp = -g \rho \ dz \qquad\qquad$ Hydrostatic equation

Since in homogeneous atmosphere ρ is constant, integrating we have

$$\int_{P_0}^{0} dp = -g\rho \int_{0}^{H} dz$$

This gives $\quad p_o = g \rho H$

or $\qquad H = \dfrac{p_o}{g \rho} = \dfrac{\rho R'T}{g\rho}$

$$H = \frac{R'T}{g}$$

Put in this equation

$$R' = 287 \ m^2/s^2 \text{ for dry air}$$

$$g = 9.8 \ m/s^2, \ T = 273 \ k^\circ$$

$$H = \frac{287 \times 273}{9.8} \simeq 8000 \ m = \ 8 \ km.$$

2. Isothermal Atmosphere

If we assume there is no change in temperature throughout the atmosphere $\left(\dfrac{dT}{dz} = 0\right)$ we get

$$p = p_o \, e^{\frac{-z}{H}}$$

where $$H = \frac{R' T}{g}$$

This shows that pressure decreases with height exponentially and there is no upper boundary of the atmosphere.

3. **Uniform Lapse Rate of Atmosphere**

If we assume that temperature decreases with height uniformly

$$\frac{dT}{dz} = \text{const} = \Gamma \ (\text{say})$$

we get $$p = p_o \left(\frac{T}{T_o}\right)^{\frac{g}{\Gamma R}}$$

where T_o = temperature at sea level

This gives pressure is a function of temperature T, while T is a function of z, leading to $p = p(z)$

4. **Adiabatic Atmosphere**

In adiabatic process there is no change in potential temperature (θ), throughout the atmosphere θ = const.

using Poisson's equation, we get

[assuming sealevel temperature 0 °C]

$$\frac{dT}{dz} = \frac{-g}{c_p} = \frac{-9.8}{1004} \frac{m/s^2}{J\,Kg^{-1}} K^{-1}$$

$$\simeq 0.0098 \ °C/m = 9.8 \ °C/km$$

i.e., the lapse rate of adiabatic atmosphere is 9.8 °C/km.

The height of the adiabatic atmosphere is roughly 27.3 km.

Standard Atmosphere

The hypothetical static atmosphere described above (four cases) do not fit as good approximation of real atmosphere. However these are used as models. Based on these models a standard atmosphere has been advocated as a reference model with the following two simple criteria. In this model lapse rate of temperature from sea level to 10 km altitude is 6.5 °C/km. This layer is called troposphere. Above the top of troposphere temperature assumed to be constant.

The NACA or US Standard Atmosphere

For various practical uses in aviation National Advisory Committee for Aeronautics in America, defined a standard atmosphere with the following specifications up to an altitude of 32 km

1. Air is assumed to be dry and obeys gas law.

2. At sea level surface air temperature 15 °C and pressure 1013.25 h Pa

3. Temperature lapse rate 6.5 °C/Km upto an altitude of 11 Km. This layer is called troposphere.

   Above 11 Km upto 32 Km, the atmosphere is assumed to be isothermal with constant temperature of –55 °C. This layer is called stratosphere.

4. The region above 11 Km elevation separating troposphere and stratosphere is called tropopause.

5. The value of g is taken to be constant $= 9.86066$ m/s^2

6. Above 32 Km temperature inversion (rise) of 1 °C/Km is taken

7. The above criteria is also used by ICAO (International Civil Aviation Organisation).

ITRA : International Tropical Reference Atmosphere

Based on US standard Atmosphere (1976), Ananthasayanam and Narasimha (1985) have proposed ITRA extending to 80 Km asl. The salient features of this atmosphere are given below.

1. ITRA is applicable between Tropic of Cancer (Lat 23½°N) to Tropic of Capricorn (Lat 23½°S)

2. Air is assumed to be dry and obeys gas law as in USSA. The molecular weight, ratios of specific heats of air, gas constant and other constants have the same values as in USSA (1976). Density of air at sea level is taken is 1.225 Kg/m^3.

3. At sea level surface air temperature 27 °C (300 °K) and pressure 1010 hPa

4. The acceleration due to gravity is g $= 9.7885$ m/s^2

5. Temperature lapse rate $\dfrac{d\,T}{d\,z}$.

   (a) 6 °C/Km from sea level to 6 Km altitude.

   (b) 6.5 °C/Km between 6 to 16 Km asl

   (c) Tropopause height 16 Km with temperature –74 °C.

   (d) – 2.3 °C/Km (inversion) between 16 to 46 Km asL

   (e) Stratosphere altitude of 46 Km and temperature – 5 °C.

(f) Isothermal (layer of lapse rate zero) between 46 to 52 Km asl with temperature of – 5 °C.

(g) 3 °C/Km between 52 to 75 Km altitude.

(h) Mesopause near 75 Km asl, with temperature – 74 °C.

(i) Above Mesopause, isothermal layer 75 to 80 Km with temperature –74 °C.

The Field Variables of Atmosphere

Fluid Particle

A fluid particle is an infinitesimal volume (strictly a geometrical point) whose linear dimensions are disregarded for the purpose of finding its velocity and acceleration. This infinitesimal volume whose size, however, is many times larger than the atmospheric intermolecular distance (order 10^{-8}m or $10^{-2}\mu$m in gases at standard condition).

An ideal fluid is a continuous fluid substance that cannot excert any shearing stress however small. Atmosphere is regarded as a continuous fluid medium (or contimum). The physical quantities pressure, temperature, density, velocity, mass, composition are assumed to have unique values at any point in the atmosphere. These field variables and their derivatives are assumed to be continuous functions of space and time.

Mathematical Equations of Motion

Newtons second law of motion may be expressed in different ways as below.

1. The first derivative of the momentum of a particle $\vec{P_i}$ (w.r.t time) is equal to the force $\vec{F_i}$ acting on the particle

$$\frac{d\,\vec{p_i}}{dt} = \vec{F_i} \quad \text{or} \quad \frac{d\,(mi\,\vec{V_i})}{dt} = \vec{F_i}$$

where $\quad \vec{P_i}$ = momentum = $mi\,\vec{vi}$

mi = mass of the particle

$\vec{V_i}$ = velocity of the particle.

2. A small (infinitesimal) change in momentum of a particle is equal to the impulse of the force acting on it

Impulse = Force × time

$$J = \vec{F} \times t$$

$$d\,\vec{p_i} = \vec{F_i}\,dt \quad \text{or} \quad d(mi\,\vec{vi}) = \vec{F_i}\,dt.$$

$\vec{Fi}.dt$ or $\vec{Fi}.\delta t$ is called an impulse of force $\vec{Fi}$.

Impulse during the time period t_1 to t_2 is given by

$$J = \vec{Fi}\ (t_2 - t_1)$$

3. The acceleration $(\vec{a_i})$ of a particle is directly proportional to the force $(\vec{Fi})$ acting on it and inversely proportional to the mass (mi) of the particle and coincides with the direction of the force.

$$\vec{a_i} = \frac{d\vec{Vi}}{dt} = \frac{\vec{Fi}}{mi}$$

4. If $\vec{Fi} = i\ \vec{F_x} + j\ \vec{F_y} + k\ \vec{F_z}$, where i, j, k are unit vectors in x, y, z directions and Fx, Fy, Fz are components of $\vec{Fi}$ in these directions.

and If $\vec{Vi} = iu + jv + kw$

where $\vec{Vi}$ = velocity vector, u, v, w, components of velocity in x, y, z directions

    $\vec{r} = ix + jy + kz$

Then $\vec{Vi} = \dfrac{d\vec{r}}{dt} = \dfrac{d}{dt}\ (ix + jy + kz)$ gives

$$u = \frac{dx}{dt}, v = \frac{dy}{dt}, w = \frac{dz}{dt}$$

$$\vec{a_i} = \frac{d\vec{Vi}}{dt} = \frac{\vec{Fi}}{m_i} = \frac{i\vec{F_x} + j\vec{F_y} + kF_z}{m_i}$$

This gives

$$\vec{F_x} = m_i\ \frac{d^2 x}{dt^2}$$

$$\vec{F_y} = m_i\ \frac{d^2 y}{dt^2}$$

$$\vec{F_z} = m_i\ \frac{d^2 z}{dt^2}$$

These are the differential equations of motion.

Streamline

A streamline is a curve drawn in a fluid such that at any instant of time, the direction of the tangent at any point of the curve coincides with the direction of the velocity of the fluid particle at that point.

The differential equations of streamline are given by

$$\frac{dx}{u} = \frac{dy}{v} = \frac{dz}{w}$$

Pathline

The path line or trajectory is a curve, which a particular fluid particle describes during its motion.

The differential equations of the pathline are given by

$$\frac{dx}{dt} = u, \quad \frac{dy}{dt} = v, \quad \frac{dz}{dt} = w$$

Note : Streamlines show how each particle is moving at a given instant, while the path line shows how a given particle is moving at each instant.

Euler's approach to the hydrodynamical or fluid dynamical problems consist in observing the changes in velocity, density and pressure as the fluid passess through a fixed point in space occupied by the fluid or local time rate of change.

Lagrange's approach to the hydro (fluid) dynamical problems consist in observing the changes in velocity, density and pressure as a particular typical particle itself moves about or individual time rate of change.

Streamlines do not cross one another, however they may join at isolated points, such as, col region, or infinite velocity. Fluid flow will be along stream line but not across it. Streamline spacing varies inversely with velocity. Convergent streamlines indicate accelerated flow in that direction.

Coordinate System

In meteorology the coordinate system is very important. Here we consider the direction of x-axis towards east, y-axis towards north and z-axis upward perpendicular to the surface level or zeinth.

Equations of Motion

In meteorology, the equations of motion given by the statement that

$$\text{acceleration} \left(\frac{d\vec{V}}{dt} \right) = \text{vector sum of all forces per unit mass}$$

$$= \sum_{i=1}^{n} F_i$$

$$= \text{Pressure gradient force} + \text{Coriolis force}$$
$$+ \text{Frictional force} + \text{gravitational force}$$

Thus,

$$\frac{d\vec{V_i}}{dt} = -\frac{1}{\rho} \nabla p - 2\vec{\Omega} \times \vec{V} + \vec{g} + \frac{1}{\rho}\frac{d\vec{\tau}}{dz}$$

where

$\vec{V}$ = wind velocity vector

∇p = Pressure gradient force

$\vec{\Omega}$ = Angular velocity of the earth or earth's rotation vector

$\vec{g}$ = gravitational force vector

$\vec{\tau}$ = Frictional force per unit area

ρ = density

Coriolis Force

The effect of earth's rotation on moving objects over the surface of the earth was explained by Gaspard Coriolis and hence it is called Coriolis force by meteorologists.

An object in motion on the surface of the earth is deflected every stage in its motion relative to the moving earth. The deflection is caused by an apparent force, is called Coriolis force.

Coriolis force deflects wind to the right of its motion in the northern hemisphere and to the left in southern hemisphere. This effect is given by the formula

$$2\,\Omega\,V\,\text{Sin}\,\phi \text{ or } fV$$

where Ω = angular velocity of the earth

         V = horizontal wind speed

         ϕ = latitude

         $f = 2\Omega\,\text{Sin}\,\phi$, is called Coriolis parameter

There is no deflection at the equator ($\phi = 0$) but increases slowly and attains maximum value at the poles. Because of Coriolis effect, near the surface of the earth, the north east and south-east trade winds have developed.

Buys Ballot's Law

If you stand with your back to the wind, the low pressure will be on your left in the northern hemisphere and to the right in the southern hemisphere.

We shall now consider some important wind models which lend support to numerical weather predictions; Geostrophic wind, Gradient wind and Thermal wind.

Geostrophic Wind

If we consider (i) frictionless flow (i.e., flow of air in free atmosphere, about 1 km amsl), (ii) further if we assume the flow is free from acceleration and constant and overlook gravity and friction, then the wind flow is termed geostrophic wind. The equation of motion reduces to

$$0 = -\alpha \, \nabla p - 2 \, \vec{\Omega} \times \vec{V}_g$$

or $$\alpha \, \nabla p = -2 \, \vec{\Omega} \times \vec{V}_g$$

where $$\vec{V}_g = \text{Geostophic wind.}$$

Definition

The horizontal wind occurring in a frictionless flow, when the pressure gradient force is in balance with the Coriolis force is called geostophic wind.

That is, the wind that occurs in atmosphere which is hydrostatic and horizontal in the absence of acceleration and friction. The following Fig. 2.4 shows the geostrophic balance.

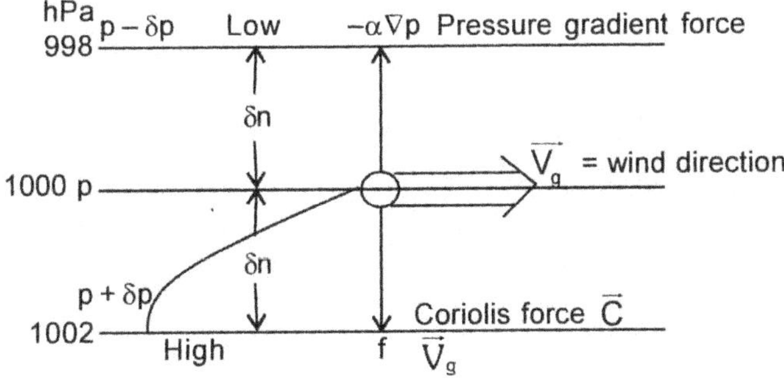

Fig. 2.4 Geostrophic balance.

If $\qquad \vec{V}_g = iu_g + jv_g$

we have

$$\alpha \frac{\partial p}{\partial x} = fv_g$$

$$\alpha \frac{\partial p}{\partial y} = -fu_g$$

Geostrophic wind as a finite difference is given by

$$Vg = \frac{\alpha}{f} \frac{\delta p}{\delta n} \quad \text{or} \quad \frac{\alpha}{f} \frac{\Delta p}{\Delta n}$$

$$Vg = \frac{9.8}{f} \frac{\partial H}{\partial n},$$

where H is measured in gpm.

Properties of Geostrophic Wind ($\vec{V}_g$)

Pressure gradient force $\alpha \nabla p$ is normal to the isobars and directed from high to low, while Coriolis force is also normal to the isobars but in the opposite direction, directed from Low to High.

Geostophic wind vector obeys Buys Ballot's law, which states that, looking downstream low pressure will be on the left in the northern hemisphere and to the right in the southern hemisphere.

Geostrophic wind is proportional to the pressure gradient and inversely proportional to the distance between the isobars, that is closers the isobars the stronger the wind and vice versa.

$$Vg \,\alpha\, \nabla p, \qquad Vg \,\alpha\, \frac{1}{\Delta n}$$

$$Vg \,\alpha\, \frac{1}{f}$$

(Geostrophic wind is inversely proportional to the Coriolis parameter)

The geostrophic approximation applies only when the isobars are straight and parallel and pressure distribution is steady and not varying rapidly with time. Geostrophic approximation is not valid in case of local winds such as land and sea breeze. Geostrophic wind is a good estimate of actual wind above friction layer.

Horizontal pressure gradient force generally balances the Coriolis force.

Atmosphere is hydrostatic and nearly in geostrophic balance, however geostrophic flow rarely occurs between Lat 15 °N, 15 °S.

These properties are used in numerical prediction models.

Note : Wind is generally measured at the standard height of 10 m agl by wind anemometers. The observed wind over land, on average is only one third of the geostrophic wind and over sea areas about two-third's of geostrophic wind . This is because of the friction effect. It is also observed that wind does not blow parallel to the isobars but blows cross isobars about an angle of $30°$ over land and $10°$ over sea areas.

The Gradient Wind

The wind occurring in a frictionless flow, when there is a balance between the pressure gradient force $(\alpha \nabla p)$, the coriolis force $(2\vec{\Omega} \times \vec{V})$ and the centrifugal force $(\Omega^2 \vec{R})$.

that is, $$\alpha \nabla p = -2\vec{\Omega} \times \vec{V} - \Omega^2 \vec{R}$$

In this case the magnitude of $\alpha \nabla p \neq \vec{C}$. Consequently gradient wind flow will be either to the left or to the right of isobars and the flow is tangential to the isobars.

Geostrophic wind is a particular case of gradient wind in which the centrifugal force is negligible as compared to pressure gradient force and Coriolis force.

In case of anticlock wise circulation (Fig. 2.5 (a))

The equation of gradient wind flow is given by

$$\frac{f V_g}{V_G^2} - \frac{f}{V_G} - \frac{1}{R} = 0$$

or $$V_G = \frac{V_g}{\frac{1}{2} + \sqrt{\frac{1}{4} + \frac{V_g}{fR}}}$$ .....(2.41)

where V_G = magnitude of gradient wind flow

V_g = magnitude of geostophic wind flow

R = radius of circular motion

From eq. (2.41) it is clear that cyclonic gradient wind is subgeostrophic $(V_G < V_g)$

In case of clockwise circulation (Fig. 2.5 (b))

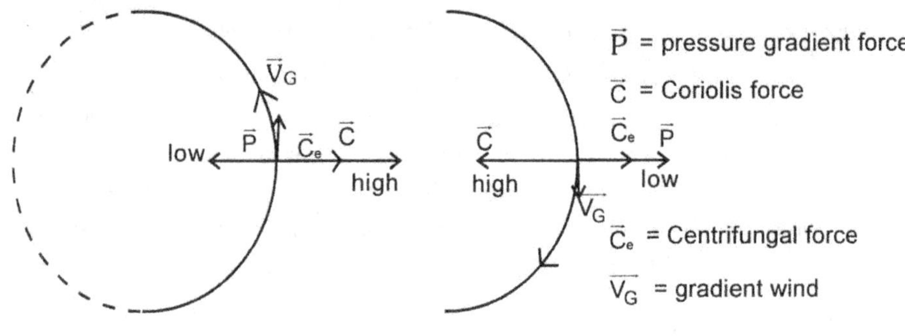

Fig 2.5

The equation of gradient wind is given by

$$V_G = \frac{2\,V_g}{1 + \sqrt{1 - \dfrac{4\,V_g}{R\,f}}} \qquad \qquad .....(2.42)$$

It follows from eq. (2.42) that anticyclonic gradient wind is supergeostrophic $(V_G > V_g)$.

Cyclonic and Anticyclonic Flow

In northern hemisphere, if the wind flow is anticlockwise it is called cyclonic flow and clockwise flow is called anticyclonic flow.

In southern hemisphere, the clockwise wind flow is called cyclonic and anticlockwise flow is called anticyclonic.

Advection of Temperature

The change in temperature in horizontal adiabatic flow is regarded as geostrophic advection or simply advection.

Veering of Wind with Height

Clockwise change of wind direction in two levels L and U is called Veering of wind.

Backing of Wind with Height

Anticlockwise change of wind direction in two levels L and U (lower and upper) is called backing of wind.

Veering of geostrophic wind with height indicates warm air advection.

Backing of geostrophic wind with height indicates cold advection.

Thermal Wind

Wind aloft at any level is considered to have two components (i) Geostrophic wind above friction layer and (ii) a wind blowing along the mean isotherms with low temperature to its left in the northern hemisphere (and to its right, in the southern hemisphere). The second component is called the thermal wind.

Consider two levels L (lower) and U (upper).

Let $\vec{V}_g L$ = Geostrophic wind vectors in lower level

$\vec{V}_g U$ = Geostrophic wind vector in upper level

The vector difference $\vec{V}_T = \vec{V}_g U - \vec{V}_g L$, is called thermal wind in the layers

L and U. $\vec{V}_T$ is parallel to the isotherms see Fig. 2.6.

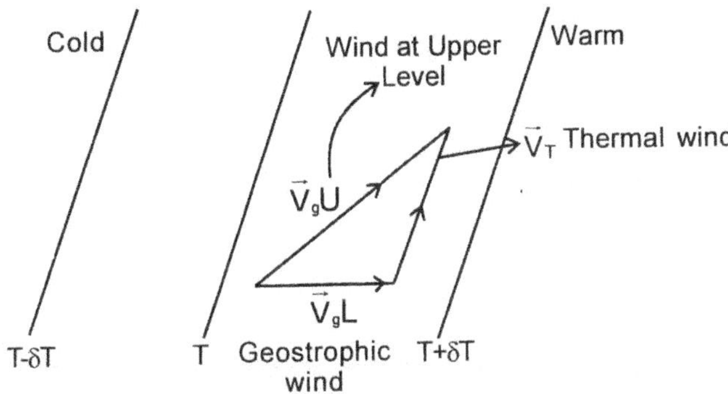

Fig 2.6 Thermal wind.

Thermal wind is a fictitious wind and it has no actual existence. However the concept is of great importance in practical weather prognosis.

1. The approximate formula for thermal wind

$$\left|\vec{V}_T\right| = V_T \text{ is given by}$$

$$V_T = \frac{g}{f\,\overline{T}} \frac{\partial \overline{T}}{\partial n} \delta z \qquad\qquad .....(2.43)$$

where $\dfrac{\partial \overline{T}}{\partial n}$ = Temperature gradient

$\overline{T}$ = Mean temperature of the layers

δz = Thickness between the layers

2. V_T is also given by

$$V_T = \frac{9.8}{f} \frac{\partial H}{\partial n} \qquad \qquad .....(2.44)$$

where $\dfrac{\partial H}{\partial n}$ = horizontal gradient of relative geopotential.

3. Geostrophic wind variation over small height 'h' is given by

$$\frac{\partial \vec{V_g}}{\partial z} = \frac{\vec{V_g}(z+h) - \vec{V}g\,(z)}{h}$$

Barotropic Atmosphere

An atmosphere in which isobaric surfaces are also surfaces of constant density.

In barotrophic atmosphere ($\vec{V_g}$) geostrophic wind is constant with height

$\vec{V_g}$ = constant.

Baroclinic Atmosphere

An atmosphere which is not barotropic.

In baroclinic atmosphere $\vec{V_g}$ is (not constant) variable with height.

1. At a given location, the thickness (δz) between two isobaric surfaces is proportional to the mean virtual temperature.

$$\Delta Z \, \alpha \, \overline{T_v}$$

2. Thermal wind is parallel to the isopleths of thickness (Δz) lines and hence coincide with the mean virtual isotherms.

3. Magnitude of the thermal wind is proportional to the magnitude of the thickness gradient. Closer the thickness lines the stronger the thermal wind and vice versa.

4. In northern hemisphere, looking downstream the cold area or lower thickness area will be on the left and warm area will be on the right. In southern hemisphere the reverse is true.

Convergence and Divergence

At surface level, in both hemispheres, loss of mass and fall of atmospheric pressure is associated with convergence of wind and gain of mass and rise in atmospheric pressure is associated with surface divergence of wind.

$$\nabla . \vec{V} > 0, \text{ divergence}$$

$$\nabla . \vec{V} \; < \; 0, \;\; \text{convergence}$$

where $\quad \nabla = i \dfrac{\partial}{\partial x} + j \dfrac{\partial}{\partial y} + k \dfrac{\partial}{\partial z} \quad \& \quad \vec{V} = iu + jv + kw$

Circulation(c)

Flow of wind round a closed curve is called circulation

$$C = \oint \vec{V} . \vec{dr}$$

$$= \oint (udx + vdy + wdz)$$

where $\quad \vec{dr} = idx + jdy + kdz, \;\; \vec{v} = iu + jv + kw$

Kelvins Circulation Theorem

The acceleration of circulation (C_{ac}) is equal to the circulation of the acceleration

$$\frac{dC}{dt} = \oint \frac{d\vec{V_r}}{dt} . \vec{dr}$$

Solenoids

A soleniod is a three dimensional body formed by the intersection of two sets of equi-scalar surfaces.

An equi-scalar surface is one on which a scalar property remains constant. e.g., isobaric surface, isothermal surface, isosteric surface etc.

Note : Presence of more number of solenoids in a vertical cross section indicates more available energy to begin circulation.

Vorticity

The velocity of rotation (angular velocity) of an infinitesimal fluid element is called vorticity.

$$\text{Vorticity} = \frac{1}{2} \nabla \times \vec{v}$$

The rotation of a vector field is given by

$$\text{rot } \vec{v} = \nabla \times \vec{v} \quad \text{or} \quad \text{Curl } \vec{v} .$$

If ϕ is a scalar function, then gradient of ϕ (denoted by grad ϕ) is given by

$$\text{grad } \phi = \nabla \phi$$

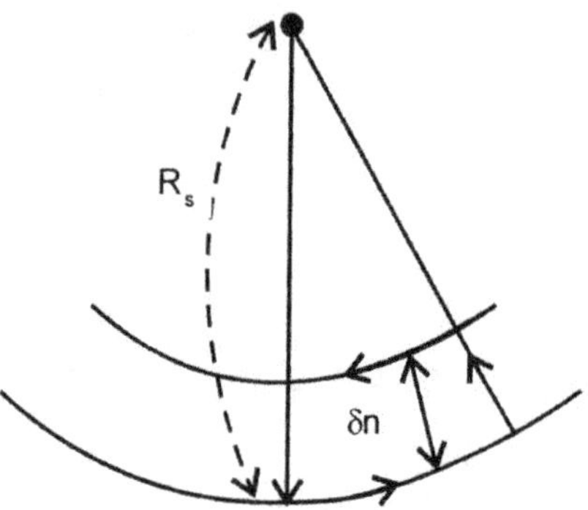

The vertical component of vorticity (ζ) is given by

$$\zeta = \frac{V}{R_s} - \frac{\partial V}{\partial n}$$

where $\dfrac{V}{R_s}$ denotes curvature

$\dfrac{\partial V}{\partial n}$ denotes shear.

R_s = radius of curvature, V = velocity along a streamline, δ_n = distance between streamlines.

Inference of Divergence, Vertical Motion and Vorticity from Synoptic (Surface and Upper Air) Charts

Analysis of weather charts enables the meteorologist to understand the current weather and to infer the future weather. The three terms divergence, vertical motion and vorticity are the key terms to understand and to unravel the weather on synoptic scale.

Fields of Divergence

The wind field is developed by the variation of wind speed along the streamlines and coming together (confluence) or going apart (difluence) of streamlines. Based on wind speed, divergence of wind is inferred as follows.

1. When the wind speed increases downstream in a field of parallel streamlines divergence occurs.

2. When the wind speed decreases downstream in a field of parallel streamlines convergence occurs. (Fig. 2.8(b))

3. Fanning out of streamlines is called difluence, converging (coming together) of streamlines is called confluence.

4. If streamlines fan out in an area of uniform wind speed along streamlines divergence occurs.

5. If the streamlines come together in an area of uniform wind speed along streamlines convergence occurs.

6. If the wind speed increases downstream and simultaneously streamlines fan out indicate divergence.

7. If the wind speed decreases downstream and simultaneously streamlines come together indicate convergence.

8. Areas of divergence indicate generation of mechanic Kinetic energy, while convergence indicate dissipation of Kinetic energy.

9. *Vertical motion of wind*

   Horizontal convergence (or divergence) of wind near the surface of the earth causes vertical motion ascent (or descent)

10. In case of low level convergence and simultaneous high level divergence at the same area leads to vertical motion. Such situations are conducive for triggering convective activity and precipitation phenomena. Contrary to this, low level divergence associated with high level convergence leads to subsidence, generally fair weather or smog/fog, inversion of temperature near the earth's surface.

A level at which neither convergence nor divergence effectively observed is called level of non-divergence.

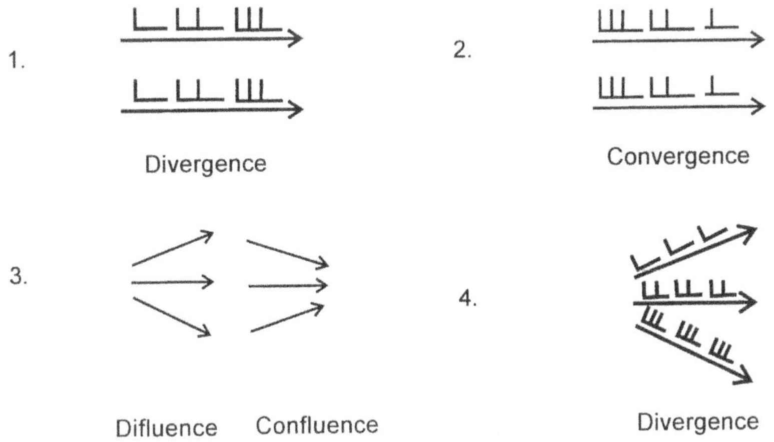

1. Divergence

2. Convergence

3. Difluence   Confluence

4. Divergence

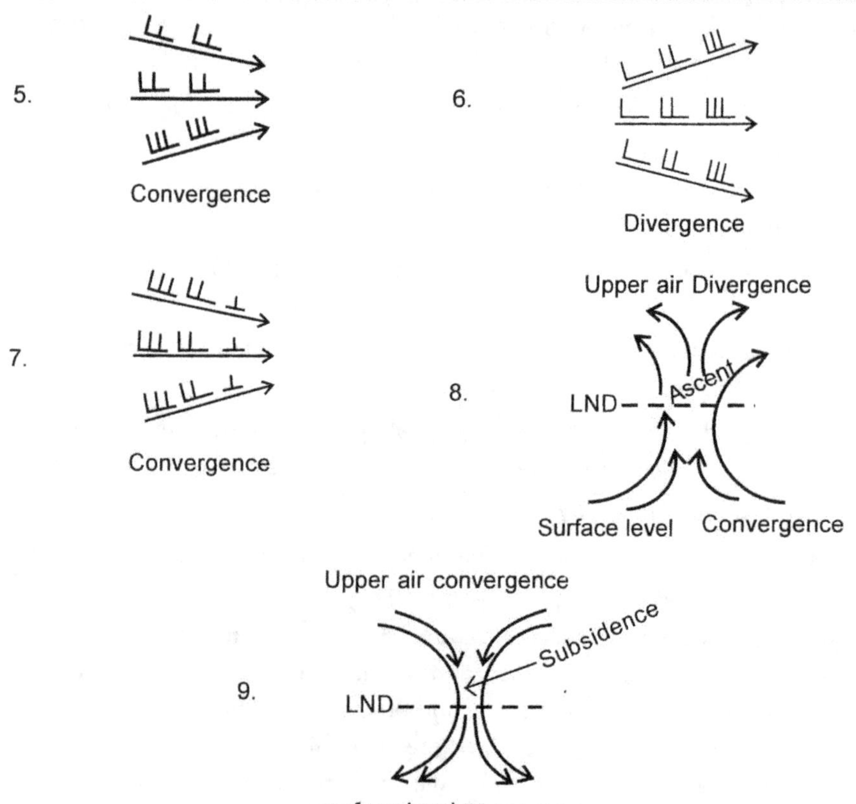

Fig. 2.7 Fields of Divergence.

Note : LND-level of Non-divergence

Fields of Vorticity

For Northern Hemisphere

1. If the flow is of uniform speed and isotaches (equal wind speedlines) are normal to the streamlines, then curvature of the flow determines the sign of vorticity (positive for cyclonic and negative for anticylonic).

2. In a straight flow, looking downstream, decreasing of wind speed to the left of the current indicates cyclonic vorticity.

3. In a straight flow, looking downstream, increasing of wind speed to the left of the current indicates anticyclonic vorticity.

4. In streamlines of cyclonic curvature decreasing of wind speed to the left of the current (looking downstream) indicates cyclonic vorticity.

5. In streamlines of anticylonic curvature, increasing of wind speed to the left of the current (looking downstream) indicate anticyclonic vorticity.

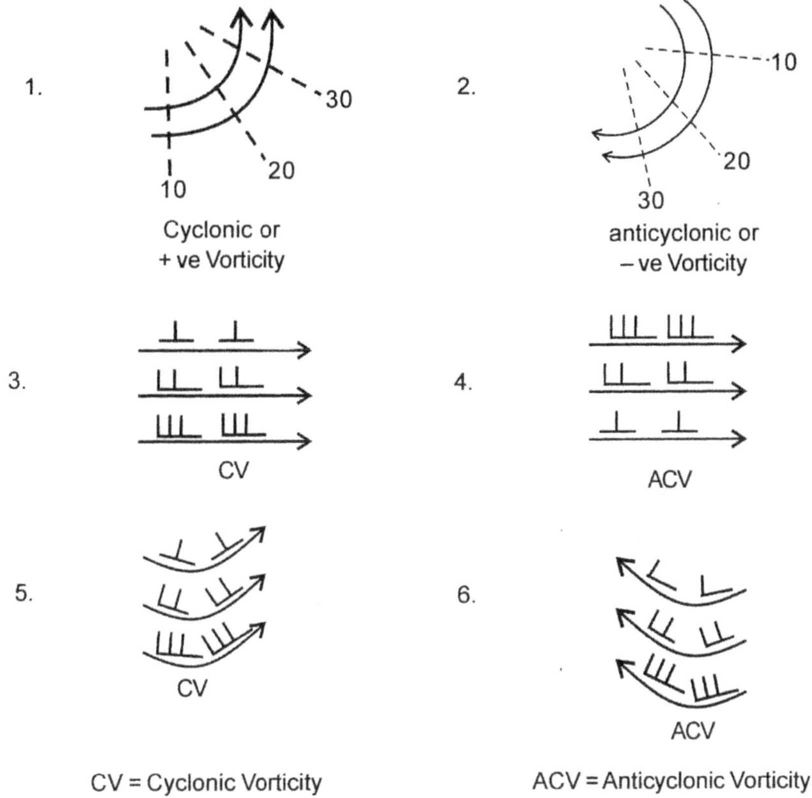

Fig. 2.8 Fields of vorticity in northern hemisphere.

Synoptic Chart

A chart or map on which meteorological data (which represents the state of atmosphere wind, pressure, temperature, cloud, present weather, precipitation, dew point etc, over a large area at a given instant of WMO/local standard times) presented and analysed is called a synoptic chart. Synoptic means having the same view point.

Surface Chart

A synoptic chart on which surface wind (direction and speed), visibility, pressure, temperature, dewpoint, rainfall, cloud (type and amount, present weather (fog, mist, rain, thunderstorms, squalls, etc) past weather (with respect to the time of observation) are plotted is called surface chart or sea level chart. Observation of these elements recorded or observed from ground. Except the station pressure which is reduced to mean sea level (msl) all other elements do not belong to a constant level.

Upper Air (aerological constant pressure) Chart

These charts are prepared for the free atmosphere for mandatory pressure levels [such as 850, 700, 500, 400, 300, 250, 200, 150, 100, 70, 50, 30, 20, 10 h Pa pressure levels]. These charts contain upper air/aerological observations of wind (direction/speed), temperature and dew point.

Synoptic Station

A weather observing station for synoptic purpose. As per WMO, the chief synoptic observation times are : 0000, 0600, 1200 and 1800 UTC (or GMT).

Subsidiary synoptic hours are 0300, 0900, 1500 and 2100 hrs UTC.

Synoptic Forecasting

Based on the analysis of synoptic (surface and upper air) charts and auxillary charts such as 24 hour change chart, vertical time section, cross-section, thickness charts, weather inference drawn. Based on this inference, climatology and experience synoptic weather forecasting is made.

In addition to the above we have numerical, statistical, climatological and remote sensing based forecasts.

WMO Recommended Network of Synoptic Station Density

Principal land based observatories should be at an intervals of not exceeding 150 Km. In sparcely populated area like deserts, hill stations the density of surface observatories should not be farther apart 500 Km and upper air land stations should not be farther apart than 1000 Km.

Surface synoptic stations may be land stations manned or automatic. There may be sea stations fixed or mobile, manned or automatic. These may be located on fixed ships or mobile merchant ships, ocean weather ships, light ships fixed or anchored platforms, buoys and ice floes.

Upper air synoptic stations may be land or sea stations where Radiosonde, Rawin and Radio wind or Pilot Balloon observations are taken.

Scales of Atmospheric Motions

Atmospheric motions exhibit different scales of time and space. They may be smallest and rapid such as molecular diffusion, sound waves. Next scale like dust devils, thunderstorms, tornadoes etc. Further large scale like frontal disturbances (frontal waves), tropical disturbances depression, cyclones and hurricanes. The largest are Rossby waves. The sizes thus vary from fraction of a centimeter to several thousand kilometers. Typical scales of motions are given in Table 2.1.

Table 2.1 Different scales of atmospheric motions.

Types of motion	Order of horizontal scale in meters
Molecualr mean freepath in atmosphere	10^{-9} m = 10^{-3} µm
Minute turbulent eddies	$10^4 - 10^5$ µm (10^{-2} - 10^{-1}m)
small eddies	10^5 µm to 1 m = 10^6 µm ($10^{-1} - 1$m)
Dust devils	1 to 10 m
Gusts of wind	10 to 10^2 m
Tornadoes	10^2 to 10^3 m
Cb cell (thunderstorm)	10^3 m = 1 Km
Fronts and squall lines	10 km – 100 Km
Cyclone/hurricanes	100 to 1000 Km
Planetary waves/Rossby waves	10^4Km
Zonal wind/jet streams	10^4 Km
Planetary-scale Horizontal length	5000 Km
Synoptic-scale Horizontal length	1000 Km
Meso-scale Horizontal length	100Km
Micro-scale Horizontal length	$\leq$ 1 Km

Time scale

    Planetary scale $\geq$ 5 days.

    Synoptic scale   1 day

    Meso-scale   5 hours

    Micro-scale $\leq$ 10 minutes

Note : Average diameter of a molecule in air is about 3.7×10^{-10} m

Number of molecules of air in one $cm^3 = 3 \times 10^{19}$

$$\text{in one } m^3 = 3 \times 10^{25}$$

Mean free path of an air molecule 6×10^{-8} m

Micro-scale

This includes the fine structure of physical process, such as diffusion, evaporation, turbulence, heat fluxes or transfers. Many of these processes can occur close to the ground surface and also at higher levels.

Meso-scale

This includes weather phenomena, which are very small to be identified or located on synoptic charts but they cause profound effects. Examples, Tornadoes, thunderstorm, cloud bursts, land and sea breeze etc. These can be studied on graphical scale between micro-meteorology and synoptic meteorology.

Synoptic Scale

This covers weather phenomena which are observed in a network observation of stations 100 to 1000 Km apart. Examples, Tropical and extra tropical cyclones, hurricanes.

Planetary Scale

This include weather systems observed on hemispherical scale, such as long Rossby Waves, mean zonal wind, jet streams etc.

Cosmic Scale

This include those weather phenomena which involve inter-planetary environment such solar wind.

Questions

1. Define ideal gas and the four laws that it obeys.

2. Using hydrostatic and gas equations assuming constant temperature derive an expression for variation of pressure with altitude.

3. Derive Poisson's equation and define potential temperature.

4. Define adiabatic process and write the properties of ideal gas in adiabatic process.

5. What are the main features in Thermodynamic diagrams, Write the properties of emagram.

6. What is $T - \phi$ gram? Derive the expression $\ln \theta = \ln T + const.$ Write the properties of $T - \phi$ gram.

7. Write the mathematical equation of continuity. What it expresses? Derive the equation of continuity for incompressible fluid?

8. Using hydrostatic and gas equations derive an expression for variation of density with height in the atmosphere?

9. State Newton's second law of motion and hence derive differential equations of motion.

10. Define a streamline and pathline and write the corresponding differential equations.

11. Write briefly on : (i) Coriolis force (ii) Buy's Ballot's law (iii) Geostrophic wind, (iv) Gradient wind, (v) Thermal wind.

CHAPTER 3

The Motion of the Earth

The earth has two motions. (a) Rotation, (b) Revolution. The earth rotates about its axis (west to eastward) with a speed of 464 m/s or 1669 km/hr. Its angular speed of rotation is about 7.29×10^{-5} radians/sec or 0.26244 radians per hour. The axis of the earth is inclined at an angle of 23½ ° to its elliptical orbit. This rotation causes day and night and diurnal variations of weather. The rotating earth revolves round the sun in an elliptical orbit (with the sun at one of its Foci) with a speed (orbital speed) of about 1.073×10^5 km/hr or 29770 m/s or 29.8 km/s. This revolution causes seasons. It has been stated earlier that the apparent path of the sun with respect to the celestial sphere is called ecliptic.

Equinoxes

When the sun apparently crosses the celestial equator (ecliptic intersects the celestial equator), all the places on the earth experience equal (length) day and nights. Every year equinoxes occur on March 21 and September 23.

Solstices

When the sun moves over the Tropic of Cancer (lat 23½ ° N, farthest north) it is called summer solstice (June 21), which is associated with the longest day and shortest night in the northern hemisphere. When the sun moves over to the Tropic of Capricorn (lat 23½ ° S, farthest south) it is called winter solstice (December 22), which is associated with the longest day and shortest night in the southern hemisphere. The two points on the ecliptic which are farthest from the celestial equator are solstices Fig. 3.1.

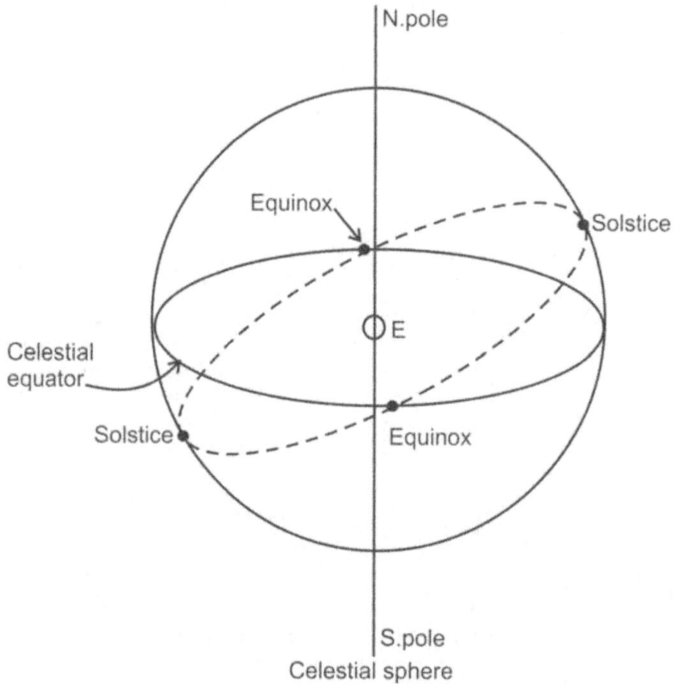

Fig. 3.1 Solstices.

Sidereal Day

The time period required for any point on the earth's surface for turning through an angle of 360° is called sidereal day. Actually a sidereal day is determined by observing the rotation of earth with respect to a fixed star instead of sun. 1 sidereal day = 23 hrs 56 m and 4.09 seconds. In scientific studies sidereal day is considered as standard time.

$$1 \text{ second} = \frac{1}{86164.09} \text{ sidereal day} = \frac{1}{86400} \text{ mean solar day.}$$

Solar Day

The time period (t_i) between two successive crossing of the sun across the meridian (over head) varies from day to day through out the year.

The average of this interval $t = \sum_{i=1}^{365} t_i/365$ is called mean solar day = 24 hrs.

Thus a sidereal day is shorter than the mean solar day by 3 minutes and 56 seconds.

Perihelion and Aphelion : According the Kepler's law, the earth is orbiting around the sun in an elliptical orbit with the sun being at one of the Focii. The earth is nearest to the sun on 1st January every year and farthest from the sun on 1st July. The nearest position of the sun is called perihelion (earth-sun distance = 147×10^6 km) and the farthest position of the sun is called Aphelion (earth sun distance = 152×10^6 km). It may be noted that the sun is nearest to the earth during winter and farthest in summer. Thus earth receives more intensity of solar radiation during winter (southern hemisphere) as compared to summer (northern hemisphere).

The Changing Lengths of Day and Night time

We know that the axis of the earth is tilted by about an angle of $23\frac{1}{2}°$ to the plane of the earth's orbit. The region between Tropic of Cancer (lat $23\frac{1}{2}°$ N) to Tropic of Capricorn (lat $23\frac{1}{2}°$S) is called Tropic. The region between Tropic of Cancer to Arctic circle (lat $66\frac{1}{2}°$N) and between Tropic of Capricorn to Antarctic circle (lat $66\frac{1}{2}°$S) are called subtropics. The regions between Arctic circle to north pole and between Antarctic circle to south pole are called frigid zones or polar regions. The length of the day (night) varies with latitudes, the approximate periods are given below Table 3.1.

Table 3.1

Lat (N/S)	0	17	41	49	63	66 ½	67	90
Max.length of day (night) hours.	12	13	15	16	20	24	1 moth	6 month

Regions which have continuous day light (24 hrs) [from lat $66\frac{1}{2}°$ to $90°$] are generally called "the land of the midnight sun". The day light varies from one day at Arctic (or Antarctic) Circle to six months at the summer poles. The night time varies from one day (24 hrs) at Arctic (or Antarctic) circle to six months at the winter pole. As the year progresses the apparent path of the sun varies from place of place and also the length of the day (or night) through out the year except at equinoxes. This variation causes seasonal changes.

At summer solstice (June 21) the sun is over head (at meridian) at Tropic of Cancer. As we proceed north wards from Tropic of Cancer, the length of the day increases (or night decreases) and within Arctic circle zone all places have 24 hrs day (within Antarctic circle, all places have 24 hrs night). North pole will be at the middle of six months period of continuous day (while south pole lies in the middle of six months period of darkness).

At equinoxes (March 21, Sept. 23) all places on the earth experience 12 hrs day and 12 hrs night.

At winter solstice (Dec. 22) the sun is over head (at meridian) at Tropic of Capricorn. As we proceed southwards (from Tropic of Capricorn to south pole) the length of the day increases (while night decreases) and within Antarctic circle zone all the places have 24 hrs light (all the places within Arctic circle will have 24 hrs night). South pole will be at the middle of six months period of continuous day (and north pole will be in the middle of six months period of darkness).

In summer hemisphere (i.e., when the sun is at Tropic of Cancer or Tropic of Capricorn) where the sun continuously shines for 24 hrs day, the apparent motion of the sun will be along the horizon (instead of sun rising and sun setting).

It may be noted that at any place, summer is warmer than winter due to direct sunshine and longer length of day.

In addition to the two motions referred to above, the earth has two other motions. 1. Precession and 2. Tides. The movement of the earth about its axis is not steady. It wobbles on its axis slowly like that of a dying top. The earth is not a perfect sphere but it has small bulge in the equatorial region. This equitorial bulge is responsible for the wobbling in the earth's axis of rotation. The period of wobbling is about 26,000 years. Because of this wobbling equinoxes are continually changing places in the earth's orbit. This effect is called precession of the equinoxes or simply precession.

The movement of earth's ocean water is called tides. In coastal areas it is observed that the sea water rises above normal level twice daily, this is called high tide. It also recedes below normal level twice daily, which is called low tide. This rise and fall of ocean water is largely due to moon's gravitational attraction, and at a smaller extent by the sun and the planets. This will be discussed separately as the lunar and the solar tides.

Questions

1. Write briefly on the cause of day and night and the seasons, equinoxes, soldtices, sidereal day, solar day, perihehon and aphelion.

2. Write briefly about the changing length time of day and night the land of the midnight seen 24 hours day, 6 months day.

3. When will be the sun closest to the earth with approximate distance and farthest to the earth. What is the mean earth-sun distance?

4. Write briefly on the Presession of the Equinoxes and Tides.

CHAPTER 4

Atmospheric Pressure

According to kinetic theory of gases, the molecules of the atmosphere (which are billions and billions in numbers) are perpetually moving at random. They are bombarding anything in their path and also colliding one another. The bombardment of molecules on any unit area constitutes pressure.

Man is living at the bottom of the ocean of atmosphere. The weight of the unit column of the atmospheric molecules at the bottom (that is at sea level) is called atmospheric pressure. Atmospheric pressure is one of the three basic state variables (the other two are temperature and specific volume) and plays a very important role in meteorology. Pressure is an internal state variable while volume is an external state variable.

Wind (atmospheric airflow) depends on the pressure difference between two neighbouring places and blows (in horizontal direction) from a high pressure region to a low pressure region, like water flow from higher region. Wind flow (in horizontal and vertical directions) ultimately results in the variation of weather phenomena. Airflow in vertical direction occurs due to (convective process) buoyancy.

Pressure is defined as the force per unit area.

$$\text{Pressure} = \frac{\text{Force}}{\text{Area}}$$

Force is a vector quantity while pressure is a scalar quantity. Atmospheric variations are associated with irregular changes in the atmospheric pressure and solar activity. The random motion of the molecules of the gases and their bombardment takes place in all directions. Consequently they exert pressure on all surfaces, irrespective of the direction of the faces. Atmospheric pressure at any place is greatest at the surface of the earth and generally decreases with altitude. The mass of the atmosphere is about 5.6×10^{18} kg, while the mass of the earth is 6×10^{24} kg and the mass of hydrosphere is 1.4×10^{21} kg.

Units of Atmospheric Pressure

At the surface of the earth the atmospheric air exerts a pressure of about 10^5 Newtons/m^2 or 10^5 Pascals, which is called one bar or one atmosphere. In meteorology one thousandth of a bar called millibar or hecto Pascal (1h Pa = 100 Pa) is widely used.

1 bar $= 10^5$ N/m$^2 = 10^5$ Pa, more accurately 1 Atm $= 1.01325 \times 10^5$ Pa

1 millibar $= 10^2$ N/m$^2 = 100$ Pa $= 1$ h Pa

The pressure of the atmospheric air will support the weight of the mercury (Hg) column of 760 mm or 30 inches. The length of the mercury column varies with temperature and gravity. Therefore, for meteorological purpose the atmospheric pressure is defined under standard condition of temperature (T = 0 $^\circ$C) and gravity (g = 9.80665 m/sec^2). The standard atmospheric pressure is 1013.25 mb (h Pa) at 0 $^\circ$C and gravity 9.80665 m/sec^2.

That is 1 Atm = 1013.25 h Pa = 760 mm Hg. column weight.

1 mm of Hg = 1.333224 h Pa. This is called standard mm of Hg. The instrument that is used for measuring the atmospheric pressure is called barometer. (In Greek "baros" meaning weight and "metron" meaning measure).

Measurement of Atmospheric Pressure

Evangelista Torricelli (1608-1647) devised a method for measuring the atmospheric pressure by mercury (Hg) barometer in 1643. In this method a long glass tube is filled with mercury and then inverted in a dish of mercury as shown in Fig. 4.1. The atmospheric pressure at any point is numerically equal to the weight of the mercury column in the tube. The space above the Hg column contains some mercury vapour and practically its pressure is negligible.

Barometer gives the atmospheric pressure p_0 as $p_0 = \rho g z$ .....(4.1)

Where ρ = density of the mercury in the column

z = height of mercury column

g = acceleration due to gravity = 9.80665 m/s^2

The height of the mercury column found to be about 76 cm at sea level when temperature is 0 °C and g = 9.80665 m/s². However this varies depending on the atmospheric conditions.

$$1 \text{ Atm} = (13.595 \text{ gm/cm}^3)(980.665 \text{ cm/sec}^2)(76 \text{ cm})$$
$$= 1.013 \times 10^5 \text{ N/m}^2 = 1.013 \times 10^5 \text{ Pa}$$

As for Fig. 4.1

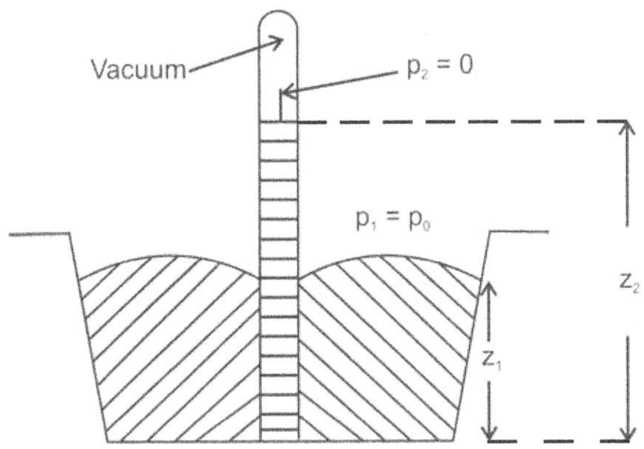

Fig. 4.1 Mercury barameter concept.

$$p_2 - p_1 = -\rho g (z_2 - z_1) \qquad \qquad .....(4.2)$$

French physicist Blaise Pascal, verified that pressure at the top of a hill is less than at the bottom and thus drawn conclusion that pressure falls with increasing altitude.

If we stock a number of pillows one above the other, we know that maximum pressure will be on the bottom pillow, and it decreases as it goes up. In a similar way the atmospheric pressure is maximum at the surface of the earth and it decreases with increasing altitude see Fig. 4.2. This decrease is not linear with height but found to be in exponential order. The approximate pressure fall with increasing altitude in given by the formula $p = p_0 \, e^{\frac{-mgz}{RT}}$

Where p = pressure at an altitude z, p_0 = Pressure at MSL

z = height of the altitude, g = acceleration due to gravity

M = Molecular mass of an ideal gas for dry air M = (28.89/mol)

R = Gas constant , T = temperature (in degrees) Kelvin or absolute scale.

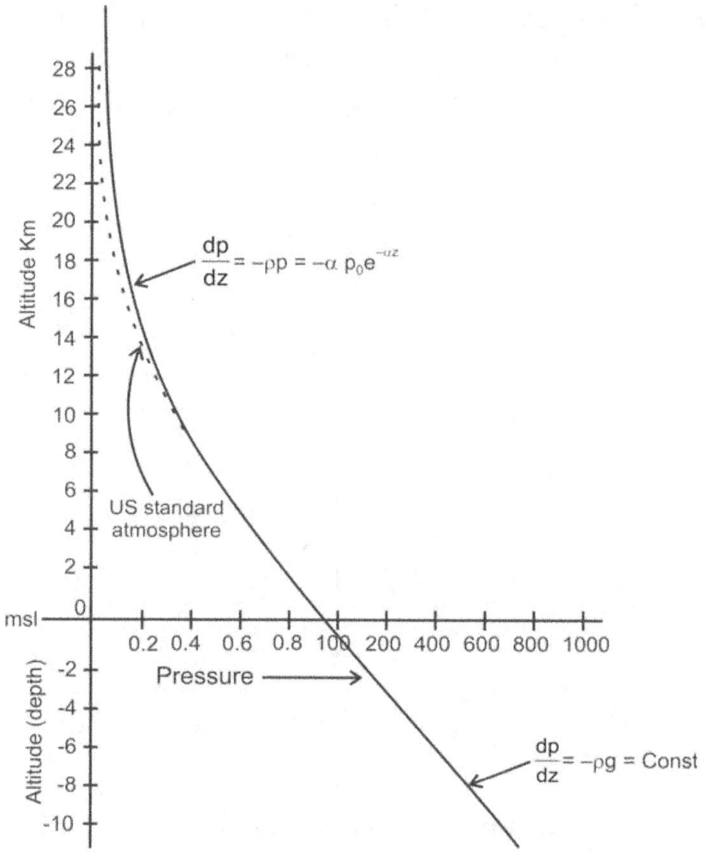

Fig. 4.2 Variation pressure with height and depth (not to scale).

The standard atmospheric pressure, temperature, density as function of geopatentral height is given in Table 4.1.

Table 4.1

Z (km)	Pressure (h Pa)	Temperature (° K)	Density (Kg/m^3)
32	8.7	228.7	0.013
28	15.9	224.7	0.025
24	29.3	220.7	0.046
20	54.7	216.7	0.088
18	75.1	216.7	0.121
16	102.9	216.7	0.165
14	141.0	216.7	0.227
12	193.3	216.7	0.311
10	264.4	223.1	0.412

Table 4.1 Contd...

Z (km)	Pressure (h Pa)	Temperature (° K)	Density (Kg/m^3)
9	307.4	229.7	0.466
8	356.0	236.1	0.525
7	410.6	242.7	0.590
6	471.8	249.1	0.660
5	540.2	255.7	0.736
4	616.4	262.1	0.819
3	701.1	268.7	0.909
2	794.9	275.1	1.007
1	898.7	281.7	1.112
0	1013.3	288.1	1.225

The approximate variation pressure with altitude in middle latitude is given in Table 4.2. and see Fig. 4.2

Table 4.2

Altitude (Km)	Pressure (hPa)
100	0.0003
80	0.0104
50	0.798
40	2.87
30	12.0
20	55.3
10	265
5.5	500
MSL = 0	1000
-50m	1020
-150 m	1030
-300 m	1050
-550 m	1080

Note : MSL pressure = 1013.25 h Pa

On an average at MSL, Nitrogen exerts pressure of about 760 h Pa, oxygen exerts about 240 h Pa and water vapour about 10 h Pa

Pressure falls roughly by 1 h Pa (mb) for every rise of 8.5 m near sea level.

At MSL (on global scale) pressure varies from less than 990 h Pa to more than 1030 h Pa. The world greatest sea level pressure of 1083.8 h Pa was recorded at Agata, Saiberia on 31st December 1968, while the lowest pressure of 870 h Pa was recorded (by drop sonde) in the eye of Typhoon Tip (lat 16.7° N, long 137.8 °E) in Pacific on 12th October 1979.

The length of the Hg column depends on temperature and gravity, consequently it is adjusted to standard conditions of temperature and gravity. These corrections are to be made for meteorological purposes. Scales are devised to read pressure directly under standard conditions in standard units. For the purpose of comparable pressure readings at different places the following corrections are made. Index error correction, temperature and gravity corrections.

Index Correction

Index correction is made to eliminate error due to calibration, residual gas in the vacuum above (mercury) Hg column and also error that would be crept due to bending of light (refraction), capillarity effect.

Temperature Correction

Pressure readings made at room temperature have to be reduced to standard temperature of 0 °C.

Gravity Correction

Pressure depends on acceleration due to gravity(g). g varies with latitudes and altitudes. Barometers are calibrated to standard g = 9.80665 m/s^2. Thus barometer readings have to be reduced to the "g" of that place.

Meteorological mercury barometers are 1. Fortin Barometer, 2. KEW pattern Barometer.

In Fortins Barometer, the mercury level in the cistern has to be raised or lowered to just touch a fixed ivory pointer before taking readings.

In KEW pattern Barometer the cistern is fixed and hence no adjustment of mercury level required. IMD uses this type of barometers in observatories.

KEW pattern Barometer consists of three essential parts. 1. A glass tube 90 cm long closed at the top and open below, 2. a cup or cistern and 3. a brass scale. The glass tube is filled with mercury and its open end is dipped in the mercury in the cistern, which prevents air from entering the tube. Above mercury column in the tube is an empty space. The mercury column in the tube is supported by the pressure of the air on the surface of the mercury in the cistern. This forms the basic principle of balance on which a barometer is constructed. As the mercury in the barometer tube rises or falls due to changes in the atmospheric pressure, the mercury level in the cistern changes in the opposite direction. This change of level in the cistern is taken into account in the graduation of the scale itself.

Aneroid Barometer

Aneroid Barometer contains corrugated walls of an evacuated small box. These walls expand (swell) when pressure decreases and contract (collapse) when pressure increases. This expansion or contraction of the walls calibrated with the help of spring indicator to read pressure directly. This principle is used in Altimeters which are used in aircrafts to indicate its flight altitude or to indicate its height above aerodrome level while take off or landing. It may be noted here that mercury barometers are more accurate but they are not portable.

Aneroid barometers are portable but not very accurate. All aneroid barometers must be calibrated with the help of mercury barometers.

Aneroid barometers are compact and portable and therefore convenient for observations over ships or in the field. However, there is disadvantage due to weakening of spring with high temperature or rapid change in temperature. This change will not indicate true pressure immediately. This is called hysteresis. Some times the spring may weaken and the instrument may show a very high pressure. There may be slow changes in the wall metal of the aneroid chamber, which are called secular changes. These changes may be corrected only comparing with standard mercury baromer.

Barograph

It is a modified form of aneroid barometer, used for continuous self recording of atmospheric pressure. In a barograph a number of aneroid cells are joined together. The expansion or contraction of these walls is magnified with the help of a lever system and is indicated by pointer, which marks on a chart wrapped over the cylindrical drum. The cylindrical drum rotates with a clock work mechanism. The cylinder completes one rotation in one week. Thus the barograph records continuous atmospheric pressure for a period of one week.

Reduction of Pressure to Standard Levels

The observed pressure at a station is called station level pressure. For comparison of pressure at various stations situated at different altitudes, first it has to be reduced to a standard level. For all practical purposes it is reduced to MSL (mean sea level) and it is called mean sea level pressure. In fact MSL pressure thus obtained is a hypothetical one. In case of stations located above msl we have to add the pressure of the unit area air column between msl and station level pressure. In case of a station located below MSL the pressure of the unit area air column between MSL and station has to be deducted from the station level pressure. In these cases, though the height of the column may be fixed but the weight of air column differs due to temperature (or density), as it is difficult to fix the msl temperature of this air column. Despite this draw back MSL pressure is widely used for synoptic pressure charts throughout the world by meteorologists. In Greek "syn" means together and "opsis" means looking. Synopsis stands for looking together. Thus pressure observations taken at standard times simultaneously throughout the world reduced to MSL and then plotted on the charts. These charts are analysed by drawing isobars (equal pressure use lines). These isobars indicate Low Pressure, High pressure regions which in turn gives first indication about the weather.

Pressure is one of the weather elements. Horizontal pressure distribution at MSL has profound bearing on weather. Some of the pressure patterns are shown in Fig. 4.3. Isobars are found to be closed, elongated or straight. If the closed isobars indicate that pressure decreasing out ward, we call it High pressure region or simply High. If the closed isobars indicate that pressure is increasing outward we call it low pressure region or Low. If the isobars are in the form of V and pressure decreasing outward, the line passing through the vertices of V is called ridge or wedge. If the isobars of V form indicate pressure increasing outward, the line passing through the vertices of V is called trough line. The region between two Highs and two Lows is called Col region. These are shown in the Fig. 4.3.

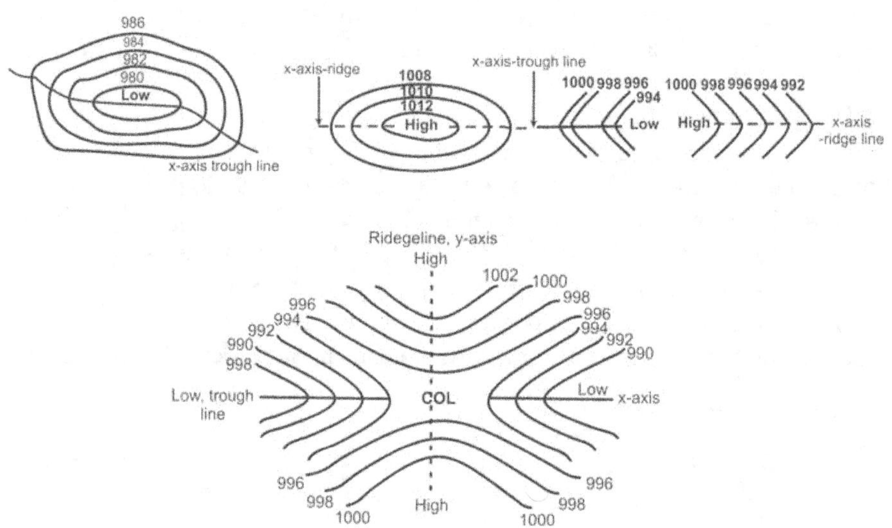

Fig. 4.3 MSL pressure.

Distribution of MSL Pressure

The horizontal distribution of MSL pressure world over indicate the following features. MSL pressure distribution depends on temperature, latitude, land and water bodies. Low pressure regions are found along equator called equatorial low or doldrums. There are regions of high pressure along latitude belts of 25° to 35° N and S. These are called subtropical High pressure belts. Low pressure belts are observed between lat 60 to 70 ° N and S, which are called sub-polar Low pressure belts. In polar regions both north and south High pressure regions are found, which are called Polar Highs Fig. 4.4. These pressure belt zones oscillate but they are semi-permanent.

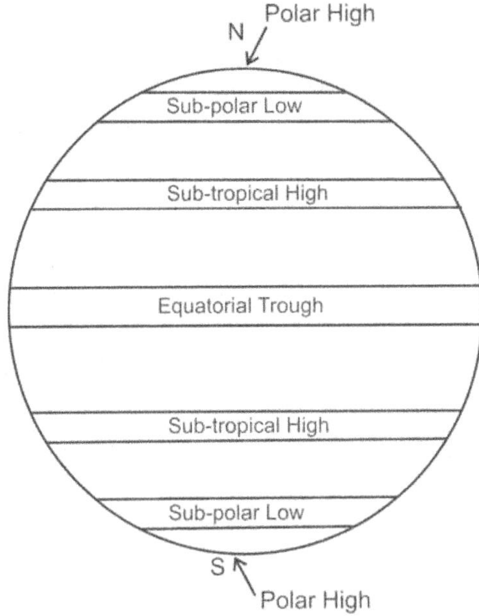

Fig. 4.4 MSL pressure distribution semi-permanant features.

Pressure Gradient

Horizontal pressure change with distance is called pressure gradient. It is measured perpendicular to the isobars and directed from High to Low. In Fig. 4.5 pressure gradient along AB is higher than along CD. If the isobars are closely spaced (more), the gradient is said to be steep or strong, if they are sparcely spaced (a few far apart) the gradient is said to be weak or flat. Wind strength depends on the pressure gradient, closer the isobars stronger the wind and vice versa.

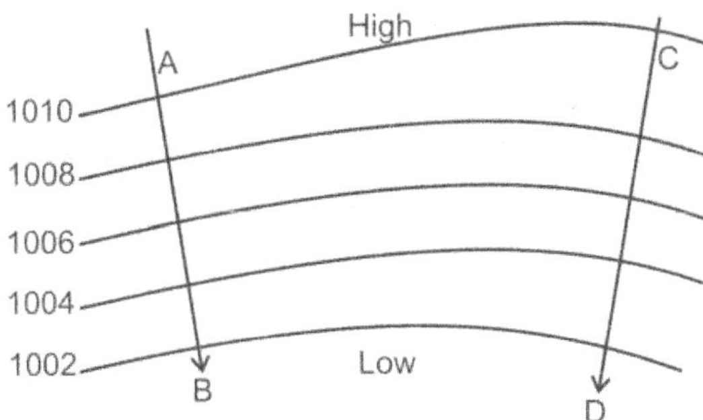

Fig. 4.5 Different pressure gradients.

Diurnal Variation of Pressure

At any given place the atmosphere will be ever changing which may be periodic (regular) or non-periodic (irregular).

Periodic variations are regular but they have different periodicity. Of all periodicity 12 hours (1/2 day) is of most important. This is called semi-diurnal variations of pressure. As the earth rotates day and nights are formed. Heating of atmosphere during day (due to insolation) and cooling at night (nocturnal cooling) causes rhythmic expansion and contraction, which results in pressure oscillations. Further it is assumed that atmosphere itself has 12 hour periodic oscillation. This is amplified by the day and night temperature variations. Because of this resonance, amplitude oscillations magnify (increase). Therefore a double sinusoidal pressure wave travels around the earth along with the sun. It is seen that a pressure maxima occurs at about 1000 and 2200 hrs local mean time and a pressure minima at about 0400 hrs and 1600 hrs (LMT) (shown in Fig. 4.6).

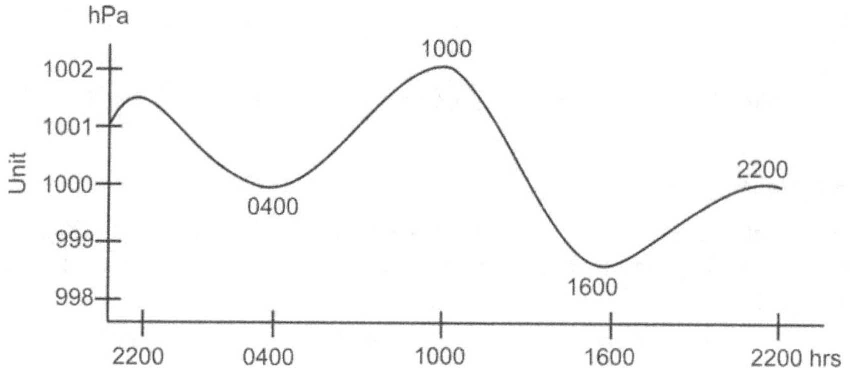

Fig. 4.6 Diurnal variation of pressure.

In tropical regions diurnal variation of pressure is clearly observed however it is not so clear on higher latitudes. Thus perodic variations of pressure are very complex and they are not perfectly symmetrical. These changes vary from place to place and have little influence on other meteorological parameters. Non-periodic variations of atmospheric pressure is mostly associated with the passages of pressure systems such as Low, Depression, Cyclone. Abrupt variations of pressure are associated with thunderstorms, tornadoes, pressure shockwaves, sudden explosions etc.

Mass of the Atmosphere

Average atmospheric pressure at the msl is about 1.013×10^5 Pa = 1.013×10^5 N, P_o = 1013 h Pa. This pressure exerts a total force of $4\pi R^2 P_o$, where P_o = Average msl pressure, R= radius of the earth.

$$F = 4 \times \frac{22}{7} \times (6400 \times 1000)^2 \times 10^5 \times 1.013 \times 10^5 \, N.$$

$$(\text{Pressure} = \frac{\text{Force}}{\text{Area}})$$

The mass of the atmosphere $M = \dfrac{F}{g}$ (Force = Mass × acceleration)

$$= [4 \times \frac{22}{7} \times (6.4 \times 10^6)^2 \times 1.01325 \times 10^5] / 9.81$$

$$= 831 \times 10^{18} \, kg$$

The mass of the earth is 5.98×10^{24} kg. This shows that the mass of the atmosphere is about (0.889×10^{-6} times) one millionth of the mass of the earth. The total mass of the water of the world is about 1.4×10^{21} kg, while the mass of the atmosphere is 831×10^{18} kg. Thus the mass of the worlds water exceeds the mass of the atmosphere by about 263 times.

Pressure Altimeters

The principle decrease of pressure with height is used in the construction of aircraft altimeters. Pressure altimeters are simply aneroid barometers with scale graduated in height instead of pressure.

Radio Altimeters

Radio altimeters use the property of reflection of radio waves from the surface below which may be land or water. The time taken for reflected pulse for to and fro journey gives the measure of the intervening height.

D-value

$D = Z - Zp$ where Z = actual height, Zp = Pressure altitude in ISA

D - is the difference between actual height Z above mean sea level of a particular pressure surface (Sp) and the pressure altitude Zp with respects to the same surface (Sp) in International standard atmosphere (ISA).

Questions

1. Define pressure, state the units of atmospheric pressure.
2. How is atmospheric pressure measured? Why is it corrections required name them.
3. Describe the MSL pressure distribution over the globe.

4. What is diurnal variation of pressure? Is it periodic? If so when do you expect pressure maxima and minima in 24 hours of a day.

5. Assuming the average atmospheric pressure as 1.013×10^5 Pa or N, radius of earth $\simeq$ 6400 Km, g = 9.81 m/s^2, find the mass of the atmosphere.

CHAPTER 5

Water in the Atmosphere

Water is present in the atmosphere in varying degrees in all levels in troposphere, but extends upto mesosphere. It is generally found in the form of water vapour. Water cycle over land and atmosphere is called hydrologic cycle, which has three important stages: (i) evaporation (ii) condensation and (iii) precipitation. The average storage of water in the atmosphere is only about 2.5 cm, but exchanges between oceans and land is large. Water enters the atmosphere through evaporation and transpiration, while it leaves the atmosphere through condensation and precipitation. The annual depth of evaporation of water over the world is 1030 mm or 1.03 m or the annual volume of evaporation over world is about 525.1×10^{12} m^3 = 525.1×10^3 km^3 [V = 4 $\times$ π $\times$ $(6370 \times 10^3)^2 \times 1.03 = 525.1 \times 10^{12}$ m^3] = 525.1×10^{12} metric tons of water. The amount of water that enters the atmosphere is received back through precipitation (pptn). The annual world average precipitation is 103 cm. The total volume of atmospheric water vapour is 14000 km^3 or 0.001% of the hydrosphere.

Volume of hydrosphere = 1454651×1000 km^3.

We know water occurs in nature in three states: solid (ice), liquid (water) and gas (water vapour), and they may change one state to another in the following ways. Solid to liquid state (ice melts into water, heat required). Solid to vapour state (Ice sublimates to water vapour, heat required). Liquid to gas state (water evaporates to water vapour, heat required). Liquid to solid state (water freezes to Ice, latent heat released). Gas to liquid state (water vapour condenses to water, latent heat released). Gas to solid (water vapour becomes ice by deposition, latent heat released). Water or moisture is a factor in humidity, cloudiness, pptn, and visibility.

The atmospheric air is a mixture of gases. Each gas exerts it own pressure, called its partial pressure. The partial pressure exerted by an individual gas is proportional to its number of molecules present in the given volume of air (mixture). Thus the atmospheric pressure is the sum of the partial pressures of all the gases present in the atmosphere including water vapour. The pressure exerted by dry atmosphere is less than moist atmosphere, since the pressure in the later case is added up due to extra pressure exerted by water vapour in comparison to dry air. The content of water vapour in the atmosphere varies with place and time. At any instant the amount of water vapour in the atmosphere over the globe $= 13000$ km^3.

The residential time of water vapour in the atmosphere

$$= \frac{365 \text{ days}}{\left(525.1 \times 10^3 \text{ km}^3/13000 \text{ km}^3\right)} \simeq 9 \text{ days}$$

Large quantities of water vapour (vapour pressure about 30 h Pa) is observed close to the earth in tropics while small quantities observed in polar regions and in particular over Antarctic region during winter. In general water vapour decreases with increasing altitude, but increase also observed on some occasions.

The Rate of Evaporation

The process of vapour formation at the free surface of water is called evaporation or vaporisation.

The rate of evaporation depends on many factors such as temperature, humidity, wind speed etc. For practical purposes, however, it can be expressed as the volume of liquid water evaporated form unit area in unit time. From the molecular point of view evaporation means the flying off of molecules from the water surfaces. The molecules of highest velocities and kinetic energy fly off first, hence the remaining water cools down due to evaporation. The rate of evaporation u (that is the amount of water transformed into vapour per second) depends mainly on the atmospheric pressure (P_o), velocities of the gaseous phase above free surface of water. It is given by the relation

$$u = \frac{KA(e_s - e)}{P_o}$$

where K = constant, A = Area of the free surface of water, e_s = saturation vapour pressure, e = pressure of vapour above free surface of water, P_o = atmospheric pressure.

The Saturation Vapour Pressure (e_s)

Consider a plane surface of water at a given temperature. The molecules of highest velocity escape from the free surface of water into the air above. A few of them return back to water while others move as gas molecules above the water surface. A stage will come when the number of molecules escaping the surface of water per second is equal to the number of molecules return to the water. In this state the space just above the water is called saturated at that prevailing temperature. If the temperature is increased, more molecules of water required to saturate the space above water and hence the partial pressure of water vapour increases. Consequently, the saturation vapour pressure increases with increase of temperature. Further it may be noted that the water holding capacity of the environment increases with the increasing temperature. Because of this tropical region has greater water vapour holding capacity compared to cold regions or polar regions. As a result saturation vapour pressure is high over tropical water bodies (oceans, lakes and rivers). The variation of saturation vapour pressure with increase in temperature is given in Table 5.1.

Table 5.1

Temp °C	e_s = saturate vapour pressure (hPa)
0	6.11
10	12.27
20	23.37
30	42.43
40	73.77

We shall now discuss the various processes that transform (three states) of water from one state to another.

The Condensation Process

If water vapour sent into a chamber containing saturated water vapour, it condenses into water. Condensation depends on the presence of various sizes of nuclei (small, Atican, large and Giant nuclei) and types (hygroscopic, non-hygroscopic). Condensation is the reverse process of evaporation.

Condensation of water vapour in the atmosphere generally takes place due to cooling. As temperature decreases less water vapour is required for saturation. As water vapour goes up (ascends) it cools, which causes saturation and further cooling by ascent results in condensation. This aspect will be considered in detail in precipitation process.

Isobaric Process

We know $PV = RT$ (ideal gas equation) where P = pressure, V = Volume, T = temperature (in Kelvin) and R = gas constant.

A physical process which proceeds, where pressure remains constant is called isobaric process. If a sample of moist air is cooled isobarically (keeping pressure constant), at certain temperature it becomes saturated. This temperature is called dew point temperature or simply dew point. If it is cooled below dew point, it condenses into water.

Adiabatic Process

A physical process in which the heat content remains constant, that is neither added nor removed, the process is called adiabatic process. If a moist air sample moved up adiabatically, it expands and cools. This results in cloud formation and condensation.

The Freezing Process

If pure water is kept undisturbed and cooled, it remains in liquid state even below $0\ ^{\circ}C$ (freezing point). This water is called super cooled. If super cooled water is disturbed or freezing nuclei/ice crystals are introduced into it, the super cooled water freezes into ice crystals. Experiments indicate that super cooled water can attain $-40\ ^{\circ}C$, but before reaching this low temperature water automatically freezes into ice crystals, irrespective of the nuclei present.

The Deposition Process

The process in which water vapour directly converts into ice without passing through the liquid state is called deposition. The conversion of ice crystals into vapour (solid to vapour state) is termed sublimation.

It has been found experimentally that the saturation vapour pressure over ice is slightly less than over super cooled water at the same temperature. This variation at different temperatures is given in Table 5.2.

Table 5.2

Temperature °C	Saturation vapour pressure (hPa) over ice	Vapour pressure (h Pa) over super cooled water
0	6.106	6.107
-10	2.597	2.862
-20	1.032	1.254
-30	0.380	0.509
-40	0.128	0.189

Frost Point Temperature

If a sample of moist air is cooled isobarically, the temperature at which it becomes saturated with respect to an ice surface is called Frost point. When moist air is cooled below frost point, the water vapour deposits as ice crystals over the container including ice surface.

Latent Heat

The heat required (or removed) to change the phase (ice to water, water to steam or vice versa) without change in temperature (in Fig. 5.1 segment bc segment de) is called latent heat.

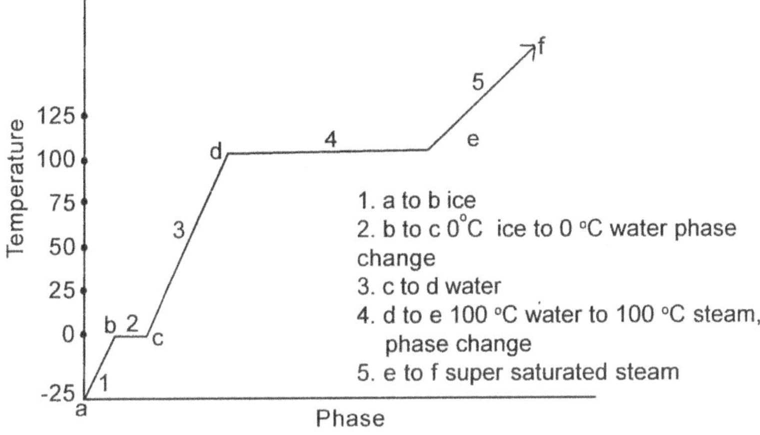

Fig. 5.1 Phase change.

The quantity of heat required to change 1 gm of ice into water at 0 °C (without change in temperature) is called heat of fusion or latent heat of fusion (335 J gm^{-1}). Similarly, the quantity of heat required to change 1 gm of water at 100 °C to vapour (with out change in temperature) is called heat of vaporisation or latent heat of vaporisation (2256 Jgm^{-1}).

Transpiration

Plants withdraw water from the soil by its roots, which moves upward to leaves. Through openings in leaves some water escapes to the atmosphere. This loss of water from the leaves and stems of vegetation is called transpiration. The combined effect of moisture loss through evaporation and transpiration from a given area is called evapo-transpiration.

Moisture Variables or Indicators

The moisture content of the atmosphere is measured besides vapour pressure by absolute humidity, relative humidity, specific humidity and mass mixing ratio, which are defined below.

Absolute Humidity (ρ_v)

The mass of water in a given volume of air is called absolute humidity. It is expressed as grams per cubic meter (g/m^3). OR, The mass of water vapour per unit of volume is called absolute humidity. Absolute humidity, in fact, is the density of water vapour.

Relative Humidity (RH)

The ratio of the mass of water vapour present in a unit volume of air to that of mass of water vapour required for saturation at that temperature.

OR

The ratio of the partial pressure to the vapour pressure at the same temperature.

Generally it is expressed as percentage.

RH (%)

$$= \frac{\text{Actual vapour pressure (or partial pressure of water vapour)}}{\text{Saturation vapour pressure (or vapour pressure) at the same temperature}} \times 100$$

$$= \frac{e}{e_s} \times 100$$

Example 5.1

Let the partial pressure of water vapour (e) in the atmosphere be 0.013×10^5 Pa (at 15 °C) and let the vapour pressure (e_s) at the same temperature be 0.169×10^5 Pa. Then the RH is given by

$$RH = \frac{0.013 \times 10^5}{0.0169 \times 10^5} \times 100 = 77 \%$$

Vapour pressure (e_s) of water at different temperatures

T(°C)	0	5	10	15	20	40	100
Vapour Pressure (10^5 Pa)	0.0061	0.00868	0.0119	0.0169	0.0233	0.0734	1.01

Note : Man becomes uncomfortable during summer when the RH is high because sweat does not evaporate, consequently body temperature does not fall. Hence man feels oppressive of heat of the environment.

Mixing Ratio (ω)

Mass of water vapour present in gms per kilogram of dry air is called mixing ratio.

$$\omega = \frac{\rho_v}{\rho_d} = \frac{\text{Absolute humidity}}{\text{Density of dry air}}$$

$$RH = \frac{\omega}{\omega_s} \times 100 = \frac{\text{Mass of water vapour in Vcc}}{\text{Mass of saturation vapour in Vcc}} \times 100,$$

where ω_s = saturation mixing ratio.

Note : When RH is 100%, air temperature = dew point temperature. RH is a function of pressure, temperature and mixing ratio.

$$RH = \frac{e_s \text{ at dew point}}{e_s \text{ at air temperature}} \times 100$$

Example 5.2

What is relative humidity (RH) on a day when the temperature is 20 $^{\circ}$C and dew point is 5 $^{\circ}$C. What is the partial pressure of the water vapour.

$$RH = \frac{\text{saturated vapour pressure at 5 } ^{\circ}\text{C}}{\text{saturated vapour pressure at 20 } ^{\circ}\text{C}} \times 100 \ .$$

$$= \frac{0.00868 \times 10^5}{0.0233 \times 10^5} \times 100 = 37.3\% \ (37.253\%)$$

$$RH = \frac{\text{Partial pressure } (e)}{\text{saturated vapour pressure } (e_s)} \times 100$$

$$\text{Partial pressure of water vapour } (e) = \frac{RH \times e_s \text{ at } 20^{\circ}\text{C}}{100}$$

(Since we found RH 37.253% at 20 $^{\circ}$C and e_s at that temperature is 0.0233×10^5Pa)

$$\therefore \quad e = \frac{37.253}{100} \times 0.0233 \times 10^5 = 0.37253 \times 2.33 = 8.679 \text{ Pa.}$$

Example 5.3

Find the RH if the actual vapour pressure (e) in air at 21°C is 20 h Pa and (e_s) saturated vapour pressure is 25 h pa.

$$RH = \frac{e}{e_s} \times 100 = \frac{20}{25} \times 100 = 80 \ \%.$$

Humidity has great effects on many organic substances. Completely degreased human hair is sensitive to variation of humidity. The length of human hair increases in length with increase in humidity and vice verse. However the increase is not linear. This property is employed in hair hygrograph. The increase in length with RH, from 0 – 100% is about $2\frac{1}{2}$% of the original length.

Between 20 to 100% RH, the increase in hair length (or Δl) is proportional to the logarithm of the change in RH (log Δ RH).

$$\Delta l = l_2 - l_1$$
$$\Delta l \; \alpha \; \log [\text{change in RH}]$$
$$\Delta l = \log [RH_2 - RH_1] \times \text{const}$$
$$\text{or} \quad \Delta l = Ln \; [(RH_2 - RH_1). \; \text{const}$$

Humidity may change without adding or subtracting moisture from the sample air. This happens by increasing or decreasing of temperature. Generally maximum RH, observed at dawn and minimum at about 2.30 PM local time. In winter at dawn saturation may occur. In such situation mist or fog may form due to condensation. As temperature increases in the later day RH decreases which disperses fog or mist.

$$\text{Specific humidity} = \frac{\text{mass of water vapour}}{\text{mass of moist air containing the vapour}}$$

$$\frac{M_v}{M_v + M_d} = \frac{\rho_v}{\rho}$$

where M_v = Mass of water vapour

            M_d = Mass of dry air contain the vapour

$$RH = \frac{\text{Actual mixing ratio}}{\text{saturation mixing ratio at the same temperature and pressure}} \times 100$$

$$RH = \frac{\omega}{\omega_s} \times 100$$

$$\omega \simeq \varepsilon \frac{e}{p}$$

$$\omega_s \simeq \varepsilon \frac{e_s}{p}$$

Where

          ω = mixing ratio

          ω_s = saturated mixing ratio

$$\varepsilon = \frac{m_v}{m_d} = \frac{\text{mole. wt of vapour}}{\text{mole. wt of dry air}}$$

          e = vapour pressure

          e_s = saturation vapour pressure and $e \ll p$.

Density of Dry Air

Density of dry air varies with temperature and pressure. For all practical purposes, at the surfaces of the earth, the atmospheric standard pressure is 1013.25 h Pa at 15 °C and density of dry air is 1.225 kg/m^3.

Note : Molecular weight of dry air implies mean molecular weight of dry air because air is a mixture of gases. Mean molecular weight is taken as standard for all practical purposes.

Density of Moist Air

The approximate molecular weight of water vapour is 18, but this is about $\frac{5}{8}$th of the mean molecular weight of dry air (28.8)

$$[z \times \frac{5}{8} = 18 \text{ or } z = \frac{144}{5} = 28.8]$$

Thus water vapour molecule is lighter than dry air, which means density of moist air is less than density of dry air.

Dew Point Temperature (T_d)

The temperature to which moist air must be cooled to make it saturated with respect to water keeping pressure and mixing ratio constant (i.e., without any addition of water vapour). OR,

If we cool a sample of moist air isobarically it becomes saturated at certain temperature, which is called dew point temperature or simply dew point.

Frost Point Temperature

The temperature of moist air to which it must be cooled to make it saturated with respect to ice is called Frost point temperature.

The Lifting Condensation Level (LCL)

The level to which unsaturated sample of air to be lifted to make it to begin condensation, OR

The level to which the moist air be lifted to make unsaturated mixing ratio to saturated mixing ratio, OR.

Lifting condensation level is that level to which unsaturated air has to be lifted dry adiabatically to produce condensation.

Wet Bulb Temperature ($T\omega$)

The temperature to which a sample of air must be cooled by evaporating water into it to make the sample of air saturated, keeping pressure constant.

(Note: in this case mixing ratio is not constant), OR

This is the lowest temperature attained by air when water is freely evaporated into it. OR

The temperature to which air may be cooled by evaporating water into it isobarically till it is saturated.

Measurement of Evaporation

Water is continuously lost from the earth's surface by evaporation. The rate of evaporation is a function of nature of soil, vegetation, temperature, humidity and wind speed. For practical use it is expressed as the volume of liquid water evaporated from unit area in unit time. Over a given area the evaporation is proportional to the depth of liquid water lost in unit time. Accordingly evaporation is measured as millimetres of water lost per day. The instrument used for measuring evaporation is called Evaporimeter. The most common evaporimeter used in IMD is the class A Pan Evaporimeter, also called open pan evaporimeter.

Class A Pan Evaporimeter

It consists of a large circular pan with a stilling well to provide an undisturbed water surface around the point of a hook gauge by breaking any ripple caused by wind that may be present in the main part of the pan. The pan rests over a white painted wooden stand.

The pan is covered with wire netting to avoid loss of water by animal and birds. The amount of water lost by evaporation from the pan during successive observations (at 08.30 and 17.30 hrs IST) is measured by adding known quantities of water to the pan with a graduated cylinder till the water level touches the reference point. The amount water added is equal to the amount of water lost by evaporation from the pan and this divided by the time interval gives the rate of evaporation. Rain falling into the pan is accounted by the nearby rain gauge record. The pan should be exposed in a relatively sheltered position to prevent out splashing caused by strong winds. It is generally installed in the observatory Stevenson screen enclosure by the side of rain gauge so that the amount of precipitation caught by the pan is represented by the measurement of pptn (=precipitation) by the rain-gauge.

The observation is made as follows. At the standard time of observation (0830/1730 hrs IST) add water to the pan with standard measuring cylinder until the tip of the fixed point coincides with the surface of the water in well. The measuring cylinder is graduated from the top to the bottom. Suppose two full cylinders of water and 9 cm, that is $20 + 20 + 9 = 49$ cm from the

cylinder have been added to the pan. The radius of the cylinder (r) is $\dfrac{1}{10}$

times that of the pan $(R = 10r)$, the reading divided by 100 $(\frac{\pi R^2}{\pi r^2} = 100)$, that is $\frac{49}{100} = 0.49$ cm = 4.9 mm is amount of water lost by evaporation from the pan provided there was no precipitation (pptn) since the last observation.

If there was pptn during the period of two observations, and if this exceeds the water lost by evaporation, water has to be removed from the pan till the level comes to the tip of the gauge. Record the amount of water removed.

Suppose the water removed is 76 cm ($\frac{7.6}{100} = 7.6$ mm) and the precipitation since lost observation is 8.2 mm. The water lost by evaporation is given by 8.2 − 7.6 = 0.6 mm.

In (Fig. 5.2) AA' indicates level of water in the pan when it is set at 0830 hrs IST (say) and CC' is the level at next 1730 hrs IST. If the amount of water removed from the pan to bring the level to AA' is 76 cm, the difference in level is 7.6 mm. If there had been no rainfall the water level would have been at BB'. To obtain the level BB', one has to imagine level CC' to be lowered by the depth of precipitation. If 8.2 mm is the amount of rainfall during 0830 to 1730 hrs interval CB or C' B' = 8.2 mm, the evaporation is then given by A'B' = AB = 8.2 − 7.6 = 0.6 mm.

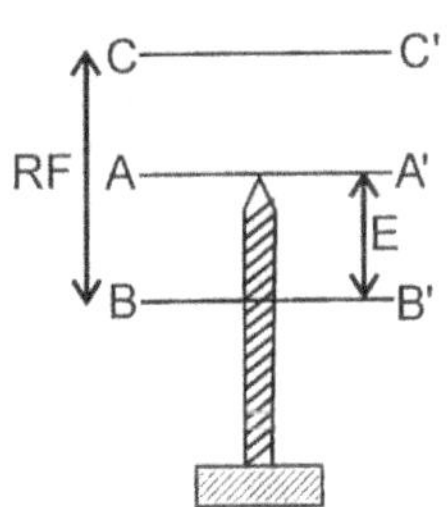

Fig. 5.2

If there is only light rain, the water, level may not rise above AA' (the setting level) as in the previous case, but may be at CC' below AA' but above BB' ([as shown in Fig. 5.3]). BB' is the level of water had there been no rain. If 42 cm of water has been added to the pan AC = A'C' = 4.2 mm and the rainfall is 2.3 mm, then the actual evaporation since last observation is 4.2 + 2.3 = 6.5 mm

Measurement of Humidity

Instruments which are used for measuring humidity (water vapour) are called hygrometers.

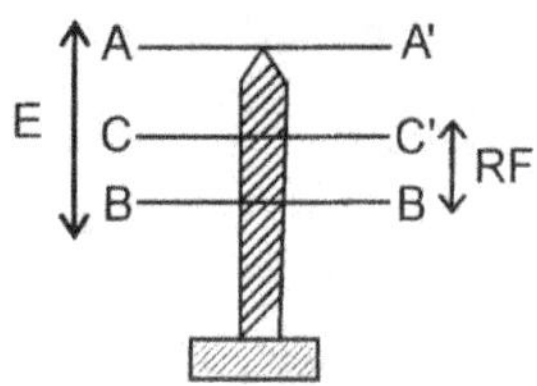

Fig. 5.3

Hair Hygrograph

Human hair is sensitive to humidity and changes its length with the variation of humidity which is discussed earlier. This property is made use of in hair hygrograph, which records atmospheric humidity continuously on a graph. The changes in the length of human hair is magnified by a lever system and operated by pen which marks on a chart attached to a rotating cylinder.

Psychrometer

It consists of two thermometers mounted side by side. One of these thermometer bulb is wrapped with muslin cloth with a few strands which are wetted with distilled water kept in a small vessel. One indicates the air temperature and the other wet bulb temperature. This mounted pair with distilled water vessel is kept in a Stevenson Screen for protection from solar radiation and ventilation. With the help of dry, wet bulb temperature readings RH and dew point is calculated using hygrometric tables.

Assmann Psychrometer

If the air is not saturated with moisture the wet bulb thermometer reads less than the dry bulb thermometer. This is because of evaporation of water from muslin cloth round the wet bulb (which cools). The wet bulb temperature depends not only on the amount of water vapour in the surrounding air but also on the speed of the air which is passing over it. RH calculated from ordinary thermometer observation is (strictly) valid for wind speeds between 4 to 10 mps (8 to 20 kt). This condition is not always satisfied for screen thermometers. These drawbacks are overcome in the Assmann psychrometer which is portable. In this instrument air is drawn past the dry and wet bulbs by means of a clock work exhaust motor. Each bulb is protected from external radiation by two highly polished coaxial tubes so that the instrument can be held even in bright (strong) sunshine without risk of solar radiation affecting the readings.

Moisten the wet bulb with distilled water or rain water, shake the instrument gently to throw out excess water, if any, round the wet bulb. Wind the clock work motor, wait for two minutes to get steady wet bulb temperature. During observation if wet bulb shows sudden rise, it indicates that the muslin cloth became dry. The muslin should be moistened again and the reading repeated. Read wet bulb and the dry bulb. Using special (hygrometric) tables Relative humidity, dew point and vapour pressure of the air can be calculated.

Whirling Psychrometer

In the whirling or sling psychrometer, the aspiration is provided by whirling or rotating the thermometers which are mounted side by side on a wooden frame for that purpose. To obtain desirable air speed (5 mps) past the thermometer bulbs, the psychrometer has to be rotated at 4 revolutions per second.

Moisten the wet bulb (by wetting the wick completely). Whirl the instrument keeping back to the sun to avoid falling of direct sunlight on the instrument, or in a sheltered place. After whirling for about 2 to 3 minutes take wet bulb and dry bulb readings, only making sure that the wetbulb temperature is steady. Again whirl for about 10 seconds and take readings. Note that these readings should be concurrent (which may not be far off).

The Theoretical Basis of Wet Bulb Thermometer

Evaporation of water from the muslin cloth takes place as long as the air is not saturated (RH 100%). When evaporation takes place latent heat of evaporation is carried away from the muslin cloth which results in fall of wet bulb temperature. The difference between the dry bulb and wet bulb temperature, called wet bulb depression, depends on the rate of evaporation, which in turn depends on the humidity of the prevailing air.

Psychrometric Tables

To calculate RH and dew point temperature, psychrometric tables have been prepared for ventilation wind speed 1 to 10 mps and pressure 1000 h Pa and 900 h Pa.

Moisture Transport

Wind transports moisture both horizontally and vertically. At any place (locality) the moisture balance equations is:

Precipitation – Evaporation = Horizontal transport of moisture into the air column.

It may be noted that local evaporation is not entirely dependent on local precipitation. It depends on moisture advection into the location.

Worlds Evaporation Depth

The average evaporation depth over land area (149×10^6 km^2) is 495 mm, over ocean area (361×10^6 km^2) is 1251 mm and over Globe

$$= \frac{495 \times 149 \times 10^6 + 1251 \times 361 \times 10^6}{510 \times 10^6} = 1030 \text{ mm}$$

The mass of the hydrosphere is only 0.024% of the mass of the earth and the mass of the atmosphere is 0.00009% of the mass of the earth.

Questions

1. What are three stages of water cycle? Write the estimated annual depth evaporation and precipitation over the globe.

2. Define absolute humidity, relative humidity, mixing ratio, saturated mixing ratio, vapour pressure, saturated vapour pressure.

3. Find the actual vapour pressure in air if the saturated vapour pressure at that temperature is 25 hPa and RH is 80%.

4. Write the relation between length of humanhair and relative humidity. Hair hygrograph.

5. Define specific humidity, dew point temperature, wet bulb temperature. What is lifting condensation level?

6. Write the theoretical basis of wet bulb thermometers and write briefly about whirling psychrometer.

CHAPTER 6

The Surface Wind

Air in motion is called wind and is expressed in terms of direction and speed. In meteorology wind refers to flow of air either at surface of the earth or in the free atmosphere. Here we are concerned with the horizontal flow of air near earth's surface. Wind velocity is a vector quantity, it has both magnitude and direction. The magnitude of the wind is called wind speed which is a scalar quantity.

Wind direction is regarded as the direction from which wind blows and speed is the rate of movement of air in its instantaneous direction. Wind direction is determined with reference to true north and is expressed in degrees and measured clockwise from geographical north in terms of the points of the compass 0-360°, where East (E) denotes 90°, South(S) denotes 180°, West (W) denotes 270°, North (N) denotes 0° or 360°. The Table 6.1 gives the wind direction in 32 points of compass with code figures used in meteorology.

Motion of air near the surface of earth is greatly affected by ground friction, heat sources and obstruction such as buildings, trees, rocks etc. Generally wind speed increases with height. A standard exposure condition and height of the wind measurement instrument is adopted. The surface wind instrument should be at an height of 10 m agl and the axis of wind wane should be vertical and exactly oriented to true north. For the convenience of aviation wind measurement is made in knots (kt = Nautical miles per hours). The conversion of knot into kmph [1kt = 1.85 kmph, 1kt = 0.51 mps (kmph = kilometers per hour, mps = meters per second)] is given in Table 6.2.

Table 6.1 Surface wind direction.

Direction of compass point	Degrees	Code
Calm	-	00
N by E	11.25	01
NNE	22.5	02
NE by N	33.75	03
NE	45	05
NE by E	56.25	06
ENE	67.5	07
E by N	78.75	08
E	90	09
E by S	101.25	10
ESE	112.5	11
SE by E	123.75	12
SE	135	14
SE by S	146.25	15
SSE	157.5	16
S by E	168.75	17
S	180	18
S by W	191.25	19
SSW	202.5	20
SW by S	213.75	21
SW	225	23
SW by W	236.25	24
WSW	247.5	25
W by S	258.75	26
W	270	27
W by N	281.25	28
WNW	292.5	29
NW by W	303.75	30
NW	315	32
NW by N	326.25	33
NNW	337.5	34
N by W	348.75	35
N	360	36
Variable	-	99

Note : Read N by E: North by East, NNE: North North East etc.

Gustiness

The irregular variations of wind speed (oscillations) is called gustiness.

$$\text{Gustiness factor} = \frac{\text{Range of wind speed}}{\text{Average wind speed}}$$

Table 6.2

I		II		III		IV		V	
Kt	Kmph	Kt	Kmph	Kt	Kmph	Kt	Kmph	Kt	Kmph
1	1.9	11	20	21	39	31	57	45	83
2	3.7	12	22	22	41	32	59	50	93
3	5.6	13	24	23	43	33	61	55	102
4	7.4	14	26	24	44	34	63	60	111
5	9.3	15	28	25	46	35	65	65	120
6	11.1	16	30	26	48	36	67	70	130
7	13.0	17	31	27	50	37	69	75	139
8	14.8	18	33	28	52	38	70	80	148
9	16.7	19	35	29	54	39	72	85	157
10	18.5	20	37	30	56	40	74	90	167
								95	176
								100	185

Wind Speed Estimation

Wind speed estimation without instruments was initially developed by Admiral Sir Francis Beaufort (1905) for ship navigation. It is still being used where wind instruments are not available. The Table 6.3 and 6.4 gives the Beaufort scale and equivalent wind speed with specification.

Beaufort Scale : Wind speed equivalents, wind speed range and mean.

Table 6.3

Beaufort Number	Description	Kt	mps	Kmph	Miles per hour
0	Calm	<1 (0)	0-0.2 (0.1)	<1(0)	<1 (0)
1	Light air	1-3 (2)	0.3-1.5(0.8)	1-5(3)	1-3(2)
2	Light Breeze	4-6 (5)	1.6-3.3(2.5)	6-11(9)	4-7 (5)
3	Gentle breeze	7-10 (9)	3.4-5.4(4.4)	12-19(16)	8-12 (10)
4	Moderate breeze	11-16 (13)	5.5-7.9(6.7)	20-28(24)	13-18 (15)
5	Fresh breeze	17-21 (18)	8.0-10.7(9.3)	29-38(34)	19-24 (21)
6	Strong breeze	22-27(24)	10.8-13.8(11.8)	39-49(44)	25-31 (28)
7	Near gale or moderate gale	28-33 (30)	13.9-17.1	50-61(55)	32-38 (35)
8	Gale or fresh gale	34-40 (37)	17.2-20.7	62-74(68)	39-46(43)
9	Strong gale	41-47(44)	20.8-24.4	75-88(82)	47-54(51)
10	Storm	48-55(52)	24.5-28.4	89-102(96)	55-63(59)
11	Violent storm	56-63 (60)	28.5-32.5	103-117(110)	64-72(68)
12	Harricane	≥ 64	32.7 and more	118 and more	73 and more

Table 6.4

B. No	Description	Speed estimation over land	Over sea
0	Calm	Calm, smoke rises vertically	Sea like a mirror.
1	Light air	Direction of wind shown by smoke-drift but not by wind vanes.	Ripples with the appearance of scales are formed, but without foam crests.
2	Light breeze	Leaves rustle, ordinary vanes moved by wind	Small wavelets, still short but more crests have a glassy appearance and do not break.
3	Gentle breeze	Leaves and small twigs in constant motion, wind extends light flag.	Large wavelets, crests begin to break. Foam of glass appearance perhaps scattered white horses.
4	Moderate breeze	Raises dust and loose papers, small branches are moved	Small waves becoming longer, fairly frequent white horses.
5	Fresh breeze	Small trees in leaf begins to sway, orested wavelets form on inland waters	Moderate waves taking a more pronounced long form, many white horses are formed (chances of some spray).
6.	Strong breeze	Large branches in motion; whistling sound heard in telegraph wires, umbrellas used with difficulty	Large waves begin to form, the white foam crests are more extensive everywhere (probably some spray)
7	Near gale or Moderate gale	Whole trees in motion, inconvenience felt when walking against the wind.	Sea heaps up and white foam; breaking waves begin to be blown in streaks along the direction of the wind (spindrift begins to be seen)
8	Gale or Fresh gale	Breaks twigs off trees, generally impedes progress	Moderate high waves of greater length; edges of crests break into spindrifts, the foam is blown in well-marked streaks along the direction of the wind.
9	Strong gale	Slight structural damage occurs (chimney pots and slates removed)	High waves; dense streaks of foam along the direction of the wind. Crests of waves begin to topple, tumble and roll over; spray may affect visibility.

Table 6.4 Contd...

B. No	Description	Speed estimation over land	Over sea
10	Strom	Seldom experienced inland; trees uprooted; considerable structured damage occurs.	Very high waves with long over hanging crests. The resulting foam in great patches is blown in dense white streaks along the direction of the wind. On the whole, the surface the sea takes white appearance. The rolling of the sea becomes heavy and shock like. Visibility is affected.
11	Violent storm or severe storm	Very rarely experienced; accompanied by widespread damage.	Exceptionally high waves (small and medium sized ships might for a time be lost behind waves). The sea is completely covered with long white patches of foam lying along the direction of wind. Every-where the edges of the wave crests are blown into froth. Visibility badly affected
12	Hurricane or super cyclone	Severe and extensive damage	The air is filled with foam and spray. Sea is completely white with driving spray. Visibility is very seriously affected

Measurement of Wind Direction

The wind direction is measured by wind vane. It is a balanced lever which freely turns about a vertical axis. In a common type of wind vane one end of the lever exposes a broad surface to the wind, while the other end, which is narrow, points to the direction from which the wind blows. Under this movable system there is a fixed rigid cross, whose arms are set to the four cardinal directions – North, East, South and West. Some wind vanes are provided with eight directions N, NE, E, SE, S, SW, W and NW.

Exposure

The wind vane must be freely exposed and be high enough above buildings and trees so that it is not affected by the eddies created by them. In a very open site the vane can be installed on steel or wooden lattice tower or mast,

6 meters high and well guyed within the observatory enclosure. Where the site is obstructed by trees, buildings etc, it may be erected on a building or a high mast so that it is at least higher than the highest obstacle (in the vicinity) by 3 meters. Alternately a separate site for the mast may be selected outside observatory, such as the roof of a building. When both anemometer and the wind vane are fixed on the same platform, they should be at least 2 meters apart

Anemometer or Cup Anemometer

Wind speed measurement instrument is called Anemometer. IMD uses (i) Cup Anemometers (ii) Pressure Tube anemometers. The cup Anemometer consists of three large semi-conical cups (diameter 127 mm) with beaded edges fixed at the ends of three rods. The cups are mounted symmetrically about a vertical axis so that the diametral plane of each cup is vertical. As the force on the concave side of any cup, due to wind is greater than that on convex side in a similar position, the cup wheel rotates. For any given anemometer, the dimensions and design are such that the speed at which the cup wheel rotates depends solely on the wind speed provided the wind speed is steady and is greater than the minimum required to set the cups in motion; the lower limit is due to the effect of friction on the bearings of the cup wheel. The rate of rotation does not depend on the direction of the wind nor to any appreciable extent on the density of the air.

The cups are attached to a central spider which is mounted on a spindle carrying a worm. The worm engages with a gearwheel and drives a revolution counter mounted in a water proof aluminium housing. The gear ratio between the cup and counter spindle is so chosen in relation to the factor of the anemometer, that the counter indicates directly the run of wind in kilometers and tenths, when the instrument is suitably exposed.

To obtain the run of the wind in km and tenths over a given period (T) the counter is read at the beginning (R_1) and at the end of the period (R_2).

The mean wind speed during this period is given by $\left(\dfrac{R_2 - R_1}{T} \right)$

where T is in minutes.

Exposure

The wind velocity near the surface of the earth varies rapidly with height and is also largely affected by the presence of irregularities in the ground or nearby obstacles such as trees, buildings. For synoptic reports of the surface wind and for general climatological records, it is therefore necessary to define the height and conditions under which the measurement should be made. The standard exposure of wind instruments over level open terrain is 10 meters above ground. Open terrain is defined as an area where the distance between the anemometer and any obstruction is at least 10 times the height of the obstruction. This ideal condition is difficult to obtain but care should be taken to select the best possible.

Wind Speed at the Hour of Observation

Surface wind speed for synoptic purposes is reported in knots. To determine the wind speed at the time of observation, take two successive readings (R_1, R_2) of the anemometer at an interval of 3 minutes. Then $20 (R_2 - R_1)$ gives the wind speed in kmph. To convert into knots multiply by 0.54 i.e., $20 \times 0.54 \times (R_2 - R_1)$ gives wind speed in knots.

Example 6.1

Let the counter reading at the beginning of the observation $R_1 = 2063.4$ and let the reading after 3 minutes be $R_2 = 2065.1$

$$\text{Wind speed in kmph} = (2065.1 - 2063.4) \times 20$$

$$= 1.7 \times 20 = 34 \text{ kmph.}$$

$$\text{Wind speed in knots} = 34 \times 0.54 = 18.36 \approx 18$$

Example 6.2

Average wind speed during the day. Let the counter reading at 0830 hrs IST on 10.11.2000 be 9985.5 and at 0830 hrs IST on 11.11.2000 be 0097.6 (which is 10097.6).

$$\text{The average wind speed in kmph} = \frac{10097.6 - 9985.5}{24}$$

$$= \frac{112.1}{24} = 4.67 \approx 4.7$$

$$\therefore \text{ Average wind speed on 10.11.2000 in kt} = 4.67 \times 0.54 = 2.5218$$

Accuracy

The anemometer is expected to have frictional and other errors less than 10%. The anemometer normally begins to rotate at wind speeds of the order of $1\frac{1}{2} - 2$ kt and between $2 - 5$ kt. Its indications depend to a large extent on the bearing friction and the state in which the instrument is maintained.

Pressure Tube Anemometer

In a pressure tube type anemometer, a wind vane at the top of the mast keeps the opening end of the tube facing the wind. As the wind blows into open and it produces pressure rise inside, which is proportional to the wind speed. This pressure rise is transmitted to the indicating device. An outer tube just below the wind vane contains small holes. The wind blowing against the holes reduces the pressure inside which again proportional to the wind speed. This effect is also transmitted to the indicator by a suction tube. The combination of these two effects forms a system which is independent of the slight pressure difference if any inside and outside the building in which the recorder is kept.

Theory

In streamline flow, the Bernoulli's equation $\frac{1}{2}\rho v^2 + p = \text{constant}$, gives that total head (velocity head + static pressure) is constant. When an open end tube is facing wind, the total pressure acting at its mouth is constant $(= C)$. If the difference between the total head and static pressure is measured, the result is proportional to v^2 (v = velocity). This is the theoretical basis of Dines pressure tube anemometer or its modification.

Anemograph is an instrument which gives continuous recorded data of wind speed and direction. A density correction has been applied by W. Ferrel when the pressure tube anemometer located at heights which are sufficiently different from the place of standardization. This is given by

$$p = \frac{0.002698}{1 + 0.004t} V^2 \frac{P}{P_o}$$

$$\begin{aligned}
\text{Where} \quad p &= \text{pressure in pounds per sq. ft} \\
P_o &= \text{standard barometric pressure (760 mm of Hg)} \\
P &= \text{Barometric pressure at observation station} \\
t &= \text{temperature in } ^{\circ}\text{C}, \ v = \text{wind speed in feet/sec}
\end{aligned}$$

At a temperature 15 $^{\circ}$C, P = 760 mm of Hg, the above formula reduces to $p = 0.00255 \ V^2$

Dines Pressure Tube Anemograph

This is a continuous recording wind speed instrument. In this device the pressure and suction are transmitted to a float chamber, which controls levels of floating cylinder in water. The float is attached to the indicator which marks on a graph chart wound on a cylinder driven by a clock device.

Squall : A squall is a sudden burst of strong wind that lasts for a few minutes. More specifically,

A sudden increase of wind speed by at least three stages on Beaufort scale (16 Kt) and the speed reaching B.F.6 or more and lasting at least one minute.

For Example

The sudden increase of wind speed from 16 kmph (BF 3) to 44 kmph (BF 6) [or 9kt to 24kt] and remains at 44 kmph for at least one minute, is a squall.

Squall line : A violent squall associated with the passing of a long line or arch of dark cloud and accompanied by thunder and lightning, rain or hail and a sudden cooling with a shift in wind direction. A line squall although of short duration may blow off trees and houses.

Gale : Wind speed of B.F. 8 or more, blowing continuously and doing damage to trees, houses etc. is called gale.

Gust : A gust is a sudden increase in wind speed relative to its mean value over a period of time which is shorter than squall duration.

Diurnal Variation of Surface Wind Speed

Clockwise change of direction of wind is called veering and anticlockwise changes of direction of wind is called backing. It is observed that wind changes during day and night both in direction and speed particularly in mountain or hill slopes, valleys, coastal areas. This change is diurnal (or daily) variation. Prominent local winds and their cause of change discussed separately. The relation between pressure gradient and wind speed discussed in Geostrophic wind. Over land, wind speed generally strengthens during midday and weakens towards evening. The maximum wind speed is observed during afternoon (at about 1400 hrs local time) and minimum at dawn. During the day time ground surface is heated up by insolation as the day advances and causes convection in the wind layer closer to the ground.

At night the air close to the ground is cooled (on earth being cooled) and tend to remain or sink at lower levels. Thus day heating strengthen the wind speed while nocturnal cooling be calms it.

It will be discussed later that in northern hemisphere Coriolis force deflects wind to the right of its direction and in southern hemisphere it deflects towards its left. Due to surface friction surface wind rarely blows parallel to the isobars. In general it flows cross isobars from High to Low pressure area at an angle varying 10 degrees over sea to 30° over land. Surface friction effect has been discussed in planetary boundary layer. In 1857 Buys Ballot, a Dutch meteorologist, gave a relationship between wind direction and isobars which is popularly called Buys Ballots law, which states as follows.

If an observer stands with his back to the wind the low pressure is on his left in the northern hemisphere, and on his right in the southern hemisphere.

Wind Veering : clock wise change of wind direction with time is called veering.

Wind Backing : Anticlockwise change of wind direction with time is called backing.

Cyclonic Circulation : In Northern Hemisphere (S.H) the anticlockwise (clockwise) spiralling in of wind (at any level) is called cyclonic circulation.

Anti Cyclonic Circulation : In Northern Hemisphere the clockwise spiralling out of wind (at any level) is called anti-cyclonic circulation.

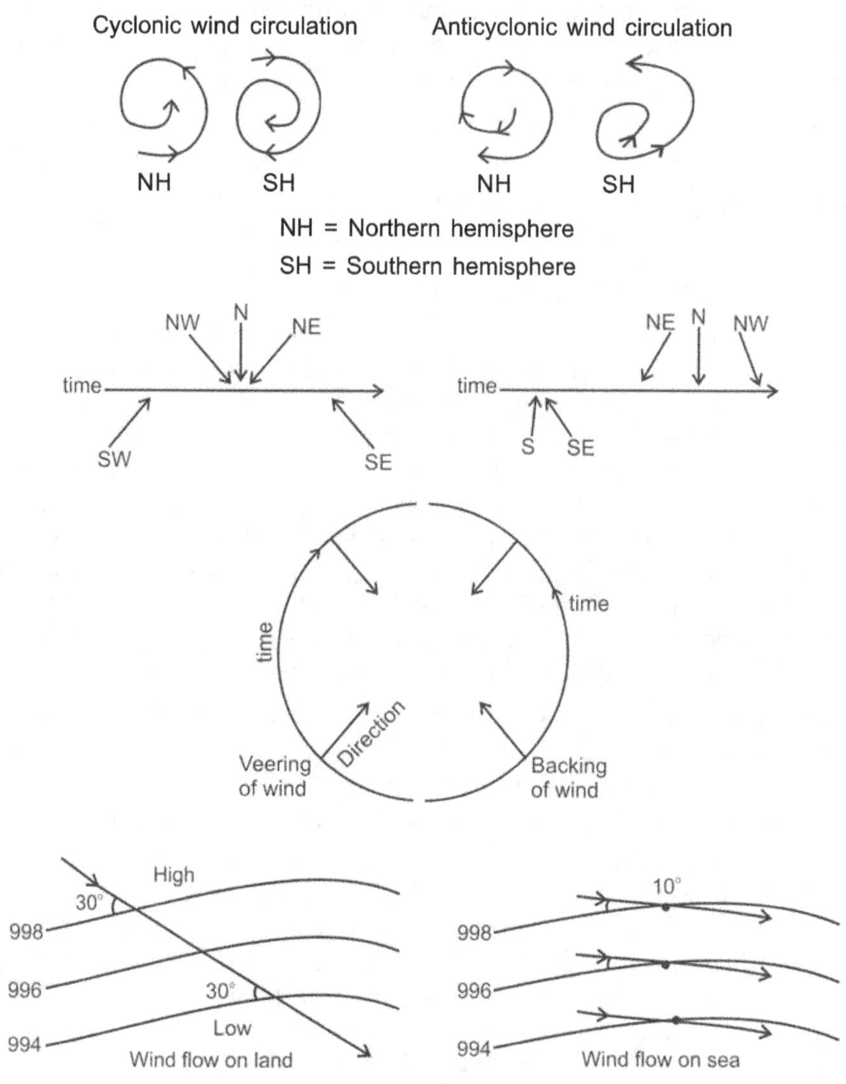

Fig. 6.1 Wind circulation.

Local Winds

Local conditions such as earth's elevation, rocky areas, forests, deserts, hills, mountain slopes, valley, lakes, reservoirs, sea coasts, towns and cities etc., have great influence on surface wind. Diurnal variations of surface wind and temperature are decisively affected by these conditions. We shall consider a few important cases.

Land and Sea Breezes

(A) Sea Breeze : A few hours after sunrise sea breeze develops on calm summer days. It has cooling effect in tropics and continues throughout the day light hours and dies down around sunset. After this seaward blowing land breeze develops which is confined to coastal area. The direction of sea breeze does not remain constant day long. It sets in with gustiness. Diurnal variation of wind in a coastal area is a striking example. Near the sea coast on shore wind generally sets in at about 9 – 10 AM and reaches peak at about 2 pm (1400 hrs local time) and then slowly weakens and dies away at about 6 – 7 pm. The strength of the sea breeze is stronger on warm days and weaker on cloudy occasions. Maximum sea breeze velocity varies (4 – 7 mps) 14 to 25 kmph. The theoretical aspects of sea breeze is considered in solenoids. The specific heat of sea water is 40% more than that of soils. During day time land gets heated up (forms local low) and sea remains comparatively cooler (forms local high). This tilts the isobaric surfaces downwards near the coast. Thus wind blows at surface level from sea to land and land to sea in lower levels of atmosphere, forming wind circulation as shown in Fig. 6.2(a). The depth of the sea breeze is about (150-1000 m) 1 km, but it thins at advancing edge. Generally sea breeze penetrates over land to about 50 km (by 2100 hrs). In tropical regions sea breeze reaches 50 to 60 km, sometimes even 125 to 145 km interior.

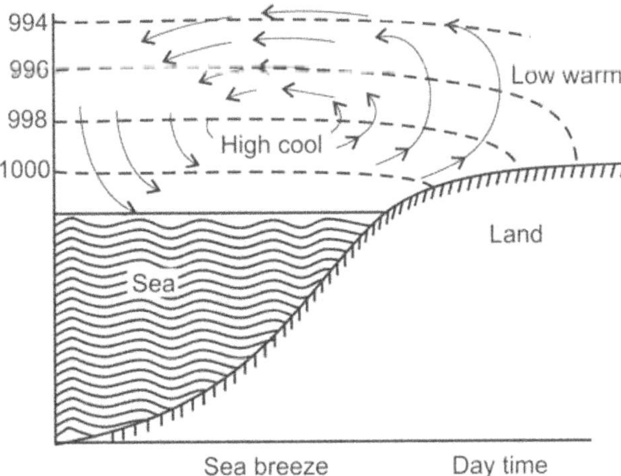

Fig. 6.2(a) Sea breeze.

A similar situation is observed (at smaller scale) near large lakes where we find lake breeze. Monsoon circulation is also caused by differential heating of land and sea, resembles to sea breeze but over a very large scale (This will be discussed in great deal subsequently). In this case the land area is a continent and sea area is ocean [The diurnal variation of temperature over land is about five times that of the diurnal variation of temperature over sea]

(B)   Land Breeze : During night the reverse of sea breeze occurs. Coastal land area cools down (local high forms comparatively) while the sea area remains warm (local low forms). As a consequence wind blows from land to sea at surface level and from sea to land in the lower levels of the atmosphere as shown in the Fig. 6.2(b). In general land breeze is not so prominent as the sea breeze. However in tropics land breeze is also marked, and in some areas thunderstorms may occur off the coast at about dawn. In this case the isobaric surfaces lifted up near the coast and wind blows from land to over sea. The maximum speed of land breeze is about (2 mps) 7 – 8 kmph and depth is about ½ km and sea ward range of land breeze is very small about 5 – 10 km only (in tropical regions extends up to 25 – 30 km).

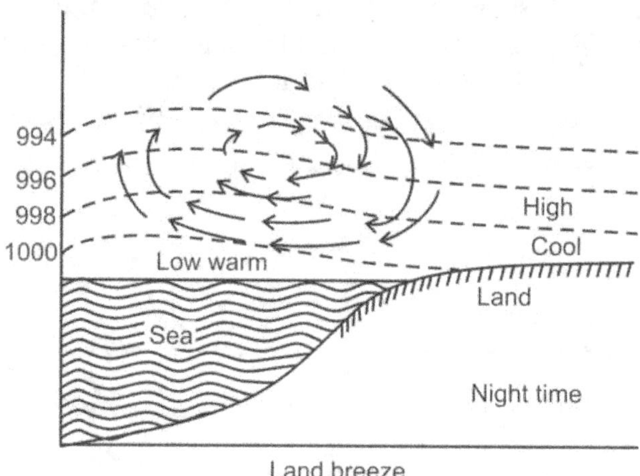

Fig. 6.2(b) Land breeze.

Note : Land and sea breeze are prominent in tropical regions with great regularity where its occurrence is 80%, of days, elsewhere its occurrence is less than 50% of days and with lesser regularity. In polar regions land and sea breeze may be hardly seen during summer sphere. In tropical monsoon regions, during summer monsoon period, land and (150 m to 1000 m) sea breeze are masked and may be seen at a small scale and strength on less than 20% of the days of the season.

Katabatic Winds (Mountain Winds)

The phenomena of daily wind change along the axis of large valley is well known in mountainous regions. During clear nights, on mountain slopes wind flows down the slopes. At night mountain ridge slope cools down and the air close to it becomes heavier (denser) and moves down into river valleys. These down slope winds are called Katabatic winds (also called mountain winds) Fig. 6.3(a). It may be noted that as the air flows down the slope it warms up due to adiabatic warming, but the cold surface counteracts and keeps it cool (compared to the open air at that height). Thus the wind continues to flow down the slope. Katabatic winds are in general light but they acquire considerable speed if the slope is quite steep and is covered by snow or ice. Down slope wind when reaches the bottom (open valley) it replaces the warm wind and attains maximum speed (5-6 mps) at about sun rise (0700 hrs), when the diurnal cooling is maximum. Thickness is about 100 – 150 m. Example, Mistral wind belongs to this category.

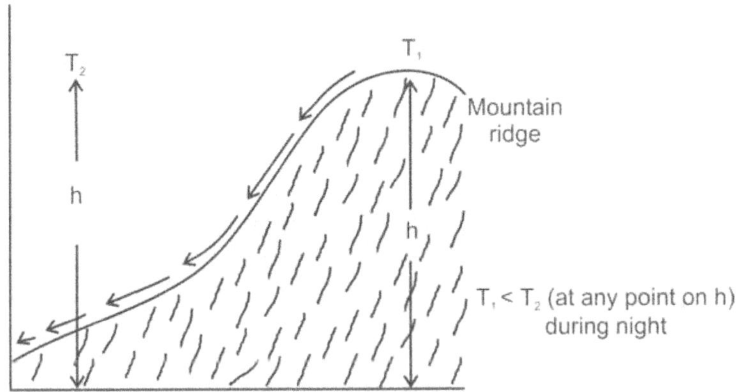

Fig. 6.3(a) Katabatic winds or mountain winds.

Anabatic Winds (Valley Winds)

On warm clear day upslope winds, a reverse process of katabatic winds occur on mountain slopes. These are called anabatic or valley winds. In Greek Katabatic means 'go down' and Anabatic means 'go up'.

On a clear warm day, the mountain slope (hill slope) gets heated up and the air close to it becomes warmer (lighter) than the free air at that level. As a result air close to the mountain slope, being lighter than the free air, ascends Fig. 6.3(b). As the wind moves upward adiabatic cooling takes place, but the cooling is countered by the warm air close to the slope. Thus, the air continues to move up slope of mountain during afternoon, at about 1400 hrs

LMT, thickness is 100 – 200 m and attain maximum wind speed of 2 – 4 mps. In general anabatic winds are lighter than katabatic winds due to friction and gravity. Mountain and valley winds (Katabatic and Anabatic winds) are well developed in the wide and deep valleys of the Alps. They are conspicuous during summer fair weather when high pressure area develops there.

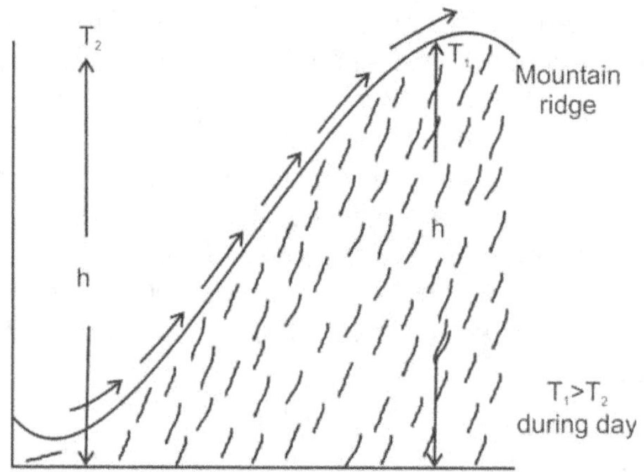

Fig. 6.3(b) Anabatic winds or valley winds.

Fohn Winds

When moist air is forced to rise a mountain barrier (due to pressure gradient), it cools adiabatically and attaining saturation it condenses into precipitation. After crossing the barrier on leeward side the dry air (mostly) moves down the mountain slope and warms up adiabatically. At the foot of the mountain, on leeward side, it blows as pleasant warm wind. This phenomenon occurs on the southern side slopes of the (mountain) Alps. These are called Fohn winds. Fohn are well developed in Alps (Europe), Switzerland, Tirol. They are seen in Greenland (from the inland ice over the mountain ranges down to the coast). In north America it is known as the Chinook. In Argentina it is called Zonda and blows down the Andees. The clouds associated with Fohn are lens-shaped (lenticularis). Any such physical process which warms the wind on lee-ward side of the mountain is now called Fohn winds. Fohn also occurs on lee of the mountains of the Caucsaus and central Asia. At Tashkhant (in central Asia) in Russia, where average temperature in winter is about 0 °C, which rises to about 20 °C due to Fohn.

In Fig. 6.4 the rising wind up to B (A to B) moves dry adiabatically (DALR = 10 °C/km), from B to C moves at saturated adiabatic lapse rate (SALR = 6 °C/km) and from C to D moves again at DALR. In between B and

C precipitation (pptn) occurs. If the ascending air is completely dry or ascending air does not precipitate to release latent heat (between B and C), then the descending air at D regains to the original temperature (T_1 = 18 °C) without warning to T_3 = 22 °C.

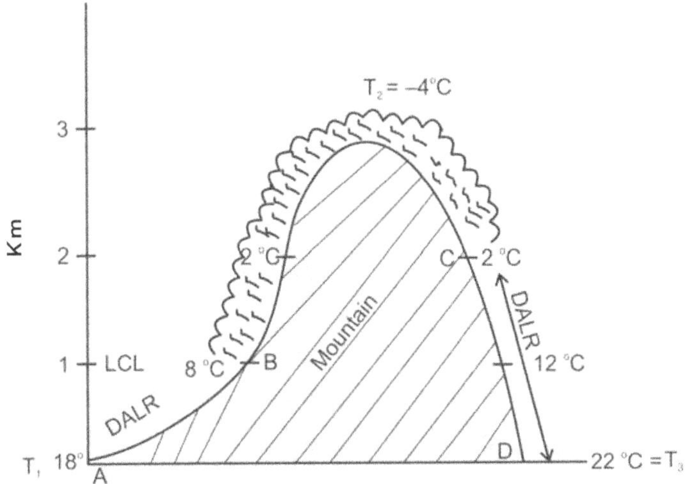

Fig. 6.4 Fohn wind model.

The Mistral

The Mistral is a Katabatic (or gravity) wind which is prominent along Mediterranean coast of France. In winter high pressure ridge frequently extends from the Azores to France. It drains from the higher lands and snow-capped mountains of the north. After draining down it is channelled by Rhone valley. It abruptly changes the climate of Riviera with coolness and dryness which otherwise is warm and pleasant during winter. Mistral may last for about a week. The annual frequency of Mistral speed exceeding 21 kt is about 105 days and 33 kt is 30 days.

Bora

Bora is another example of Katabatic wind which flows down from the plateau of Yugoslavia on to the northern coast of Adriatic sea and northern Aegean sea. In winter very cold continental wind from Russia, travelling over Dalmatia mountain arrives to the coast of Adriatic sea as cold wind and replaces the normally existing warm wind. This cold gusty wind causes damage. There are lulls between strong gusts of about 50 mps, called Reffoli, which are associated with pressure variation of about 4 mm of Hg. Bora does not extend to a long distance over the sea, but it causes a cloud of mist over the sea which is locally called Fumarea. The Bora of the Adriatic sea has marked diurnal variation of strength and frequency. It attains maximum intensity at about 0700 (LMT) and minima at about midnight (2400 hrs). It has maximum frequency of occurrence at about 0600 hrs in the morning and minimum frequency in the afternoon at about 1400 hrs.

Harmattan

In certain regions winds blow with some regularity and characteristics of the region in association with pressure systems. Harmattan is one such wind which is cool, dry and dusty. In winter it comes from Sahara and flows on to Guinea coast of north Africa and replaces humid tropical air. Due to its cooling effect locally it is called Doctor.

Scirocco

Hot, dusty and dry winds of continental tropical origin blows in north Africa in summer. These winds locally called by various names. In Algeria and the Levant they are called by Scirocco, in south-east Spain Leveche, in Egypt Khamsin. During summer depressions in this region move west to eastward and these local winds move ahead of the depressions in a northerly direction.

Chinook

Like Fohn, Chinook is a warm dry air which blows on the leeward side of Rocky mountains. These winds have the origin over Pacific ocean. While crossing over the Rockies they loose their moisture in precipitation over the mountain. In Canada Chinook's effect is found over a large distance. It has been recorded a variation of about 22 °C temperature during the passage of Chinook in just five minutes. Winter maximum temperature records during the passing of Chinook. In local Red Indian language Chinook means snow eater. In winter months at Medicine Hat the lowest minimum temperature goes to about –45 °C while the highest maximum is associated with Chinook goes to about + 20 °C. Similarly western suburbs of Calgary records about 0 °C where Chinook effects while the eastern suburb records about – 15 °C (where there is no Chinook effect).

Questions

1. What are the standard exposure conditions of wind measurement instrument and units of measurement?

2. Define gustiness factor, and squall. Using Beaufort scale number and knots define (i) Breeze (ii) Gale, (iii) Storm, (iv) Hurricane. What are its effects on trees, telegraphwires and human beings and damages.

3. Describe the working of Cup Anemometers and pressure tube anemometer.

4. What is the theoretical basis of Dines pressure tube anemometers. What is the output of Dines pressure tube anemograph.

5. Write briefly about the diurnal variation of surface wind speed.

6. Write briefly on (i) Land breeze (ii) Sea breeze, (iii) Katabtic winds (iv) Anabatic winds (v) Form winds, (vi) Chinook.

CHAPTER 7

Cloud Classification

Definition of a Cloud

According to WMO, a cloud is a visible aggregate of minute particles of water, ice or both in the free atmosphere. In simplest terms, a cloud is a fog high in the air. From micro-physical point of view, clouds are the result of the action of atmospheric nuclei, from a macro-physical point of view clouds are the result of the cooling of moist air caused by the expansion of air during vertical motion, contact with colder air masses or by contact with colder land or water surfaces.

Clouds are formed in the atmosphere when the moist air from the earth's surface lifted and then condensed on nuclei due to cooling. The level at which moist air begins to condense in called lifting condensation level (LCL). Clouds are continuously in a process of formation or dissipation depending on the environmental conditions and hence appear in an infinite variety of forms. There are, however, some characteristic forms frequently observed all over the world. Clouds are classified based on their height, shape, reflection or refraction of light and colour. The WMO (International Atlas) classification of clouds follows. It is based on Genera (10 main groups, Latin names), species (sub-divisional genera based on internal structure and shape), varieties (special characteristics such as arrangements and transparency), supplementary features (attached to the main clouds such as udder like, hanging, protuberance etc), Mother clouds (a part of the cloud may develop into pronounced extension from the main cloud, which is called mother cloud).

Genera (Average height of cloud base)	Species	Varieties	Cloud temperature	Supplementary features	Mother cloud Genitus
Cirrus (Ci) (5000 – 13700 m)	Fibratus Uncinus Spissatus Castellanus Floccus	Intortus Radiatus Vertebratus Duplicatus	– 20 to – 60 °C	Mamma	Cc, Ac, Cb
Cirrocumulus (Cc) 5000 – 13700 m	Stratiformis Lenticularis Castellanus Floccus	Undulatus Lacunosus	– 20 to – 60 °C	Virga mamma	– –
Cirrostratus (Cs) 5000 – 13700 m	Fibratus nebulosus	Duplicatus Undulatus	– 20 to – 60 °C	–	Cc Cb
Altocumulus (Ac) 2000 – 7000 m	Stratiformis Lenticularis Castellanus Floccus	Translucidus perlucidus Opacus duplicatus radiatus lacunosus	+ 10 to – 30°C	Virga mamma	Cu, Cb
Altostratus (As) 2000 – 7000 m	–	Translucidus Opacus duplicatus undulatus radiatus	+10 to – 30 °C	Virga praecipitatio pannus mamma	Ac, Cb
Nimbostratus (Ns) 900 – 3000 m	–	–	+10 to – 15 °C	Praecipitatio Virga pannus	Cu, Cb
Stratocumulus (Sc) 460 – 2000 m	Stratiformis lenticularis castellanus	Translucidus perlucidus opacus duplicatus Undulatus lacunosus	+15 to – 5	Mamma virga praecipitatio	As, Ns, Cu, Cb
Stratus (St) surface to 460 m	Nebulosus fractus	Opacus translucidus undulates	+ 20 to – 5 °C	Praecipitatio	Ns, Cu, Cb
Cumulus (Cu) 460 – 2000 m	Humilis, mediocris, congestus, fractus	Radiatus	+15 to – 5 °C	Pileus velum, virga, praecipitatio, arcus, pannus, tuba	Ac, Sc
Cumulonimbus (Cb) 460 – 2000 m	Calvus capillatus	– –	+15 to – 5 °C	Praecipitatio virga, pannus, incus, mamma, pileus, velum, arcus, tuba	Ac, As, Ns, Sc, Cu

See Figs. 7.1, 7.2, and 7.3

Note: Cloud base height and cloud base temperatures are averages. They differ widely form equatorial regions to polar regions.

Table 7.1

Etage	Polar regions km	Temperate regions km	Tropical regions km
High (Ci, Cs, Cc)	3 -8	5 – 13	6 – 18
Middle (Ac, As, Ns)	2 – 4	2 – 7	2 – 8
Low (Sc, St, Cu Cb)	From the earth's surface to 2 km	From the earth's surface to 2 km	From the earth's surface to 2 km

(A)　Definitions of Clouds

WMO (International Atlas) defined the ten cloud genera (10) is given below. Fig. 7.1 to 7.3

1. **Cirrus (Ci)** : Detached clouds in the form of white, delicate filaments or white or mostly white patches or narrow bands. These clouds have fibrous (hair-like) appearance, or a silky sheen (lustre) or both.

2. **Cirrocumulus (Cc)** : Thin, white patch, sheet or layer of cloud without shading, composed of very small elements in the form of grains, ripples etc., merged or separate, and more or less regularly arranged; most of the elements have an apparent width of less than one degree.

3. **Cirrostratus (Cs)** : Transparent, whitish cloud veil of fibrous or smooth appearance, totally or partly covering the sky, and generally producing halo phenomena.

4. **Altocumulus (Ac)** : White or grey or both white and grey, patch, sheet or layer of cloud, generally with shading, composed of laminae, rounded masses, rolls etc., which are sometimes partly fibrous or diffuse and which may or may not be merged; most of the regularly arranged small elements usually have an apparent width of between one and five degrees.

5. **Altostratus (As)** : Greyish or bluish cloud sheet or layer of striated (small channeled), fibrous or uniform appearance, totally or partly covering the sky, and having parts thin enough to reveal the sun at least vaguely, as through ground glass. Altostratus does not show halo phenomena.

6. **Nimbostratus (Ns)** : Grey cloud layer, often dark, the appearance of which is rendered diffuse by more or less continuously falling rain or snow, which in most cases reaches the ground. It is thick enough throughout to blot out the sun.

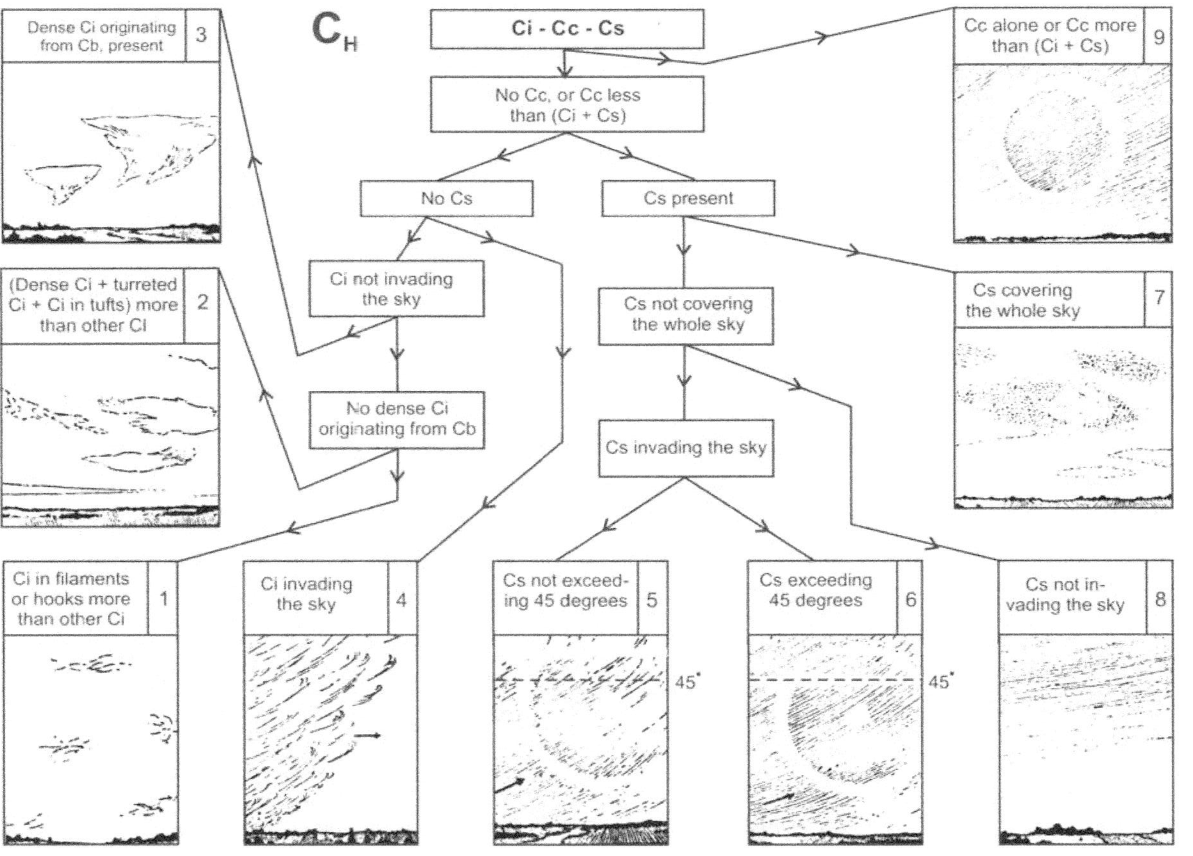

Fig 7.1 Pictorial guide for C_H- clouds.

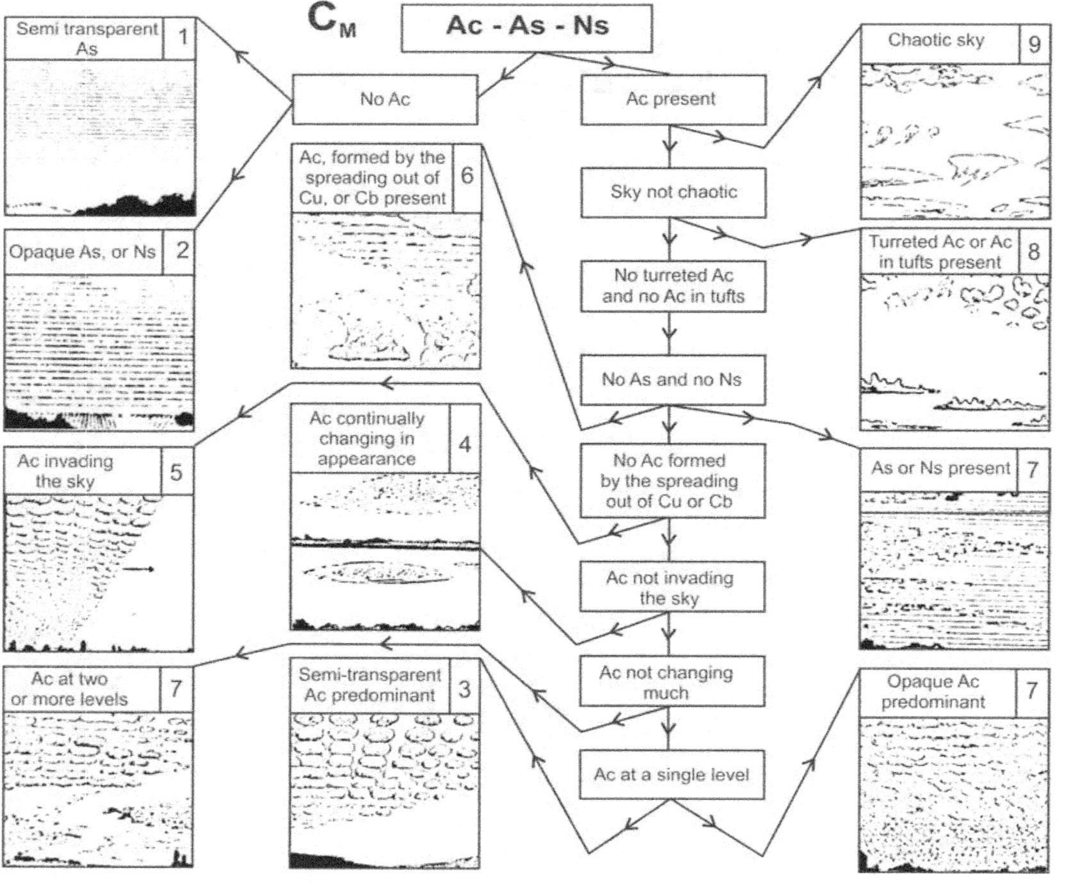

Fig 7.2 Pictorial guide for C_M - clouds.

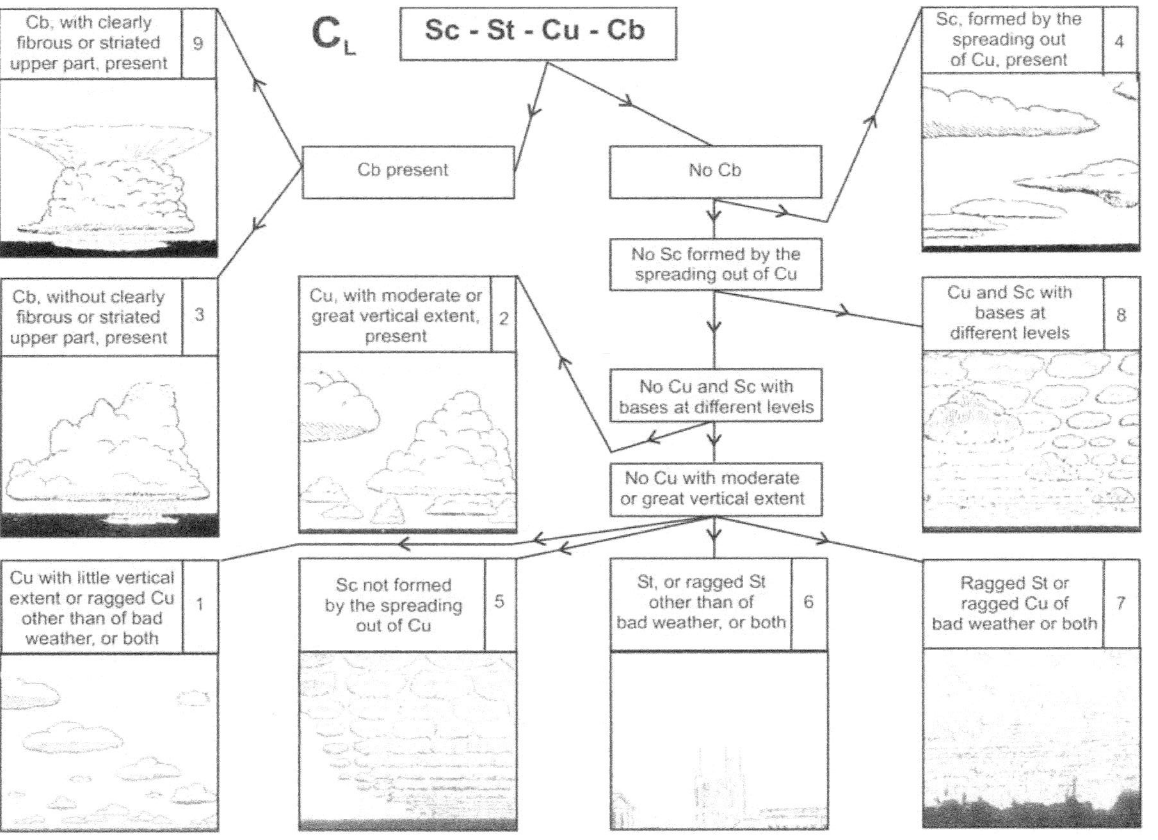

Fig 7.3 Pictorial guide for C_L - clouds.

Low ragged clouds frequently occur below the layer, with which they may or may not merge.

7. **Stratocumulus (Sc)** : Grey or whitish or both gray and whitish, patch, sheet or layer of cloud which almost always has dark parts, composed of tessellations, (form into squares or checkers) rounded masses, rolls, etc., which are non-fibrous (except for virga) and which may or may not be merged; most of the regularly arranged small elements have an apparent width of more than five degrees.

8. **Stratus (St)** : Generally grey cloud layer with a fairly uniform base, which may give drizzle, ice prisms or snow grains. When the sun is visible through the cloud, its outline is clearly discernible. Stratus does not produce halo phenomena except, possibly, at very low temperature.

    Sometimes stratus appears in the form of ragged patches.

9. **Cumulus (Cu)** : Detached clouds, generally dense and with sharp outlines, developing vertically in the form of rising mounds, domes or towers, of which the bulging upper part often resembles a cauliflower. The sunlit parts of these clouds are mostly brilliant white; their base is relatively dark and nearly horizontal. Sometimes cumulus is ragged.

10. **Cumulonimbus (Cb)** : Heavy and dense cloud, with a considerable vertical extent, in the form of a mountain or huge towers. At least part of its upper portion is usually smooth, or fibrous or striated, and nearly always flattened; this part often spreads out in the shape of an anvil or vast plume.

    Under the base of this cloud which is often very dark, there are frequently low ragged clouds either merged with it or not, and precipitation sometimes in the form of virga.

(B) Definition: Cloud Species (14)

1. **Fibratus** : Detached clouds on a thin cloud veil, consisting of nearly straight or more or less irregularly curved filaments which do not terminate in hooks or tufts.

    This term applies mainly to Cirrus and Cirrostratus.

2. **Uncinus** : Cirrus often shaped like a comma, terminating at the top in a hook, or in a tuft, the upper part of which is not in the form of a rounded protuberance.

3. **Spissatus** :Cirrus of sufficient optical thickness to appear greyish when viewed towards the sun.

4. **Castellanus** : Clouds which present, in at least some portion of their upper part, cumuliform protuberances in the form of turrets which

generally give the clouds a crenelated (loopholes) appearance. The turrets, some of which are taller than they are wide, are connected by a common base and seem to be arranged in lines. The castellanus character is especially evident when the clouds are seen from the side.

This term applies to cirrus, cirrocumulus, Altocumulus and Stratocumulus.

5. **Floccus :** A species in which each cloud unit is a small tuft with a cumuliform appearance, the lower part of which is more or less ragged and often accompanied by virga.

   This term applies to Cirrus, Cirrocumulus and Altocumulus.

6. **Stratiformis :** Clouds spread out in an extensive horizontal sheet or layer.

   This term applies to Altocumulus, Stratocumulus and, occasionally, to cirrocumulus.

7. **Nebulosus :** A cloud like a nebulous (cluster or vague) veil or layer showing no distinct details.

   This term applies mainly to Cirrostratus and Stratus.

8. **Lenticularis** : Clouds having the shape of lenses or almonds, often very elongated and usually with well defined outlines; they occasionally show irisation . Such clouds appear most often in cloud formations of orographic origin, but may also occur in regions without marked orography.

   This term applies mainly to Cirrocumulus, Altocumulus and stratocumulus.

9. **Fractus :** Clouds in the form of irregular shreds, which have a clearly ragged appearance.

   This term applies only to Stratus and Cumulus.

10. **Humilis :** Cumulus clouds of only a slight vertical extent; they generally appear flattened.

11. **Mediocris :** Cumulus clouds of moderate vertical extent, the tops of which show fairly small protuberances

12. **Congestus :** Cumulus clouds which are markedly sprouting and are often of great vertical extent; their bulging upper part frequently resembles a cauliflower.

13. **Calvus** : Cumulonimbus in which at least some protuberances of the upper part are beginning to lose their cumuliform outlines, but in which no cirriform parts can be distinguished. Protuberances and sproutings tend to form a whitish mass, with more or less vertical striations.

14. Capillatus : Cumulonimbus characterised by the presence, mostly in its upper portion, of distinct cirriform parts of clearly fibrous or striated structure, frequently having the form of an anvil, a plume or a vast, more or less disorderly mass of hair. Cumulonimbus capillatus is usually accompanied by a shower or by a thunderstorm, often with squalls and sometimes with hail; it frequently produces very well -defined virga.

(C) Definition: Cloud Varities (9)

First six refer to arrangement of macroscopic elements and the last three the degree of transparency.

1. **Intortus :** Irregularly curved cirrus filaments.
2. **Vertebratus :** Cloud elements arranged like vertebrae, ribs or a fish skeleton. This applies to mainly cirrus.
3. **Undulatus :** Clouds in patches, sheets or layers showing undulations (wave like pattern).
4. **Radiatus :** clouds in parallel bands seem to converge towards a points on the horizon or towards two opposite points on the horizon called "radiation point(s)". This mainly applies to Ci, Ac, As, Sc and Cu.
5. **Lacunosus :** Cloud patches, sheets or layers (generally thin) arranged in a manner that appears as a net or a honey-comb. This mainly applies to Cc, Ac and on rare occasions to Sc.
6. **Duplicatus :** Superposed cloud patches, sheets or layers, at slightly different levels or partly merged. This mainly applies to Ci, Cs, Ac, As and Sc.
7. **Translucidus :** Clouds in an extensive patch, sheet or layer, the greater part of which is translucent to reveal the position of the sun or moon. This mainly applies to Ac, As, Sc and St.
8. **Perlucidus :** An extensive cloud patch, sheet or layer, with small spaces between the elements. The spaces allow the sun, moon etc., or overlying clouds to be seen. This mainly applies to Ac and Sc.
9. **Opacus :** An extensive cloud patch, sheet or layer, the major part of which is opaque to mask the sun or moon. This mainly applies to Ac, As, Sc and St.

(D) Definition: Supplementary Features of Clouds (6)

1. **Incus :** The upper part of Cb cloud spreading into anvil with a smooth, fibrous or striated appearance.
2. **Mamma :** Hanging protuberances, like udders, on the under surface of a cloud. This mainly applies to Ci, Cc, Ac, As, Sc and Cb.

3. **Virga :** Vertical or inclined trails of precipitation (fall-streaks) attached to the under surface of a cloud, which do not reach the earth's surface. This occurs mostly with Ce, Ac, As, Ns, Sc, Cu & Cb.

4. **Praecipitatio :** Precipitation (rain, drizzle, snow, ice pellets, hail etc.,) falling from a cloud and reaching the earth's surface. This feature is accompanied with As, Ns, Sc, St, Cu and Cb.

5. **Arcus :** A dense, horizontal roll with more or less tattered edges (lace like edging) situated on the lower front part of certain clouds with the appearance of a dark, menacing arch. This is associated with Cb and sometimes with Cu.

6. **Tuba :** Cloud column or inverted cloud cone, protruding from a cloud base; showing more or less intense vortex. This is associated with Cb and sometimes with Cu.

(E) Definition: Accessory Clouds (3)

1. **Pileus :** An accessory cloud of small horizontal extent, in the form of a cap or hood above the top or attached to the upper part of a cumuliform clouds which often penetrates it. Several pileus clouds may fairly often be observed in superposition. Pileus principally occurs with Cu and Cb.

2. **Velum :** An accessory cloud veil of great horizontal extent, close above or attached to the upper part of one or several cumuliform clouds which often pierce it. Velum occurs principally with Cu and Cb.

3. **Pannus :** Ragged shreds, sometimes constituting a continuous layer, situated below another cloud and sometimes attached to it. Pannus mostly occurs with As, Ns, Cu and Cb.

Definition: Mother Clouds

Clouds may develop in either of two ways: (a) They may form in clear air (b) They may form from other clouds, called mother clouds.

Non-instrumental Observations

Observations of clouds, visibility and weather (and wave observations at coastal stations) fall under the non-instrumental meteorological observations. They are naked eye observations. Of these observations the most interesting is that of the growth and dissipation of clouds in the sky. To identify clouds it requires a great deal of experience and thorough knowledge of general meteorology is necessary. A cloud observation consists of (a) identification of cloud form, (b) estimating its amount, (c) estimating the height of its base above the station level and (d) determining the direction of its movement.

We have already dealt with the cloud identification of various form. We shall briefly discuss with the other three aspects of cloud observations.

Cloud Amount

The international unit for reporting cloud amount is okta or 1/8 of the sky. When the sky is completely cloudless the cloud amount is zero, when it is completely over cast (without gaps or openings) the amount is 8 oktas. When the sky is not discernible due to thick fog, the cloud coverage is coded as 99.

Cloud Height

By cloud height we mean the vertical distance of the base of cloud from the ground level. If the cloud base is diffuse and irregular as in case of ragged low clouds of bad weather, the height of the lowest patch of such clouds is reported. Cloud height can be measured with the help hydrogen filled balloons (called ceiling balloons) or by cloud search-light at night. Observatory stations which are not provided with these facilities, cloud height are estimated and reported. In hilly or mountainous areas this may be done fairly accurately by comparing the level of the cloud base with the heights of well marked topographical features but in level country the observer should use his judgement taking into consideration the form and general appearance of the cloud.

Direction of Movement of Cloud

A few departmental observatories have been provided with an instrument called "Nephoscope" for measuring the direction of movement of cloud. Instruction for its use are supplied along with the instrument. At other observatories, the direction of movement of cloud is visually estimated nearest to the eight points of the compass (namely N, NE, E, SE, S, SW, W and NW) from which the predominant low cloud and the predominant medium or high cloud are coming. This is done best by observing the movement of the cloud against a fixed point like a pole erected in an open space, corner of a building or stars at night time.

Questions

1. Define a cloud, LCL. Write the average height of Low, Medium, High clouds in polar, temperate and Tropical regions.

2. Name the WMO classification of cloud genera with cloud abbreviations.

3. Define or explain mother clouds, cloud amount, cloud height, direction of cloud movement.

CHAPTER 8

Meteors

Definition

According WMO, a meteor is a phenomenon, other than cloud, observed in the atmosphere or on the surface of the earth, which consists of a precipitation, a suspension or a deposit of aqueous or non-aqueous liquid or solid particles, or a phenomenon of the nature of an optical or electrical manifestation.

Note : Astronomical meteors that enter the atmosphere such as shooting stars are different from meteorological meteors.

Meteors are classified into four categories, They are (i) hydrometeors. (ii) litho-meteors, (iii) photometeors and (iv) electrometeors.

1. **Hydrometeors :** *Definition* A hydrometeor is a meteor consisting of an ensemble of liquid or solid aqueous particles, falling through or suspended in the atmosphere, blown by the wind from the earth's surface or deposited on objects on the ground or in the free air (Hydro means water).

   The most common hydrometeors that are frequently referred in meteorology are given below.

   Precipitation : This refers to falling of hydrometeors that finally reach the ground. This will not include a virga, which is seen streaming from the bottom of a cloud, but does not reach the earth's surface.

   Rain : Precipitation of liquid water particles, either in the form of drops of more than 0.5 mm diameter, or of smaller widely scattered drops.

Shower : liquid precipitation of big size usually marked with sudden onset. The falling drops are more widely scattered than rain. The diameter of the drop size is more than 0.5 mm. Showers generally fall from convective clouds (Cu or Cb).

Freezing rain : Rain, the drops of which freeze on impact with the ground or with objects on the earth's surface or with aircraft in flight.

Drizzle : Fairly uniform precipitation composed of exclusively of fine drops of water, whose diameter is less than 0.5 mm, and that these drops fall very close to one another. The impact of the drops on water surface is imperceptible.

Freezing drizzle : Drizzle, the drops of which freeze on impact with the objects on the earth's surface or ground or with aircraft in flight.

Snow : Precipitation of ice crystals, most of which are branched (or star shaped).

Sleet : Rain and snow falling together or snow melting as it falls.

Snow pellets: Precipitation of white and opaque grains of ice. The grains are spherical or conical with diameter ranging 2 to 5 mm.

Snow grains : Precipitation of very small white and opaque grains of ice. These grains are fairly flat or elongated and have diameters less than 1 mm.

Ice pellets : Precipitation of transparent or translucent pellets of ice, whose shapes are spherical or irregular, but rarely conical, diameter less than or equal to 5 mm. These ice pellets are of two types: (a) Frozen rain drops or largely melted and re-frozen snowflakes, (b) Pellets of snow encased in a thin layer of ice.

Hail : Precipitation of small balls or pieces of ice (called hailstones) diameter ranging 5 to 50 mm (or sometimes more), falling either separately or agglomerated into irregular lumps.

Ice prisms : A fall of unbranched ice crystals in the form of needles, columns or plates, often so small that they appear to be suspended in the air. These crystals may fall from a cloud or from a cloudless sky.

Fog : A suspension of very small water droplets in the air, mostly reducing the horizontal visibility at the earth's surface to below one kilometer (1000 m). Fog normally occurs when the surface wind is calm or very light and relative humidity 75% or more.

Ice Fog : A suspension of a large number of minute ice crystals in the air, that reduces the horizontal visibility at the earth's surface to less than one kilometer. Ice fog creates luminous pillars, small haloes etc.

Mist : A suspension of microscopic water droplets or wet hygroscopic particles that reduces the horizontal visibility at the earth's surface to (1 – 2 km) less than 2 km but more than one kilometer. Relative humidity is generally less than 100% but at least 75%.

Drifting snow and blowing snow : An ensemble of snow particle raised from the ground by a sufficiently strong and turbulent wind, that is snow blown off the ground into air after it has already fallen.

(a) *Drifting snow :* An ensemble of snow particles raised by strong wind to a height less than 2 meters above ground, but the horizontal visibility is not deteriorated to mist or fog level.

(b) *Blowing snow :* An ensemble of snow particles raised by strong wind to moderate, to great heights above ground level. Horizontal visibility below 2 meter height sufficiently deteriorated to below 2 kilometre.

Spary : An ensemble of water droplets torn by the wind from the surface of an extensive water body, generally from the crests of waves and carried up a short distance into the air.

Dew : A deposit of water drops on objects at or near the ground, produced by the condensation of water vapour from the surrounding clear air.

OR

Moisture condensed from atmosphere on exposed surfaces. It is a common phenomena of night or early morning with calm and clear sky.

White dew : A deposit of white frozen dew drops.

Hoar frost : Water vapour of the clear air directly transformed into ice crystals of the form scales, needles, feathers or fans.

Frost : Crystalline ice deposit formed on objects near the ground similar to dew when the temperature is below freezing point.

Rime : A deposit of ice, composed of grains more or less separated by trapped air, sometimes adorned with crystalline branches.

Glaze (clear ice) **:** Generally a homogeneous and transparent deposit of ice formed by the freezing of supercooled drizzle droplets or raindrops on objects, the surface temperature of which is below or slightly above 0°C.

Spout **:** A phenomenon consisting of an often violent whirl-wind, revealed by the presence of a cloud column or inverted cloud cone (funnel cloud), protruding from the base of a Cb and of a 'bush' composed of water droplets raised from the surface of the sea or of dust, sand or litter, raised from the ground.

2. **Lithometeors**

A lithometeor is a meteor consisting of an ensemble of particles most of which are solid and nonaqueous. The particles are more or less suspended in the air or lifted by the wind from the ground. The common lithometeors are described below.

Haze : A suspension in the air of extremely small, dry particles (smoke, dust etc but not of hydrometeors) invisible to the naked eye but sufficiently large number and cause the air an opalescent appearance. Visibility deteriorates (2 – 10 km) to 10 km or less but more than 2 km.

Dust haze : A suspension in the air of dust or minute sand particles, raised from the ground prior to the time of observation by a dust storm or sand storm, reducing the horizontal visibility 2 to 10 km and relative humidity less than 75%.

Smoke : A suspension in the air of small particles emanated by combustion.

Smog : A combination of smoke and haze (or dust haze) that reduces the horizontal visibility 2 to 5 km

Note : Combination of smoke and fog is also called smog. Generally observed around industries and cities in the morning.

Drifting and Blowing Dust or Sand

An ensemble of particles of dust or sand raised from the ground, at or near the station to a small or moderate heights (20 to 30 m) by strong and turbulent wind. In drifting dust / drifting sand, the visibility is not sensibly deteriorated at the eye level (or 1.8 m height). In blowing dust / blowing sand the horizontal visibility at eye level is appreciably reduced (2 to 5 km).

Dust Storm or Sandstorm

An ensemble of particles of dust or sand, energetically lifted to great heights (about 100 m or more) by a strong and turbulent wind causing horizontal visibility to less than 1000 m.

The further classification of dust/sand storm are :

Slight (wind speed 20-49 Kmph, visibility less than 1000 m)

Moderate (wind speed 39-74 Kmph, visibility 500 m or less)

Severe (wind speed 75 kmph or more, visibility 200 m or less)

This is defined on the basis of wind speed and visibility.

Dust Whirl or Sand Whirl (Dust Devil)

An ensemble of particles of dust or sand, sometimes accompanied by small litter, lifted from the ground in the form of a whirl column of varying height with a small diameter and an approximately vertical axis. OR

A narrow columns of whirling dust or sand going up in a spiral form.

3. Photometeors

A photometeor is a luminous phenomenon produced by reflection, refraction, diffraction or interference of light from the sun or the moon.

The common photometeors are: (i) Halo phenomena, (ii) Corona, (iii) Irisation, (iv) Glory, (v) Rainbow, (vi) Bishops ring (vii) Mirage, (viii) Shimmer (ix) Scintillation, (x) Green flash, (xi) Twilight colours. These are briefly described below.

Halo phenomen : A group of optical phenomena in the form of rings, arcs, pillars or bright spots produced by the refraction or reflection of light by ice crystals suspended in the atmosphere (cirriform clouds, ice fog etc).

Corona : One or more sequence of coloured rings of relatively small diameter, centred on the sun or moon.

Solar Halo : A ring of light round the sun by thin veil of cirrus cloud. It is often white but sometimes red near the sun, then orange, then yellow. In most cases the ring has a radius of $22°$ surrounding the sun.

Lunar Halo : A circle round the moon similar to solar halo.

Solar Corona : A ring of light round the sun, much smaller than halo. Its inner edge is brownish red, while the sky between the ring and the sun has a distinct bluish white colour. The radius of the ring varies from $5°$ to $8°$.

Lunar Corona : A ring round the moon similar to the solar corona.

Irisation : Colours appearing on clouds, sometimes mingled, sometimes in the form of bands nearly parallel to the margin of the clouds. Green and pink predominate, often with pastel (delicately tinted or coloured crayon) shades.

Glory : One or more sequence of coloured rings, seen by an observer around his own shadow (i) on a cloud consisting mainly of numerous small water droplets, (ii) on fog or, (iii) very rarely on dew.

Rainbow : A group of concentric arcs with colours ranging from violet to red (VIBGYOR – Violet, Indigo, Blue, Green, Yellow, Orange and Red) produced on a screen of water drops (raindrops, droplets of drizzle or fog) in the atmosphere by light from the sun or the moon.

The primary rainbow shows violet on the inside (radius $40°$) and red on the outside (radius $42°$).

Secondary rainbow is much less bright, shows red on the inside (radius $50°$) and violet on the outside (radius $54°$).

Note : If the size of a drop exceeds a certain size, its shape in no longer spherical and hence rainbow is not observed (as dispersion does not take place). Similarly if the droplets are too small, even though droplets are spherical, rainbow may not occur. The fog bow is a primary rainbow consisting of a white band which appears on a screen of fog or mist; it is usually fringed with red on the outside and blue on the inside.

Bishop's Ring : A whitish ring, centred on the sun, or moon with a slightly bluish tinge on the inside and reddish brown on the outside.

Mirage : An optical phenomenon consisting mainly of steady or wavering, single or multiple, upright or inverted, vertically enlarged or reduced images of distant objects. This occurs due to total internal reflection of light.

Shimmer : The apparent fluttering of objects at the earth's surface when viewed in the horizontal direction.

Scintillation : Rapid variations, often in the form of pulsation of the light from stars or terrestrial light sources.

Green flash : A predominantly green colouration of short duration, often in the form of a flash, seen at the extreme upper edge of a luminary (Sun, moon or sometimes even a planet) when disappearing below or appearing above the horizon.

Twilight colours: various colourations of the sky and of the peaks of mountains at sunset and sunrise. A number of these colour effects have been given special names.

Note:

1. Astronomical twilight ends in the evening (or begins in the morning) when the sun's centre is $18°$ below the horizon.
2. Civil twilight ends in the evening (or begins in the morning) when the Sun is $6°$ below the horizon. Twilight begins (ends) when the sun just receds (rises) below (above) the horizon in the evening (morning).
3. The duration of twilight varies with the latitude.
4. Twilight means half light after sunset or before dawn (sunrise)

4. Electrometeors

An electrometeor is a visible or audible manifestation of atmospheric electricity.

Some important electrometeors are described below.

Thunderstorm : One or more sudden electrical discharges manifested by a flash of light (lightning) and a sharp or rumbling sound (thunder).

Lightning : A luminous manifestation accompanying a sudden electrical discharge which takes place from or inside a cloud or less often from high structures on the ground or from mountains.

Thunder bolt or ground discharge which take place between the thunder cloud (Cb) and ground.

Sheet lightning or cloud discharge which takes place within the thunder cloud (Cb).

Air discharge which pass from a thunder cloud (Cb) to the air but do not strike the ground.

Thunder : A sharp or rumbling sound which accompanies lightning.

Saint Elmo's fire : A more or less continuous luminous electrical discharge of weak or moderate intensity in the atmosphere, emanating from elevated objects at the earth's surface (lightning conductor, wind wanes, masts of ships) or from aircraft in flight (wing tips, propellers etc.).

Polar Aurora : A luminous phenomenon which appears in the high atmosphere, in the form of arcs, bands, draperies or curtains. More will be discussed in optical phenomena of atmosphere.

Note :

Fine weather : Sky cloudless or with isolated cirrus floating in the blue sky and showing signs of dissolving or with a small amount of pure stratiform cloud at a fixed level, but with no clouds having vertical development.

Fair weather

Thin cirrus covering considerable part of the sky but not increasing or forming a continuous layer or sky with "Fair weather cumulus" or Altocumulus with characteristic changes in the courses of the day.

Fair weather cumulus appears in patches in the afternoon and disappears in the evening. It never develops into cumulonimbus.

Precipitation Associated with Cloud Genera

Hydrometeors are closely related with certain cloud genera. The following Table 8.1 gives various precipitations associated with cloud genera.

Table 8.1

Hydrometeors	Cloud genera					
	As	**Ns**	**Sc**	**St**	**Cu**	**Cb**
Rain	+	+	+		+	+
Drizzle				+		
Snow	+	+	+			+
Snow pellets			+			+
Snow grains				+		
Ice pellets	+	+				+
Ice prisms				+		
Hail						+

An aircraft observer may encounter ice prisms under Ci, Cc, Cs Ac clouds.

The Rate of Fall of Water Droplets

Any objects falling through the atmosphere is subjected to three forces. (i) The force of gravity, (ii) The upthrust due to buoyancy and 3. the air resistance due to its motion.

A freely falling object in the atmosphere, first it accelerates due to force of gravity, overcoming the upward effect of thrust and air resistance. However, after a while the forces are balanced and it then comes down with a constant velocity. This is called terminal velocity. A parachutist at first accelerates downwards, but after a while when his parachute opens he descends at a constant speed. Similarly falling rain drops initially accelerate downwards and after a while achieving certain maximum speed they come down with a constant velocity called terminal velocity. The terminal velocity of rain drops depends on its size which are given in the following Table 8.2.

Table 8.2 Terminal velocity of rain drops (water droplets).

Diameter (µm = micron meters)	Terminal velocity (mps)
2	0.12×10^{-3}
8	0.192×10^{-2}
100	0.27
200	0.72
500	2.06
1000	4.03
2000	6.49
5000	9.09
Drizzle r < 250mm	1 mps
Rain drop 0.1 – 2.2 mm	9 mps
Hail (r) 0.5 – 5 mm	20 – 50 mps

where r = radius

It may be noted here that the shape of the rain drops are deformed as they fall through the air. If the drops are larger in size, and if they are spherical initially they loose their spherical shape as they fall down and they may break down into smaller drops. This property accounts in non formation of rainbow.

Questions

1. Define meteors, hydrometers, lithometeors, photometeors and electrometers. Give examples of each.

2. Write briefly on : (i) Precipitation, (ii) Drizzle, (iii) Hail (iv) Fog (v) Mist (vi) Dew (vii) Haze.

3. Define dust/sandstorm and classify them.

4. What is Halo phenomena? Write briefly on solar-halo, lunar-halo, solar coronia, lunar corona.

5. Write briefly on Rainbow. What is the difference between primary and secondary rainbow?

6. Differentiate between astronomical and civil twilight?

7. What is terminal velocity of rain drops? Give a few examples of terminal velocities of different sizes of rain drops.

CHAPTER 9

Visibility and Fog

Visibility

In meteorology visibility denotes the transparency of the atmosphere with respect to normal human vision, and it is expressed as certain distance. Meteorological visibility refers to the greatest distance at which a black object of suitable size can be seen and recognized against horizon sky. The object should subtend an angle of 0.5 degrees, both horizontal and vertical directions, at the observers eye. However the dimensions of the object should not subtend an angle more than 5 degrees (both horizontal and vertical directions).

For meteorological purposes visibility is defined as the farthest horizontal distance at which a person with normal vision can see an object under normal day light condition (such as tree or building) distinctly enough to recognise. It is common experience that a dark building on the skyline (horizon) is easily recognised than a sheep at much less distant bare paddock (small field). Visibility during the night may be identified as the longest distance up to which lights of moderate intensity can be identified as such. The criterion used for day light visibility cannot therefore, be used for night measurements. It is often possible to see that there is "something" without being able to distinguish what it is, in such case the object is not visible according to the above definition. The first step in estimating visibility is to choose some prominent object (called visibility land marks) situated at standard distance as laid down in the visibility code or as near to them as possible, preferably within 10% margin. An ideal visibility landmark should be an object intrinsically dark in colour and so placed that it can be viewed against the horizontal sky or other light background. Minarets, towers, factory chimneys and like objects can serve as good landmarks for shorter distance, while for longer distances larger objects like hills and large buildings would be better, as they can be easily seen without straining eye.

A list of credibility land marks showing their respective distances and directions from the observatory with corresponding code figures for reporting visibility be prepared and kept at a fixed place which may be readily available. It may not be possible to fix visibility landmarks for all code figures in all directions. It is therefore necessary for visual interpolation or extrapolation depending on the location of observatory with its surrounding. Visibility observations then consists in seeing which is the farthest object that is visible and the nearest object not visible. The distance of the later (and corresponding code figure) gives the range of visibility at the time of observation as shown in Table 9.1(a).

Table 9.1(a) A day light observations visibility code.

Landmark	Distance form the place of observation	Direction	Observation as to which landmark visible and which is not	Code fig
A	50 m	( )	A is not visible	90
B	200 m	( )	A is visible but B is not visible	91
C	500 m	( )	B is visible but C is not visible	92
D	1 km	( )	C is visible but D is not visible	93
E	2 km	( )	D is visible but E is not visible	94
F	4 km	( )	E is visible but F is not visible	95
G	10 m	( )	F is visible but G is not visible	96
H	20 km	( )	G is visible but H is not visible	97
I	50 km	( )	H is visible but I is not visible	98
	Objects visible at 50 km or more	( )	I or further objects visible	99

Note : At greater distance 100 watts lamp is not visible.

Visibility at night may be determined in the same manner as during the day. Night observation visibility code is given in the Table 9.1(b).

Table 9.1(b) Night observations visibility code.

Landmark	Distance form the place of observation	Direction	Observation as to which landmark visible and which is not	Code fig
A	100 m	( )	A is not visible	90
B	330 m	( )	A is visible but B is not visible	91
C	740 m	( )	B is visible but C is not visible	92
D	1.34 km	( )	C is visible but D is not visible	93
E	2.3 km	( )	D is visible but E is not visible	94
F	4 km	( )	E is visible but F is not visible	95
G	7.5 km	( )	F is visible but G is not visible	96
H	12 km	( )	G is visible but H is not visible	97

Note : At greater distance 100 watts lamp is not visible.

It many observatories it will not be possible to get landmark lights for code figs above 95. In such case, the observer should estimate the visibility by noting the brightness of the light corresponding to the last code figure, and making use of his personal knowledge of any fixed lights in his locality, whose distances are known to him for higher code figures. Apart from the use of lights, a careful observer can make a fairly good assessment of visibility from a general inspection of the clearness of the atmosphere and his surroundings. For example, even on a fairly dark night one may spot a distant tower, a range of hills or a long road indicating that in day light an object at the distance would be clearly visible.

Visibility in Different Directions

When visibility is different in different directions, the lowest (code) figure should be reported with making entries of visibility in different directions. e.g., 95E, 96 NW. This means that visibility in the east corresponds to code 95, visibility in the North west corresponds to code 96, 95 being the lowest it should be reported.

In airports visibility meters (skopograph and transmissometer) are used for measuring runway visibility.

Visibility is affected by the following factors. 1. Precipitation 2. Haze, mist, fog 3. Wind blown spray from sea 4. Evaporation of oils, 5. Smoke 6. Dust and sand and 7. Salts.

In precipitation visibility deteriorates due to the presence of water droplets or ice particles. Reduction of visibility depends on the intensity of rain and size of rain drops. Light rain effects little, moderate rain reduces to 10 to 3 km. While heavy rain reduces to 500 m to 50 m or less. Similarly, light snow reduces to less than 5 km, moderate snow to less than 1 km, while heavy snow reduces to 200 to 50 m or less.

Fog

Fog is defined as a suspension of very small water droplets (radius 1 to 70 μm or more) that remain suspended in the air in such concentration that it reduces the horizontal visibility near earth's surface to less than 1000 m. In case if the visibility is 1 to 2 km it is called mist. In fact fog is cloud (stratus) on the surface of earth. Lifting of fog due to insolation and turbulence results into stratiform cloud.

Fog is formed in the presence of sufficient number of condensation nuclei and moisture saturation. Saturation of moisture occurs either by cooling of air or evaporation of water into it. Based on these processes fogs are classified as: (i) Cooling fog, (ii) Evaporation fog.

Cooling Fogs are : (a) Radiation fog, (b) Advection fog, (c) Upslope fog and (d) inversion fog.

Evaporation Fogs are : (a) Frontal fog, (b) Steam fog.

Cooling Fogs

(a) **Radiation fog :** Radiation fog (is also called ground fog) is formed by the cooling of the ground by nocturnal radiation (long wave radiation). It generally develops when the night sky is clear (cloudless), relative humidity is more than 75% (high dew point), wind is calm or light. When the sky is clear, the nocturnal cooling becomes prominent, which results in fall of surface temperature and surface inversion of temperature. This causes turbulent free (stable) layer near the ground and condensation of moisture in the early morning. On sunrise minute stirring (turbulent mixing) takes place near ground which causes increase of fog (spreading vertically) in the morning. When insolation effect becomes sufficiently strong (turbulence increases) fog slowly lifts from the surface of the earth develops into stratus cloud.

Radiation fog does not form over sea area because the diurnal variation of temperature is insignificant. On the contrary in polar regions when surface temperature is below − 30 °C, radiation fogs contains minute ice crystals and are called ice fogs or diamond dust because of glittering in the light.

(b) **Advection fog :** When moist air is transported over a cold surface area, where the surface temperature is below dew point, moisture condenses into fog near the ground. This type of fog is called advection fog. Such fog occurs over coastal areas by the advection of moist air from sea area to over coastal area. In a similar way if cold air advects over a moist area (which is relatively warm) condensation takes place resulting into fog. Sea fogs are developed in this manner by the advection of cold continental air to over sea area or big lakes or reservoirs. In winter when moist air from tropical or sub-tropical area advects over to a cold middle latitude regions, advection fog or advection ice fog occurs.

(c) Upslope fog : When moist air is lifted up (to lifting condensation level) due to orography along a sloping plain or hill, air cools down adiabatically. On saturation upslope fog develops. In this case conditional stability of air mass takes place. This type of fog is frequently observed on east of Rocky mountains. In Wyoming it is locally called "Cheyenne fog". Before formation of this fog light precipitation is observed over the area.

Upslope fog metamorphosises into stratus cloud on further heating by insolation (convection) and turbulence, generally by midday.

(d) Inversion fog : Fog that forms at the top of a moist layer under subsidence is called inversion fog. In an anticyclone region, when a cold air mass under subsidence comes in contact with warm water bodies inversion fog forms. In this region ground inversion of temperature is observed besides inversion stratus and light precipitation.

Evaporation Fogs

Frontal Fog

The sloping boundary surface between the contrasting air masses is called front. The leading boundary of cold air mass that penetrates and lifts the warm air which was existing is called cold front. When warm (comparatively) rain occurs in this frontal surface, the rainfall evaporates in the cold air and results into fog at the ground surface (You might have observed such phenomena while taking hot water bath during chill winter). This type of fog is called frontal fog. Similarly when rain occurs in a warm front area it evaporates and fog forms because of unstable layer at the ground surface and stable layer above it.

Steam Fog

Intense evaporation of water in cold air that leads to fog is called steam fog. Steam fog is not caused by any cooling of air. When water vapour enters cold air it condenses, and saturation leads to the formation of fog. This type of fog is unstable and observed during autumn in middle and high latitudes in and around water bodies (lakes, rivers, etc). The mixing of cold Labrador current with warm Gulf stream results into thick steam fog. In tropics steam fog is sometimes observed just after or during thunder shower which was preceded by hot and bright sunshine. In this case air cools down after thunderstorm because the ground surface cools down due to evaporation.

Fog in general is an impediment for aviation. Thick fog reduces visibility drastically and it effects all colours of visible light equally. Similarly steam fog over sea is an impediment for maritime navigation. Thick Fog also equally effects road and rail transportation services. With all this winter fog is still welcome to crops which helps for their growth in north India.

Note : The droplets of fog diameter ranges 25-75 µm with average of 45 µm. Mean liquid water content in air is 0.13 gm/m^3 with a range of 0.01 to 0.30 gm/m^3.

Importance of Fog in Aviation

Fog presents impediments to pilots during take-off and landing phases of an aircraft flight. Fog may occur in patches or as a continuous thick or shallow layer over an extended area. It was explained how visibility varies and the meteorological visibility is the lowest visibility in any direction. Therefore, it is important to make frequent observations along runways both during deterioration and improvement of visibility with respect to fog. When visibility is less than one kilometre it is called fog and when it is 1 to 2 km is it is called mist. When visibility is less than 1500 m Runway visibility called RVR (Runway Visual Range) is reported.

RVR is the maximum distance in the direction of take off or landing, at which the runway or the specified lights or markers delineating it can be seen from a portion above a specified point on its central line, at a height corresponding to the average eye-level (5m) of pilots at touch-down.

In daylight, the objects are markers placed along side the runway. At night, runway lights are used. In certain circumstance, the range of the lights by day is greater than that of the markers. In such situations, the range of the runway lights is reported. RVR is determined by TV cameras or videometers. In these cases atmospheric (trubidity) obstruction is measured by the back scatter of light from the particles in suspension of air.

Transmissometer

It measures the transmission of light through a portion of the atmosphere (runways).It consists of an intense source focussed light at one end and a photocell at the other end of runway. The instrument is used at both day and night with modulated light and tuned telephotometer. This gives a continuous record of runway visibility. The output of photocell depends on the turbidity (water particles and dust) of the atmosphere (on runway) between the source and the receiver and gives a measure of transparency of air on the runway.

Fog Dispersal

Cold fog over runways may be dispersed by seeding dry ice or by releasing propane gas through expansion nozzle which produces freezing and precipitation. However the dispersal of warm fog (having droplet temperature above freezing point) is comparatively difficult. Warm fog dispersal was attempted by seeding with hygroscopic particles which grow by condensation and then settle down by gravity or by precipitation. It was calculated that hygroscopic particles of size 30 μm (diameter) dispersed at the rate of 5 gm/m^2 may improve the visibility from about 100 m to 1000 m in about ten minutes. In practice it was observed that visibility again deteriorated after a while with the reformation of fog. Consequently this method was abandoned. In western countries particularly in America and Europe, warm fog dispersal is made by ground based heating. During second world war this brute force of dissipating fog, called FIDO, was used in 15 Airfields in England successfully. This system is however expensive. In FIDO direct application of heat is made by burning of oil in long lines of burners on either side of the runway. Fog is dispersed by dropping electrified sand (electrically changed particles) through fog.

Questions

1. What is meteorological visibility? Write day light observations visibility code.

2. Name the factors that effects the visibility. In airports how is visibility measured. If visibility is different in directions which visibility code figured will be reported.

3. Define fog. What are the type of fogs?

4. Write briefly on (i) Radiation fog, (ii) Advection fog, (iii) Upslope fog, (iv) Inversion fog, (v) Stream fog, (vi) Frontal fog.

5. Write the importance of fog in Aviation.

6. Write briefly (i) the working of transmissometers (ii) Dispersal of fog.

CHAPTER 10

Vertical Stability of the Atmosphere

Atmosphere is said to be stable (or unstable) at any point, $P(X_0, Y_0, Z_0)$ at any instant of time t, when a particle or parcel of air displaced form its initial equilibrium position acquires a relative motion in the environment is returning back (or going away) from its initial position. See Fig. 10.1

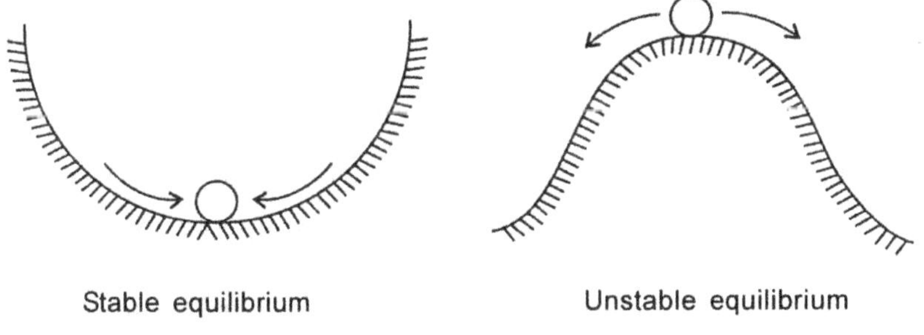

Stable equilibrium   Unstable equilibrium

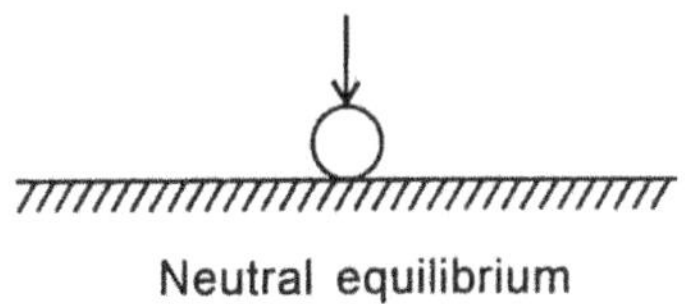

Neutral equilibrium

Fig. 10.1 Stability of a parcel.

Hydrostatic equation $dp = -g\rho\,dz$ is an excellent approximation for vertical direction stability and instability of the atmosphere. This classical criteria of stability and instability of air is applicable only in case of vertical displacements under the action of vertical forces (namely gravity and pressure gradient).

The horizontal wind velocity scale is 10 mps, while the vertical velocity scale is 0.01 mps (1 cm/sec). Though vertical scale is quite small (three orders small), yet when it is widespread its effects are very important. It may be noted here that different weather phenomena results from vertical motion of air. Vertical motion is a function of vertical stability of the atmosphere.

Adiabatic Lapse Rate

An adiabatic process is defined as a thermodynamic process in which no heat exchange (heat gain or heat loss) takes place with its surroundings. Thus in an adiabatic process vertically displaced parcel of gas may change in its volume or pressure but no flow of heat into or out of the parcel takes place. Most of the atmosphere pressure changes on small scale of time may be regarded as adiabatic for three reasons. (i) Air is a poor conductor of heat, (ii) Air parcel mixing with the environment is slow and (iii) effect of radiative processes is very small in short periods of time.

When an air parcel moves up adiabatically, the rate of fall of temperature with height is called adiabatic lapse rate. When an air parcel, dry or unsaturated, ascends without condensation of moisture (or without saturation), the rate of fall of temperature is 9.8 °C/km ($\cong$ 10 °C/km or 3 °C/1000 ft). This is called dry adiabatic lapse rate (DALR) generally denoted by Γ_d. Lapse rate greater than DALR is called super adiabatic lapse rate.

When a saturated air parcel moves up adiabatically some moisture will be condensed due to expansion cooling and thus releases latent heat (about 575 cal/gm) which counter acts cooling. Thus the saturated adiabatic lapse rate (SALAR) generally denoted by Γs, is less than the DALR. At sea level the SALR is 5 °C/km. The SALR is not uniform but varies with height. At temperatures below –40 °C, SALR is equal to DALR (since almost all water vapour condenses before reaching this temperature, and hence behaves like dry air). The SALR depends on pressure and temperature because the moisture holding capacity of air is high at higher temperatures (More latent heat is released at higher temperatures on condensation).

The Table 10.1 gives the SALR at different temperatures.

Table 10.1

Temperature °C	SALR °C/km	
	Pressure h Pa	
	1000	500
30	3.6	-
20	4.5	-
10	5.6	4.2
0	6.9	5.4
-10	8.1	6.8
-20	8.8	8.4
-40	9.8	9.8

Environmental Lapse Rate (ELR)

The vertical temperature distribution in atmosphere at any locality varies from day to day. We know temperature decreases with increasing altitude, on an average 6.5 °C/km in troposphere. Because of daily variations, main meteorological observatories all over the world take two Radiosonde ascents (Launching of hydrogen filled balloons which carry instruments to measure temperature, dew point at standard levels of pressure) to know the vertical temperature distribution. These ascents provide the environmental lapse rate of temperature at that particular hour (0000 UTC or 1200 UTC) and place. With the help of ELR and moisture meteorologists gather information of stability or instability of the local atmosphere as described below.

Parcel Method

Consider a parcel of air (for discussion, insulated) given an initial displacement which is at a state of equilibrium (that is, the parcel was under the action of equilibrium of forces initially). After the initial displacement, if the parcel sinks back, then the layer of atmosphere is said to be stable. If the parcel remains at the place where it is left, then the layer of atmosphere is said to be neutral. But if the parcel continues to move upward in the atmosphere then the layer of atmosphere is said to be unstable. In each of these three cases the vertical motion of the parcel of air depends on the environmental temperature which is higher (parcel sinks), equal (parcel remains where it is left) or lower (parcel moves upward). Thus the atmosphere is stable, neutral or unstable depends on the temperature variation of atmosphere with height. We shall consider these cases with respect of unsaturated or dry parcel and saturated air parcel.

A Vertical Motion of Unsaturated Air

Case I

Consider a layer of unsaturated atmosphere (where RH is much below 100%) with surface temperature 27 °C and the environmental lapse rate 7 °C/km as shown in Fig 10.1. Suppose an air parcel at the ground (temp

27 °C) displaced upward. The parcel will attain a temperature of 17 °C at 1km (7 °C at 2 km) height as the parcel expands and cools at DALR. (It is presumed parcel remained unsaturated or no condensation took place). The environment temperature at 1 km height is 20 °C while the parcel attained 17 °C (parcel is cooler than the environment), hence it sinks, that is the environment is stable. Consider a parcel of air at 4 km height where environment temperature is -1 °C. Suppose the parcel is pushed down, which will warm up due to adiabatic compression and attain temperature of 9 °C at 3 km height. The environment temperature at 3 km height is 6 °C (which is cooler than parcel). The pushed parcel returns back to 4 km height being warmer (lighter) than the environment, that is the environment is stable. From the above example it follows that: For unsaturated air the atmosphere is stable if ELR < DALR (as long as the air parcel remains unsaturated)

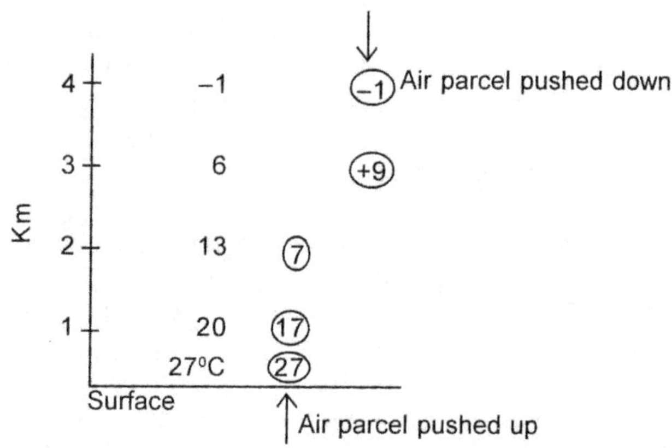

Fig. 10.1 Stable atmosphere for dry/unsaturated air.

Case II

Consider a layer of atmosphere with surface temperature 27 °C and lapse rate 10 °C/km as shown in Fig. 10.2. Suppose an air parcel at ground (temperature 27 °C) is displaced upward. It will attain a temperature of 17 °C at 1km, (7 °C at 2 km) height as the parcel expands at DALR. The environment temperature at 1 km height is 17 °C while the parcel also attained 17 °C hence the parcel remains there without movement indicating neutrality of the atmosphere. Now consider a parcel at 4 km height with temperature – 13 °C. when it is pushed down it attains a temperature – 3 °C at 3 km height. The environment temperature at 3 km height is also –3 °C. Consequently the parcel remains there without farther movement, indicating the neutrality of the atmosphere. It follows from this example that: For unsaturated air the atmosphere is neutral if ELR = DALR

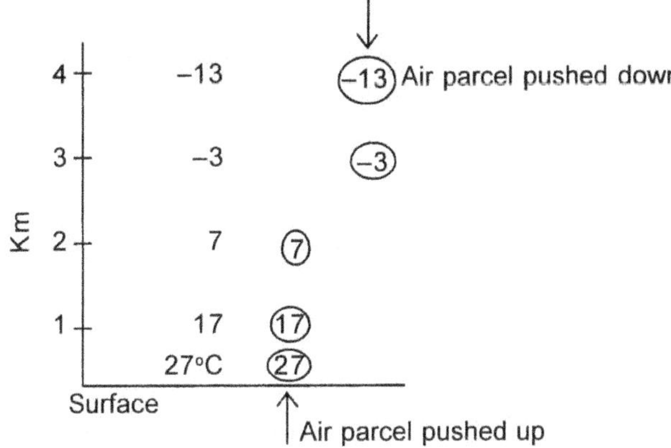

Fig. 10.2 A neutral atmosphere for unsaturated/dry air.

Case III

Consider a layer of atmosphere with surface temperature 27 °C and lapse rate 11 °C/km as shown in Fig 10.3. Suppose an air parcel at the ground (temp 27 °C) is displaced up wards. The parcel will attain a temperature of 17 °C at 1 km height (7 °C at 2 km height) as the parcel expands and cools adiabatically (DALR). At 1km height the environment temperature is 16 °C, therefore the parcel is warmer than the enviro:iment and continues to move upwards. It attains temperature of 7 °C at 2 km height while the environment temperature at that height is 5 °C and thus parcel being warmer, lighter continues to move upwards. This indicates instability of the atmosphere. Consider a parcel of air at 4 km height with temperature −17 °C is pushed down. The parcel will attain a temperature of -7 °C at 3km height (3 ° at 2 km height) under adiabatic compression (DALR). At 3 km height the environment temperature is − 6 °C, that is the parcel is cooler, heavier than the environment continues to move down wards. At 2 km the parcel attains a temperature 3 °C while the environment

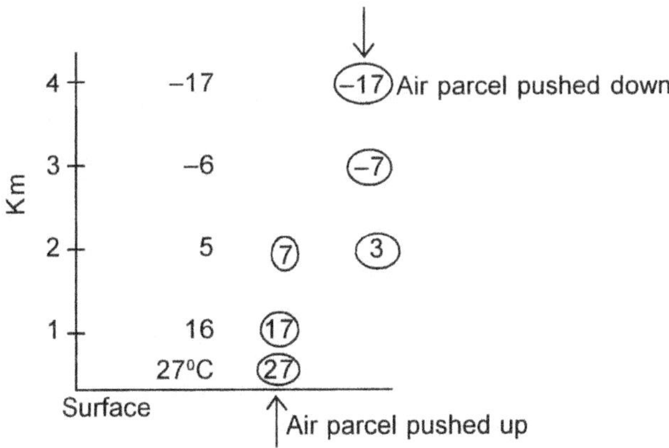

Fig. 10.3 An unstable atmosphere for unsaturated air.

temperature is 5 °C and hence continues to move down indicating the instability of the atmosphere. It follows from this example that: For unsaturated air the atmosphere is unstable if ELR>DALR

B The Vertical Motion of Saturated Air

Consider a layer of saturated atmosphere from surface to 3 km height (where RH. 100%). Let the temperature of the air at surface be 27 °C and lapse rate be 4 °C/km as shown in Fig. 10.4. Suppose an air parcel at the surface (temp 27 °C) is displaced upward. The parcel being saturated attains the temperature of 22 °C at 1km (17 °C at 2 km) height as it expands and cools adiabatically (at SALR 5 °C/km). At 1km height the parcel (temp 22 °C) is cooler and heavier than the environment (temp being 23 °C) sinks down indicating stable atmosphere. Consider an air parcel at 3 km with temperature 15 °C being pushed down. The parcel attains the temperature 20 °C at 2 km under saturated adiabatic compression. At 2 km height the air parcel (temp 20 °C) is warmer lighter than the environment (temp 19 °C) returns back to 3 km indicating stability of atmosphere. From the above example we draw conclusion that: For saturated air the atmosphere is stable if ELR<SALR.

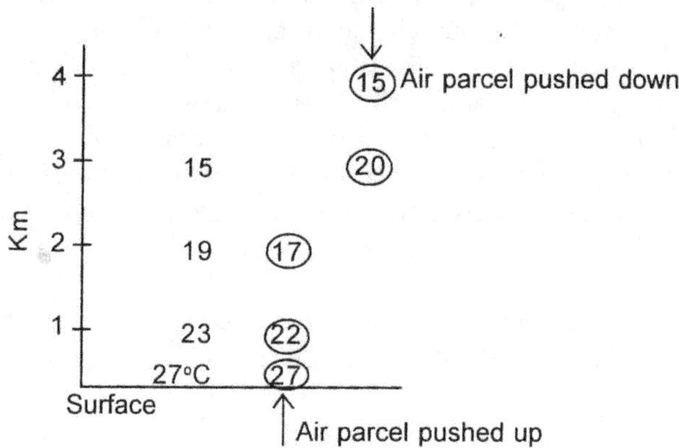

Fig. 10.4 Vertical motion of saturated air.

Similarly, as above, we arrive at the conclusion that: For saturated air, the atmosphere is neutral if ELR = SALR and unstable if ELR>SALR.

Conditional Instability

Consider a case of ELR of 8 °C/km. We know from above discussion that:

For dry air, ELR<DALR, the atmosphere is stable (since ELR = 8 °C/km and DALR = 10 °C/km)

For saturated air, ELR>SALR, the atmosphere is unstable (since ELR = 8 °C/km and SALR = 5 °C/km)

It follows from discussion that for unsaturated air the atmosphere is stable while for saturated air the atmosphere is unstable. Such condition is called conditional instability for which SALR<ELR<DALR.

From the results of three cases of dry air and three cases of saturated air we have:

(i)   For day air

If ELR<DALR, the atmosphere is stable

If ELR=DALR, the atmosphere is neutral

If ELR>DALR, the atmosphere is unstable

(ii)   For saturated air

If ELR<SALR, the atmosphere is stable

If ELR=SALR, the atmosphere is neutral

If ELR>SALR, the atmosphere is unstable

And if

SALR<ELR<DALR, the atmosphere is conditionally unstable (ie. Stable with dry air and unstable with respect to saturated air).

From these results we draw some more conclusions that: if ELR<SALR, the atmosphere is absolutely stable.

If SALR < ELR < DALR, the atmosphere is conditionally unstable and if ELR>DALR, the atmosphere is absolutely unstable.

Entrainment

The vertical movement of air is presumed to be adiabatic for reasons that air is poor conductor and the mixing of the parcel with the surrounding is slow. However in case of cumulus congestus and cumulonimbus clouds, the vertical currents (updrafts) in the cloud suck in the surrounding air. This sucking of air into the cloud is called mass entrainment. The assumption of adiabatic cooling (warming) in ascent (descent) is not valid in case of mass entrainment into the cloud, as moisture and heat content mixes up into the parcel. We have assumed that the parcel of air was in equilibrium state initially (that is at rest, this is called static stability).

Questions

1.   Define adiabatic process, lapse rate, dry adiabatic lapse rate and saturated adiabatic lapse rate.

2.   Discuss the vertical motion (of stability) of unsaturated air parcel with respect to different environments.

3.   Discuss the vertical motion stability of saturated air parcel.

4.   What is entrainment? Discuss the conditional instability with some example.

CHAPTER 11

Atmospheric Aerosols

An aerosol is a suspension of solid or liquid particulate matter in atmosphere or gaseous medium.

We know the atmospheric air is a mixture of gases, solid and liquid particulate matter. The important constituents of atmospheric aerosols are Silicates, Salts [Nacl, $NH_4 NO_3$, $(NH_4)_2 SO_4$, NH_4Cl, NA_2SO_4, $NaNo_3$, $MgCl_2$, $Mg SO_4$], Acids [$H_2 So_4$, $H No_3$], Metal oxides, Organic combustion products, Biological material, Volcanic material, Meteorites dust (extra terrestrial matter of metallic and stony compounds). The radius of aerosols vary from 6×10^{-4} μm to 150 μm (1 μm = 10^{-6}m).

Atmospheric ions (electrically charged particles) are classified into four groups according to their mobility B.

Small ions $\quad 1 > B \geq 10^{-2} cm^2 v^{-1} sec^{-1}$ ($6.6 < r < 78 \times 10^{-4}$ μm)

Large ions $\quad 10^{-2} > B \geq 10^{-3} cm^2 v^{-1} sec^{-1}$ ($78 < r < 250 \times 10^{-4}$ μm)

Langevion ions $\quad 10^{-3} > B \geq 2.5 \times 10^{-4} cm^2 v^{-1} sec^{-1}$ ($250 < r < 570 \times 10^{-4}$ μm)

Ultra large ions $B < 2.5 \times 10^{-4} cm^2 v^{-1} Sec^{-1}$ ($r > 570 \times 10^{-4}$ μm).

Electrically neutral particles are classified into three groups. (r = radius of particle).

The Aitken nuclei ($r < 0.1$ μm).

The Large nuclei ($0.1 < r < 1$ μm).

The Giant nuclei ($r > 1$ μm)

The density (or the number concentration) of particles in the atmosphere is highly variable. Near the earths surface, the small ions vary $100 - 1000$ cm^{-3}; the large ions $1000 - 80,000$ cm^{-3} in polluted air.

Near the earth's surface (a) Over land the density of Aitken nuclei vary few hundred per cubic cm (in polluted air about $10^6/cm^3$), (b) over sea $10^3 - 10^5$ per cubic centimeter.

The concentration of aerosol particles decrease with increasing size. (a) Over land: $10^3/cm^3$ ($r = 0.1$ μm), $1/cm^3$ ($r = 1$ μm), $1/liter$ ($r = 10$ μm), $1/m^3$ ($r = 100$ μm). (b) Over sea: hygroscopic particles (including sea salts): $100/cm^3$ (mass 10^{-16} gm), $10/cm^3$ (mass 10^{-14} gm), $10/cm^3$ (mass 10^{-12} gm), $1/cm^3$ (mass 10^{-9} gm).

The life period of small ions in the atmosphere varies from few minutes (in clear air, to few seconds in heavily polluted air). Intermediate size particles in lower troposphere 2 days to 2 weeks, in upper troposphere 2 weeks to 4 weeks.

The life period of atmospheric gases such as N_2, O_2 ,He, Ne, A, Kr, Xe and H_2 have long residence, that of CO_2, O_3, N_2O (Nitrous oxide), CH_4 (methane) few to many years, while that of water vapour (H_2O), Nitrogen dioxide (NO_2), Nitric oxide (NO), Ammonia (NH_3), SO_2, H_2S, CO, HCl, I_2 are highly variable both in time and space (few days to few weeks).

Aerosols play an important role, particularly large nuclei, in the formation of clouds and precipitation. The concentration of giant nuclei in the atmosphere is very small (few in numbers), consequently their role is limited. The concentration of Aitken nuclei is high, but they require super saturation for condensation. Thus the role of large nuclei in condensation process is very important. Aerosols also play considerable role in absorbing and scattering of solar radiation. The average radii of cloud droplets over land vary from 2 to 20 μm, over sea 3 to 22 μm, while the average radii of the rain drops exceed 100 μm.

Formation of Aerosols

Aerosols are formed mainly by three processes. Viz., Dispersion, combustion and photochemical action. Breaking of large particles or drops into smaller units which remain suspended in the atmosphere. On combustion the volatile components of fuels particularly carbon, and from mineral inorganic compounds aerosols enter into atmosphere. The volatile components on cooling mix up with air and form very minute aerosols of diameter less than 0.1 μm. and they are found in great concentration. The ultraviolet rays from the sun dissociates some of the constituents of the atmosphere into ions while atmospheric lighting in the lower atmosphere produces nitrates (fixation of nitrogen) and other acidic salts. Photochemical action produces new chemical components but their concentration is small.

The diameters (μm = micrometer meters) of some particles are given in Table 11.1.

Table 11.1

Smog	0.01 - 1 (μm)
Tobacco smoke	0.01 – 1 μm
Thick fog and cloud	3 – 80 μm
Coal dust	1 – 100 μm
Metallurgical dust and fume	0.01 – 100 μm
Sulphuric acid mist	0.5 – to 5 μm
Bacteria	0.5 – 50 μm
Human hair	50 – 150 μm
Soot	0.01 – 0.5 μm
Cement dust	0.7 – 10 μm
Powder milk	0.1 – 10 μm
Atmospheric dust	0.01 – 30 μm
Drizzle drop	50 – 500 μm
Rain drop	500 – 2500 μm

Note : (i) when RH exceeds 100%, the air is termed super saturated (ii) Aerosols which have affinity to water are called hygroscopic nuclei.

(i) Mobility B

The mobility of the particles play an important role. The mobility B for particles equal or smaller than the length of mean molecular free path

$(l = 0.1$ μm$)$ is given by $B = \dfrac{1 + Al/r}{6\pi\eta r}$ where B = mobility (sec/gm)

l = length of mean free path (cm)

r = radius of particle (cm)

A = numerical factor 0.9

η = viscosity of air (gm cm^{-1} sec^{-1})

The fall velocity of the particle is $V = BMg$

where B = mobility, M = mass of the particle (gm), g = gravitational attraction (cm sec^{-2})

(ii) The diffusion coefficient (valid for adhesion of particles to surfaces

of all kinds) in $D = \dfrac{RT}{N} \cdot B$

where   R = gas constant (erg $\deg^{-1}$ mol^{-1})

   T = absolute temperature (deg Kelvins)

   N = Loschmid number.

For coagulation : the decrease in the number of particles n per unit of time

and volume in homogeneous aerosol is $\dfrac{dn}{dt} = -\dfrac{8\pi RT}{N}\, B\, r\, n^2$

$$= -\dfrac{4RT}{3\eta N}\,(1+\dfrac{Al}{r})n^2\,.$$

Small particles coagulate more rapidly.

Condensation Nuclei

Condensation is a reverse process of evaporation in which water vapour (gas state) changes to liquid water. Condensation occurs in the presence of minute particles (aerosols) in the atmosphere which are termed condensation nuclei or cloud condensation nuclei. In a clear air (dust free), condensation of moisture takes place with great difficulty.

Different Types of Nuclei

(i)   There are particles which are insoluble in water and unwettable. These need supersaturation for condensation. These play very small role as condensation nuclei.

(ii)   Particles which are insoluble in water but wettable. These particles are irregular in shape and flaky in structure. Condensation occurs on these particles even before saturation (due to capillary condensation in cavities or absorption of water in pores). These play important role in condensation process.

(iii)   Particles soluble in water or droplets of solution. These particles (due to dissolved substance) require one third supersaturation as compared to first and second type particles.

As a result of coagulation of these wettable soluble and insoluble particles, there are a large number of particle types between type 2 and type 3. At super saturation it is found only 20% of total nuclei contains as condensation nuclei which help in the cloud formation at supersaturation.

Generally condensation begins on nuclei when RH 100%. However there are large nuclei, particularly hygroscopic, on which condensation begins even before air becomes saturated. On sodium chloride aerosol particles condensation begins at 78% RH. In contrast to this in a purified air (free from aerosols) water vapour may not condense even at supersaturation of 200% or more. Condensation on Aitken nuclei may occur at a supersaturation of 400 – 500% RH. Condensation on Aitken nuclei of radius r = 0.001 µm may begin at a supersaturation of 300% RH, on nuclei of radius r = 0.1 μm it may begin at a little more than 101% RH. Generally in a cloud super saturation does not exceed 1%.

The general drop size of a few cloud (genera) is given in Table 11.2 below.

Table 11.2

Genera of cloud	Peak diameter (μm)	Range (μm)
Dense Cu	14.5	3-40
Fair weather Cu	8.5	2-20
Sc	7.9	2-25
Ns	13.2	2-45
St	12.9	2-45
As	10.6	2-30

Freezing Process of Water in the Atmosphere

Water droplets in the atmosphere need not compulsorily solidify when they are cooled to 0 °C. Droplets which are not solidified at below 0 °C are called super cooled droplets. It is found even at temperatures of –25° to –35 °C cloud droplets remain in liquid state.

If pure water, without disturbing, is cooled to 0 °C, it remains in liquid state, and is called supercooled water. Supercooled water can be solidified by introducing ice particles into it. Experimentally it is established that pure water remains in liquid state even when it is cooled to –40 °C, when it is further cooled to temperatures below –40 °C it solidifies without the presence of any nuclei.

Turbulence in Atmosphere

Atmospheric wind is rarely steady in direction and speed, undergoes rapid changes. These changes indicate the turbulent and eddy flow air near the earth's surface. The cause of turbulence is mostly friction (roughness of the earth) and convection currents. The former is called mechanical turbulence and the later is termed thermal turbulence. Turbulence is mostly responsible in mixing of air from surface level to upper layers. It is also the cause for transport of moisture, particulate matter from surface to higher altitudes, and high density area to low density area. It is the main player of weather phenomena in (PBL) Planetary Boundary Layer.

Formation and Dispersal of Clouds

Water is present in the atmosphere at all levels in the troposphere in varying degrees. Water vapour is not visible in the atmosphere but the formation of clouds gives a visual evidence of presence of water (in the atmosphere). We know that moist air is a mixture of moisture and dry air and clouds are the result of cooling of moist air. The formation and dispersal of clouds is a very complex process but an elementary notion of the process is described here.

Condensation Nuclei

When moist air is cooled below dew point , the water droplets condense on the nuclei that are present in the atmosphere. These nuclei are called condensation nuclei. As described earlier cloud droplets remain super cooled up to –20 to –35 °C but they freeze to ice crystals before reaching –40 °C, even without nuclei.

Freezing Nuclei

Certain particles in the atmosphere help freezing supercooled droplets about itself to ice crystals at much higher temperature (–15 to –20 °C) than –40 °C. These nuclei are called freezing nuclei.

Deposition

Change of water vapour directly into a solid state (ice crystals) without transforming into liquid state is called deposition or sublimation. The particles upon which ice crystals form by deposition are called sublimation nuclei. It has not been proved experimentally that sublimation nuclei is different from freezing nuclei of the atmosphere. In meteorology the term freezing nuclei is used for ice-forming nuclei, which is also called ice nuclei. The presence of condensation nuclei is common in the atmosphere follows from the fact that water droplets in the atmosphere are frequently observed below 0 °C, while the ice nuclei is rare. The majority of freezing nuclei seems to be particles of certain clay mineral which were lifted into high atmosphere from the surface of the earth by turbulent mixing.

Methods of Formation of Clouds

Clouds are mainly formed by the vertical lifting of moist air. The various process that are involved in the vertical lifting of air are:

1. Mechanical or frictional or forced turbulence.
2. Thermal turbulence or convection
3. Orographic ascent
4. Slow widespread ascent in association with a low pressure system.

1. Mechanical Turbulence

The flow of air near the surface of earth is affected by the friction which is associated with the roughness of the earth. Turbulent wind motion is further triggered by tall buildings, trees, big rocks, hills etc. In the friction layer air is well mixed by mechanical turbulence and as a consequence DALR may be established as long as air is unsaturated. If there is sufficient moisture in the lower level, and the air is mixed up by mechanical turbulence, it may become saturated at some height below top of the friction layer. As a result of this, condensation occurs at a height above ground. This height is called mixing condensation level (MCL). MCL represents the base of the cloud. When cloud is formed in this manner, air will obey DALR up to MCL and then follows SALR. In this process stratus cloud is formed initially and subsequently it may develop like a wave form and transform into stratocumulus. Turbulence clouds may develop below precipitation cloud such as Ns, As and Cb. Turbulence clouds that form may be Fractostratus or Fractocumulus of bad weather. (Fracto-means broken). In a moist layer, above friction layer, if wind changes with height, high Sc or Ac may form due to turbulent motion.

2. Thermal Convection

The transfer of heat by means of actual motion material is called convection. Hot water heating is an example of this convection. If a material is forced to move (by a blower or pump), the process is called forced convection. If the material flows due to density difference (caused by thermal expansion), the process is termed free convection (or natural convection). In meteorology all the three processes of convection are involved.

In the morning as the day advances, the surface of earth gets heated up first due to insolation, then the air close to it. This develops convection currents in the atmosphere. This process is called thermal convection or thermal turbulence. As long as air remains unsaturated ELR tends to be DALR. This lapse rate (ELR = DALR) may extend upto condensation level, which is called in this process convection condensation level (CCL). Clouds begin to from at this level (CCL). Upward development of cloud depends on the ELR above CCL. If ELR >SALR the atmosphere is unstable for saturated air parcel, which continue to move up. This leads to the formation of cumulus. Thickness of cloud may vary 1 to 10 km or more depending on moisture and ELR. Isolated convection cell of moderate height is called fair weather cumulus, which will not give rain. If there is marked inversion above CCL, the vertical currents end up at CCL and cloud tops spread out below inversion level. In this case Stratocumulus forms.

If the tops of the cloud extends to the height where ice crystals form, in such case Cumulonimbus (Cb) cloud develops. The veil of floating ice crystals at the top of Cb gives a fibrous (cirrus like) appearance, called false cirrus. This distinguishes Cb form Cu. Depending on the degree of instability and height of cloud slight to heavy precipitation may occur from the cloud. Base of cloud may be less than one kilometre, while its thickness may be more than 10 km. During showers ragged turbulence cloud may develop below the cloud base which may even reach to the surface of earth. When the cloud top reaches a stable or inversion layer then it spreads out horizontally which is called anvil shape of Cb. The tops of Cb may reach to tropopause, but in some intense instability cases it may extend beyond troppause. Such cases are generally noticed in tropical regions. The phenomenon of convection clouds is of great importance. We shall study about Cb and associated its weather separately.

3. Orographic Ascent

Orographic ascent is mainly concerned with mountains and hilly barriers. When wind flow reaches mountain/hilly barriers, air is forced to ascend at all levels right from the foot of the barrier. This causes vertical ascent of air and associated adiabatic cooling and cloud formation. Depending on moisture content, orographic ascent and other factors St or Cu may develop. If St is formed in this manner it will have a flat base and little thickness of cloud. On leeward side of the mountain, the descent of air acquires adiabatic compression warming leading to dispersal of cloud. In general clouds form on the windward side of the mountain/hilly barrier and dissolve on leeward side. On leeward side at the top of the mountain barrier Lenticlular clouds are formed which appear to be stationary and associated with rain. On leeward side foot of the range, standing waves may form which affect the flight performance of aircraft.

4. Slow Widespread Ascent with a Low Pressure System

Vertical wind motion is caused by a large scale low pressure systems such as depression, cyclones and anticyclones. In association with a low pressure system slow wide spread ascent occurs due to low level wind convergence and high level divergence over the area. As a result a lot of convective clouds and precipitation occurs. Generally widespread ascent begins due to upper level divergence (in the upper troposphere) which follows low level convergence, surface low, depression/cyclone. In frontal areas frontal depressions are formed with widespread ascent. In contrast to this, in association with anticyclones, slow widespread subsidence occurs due to low level divergence and high level convergence. Because of subsidence inversion of temperature, fair weather or widespread fog/mist/smog occurs.

Temperature Inversion

We know in troposphere temperature generally decreases with increasing height, this is because air is heated up by the underlying ground surface. However in certain conditions temperature rises with increasing height in some layers or at ground level. This is termed temperature inversion or simply an inversion.

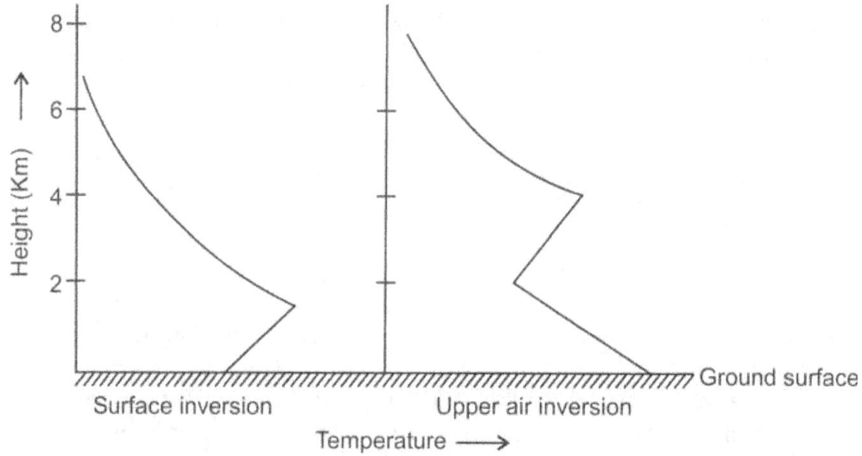

Fig. 11.1 Temperature-height inversion curves.

If the inversion occurs at the surface (ground) level, it is called surface inversion. If the inversion occurs in the air above ground, it is called upper air inversion see Fig. 11.1. The inversions are generally caused by (i) Radiation, (ii) Turbulence, (iii) Subsidence and (iv) Frontal development.

In winter radiation inversion occurs mostly in the early morning due to nocturnal cooling under calm, clear sky conditions. If in addition RH is more than 75% mist, or fog develops, in some cases ground frost may occur.

Mechanical turbulence may lift cold air from below the surface inversion to form an upper air inversion. In case of strong turbulence the inversion will be annulled.

During strong subsidence, associated with anticyclone, inversion develops in the air or at the ground. This is called subsidence inversion.

In a frontal zone when the warm air mass glides over the cold air mass inversion develops. This is called frontal inversion.

Inversion in general, indicates stability of the air, subsidence and nearly absence of turbulence. In general fog, mist, smoke and dust particles spread out below inversion level and becomes the cause of poor visibility. Inversion affects transmission of radio waves of wave length less than 10 m.

Cloud Dispersal

Cloud dispersal is a reverse process of cloud formation. Any process of cloud formation ceases to operate, cloud dissipation begins. With the warning of cloud air mass, precipitation, and mixing of drier air from surrounding, cloud dispersal takes place. Stable conditions at the surface level [that is ELR < SALR] inihibits cloud formation and aids dispersal of cloud. Low level divergence with assciated high level convergence, surface High or Anticyclone or upper air subsidence favours dispersal of cloud. Warming or insolation generally aids dispersal of cloud. Fair weather cumulus which forms by insolation, develops in the morning, reaches maximum height in the afternoon and dissipates towards evening. Entrainment of dry air/mixing of dry air from the surrounding, leads to fair weather cumulus dissipation.

Questions

1. What are the main atmospheric aerosols? How are they categorised? Discuss their concentration and life period?

2. How are aerosols formed? What is their role in cloud formation and precipitation? Write briefly about the dispersal of cloud?

3. Write briefly about the freezing process of water in the atmosphere?

4. Write briefly on : (i) Formation and dispersal of cloud (ii) condensation nuclei, (iii) Freezing nuclei.

5. Write the various processes involved in vertical lifting of air and write briefly on each of them?

6. What is temperature inversion? What is its cause?

CHAPTER 12

Precipitation Processes

Precipitation is defined as the falling hydrometeors which reach the ground finally. It does not include a virga. Thus precipitation process refers to the return of hydrometeors to the ground. Precipitation process involves four steps. Condensation nuclei, initiation of condensation, growth of condensation products and precipitation. The first two steps were dealt in aerosols of atmosphere and cloud condensation nuclei. We shall now deal with the other two steps.

In cloud condensation the cloud droplets grow to the size of about 15 μm (range 3-40 μm) while the rain drops diameter ranges 500 μm to 1000 μm or more. In condensation, initially droplets grow from nucleus size to visible size (cloud) rapidly but there after the growth of droplets slow down. Condensation alone is unlikely to produce droplets of average size larger than 30 μm. This hints that precipitation process differs from that cloud condensation process. In cloud condensation process minute nuclei (Aitken nuclei) transforms to higher concentration of small particles. In precipitation process large number of small particles convert into a small number of much larger particles which comedown to earth under gravity as precipitation [drizzle, rain, snow etc]. We have studied that drizzle drops, raindrops, hail size vary from 10–250 μm, 100–2200 μm and 500 to 5000 μm, and acquire a fall speed of 1 mps, 9 mps and 20 to 50 mps respectively.

Radius of drizzle drops 0.05–0.5 mm and terminal velocity 0.7 to 2.0 mps, Radius of rain drops 0.5–2.5 mm and terminal velocity 3.9 to 9.1 mps.

The condensation nuclei concentration over land is found to be much larger than over sea. On an average the cloud droplets are found to be of smaller radii ranging 2 to 10 µm over continents while in maritime clouds droplet size vary from 3 to 22 µm. The concentration of giant salt nuclei (hygroscopic nuclei) is very small, one per litre both on continents and over oceans, which produce cloud droplets of size radius 20–30 µm or more. Due to ascending currents in the clouds, droplets may acquire a size more than 100 µm radius and fall as rain drops with a terminal velocity of about 1 mps under gravity. Here we shall study about the growth of cloud droplets to precipitation size.

Clouds whose tops lie entirely below freezing level (0 °C) are called **warm clouds.** They consist of fully water droplets.

Clouds whose base lies entirely above freezing level are called **cold clouds**. Cold clouds contain super cooled droplets and ice crystals.

Clouds whose base lies below freezing level and whose tops lie above freezing level are called, **mixed cloud**.

Mixed clouds contain, water droplets, supercooled droplets and ice crystals. Generally clouds appear in the mixed form.

In warm clouds (cloud) droplets grow to precipitation size by collision and coalescence process, while in cold and mixed clouds droplets grow to precipitation size by Bergeron (ice crystals) process. These processes are described below.

The growth of cloud droplets depend mainly on moisture (RH), nature of nuclei, effects of surface tension and the amount of heat transferred to surroundings (by way the latent heat of release). Studies indicate that more than 80% of clouds containing liquid water were warmer than –10 °C, half of them were mixed with liquid and ice crystals. By –20 °C, only 10% contain liquid clouds and 30% contained both supercooled droplets and ice crystals.

The coalescence process

In this process the cloud droplets grow to rain drop size by direct collision and coalescence (that is fusing together) of water droplets. Droplets lifted by updraught (ascending currents) fall relative to the rising air and by this larger (heavier) droplets fall faster relative to smaller droplets. When falling droplet's size exceeds 18 µm, they overtake and collect smaller droplets in their path, thereby they become larger, fall faster and sweep up small droplets more rapidly. In warm clouds collision and coalescence process alone occurs for droplet's growth and precipitation.

Similarly in cold clouds supercooled droplets may grow by colliding and coalescing.

Calculations show that it requires about 50 minutes for a droplet size of 30 μm radius to grow by coalescence to the size of 300 μm in a typical cloud of average droplet size of 8 μm and liquid content 1 gm/m^3. The Table 12.1 gives the growth of initial droplet size of 30 μm falling through a homogeneous cloud of average drops of 20 μm containing 1gm/m^3 liquid water.

Table 12.1 Droplet growth.

Diameter (μm)	Time (cumulative) minutes	Falling distance (cumulative distance) meters
30 (initial)	0	0
40	45	65
60	75	165
100	90	320
200	105	650
500	116	1475
1000	123	2675

The Bergeron (Ice Crystals) Process

This process deals with the initial growth of ice crystals in mixed clouds.

In 1933, a Swedish meteorologist propounded the basic principles of growth of ice crystals in mixed clouds. It is assumed that ice crystals develop initially by freezing of supercooled water over ice nuclei. When ice crystals exist in the presence of supercooled water droplets, the system becomes unstable because the saturation vapour pressure over ice is less than over water at sub-zero temperatures. As a result water droplets evaporate and transform directly to (sublimates) ice and is deposited on the ice crystals. Thus ice crystals grow at the expense of the water droplets. This transformation is not evident at slightly below 0 °C , but it is conspicuous at –10 °C, and at –15 °C the transformation of vapour to ice crystal is the most efficient. However the growth rate decreases as crystals become larger. Thus the Bergeron process essentially deals with the initial growth of ice crystals.

Once the ice crystals grew appreciably (become heavier than water droplets) they begin to fall relative to the water droplets and collide with them. Collision of ice crystals with supercooled water droplets leads to the freezing of water droplets on ice crystals. This process is called accretion. Accretion of supercooled water droplets on ice crystals results into formation of rime, glaze and snowflakes. Finally when water droplets or ice crystals grow to a size which the air currents in the cloud may not be able hold, then they fall under gravity colliding with different size particles and descend below the cloud base as precipitation. In some cases the precipitation particles (hydrometeors) below the cloud may evaporate (virga), or ice crystals melt or sublimate. After survival of this process, the hydrometeors reaches the surface of earth as precipitation viz drizzle, rain, shower, snow or hail.

Note : Fog, dew and frost are also involved in the transfer of moisture from atmosphere to the surface of the earth but they are not considered as precipitation.

Summary of Precipitation Process

1. Water vapour transforms into cloud droplets above lifting condensation level, which may be liquid or ice. If these cloud droplets reach to a height where temperature is below –35 °C, all droplets convert into ice crystals, but if it reaches/falls to a height where temperature is above freezing level (> 0 °C), all droplets may convert into liquid droplets.

2. (a) Either by accretion process or auto conversion cloud liquid droplets grow and come down as rain or (b) by accretion and Bergeron process grow as hail or rain, (c) cloud ice crystals by accretion process may grow as hail and comedown to earth as hail or rain water. (d) when the ice crystals grow as hail and come to a level, where temperature is above 0 °C, they may melt and come down to the earth as rain. (e) when the rain drops in the cloud passes through a level, where temperature is below 0 °C, some of which by accretion or big drops freezing may become hail and come down to earth, (f) when rain drops or hail passes through the atmosphere below condensation level, where temperature is more than 0 °C, they may evaporate and virga may occur.

   "Nature rewards every virtuous man's deed, if there is a good man amidst us, it will rain indeed"

Precipitation Measurement

Precipitation (pptn) is expressed as the depth to which it would cover a horizontal projection of the earth's surface, in the absence of no losses by evaporation, run-off, infiltration and the part of precipitation that fallen snow or ice melted.

The average depth of precipitation over the earth's land surface (area about 149 million km^2) is 760 mm, over the oceans (area about 361 million Km^2) 1140 mm and over the entire globe (area 510 million km^2) is

$$1030 \text{ mm } [= \frac{760 \times 149 \times 10^6 + 1140 \times 361 \times 10^6}{510 \times 10^6}]$$

Note : Average evaporation depths on land 495 mm, over oceans 1251mm

$$\text{and over entire globe } 1030 \text{ mm } [= \frac{495 \times 149 \times 10^6 + 1251 \times 361 \times 10^6}{510 \times 10^6}].$$

The principal factors that determine the average amount of precipitation received at any point on the earth's surface are: Latitude, Altitude, Distance from the source of moisture position, prevailing winds, relation to mountain ranges (windward or leeward) and relative temperatures of land and neighbouring oceans.

The simplest way of measuring precipitation is by installing rain-gauges with horizontal circular aperture of standard area, collecting and measuring at regular intervals the precipitation collected in them. The amount of precipitation collected in the gauge is representative of certain area around the point of measurement. The choice of the instrument and the site, the form and exposure of measuring gange, the prevention of loss of precipitation by evaporation and the effects of wind and splashing are some of the points to be viewed in the correct measurement of precipitation. The minimum density of raingange (RG) net work recommended by WMO are : (i) For Temperate, Mediterranean and Tropical zones: (a) one RG for every 600-900 km^2 in respect of Flat areas and (b) for every 100-250 km^2 one RG for mountainous areas. (ii) For Arid and polar zone: one RG for every 1500-1000 km^2. (iii) For small mountains and islands: one RG for every 25 km^2.

Measurement of Rainfall

The amount of rainfall at a station is measured by a Rain gauge (RG). The WMO recommended RG is made of Fibreglass Reinforced Polyester (FRP) and used as a standard. The essential parts of RG (see Fig. 12.1) are: (i) A collector with a gunmetal rim of truly circular shape of area 100 or 200 cm^2, (ii) Base, (iii) polythene bottle and (iv) a measure glass. Both collector and the base are made of FRP. The collector contains a deepest funnel and the complete RG has a slight tapper with narrow portion at the top and bottom a little wide. The aperture of collector designed to suite both 100 and 200 cm^2 area (interchangeable). The smaller collector diameter is 112.9 mm corresponding to 100 cm^2 and the bigger one has a diameter of 159.6 mm corresponding to 200 cm^2. There are also two types of interchangeable bases, the smaller base being used for all types of receivers except the largest. The rain falling into the funnel collects in the bottle kept inside the base and measured by means of special measure glass which is appropriate to the area of aperture and which is graduated in tenths of millimeter. This has usually a capacity of 20 mm of rain.

In 1 cm height, volume collected would be

$$V = \pi r^2 h = \text{Base area} \times \text{height}$$

$$V_1 = \pi\left(\frac{159.6}{20}\right)^2 \times 1 = 200 \text{ cm}^3$$

$$V_2 = \pi\left(\frac{112.9}{20}\right)^2 \times 1 = 100 \text{ cm}^3$$

In case of V_1, the base area 200 cm^2 and in case of V_2, the base area 100 cm^2.

Note :

(i)   correct type of measure glass, appropriate to the type of rain gauge funnel in use should be used for measuring the amount of rain, otherwise completely erroneous results will be obtained.

(ii) On measure glass it will be seen (a) 100 cm^2 rain gauge or Rain measure for 100 cm^2 collector (b) 200 cm^2 rain gauge or Rain measure for 200 cm^2 collector.

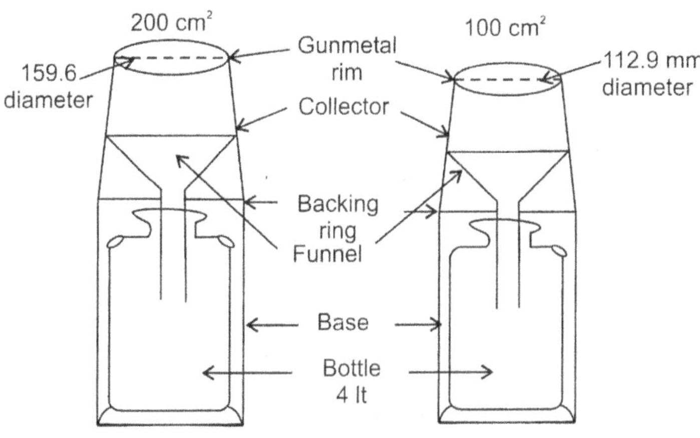

Fig. 12.1 FRP rain gauge.

Standard Exposure of RG

The amount of pptn collected in a RG depends to a considerable extent on its exposure and great care must be taken for selecting a suitable site. RG should be on a level ground away from trees, buildings and other obstructions and not upon slope or terrace. The distance between RG and object (obstacle) should be as far as possible 4 times the height of the object (obstacle). In any case the distance between the RG and the nearest object should not be less than twice the height of the object above the rim of gauge.

The rim of the RG should be exactly horizontal and remain at a height of 30 cm above ground level.

Measurement of Snowfall

Snowfall is measured as the depth which has fallen in a stated period, or melted and measured as water. The depth of snow is usually measured in mm and its water equivalent in mm and tenths obtained by dividing snow depth by 10, assuming the density of snow as 0.1. This value is however, only a rough approximation and varies very much with the depth and texture of the snow. For an accurate measurement of pptn at stations where snowfall is likely, special snowgauges are used. A snowgauge consists of a cylindrical receiver 203 mm or 127 mm in diameter, mounted on an iron stand at such a height as to be well above the average snow level at the station and provided with wind shields. Unshielded gauges are quite unreliable in strong winds because their catch may either be increased by drifting snow or blown off by the wind eddies around the mouth of the gauges. At stations where snowgauges are not available, the snow is measured with the ordinary RGs if the snowfall is light. In heavy snowfalls, the depth of snow is measured with snow-poles and water equivalent of snow determined; or cut samples of snow taken, melted and measured as water in the ordinary 127 mm measure glass.

Measurement of Snowfall at IMD Observatories

Remove the receiver from the wind-shield and pour into it a measured amount of very warm water. Since a large excess of water increases the error due to decrease in volume of water with fall of temperature, only sufficient water to melt snow in the receiver should be added. The warm water when first measured (V_1) into the measure glass, should be poured slowly so as not to crack the glass. When snow is completely melted, measure out the total amount of water (V_2) in the receiver into the empty measure glass. Then (V_2-V_1) gives the amount of pptn in mm. Suitable remarks to be made whether the pptn consisted of snow or rain or both.

Measurement of Snowfall with 100 cm^2 RG

(i) During light snowfalls the amount of snow collected in the RG should be melted by adding warm water and the quantity of water measured and recorded as described above. (ii) During heavy snowfalls the readings from the RGs will be unreliable because the gauge may be entirely buried in snow, or on the other hand, the snow may have been blown out by wind or eddies generated by winds. During periods of heavy snowfall, the observer should make one of the following three measurements.

(a)   With 100 cm^2 RG

When the pptn is in the form of sleet or snow. (i) Any snow immediately above the funnel should be separated and pressed into the funnel. This snow should be melted as described above and the water equivalent obtained. (ii) Measure the amount of water (actual rain or thawed snow) already in the receiver. The sum of these two (in mm and tenths) will give the amount of pptn.

(b) **Depth of Fresh Undrifted Snow, when the pptn has Occurred as Snow**

The depth of snow may be obtained by taking a mean of several vertical measurements in places where there is no drifting snow, using a graduated ruler or scale. Care should be taken to measure only the depth of snow which has fallen since the previous hour of observation. Special care should be taken not to measure any old snow. This can be achieved by selecting a suitable material (wooden boards or white painted metal sheets about 1 m^2) and placing it on top of the snow surface after taking the earlier observation and measuring the depth down to this material at present. On a sloping surface, measurements should still be made with the measuring rod verticals. Divide the depth of snow thus obtained by ten and record this amount in mm and tenths.

(c) **Cut Sample Method**

When there is no sleet or rain and the snow has not melted this method has to be used.

A cut sample of snow is obtained by inverting the body of the RG over the snow where its depth appears to be uniform and about the average amount. To avoid collecting old snow, small wooden or stone floors may be used into the snow till its edge touches that floor. Slip a thin plate under the body remove the cylinder of snow thus cut out, melt it by adding a measured quantity of warm water from the measure glass and record the water equivalent as described earlier. This method can be adopted only when all the pptn has occurred in the solid form.

Definition of Intensity of Rainfall in 24 hrs

A rainy a day is defined when 24 hr rain is 2.5 mm or more.

Table 12.2 Rainfall intensity.

Rainfall	Description
0.1 – 2.4 mm	Very light rain
2.5 – 7.5 mm	Light rain
7.6 – 34.9 mm ⎫ 35.0 – 64.9 mm ⎭	Moderate rain
65.0 – 124.9 mm	Heavy rain
125.0 mm or more	Very heavy rain

Record Rainfall in India

Mumbai recorded 944 mm of rain (in 21 hrs) at 0830 hrs on wednesday the 27th July 2005, which surpassed the earlier record of 838 mm recorded in 24 hrs at Cherrapunji in 1910. The earlier record of Mumbai was (24 hrs) 575.6 mm in 1974.

Questions

1. Define precipitation? Does condensation alone leads to precipitation in clouds? Give the normal sizes of drizzle, rain drops and their terminal velocity. Clarify whether fog, dew, frosts are precipitations?

2. Explain coalescence process of clouds droplet growth. In a typical cloud of average droplets size and water content in coalescence process how much time would require a droplet size of 30 μm radius to grow to the size of 300 μm?

3. Describe the Bergeran (Ice crystal) process of growth of particles in mixed clouds?

4. Write the WMO recommended minimum density of rain gauge network.

5. Describe WMO recommended FRP rain gauge and standard exposure condition?

6. Describe the measurement of snowfall at IMD observatories?

7. Describe the cut sample method of snowfall measurement?

CHAPTER 13

Air Masses and Fronts

A large body of air, which has uniform characteristics horizontally level by level in respect of temperature, moisture, lapse rate is called an **air mass**. In an air mass isobaric, isothermal surfaces are parallel and isobaric – isosteric (constant densety) surfaces do not intersect. The atmosphere with these characteristics is called barotropic atmosphere. In a baroclinic atmosphere isobaric, isosteric surfaces intersect. The region of an air mass is stable and non-turbulent.

Formation of an Air Mass

When air is calm or under subsidence it remains over the same area for a long time (about a week or more). It acquires a state of balance (or equilibrium) with the underlying surface below in respect of temperature and moisture. This balance extends vertically a few kilometres over a period of time. An air mass retain these characteristics for some time more when it is displaced from its source region.

Front

When two air masses of different characteristics meet, the interface region of these two is called a Front or Frontal region. It is similar to that of an army front of two enemy battalions.

Classification of Air Masses

Considering the general circulation of the atmosphere a single front called the polar front, may be defined in each hemisphere. Polar front circles in each hemisphere (in middle latitudes) meanders equator wards and pole wards. In northern (southern) hemisphere south (north) of the front there is warm equatorial or tropical air mass and north (south) of the front there is polar air mass. Tropical air mass includes air from the equatorial and sub-tropical regions while polar air mass includes air from polar and sub-polar regions. This simple classification may be true to some extent in middle and upper troposphere. However in the lower troposphere it is much complicated due to semi-permanent Lows and Highs and due to continental and maritime air. Some meteorologist prefer to use source region classification as under. 1. Equatorial air mass, 2. Tropical air mass, 3. Polar air mass and 4. Arctic (or Antarctic) air mass.

Air Mass Symbols

Based on source region, temperature, moisture the following simplest classification is described below. Tropical (T), Polar (P) air masses are considered as warm and cold air masses. Maritime (m) and continental (c) air masses as moist and dry respectively. The following notations are used to distinguish them.

$$Tm \quad - \quad \text{Tropical maritime}$$
$$Tc \quad - \quad \text{Tropical continental}$$
$$Pm \quad - \quad \text{Polar maritime}$$
$$Pc \quad - \quad \text{Polar continental.}$$

Cold Air Mass

In northern hemisphere the cold air mass regions are located over Anticyclones of Siberia, northern Canada and over High of Arctic region. These cold sources have profound effect on the air in the lower levels due to low moisture content (0.1 – 0.5 gm/kg) near the surface. Because of subsidence of air over these areas the air is stable with surface inversion extending to about (850 h Pa) 1.5 km in winter. During summer this inversion disappears due to ground heating and convection particularly over Canada and Siberia, while inversion depth becomes shallow over Arctic region.

Warm Air Mass

In both hemispheres warm air mass regions are located over Sub-tropical High pressure regions, particularly in summer. In northern hemisphere it is located over central Asia, while in southern hemisphere over north Africa. The warm regions have profound effect on the air in the lower levels (due to ground heating) with steep lapse rate of temperature. However convective clouds are not prominent in these areas due to low humidity. During winter startiform clouds form over north Africa under the influence of maritime air mass while over central Asia no such effect is observed due to lack of moisture (only clear skies prevail).

As the air masses change their characteristics once they move away from their source regions, marked effects may occur on moving over to a warmer or colder surface. Based on this we have sub-classification as:

1. Air mass cooler than the surface over which it moves (K)
2. Air mass warmer than the surface over which it moves (W).

Combining these symbols with the source region symbols we have the following.

K Tm –   Tropical maritime air mass moved over to a warmer surface

K Tc –   Tropical continental air mass moved over to a warmer surface

K Pm –   Polar maritime air mass moved over to a warmer surface

K Pc –   Polar continental air mass moved over to a warmer surface

In all these cases thermal instability may develop in the lower level, Cu, Cb may develop.

W Tm – Tropical maritime air mass moved over to a colder surface

W Tc – Tropical continental air mass moved over to a colder surface

W Pm – Polar maritime air mass moved over to a colder surface

W Pc – Polar continental air mass moved over to a colder surface.

In all these cases thermal stability may develop in the lower levels. Inversion of temperature develops. Fog, mist, smog may form.

Classification of Fronts

A sharp interface or boundary between two air masses is next to impossible, however on synoptic charts a line is drawn to demarcate them and is called a front. In a vertical direction a front is not a vertical plane but a curved surface.

Air mass Fronts are classified as (a) cold Front, (b) Warm Front and (c) Occluded Front.

Cold Front

If a cold (dense) air mass advances in to a warm air mass, the later is lifted up and the cold air mass under cuts with its sloping surface (as shown in Fig. 13.1) is called cold front. Cold front is not a vertical plane but a steep curved surface with an average slope (1/75) of one in seventy five. Convective clouds (Cu, Cb), turbulent wind, heavy precipitation are common features. Cloud bands and pptn occur in a narrow band.

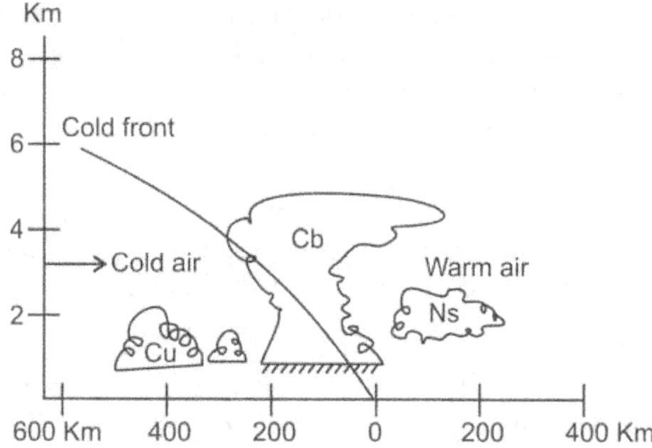

Fig. 13.1 Vertical cross-section of a cold front.

Warm Front

If a warm (light) air mass advances into a cold air mass, it will glide over the cold air mass (as shown in Fig. 13.2) is called a warm front. On an average the warm front slope (1/100) is one in hundred. In general cold front slope is steeper than the warm front. However in some cases the slope of warm front may be more than cold front.

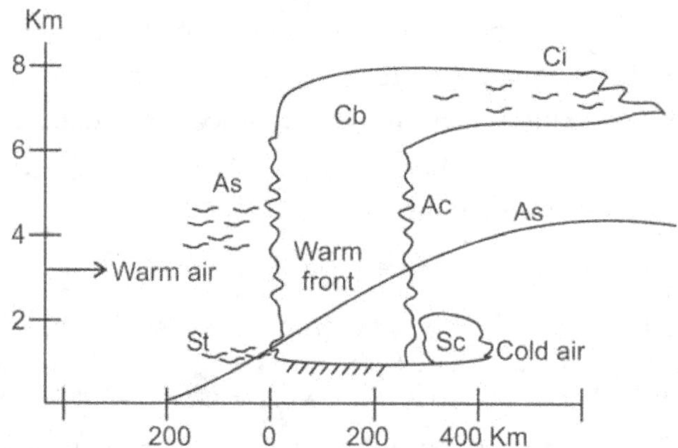

Fig. 13.2 Vertical cross-section of a warm front.

If warm front is sufficiently moist, the approach and passage of warm front may have the sequence of clouding, first Ci, Cs, Cc, As, Ac followed by precipitation and Cb, Nimbostratus. In some cases virga may be seen.

Occluded Fronts

The cold front generally moves faster than the warm front and if over takes it, the warm air sector of the two fronts mixes and forms an occluded front.

Let Tcc be the temperature of the cold air sector of the cold front and Tcw be the temperature of cold air sector of the warm front. If Tcc < Tcw (i.e., cold air of the cold front is cooler than the cold air of the warm front) the cold front under cuts the warm front. This is called cold fornt occlusion [*see* Fig. (13.3(a))].

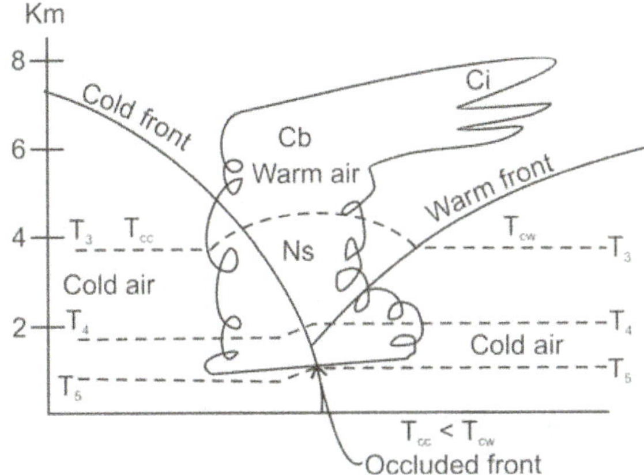

Fig. 13.3(a) Cold front occlusion.

If Tcw < Tcc (i.e., Cold air under warm front sector is cooler than the cold air of the cold front) the warm front undercuts the cold front. This is called warm front occlusion [*see* Fig. (13.3(b))]

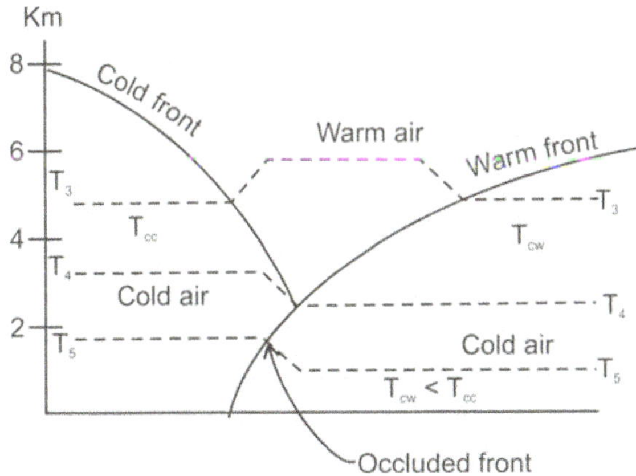

Fig. 13.3(b) Warm front occlusion.

In both types of occlusions cold air lies at the surface and warm air above it. This generally leads to cloud formation and precipitation in warm air. In case of extra tropical cyclones the system dissipates.

Table 13.1 Weather at a station as the front moves over it.

Element	Cold front			Warm front		
	Ahead	At	Rear	Ahead	At	Rear
Tempera-ture	Slight fall	fall	fall will continue	Slight rise	rise	no change
Pressure	Slight rise	rise	rise will continue	Slight fall	fall	no change
Relative humidity	Slight fall	fall	fall will continue	Slight rise	rise	no change
Wind	Veering	no change	Veering	Backing	Calm or variable	veering
Cloud	Sc, Cu	Cb, Ns	marked clearance	Ci, Cs, Cc Ac, As	As, Ns	Less of cloud or cloud clearing
Hydrome-teors	Rain or showers	Heavey showers /thunder showers	Clearing	Rain	Heavy rain	Less pptn or clearing
Visibility	Poor	Poor	Poor	Generally good except in pptn	Poor	improve-ment

Frontogenesis and Frontolysis

The development of a wave form between cold and warm air masses is the beginning of a front similar to the enemy army forming a strategy in the war. The formation of wave creates and demarcates the air masses of different characteristics and develops a zone in which extratropical depression/cyclone develops. This development of frontal waves and then depression/cyclone is called Frontogenesis. Frontolysis is the reverse of this in which frontal decay or dissipation occurs, not necessarily by frontal occlusion. Frontolysis may occur due to mutual stagnation of air masses over a similar surface or due to the movement of both air masses with same speed and direction.

According to polar front theory extratropical depressions/cyclones develop on the polar front. However it has been found that the extra-tropical depressions/cyclones may develop from a wave like twist on any front. These frontal waves are of two types, (i) stable, (ii) unstable (dynamically). Stable waves do not undergo any change, that the atmosphere is barotropic. These waves finally fade out. Unstable waves undergo rapid changes with atmosphere becoming baroclinic. The amplitude of the waves increase so much so that the air masses mix up thoroughly and look like breaking of ocean waves. The various stages of wave changes are shown in Fig. 13.4(a) and 13.4(b) for North and South Hemispheres respectively.

Life Cycle of a Wave Cyclone in Northern Southern Hemisphere

The Fig. 13.4(a) i or 13.4(b) i shows a kink of wave in cold and warm air masses, (ii), (iii) shows development of wave into depression/cyclone and (v) shows decay of cyclone, while (iv) shows occlusion. Extra–tropical cyclones (ETC) may develop either due to unstable frontal wave or due to unstable growth of upper wave trough. A combination of these two is found to be most favourable for formation and intensification to an ETC. ETC do not occur as isolated entities but they form as a family of three or four in succession at the trailing of first, second and third. In northern hemisphere the second depression forms at the south rear of the first, third depression forms at the south rear of the second and so on. Normally the family moves in a north easterly direction.

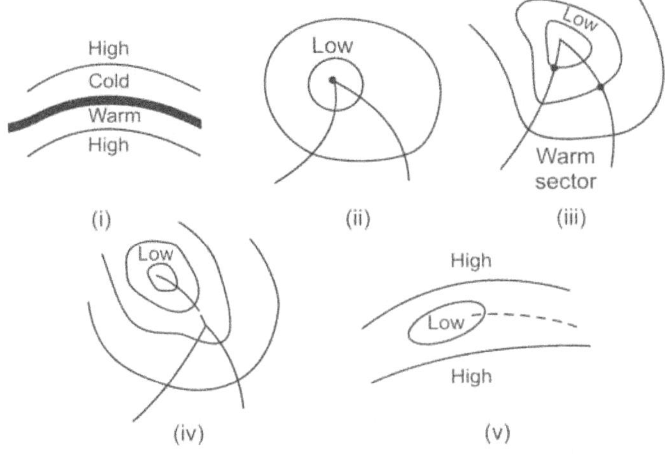

Fig. 13.4(a) Life cycle of a wave cyclone in Northern Hemisphere.

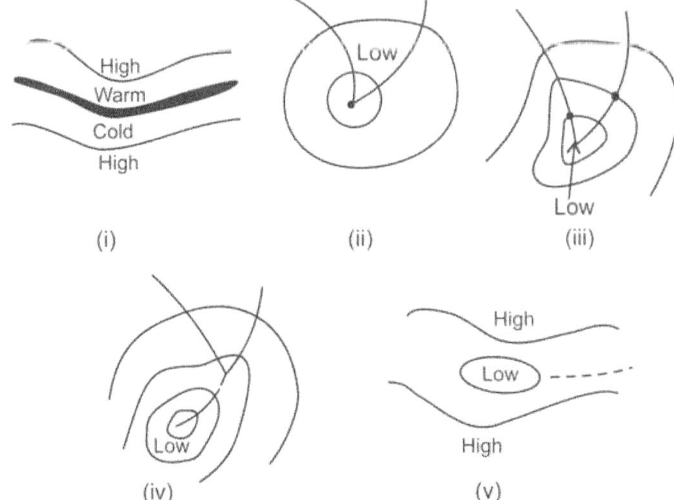

Fig. 13.4(b) Life cycle of a wave cyclone in Southern Hemisphere.

Representation of Fronts on a Synoptic Chart

The WMO accepted notation of fronts on synoptic charts are given in Table 13.2.

Table 13.2 Notation of fronts.

(i) Cold front	Cold air → Warm air direction of movement	
(ii) Warm front	Warm air → Cold air direction of movement	
(iii) Stationary front	Cold air — Warm air	
(iv) Frontal wave	Warm air (a) Northern Hemisphere	Warm air (b) southern Hemisphere
(v) Occluded front	Direction of motion →	Alternate barbs ▲, semicircles ● pointing in the direction of motion.

Questions

1. Describe the characteristics of an air mass, and classify the air masses?

2. Describe cold air mass and warm air mass with locations?

3. Define a front (or frontal region) and classify the fronts?

4. Describe the characteristics of cold front, warm front and occluded front with types of cloud and precipitation?

5. Describe the weather sequence as cold front, warm front passes over a station?

6. Explain frontogeneses and frontolysis. Describe the life cycle of a frontal wave cyclones?

CHAPTER 14

Convection Clouds

Mainly cumulus and cumulonimbus are treated as convective clouds. Convective cells develop in (i) unsaturated air if $\Gamma >> \Gamma_d$ and (ii) in saturated air if $\Gamma > \Gamma_s$ [where Γ = environmental lapse rate of temperature, Γ_d, Γ_s are dry and saturated adiabatic lapse rates]. This condition implies that the environment is unstable. The presence of cumulus clouds is an indication that the (i) cloud air mass is saturated and (ii) environment is unsaturated. In any locality/place instability and pptn may develop by the following processes [for development of cumulus or convective cloud]. (a) cumulus may develop in the afternoon by the heating of the lower atmosphere through its contact with the hot ground surface. Or (b) if a cold air mass over runs a warm ground surface. (c) The cooling of the middle atmosphere (4-10 km asl) may initiate instability in the lower atmosphere (0-4 km asl). (d) Cumulus clouds may persist in the presence of high humidity and large horizontal extent of cloud. (e) Vertical development occurs in the presence of steep lapse rate. (f) Showers may occur frequently in a region where lapse rate is 6 °C/km in the presence of large humidity.

If wet-bulb potential temperature decreases with height and the layer becomes saturated then convective instability develops which favours cumulus development in the layer.

In general low level convergence and high humidity associated with upper level divergence is a most favourable condition for the cumulus development and precipitation. An inversion of temperature at the top of the moist boundary layer is favourable for the development of severe thunderstorm. Vertical wind shear in lower level aids convection clouds. Upper level divergence exists in association with Jet stream. Lines of discontinuity or dry lines also aids convection clouds.

Note : A line of discontinuity is a vertical transition line which represents a divide between dry and moist region without any significant difference of temperature.

Thunderstorms

We have studied that an electrometeor is a visible or audible manifestation of atmospheric electricity. A thunderstorm is therefore an electrometeor.

Definition

According WMO, a thunderstorm is defined as, one or more sudden electrical discharges, manifested by flash of light (called lightning) and a sharp or rumbling sound (which we call thunder).

Cloud is invariably cumulonimbus and generally associated with precipitation. The sound of thunder is due to the electrical discharge. In electrical discharge enormous heat develops which heats the air channel between the cloud and earth. The conducting channel is heated up to 30000 °K, which is five times the temperature of the surface of the sun (6000 °K). Because of sudden heating of the conducting air channel in a very short period, the diameter of the channel expands abruptly from a few millimeters to a few centimeters (ten folds). This sudden expansion of conducting channel generates shock waves, which spread out as sound of thunder (A shock wave is a wave surface that spreads in solid or fluid medium, at which there is abrupt pressure increase accompanied by variations in density, temperature and velocity of the medium). Thunder is generally heard up to 10 –15 km and rarely beyond 25 km. At any instant there are about 2000 to 3000 active thunderstorms around the globe, most of these occur in tropics. Simultaneously at any moment there are about 100 lightning flashes per second.

A thunderstorm day is defined as a local calendar day on which thunder is heard, irrespective of the actual number of thunderstorms on that day. The highest annual average of 242 thunderstorm days was recorded at Kampala, Uganda (0 ° 20′ N, 32° 30′ E) in Africa.

Thunderstorm consists of several convective cells each having a distinctive convective circulation. The dynamical building block of a thunderstorm is the cell. A cell is a large compact region having updrafts (about 10 mps). Most of the thunderstorm cells have short period of life. At any instant of a thunderstorm, it consists of cells which are at different stages of evolution and each interacts with neighboring cells or outside environment. Thunderstorms are local affairs and rarely exceed diameter of 10 km (100 km^2 area) and duration 1-2 hours.

The Life Cycle of a Thunderstorm Cell

Thunderstorm cell has three stages of life cycle, (i) Cumulus or growing stage, (ii) Mature stage and (iii) Dissipating stage (Fig. 4.1).

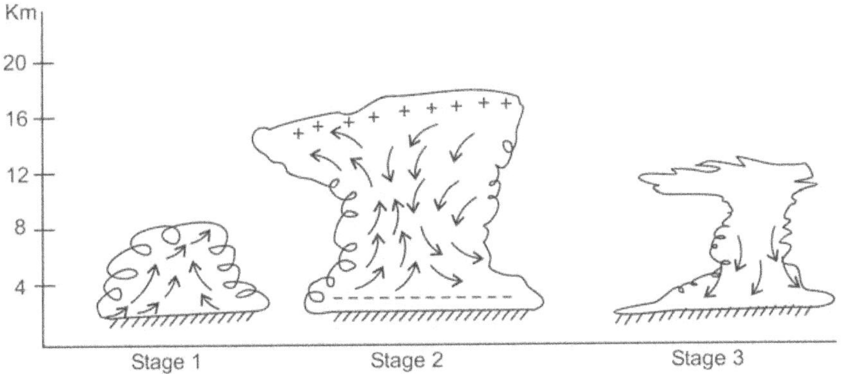

Fig. 14.1 Life cycle of thunderstorm cell.

Cumulus Stage

This stage is characterized by updrafts throughout cloud, height grows (8-10 km) and takes the shape of cauliflower type domes on the top and cloud top temperature may be about –20 to –30 °C. Unsaturated air mass entrainment into the cloud takes place at all levels of the cloud and in all stages. Marked temperature difference ($\Delta t = Tu - Te$) is observed between strong updraft (Tu) temperature and the ambient environment temperature (Te) of the cloud, Te being comparatively warmer in updrafts. In fact the temperature difference (Δt) is the measure of the strength of updrafts (Vu). Updrafts of the order 10 mps are common. When Δt exceeds 4 °C, the updrafts in the cell may be of the order 30 mps or more. Below the cloud base, horizontal convergence of wind may be of the order $5 - 10 \times 10^{-4}$ s^{-1} and wind speed 3 – 4 mps observed. The life period of cumulus stage may be 10 – 15 minutes. Light rain may occur at the base of the cloud.

Mature Stage

This stage is characterized by both updrafts and downdrafts. Downdrafts are prominent below freezing level. The lowest temperature in the cloud are observed in downdrafts. At the top of the cloud false cirrus anvil develops which mostly contains ice crystals. Downdrafts generally begin in the cloud at a level where cloud temperature (T) is equal to the ambient air temperature (Te = T). Severe turbulence occurs in this stage and downdrafts extend below the cloud base. The temperature of the air in the downdrafts (T_D) attains minimum value compared to the environment (Te) ($T_D < Te$). Squall wind speed is a measure of downdraft. This stage lasts for about 30 minutes, when it attains maximum height (12 to 18 km). In this stage severe lightning, precipitation and squalls occur. The strength of the updrafts may be 10 – 30 mps and downdrafts 5 to15 mps. Downdrafts below the cloud base produces divergence of cold air and in the wake of this cold divergence new cells develop and thus trigger chain action of thunderstorms. The life period of the cell in this stage is about 0 – 75 minutes.

Dissipating Stage

This stage is characterized by only downdrafts (speed 10 mps) throughout the cell. Squall speed at ground 10 – 15 mps. In this stage entire cloud mass temperature becomes comparatively cooler than the ambient air temperature. No more condensation occurs. The height of the cloud dissipates with dissipation of anvil. Cold air divergence commonly spreads below the cloud and downdrafts cease to exist. Average life of the cell in this stage is about 10 – 30 minutes.

Conditions Favourable for the Occurrence of a Thunderstorm

1. Conditional and convective instability in the atmosphere.
2. Adequate supply of moisture in the lower troposphere.
3. Suitable synoptic situation to cause low level convergence and upper level divergence. This will act as a trigger and release the instability present in the air mass.
4. Suitable upper air flow which may advect warm moist air in the lower troposphere and cold dry air in the upper troposphere. This increases the instability.

Conditional and Convective Instability

The presence or absence of conditional and convective instability is deduced from the thermodynamic diagram, called T - ϕ gram. Baring for minor changes in the lowest level temperature profile (caused by surface heating), it is difficult to assess the other changes objectively from the T - ϕ gram. Instability in the atmosphere is quantified by (empirical relation) different stability indices. A few are given below.

In a thermodynamic diagram of T - ϕ gram, T stands for temperature axis and ϕ stands for entropy axis. We have learnt that entropy is a thermodynamical property which remains constant during an adiabatic process.

Normands theorem

According to this theorem, if we produce the dry adiabat from dry bulb temperature (T) point and the isohygric from the dew point (Td) and the saturated adiabat from the wetbulb temperature (Tw) all the three meet at a point. This is LCL or Normand point. Temperatures at different heights are obtained by interpolation from $T - \phi$ gram.

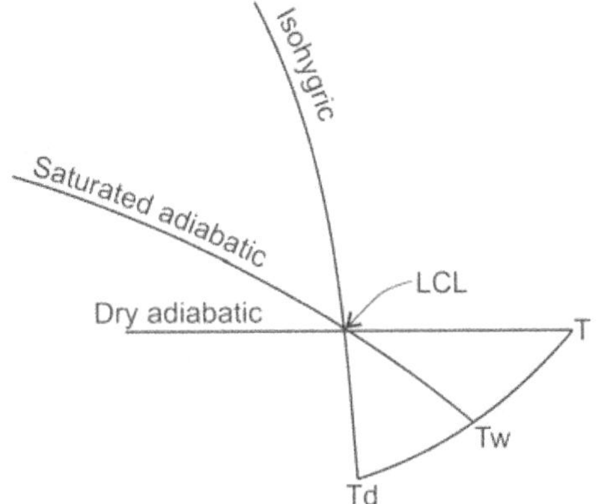

Fig. 14.2(a) Normand point.

Showalter Index (SI)

This index is the difference between the observed 500 hPa temperature (T_{500}) and the temperature (T^*) that an air parcel at 850 h Pa would acquire if it were lifted dry adiabatically to its LCL (Lifting Condensation Level) and then moist adiabatically to 500 hPa.

$$SI = T_{500} - T^*.$$

If SI >0, the atmosphere is stable

If SI <0, the atmosphere unstable.

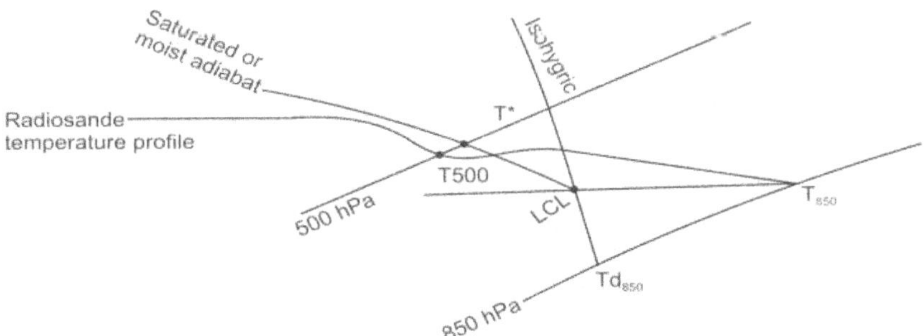

Fig. 14.2(b) Showalter Index.

Total Totals Index (TTI)

This index is the sum of the following indices.

1. Vertical total Index (TvI) $= T_{850} - T_{500}$

$$2.\ \text{Cross total Index (Tcl)} = T_{d850} - T_{500}$$
$$TTI = TvI + TcI$$
$$= T_{850} + T_{d850} - 2\,T_{500}$$

Ramalingam Index

The critical condition is given by

$$RI = \frac{\log(P_1/P_0)}{\log(\theta_0/\theta_1)} > 19 \text{ for the occurrence thunderstorm}$$

where p_0 = Pressure at the surface

$\quad p_1$ = Pressure at 500 or 600 h Pa level

$\quad \theta_0$ = Potential temperature (in absolute scale) at surface

$\quad \theta_1$ = Potential temperature (in absolute scale) at 500 or 600 h Pa level.

The Lightning Discharge

In a thunderstorm the lighting stroke starts from the cloud in the form of stepped leader, with average step length of 50 m. The time period of each successive step is about 50 μ seconds. The average velocity of the individual step is about 5×10^4 km/sec and the velocity of total step is about 1.5×10^2 km/sec. The total time for required for stepped leader to reach the earth is about 0.01 sec. After leader reaches the earth, a reverse stroke starts from the earth to the cloud through the channel made by the leader with an average velocity of 15×10^4 km/sec. After first discharge there may be other discharges which start from the cloud but little different from that of stepped leader. It will be a continuous leader, also called dart leader, with average velocity of about 2×10^3 km/sec.

When a lightning flash strikes near an observer, about 100 m of radius, the sound consists of (i) first a chick, (ii) then a crack (like that of a whip) and (iii) finally the characteristic continuous rumble of thunder. The click is caused by a discharge streamer directed upward from the ground (towards the leader stroke before commencement of the return stroke). The crack is caused by the intense return stroke in the lightning channel nearest to the observer. The rumble comes from the multiplicity of sound sources distributed along the lightning channel. When the lightning strikes away from the observer (several hundred meters) the first sound heard like the tearing of cloth.

The Electrical Field of the Atmosphere

In a fair weather the atmosphere carries a net positive charge everywhere on earth, the average potential gradient being 150 v/m. This implies that there is negative charge on the ground and it is assumed that the electrical potential of the ground is zero. The rate of change of potential with height is called potential gradient. The gradient decreases with altitude in free air, rapidly in the first one kilometer and slowly there after. The gradient at the surface varies during the day, being maximum at 1900 UTC and minimum at 0400 UTC, and the durinal range being about 35% of the daily mean. The potential gradient near the earth in fog is about 2000 v/m. In such conditions electric sparks (or electric discharges) may take place from the extremities of metallic conductors connected to the earth.

Critical Potential Gradient

An electric discharge (spark) takes place in the atmosphere when the electrical potential between two points reaches a certain value. In clear air of normal density the critical potential gradient is about 3×10^6 v/m. Dry air is a very poor conductor of electricity and requires a large potential gradient for eletrical discharge . However the presence of water droplets in the atmosphere increases the conductivity of the atmosphere. In cloud, the lightning discharge occurs at a potential gradiant of 1.0×10^6 v/m. It is found the average field strength within a thunderstorm cell is about 10^5 v/m.

Lightning

In 1600 AD, William Gilbert, the physician of Queen Elizabeth, studied about static electricity. He showed that glass/amber rod when rubbed with silk/fur attract small pieces of paper. He named this phenomena electric. In 1764, Benjamin Franklin found out positive and negative charges. In 1752 he showed, with kite experiments, that thunderstorms have electricity. In 1753 he found that thunder clouds have negative charges, but later in one experiment he also found positive charge in the thunderstorm cloud. From his experiments Franklin propounded that a majority of thunder cloud base possess negative charge but rarely positive charge may be seen. In 1756 he developed the lightning conductor for the protection of buildings, which is still in use.

Charge Separation in Clouds

In convective clouds charge is observed from the formative stage. The positive and negative charges formed in the cloud are separated by the following mechanisms. (i) By the breaking of big rain drops, (ii) by capture of ions by rain drops and ice particles in the atmosphere, (iii) by collision of ice particles at different temperatures, (iv) by freezing of supercooled water droplets.

In the cloud the charges are separated by the combination of the above mechanisms.

In a well developed convective cloud (matured stage) updrafts transport positive charge to the top of the cloud, while downdrafts and falling rain drops transport negative charge to the base. Small ice particles or ice nuclei carry positive charge to the top of the cloud while big ice particles, soft hail carry negative charge to the base of the cloud. Generally positive charge is accumulated at the top of Cb and negative charge at the base. In contrast to this classical dipole model, modern theory assumes that there can be much lateral displacement between two charges. Most lightning flashes either transfer the lower negative charge to the ground or achieve effective partial neutralization of the cloud electrification by internal cloud discharge.

Types of Lightning

Lightning occurs in various forms. More than 60% of the occasions lightning discharge takes place within the cloud, or cloud-to-cloud between positive and negative charges. The second type of discharge is between cloud and the earth. This is commonly called thunderbolt. The negative charge at the base of the thunder cloud induces positive charge on the earth in the vicinity of the cloud base, which repels the negative charge on the earth. The discharge between cloud and the earth takes place at a critical potential gradient.

Ribbon Lightning

This occurs between the cloud and the earth when the wind is very strong (speed 30 kmph or more) which displaces the conducting air channel into a ribbon form.

Sheet Lightning

Sheet lightning is observed due to distant thunderstorms at the horizon, which illuminates the sky by its lightning flashes.

Air Discharge

This occurs between the cloud base and the air below but it fails to reach the ground. Air discharge generally takes place in deserts.

Bolt from the Blue

The electrical discharge which begins as an air discharge but after traveling a distance away from the cloud, the leader stroke reaches the ground. The discharge may reach the ground in some cases about 15 km away from the cloud, where the sky may be completely clear. That is why it is termed Bolt from the Blue.

St. Elmo's Fire

This form of discharge generally takes place at the top of high masts or elevated objects. When the thunderstorm cell passes over elevated masts a positive glow will be seen at the top of the masts, while the negative charge flows down. This type of discharge observed over ship masts, on aerial masts of flying aircrafts etc.

Ball Lightning

There is no scientific explanation about ball lightning. Ball lightning occurs in the form of electric balls of diameter ranging 1 cm to about 2 meters. These lightning balls may fall from thunderstorm clouds and explode or the balls may roll down a hillock and strike against some objects and then explode. The electric balls appear like soap bubles. These balls sometimes may roll down into rooms through windows, doors or electrical discharges.

Lightning Discharge Between Cloud and Earth

An ionized air column that is formed by the accelerated electrons in a thunderstorm electric field is a leader stroke. The stroke constitutes air channel in the form of progressive steps of 50–100 m from the base of cumulonimbus cloud to the earth.

Generally the first leader stroke is not visible but the first one which steps on the earth is visible. This looks like an illuminated river with tributaries. The speed of the leader stroke may range 160 km/s to 1600 km/s. The return stroke which goes (in the reverse direction) from the earth to the cloud base appears as a flash. This we call lightning. The speed of return stroke may be 14×10^4 km/s which is approximately half the speed of light (speed of light $C = 3 \times 10^8$ m/s). The period of lightning discharge may vary from a few micron seconds to a full second. If T seconds denotes the duration of time between the flush and thunder, then the distance of thunderstorm from the observer is approximately T/3 km, that is every three seconds is equal to one km.

Lightning

An electrically active convective cell within a thunderstorm has a life period about 30 minutes to 1 hour. During this period flashings may occur 1 to 10 per minute with maximum of 20 flashes per minute after first flash. During the life of the cell the mean flashing rate is about 3 per minute.

Some Effects of Electric Shock

Electricity produces its immediate effects on the body. It causes disturbance in body function but they do not always cause structural changes. Two principal mechanisms that cause death are: 1. Ventricular fibrillation and (ii) Respiratory arrest.

Ventricular Fibrillation

The human heart has two main pumping chambers for blood circulation. One chamber to pass blood around the body (the left ventricle) and the second to pump it through the lungs (the right verticle). The thick walls of the ventricle consists almost entirely of muscle. It is the simultaneous contraction of all the individual muscle fibres establishes a pressure within the ventricle which is sufficient to circulate blood. An electric current passing through the heart disturbs the coordination of these individual muscle fibres so that instead of contracting simultaneously they contract individually each at its own rate. A head pressure is no longer established in the ventricles and blood circulation ceases so that death ensues within about 4 minutes. If the ventricles are viewed in this state when the muscle fibres are contracting individually, the ventricles, instead of showing forceful regular contractions "heart beats" are seen to be lying in a flacid state, with irregular twitchings, 'fibrillation'.

Respiratory Arrest

An electric shock might affect respiration in two ways. It may cause enduring arrest of respiration persisting after the shock current has ceased to flow or the path of the current may cause the chest muscles to contract and thereby preventing respiratory movement. In the later case, the effect lasts as long as the current flows and because a lightning current flows only for a very short time (a few tenths of millisecond), the effects caused by very short period respiration arrest is negligible. On the other hand, the first way in which the arrest of respiration persists even after the shock current has ceased, deserves further consideration.

These two ways (circulatory arrest or respiratory arrest) in which death might be caused are likely to be the most common. Though both cases result from changes in function unaccompanied by alteration in structure, there is nothing to prove (in post-mortem reports) that either or both had occurred. Therefore, man may die due to direct lightning stroke either by respiration arrest or due to malfunctioning of cardiac nerves. On such occasions the life of a man can be saved by heart massage or by mouth-to-mouth resuscitation.

In all types of lightning strokes only one-third of the victims loose their lives while the rest two-thirds recover partially or completely.

Types of Lightning Stroke on a Human Being

Injury/death from lightning may occur either by direct strike, by side flash over from an adjacent struck object or by ground currents. Direct strikes are the most severe which often leads to death. If the victim is struck directly, he will be initially (at least), conduct the whole current and is said to receive direct strike. His body resistance to earth may exceed that of the surrounding air, so that a flash over to the earth occurs. If the victim is near to another object which is struck, one of three things can happen. If he is actually in touch with the conducting object which is struck, he is then subject to contact voltage. If he is standing nearby, a part of the current may cross the air gap and discharge to the earth through him, in that case he is the victim of a side flash. If he is at a relatively long distance from the object which is struck, he may receive a side flash if a step voltage generated in the ground near the strike. A person in contact, generally by his feet, with two points at differing potential on the ground will receive an electric shock through the body between the two points of contact.

Favours and Frownings of Thunderstorms

(a) Favours

Thunderstorms, are also called atmospheric hot towers, are one of the main natural agency which transports sensible and latent heat energy from the surface of the earth (ocean surface or water bodies) to the atmosphere. A consequence of this is the transport of heat from equatorial region to polar regions.

Thunderstorms maintain the electrical field of the earth's atmosphere.

Fixation of atmospheric Nitrogen is mainly achieved by thunderstorms. By thunderstorm Nitrogen fixation, atmospheric Nitrogen is converted into nitrogen compounds which are brought down to earth by rain, which is utilized by the plant kingdom and aids agricultural production. Thunderstorm lightning produces about 30% to 50% nitrogen compounds in the atmosphere. In arid regions, the lower atmosphere is very dry and that the light rain or drizzle droplets which fall from clouds completely evaporate before reaching the ground (called virga). Only large drops can survive and reach the ground. Such large drops are effectively produced by thunderstorms.

(b) Frownings or Hazards

Lightning

Lightning causes death to human beings and animals and damages property by fire. To protect houses and tall buildings lightnings conductors are fitted.

Hail

It causes damage to life and property, particularly crops, fruits and fruit trees. Hail storm is a severe hazard to aviation and aircrafts.

Thunder Squalls

Strong winds associated with thunderstorms are called thunder squalls. These cause damage to buildings, structures, parked and moored aircrafts, electric poles, huts and uproot even big trees.

Dust Storms

Over dry area with loose soil, thunder squalls raise lot of dust/sand into atmosphere, which may be suspended in the air for days and sometimes weeks. These are locally called Andhi (meaning blinding). This is hazardous for air and surface transportation services.

Heavy Rain and Flash Flooding

Heavy to very heavy rains are generally associated with thunderstorms, even during cyclones. These heavy falls cause flash flooding in urban areas and in small rivulets and cause havoc to life and property.

Aviation Hazard

Thunderstorm area is an aviation hazard not only for parked and moored aircraft in hangers but also for aircraft in flights due to steep wind shears, poor visibility (this is hazardous during landing and take off of aircrafts). Lightning causes radiostatic, interruption in communication system. If an aircraft enters into a thunderstorm cloud, it will be tossed violently up and down (in severe cloud turbulence), causes severe icing on aircraft frames and fatigue to airframes. On small scale, severe local thunderstorms (known as Kalbaisakis) generate Tornadoes on land, and Water spouts on Water bodies. These cause devastating spell on life and property.

Safety Measures from Lightning Hazards during Active Thunderstorms

1. All tall buildings, masts (minarets) must be fitted with lightning rods on their tops to discharge electric energy directly to the ground without touching beams, pillars etc.

2. **Indoors**

   During active thunderstorm it is safer to be inside the house on an insulated bed. Keep off from electrical conductors, telephone wires, T.V. Antennas and radio aerials. Switch off radio, TV, fans, grinders, gas stoves etc. One should not touch any conducting material such as telephones, sewing machines, iron bars (picklocks, doors, grills, chains), water taps, pumps etc. Avoid smoking, better remove wrist watch, key chains etc.

3. **Out doors**

   In open fields/areas one should not seek shelter under trees (particularly tall trees like palmer, coconut etc). It is better to lie down on the ground drenching in rain rather than seeking shelter under trees. One should avoid elevated places, rocks etc., which are the favorable places of lightning discharge. During active thunderstorm, in corn field one should not seek shelter even in a sheaf of corn. It is safer to set up two sheaves some 30 meters apart and then sit down half way between them. One should not ride on a horse/camel back, instead dismount from the horse/camel back and walk slowly.

Energy Release in a Thunderstorm

In a thunderstorm energy released mostly in the form of latent heat. We know that one gram of water vapors on condensation to liquid water releases about 600 calories of heat. This means that one kilogram of rain water reaching the ground releases about 6×10^5 calories of heat into the atmosphere.

$$\text{Average energy of a thunderstorm} = 10^{-8} \times \text{solar energy received by the earth per day}$$

$$1 \text{ cm of rain over an area of } 1 \text{ km}^2 = 1 \times [1000 \times 100]^2 \text{ gm of water}$$
$$= 10^{10} \text{ gm of rain water}$$
$$= 10^4 \text{ kilo tons of water}$$

suppose the average rainfall over an area of 50 km^2 is 2 cm (which is common in thunderstorm). This means the amount of heat released into the atmosphere is

$$2 \times 50 \times 10^{10} \times 600 \text{ calories}$$
$$= 6 \times 10^{14} \text{ calories}$$
$$\approx 6 \times 4.186 \times 10^{21} \text{ or } 25 \times 10^{21} \text{ ergs}$$
$$(\because 1 \text{ cal} = 4.186 \times 10^7 \text{ ergs})$$
$$\approx 31 \text{ Hiroshima atomic bomb explosions}$$

Note : The atomic bomb which exploded on 6th August 1945 over Hiroshima released energy $= 8 \times 10^{20}$ ergs]

According to C.E.P. Brooks there are about 44,000 thunderstorms per day, 1800 simultaneously at any time, and 100 flashes of lightning per second. These discharges equivalent to a continuous current of about 2000 amperes. Assuming the potential difference 10^8 volts in the vicinity of thunderstorm area, the lightning discharges world over continuously transfers energy approximately 268×10^6 horse power.

Climatology of World Thunderstorms

Europe and Australia have the minimum frequency of thunderstorms about 20 thunderstorm days annually. In Asia, the frequency is about 60 thunderstorm days in the south east Asia comprising around Bangladesh; (Bangladesh, India and Myanmar). South America, Africa and Indonesia have maximum thunderstorm days. Tropical oceanic regions around lat 20° north and south and semipermanenet high pressure regions are relatively free from thunderstorms.

Note : (polar regions are virtually free from thunderstorms). During Northern Hemisphere winter (Dec, Jan, Feb) a few thunderstorm days may occur north of lat 10 °N, while there will be maximum thunderstorm in the Southern Hemisphere in ITCZ.

During Northern Hemisphere summer (June, July, Aug) a few thunderstorms may occur south of lat 10 °S, while the maximum thunderstorms occur over central parts of north America, central Africa, Indonesia.

The accepted record 242 thunderstorm days per year recorded over a period of 10 years at Kampala, Uganda 0° 20′ N, 32° 36′ E (north of lake Victoria) in Africa. Mbarara, on the west side of lake Victoria, experiences only seven thunderstorm days per year. Bogor, in Indonesia, is sometimes referred as recorded 322 thunderstorms per year (1916-1919). This figure seems, actual number of thunderstorms and not thunderstorms days. There are only a few stations around the world which recorded 200 or more thunderstorm days per year which are given in Table 14.1.

Tabe 14.1 Stations with more than 200 thunderstorm days per annum.

Name of the station	Country	Lat 0°20'	Long 32° 36'	Average no. of thunderstorm days per year.
Kampala	Uganda	0 ° 20' N	32° 36' E	242
Buma	Zaire	1 ° 34' N	30° 13' E	228
Kamembe	Rwanda	2 ° 27' S	28° 54' E	221
Bandug	Indonesia	6 ° 54' S	107° 34' E	218
Calabar	Nigeria	4 ° 57' N	8° 21' E	215
Entebbe	Uganda	0 ° 02' N	32° 37' E	206
Carauri	Brazil	4 ° 53' S	66° 54' W	206
Mamfe	Cameroon	5 ° 46' N	8° 20' E	201

Java (Indonesia) in southeast Asia is reported to hold the pride of place for maximum number thunderstorm days but seems doubtful. Kula Lumpur in Malaysia recorded 180 TS days, K long Yai (Thailand) recorded 159 T.S days per year. In India the maximum of 108 T.S. days recorded at Sibsagar (Assam), Krishna nagar (west Bengal) and there are very few stations which have annual frequency of 100 T.S days or more.

Distribution of Thunderstorms Over Sea Areas

According to world map of thunderstorms, the incidence of thunderstorms over sea area is less than over land. This is clearly reflected in tropics. In tropical sea regions of Panama, Ecuador, Columbia, over Gulf of Guinea along African coast, along the sea areas of western coast of Indonesia and Malaysia the thunderstorm frequency is high. High frequency also seen in the ITCZ zone of North Atlantic and North Pacific. Minimum frequency TS observed in the Trade wind zone between lat 10° and 30°.

Questions

1. Define a thunderstorm, day. What are the favourable environmental conditions for development of a thunderstorm?

2. Describe briefly the life cycle of a thunderstorm cell?

3. Write short notes on (i) Normands theorem (LCL), (ii) Showalter index, (iii) Ramalingam index.

4. Write short notes on (i) Electrical field of the atmosphere, (ii) Critical atmospheric potential gradient, (iii) Lightning, (iv) Electrical charge separation in clouds.

5. Write briefly the different types of lightning?

6. Write briefly Ventricular fibrillation, Respiratory arrest, the electrical shocks on human beings?

7. Write the favours and hazards, including in Aviation associated with thunderstorm.

8. Name the hazard safety measures to be observed during an active thunderstorm both indoors and out doors?

9. Write briefly about a thunderstorm energy and climatology of global thunderstorms?

CHAPTER 15

Radar Detection of Clouds and Thunderstorms

The word Radar is derived from Radio Detection And Ranging. Radar was developed during second world war to detect enemy planes, ships in all weathers and in day and nights. Now radar is extensively used for detection of thunderclouds, Tornadoes, Water Spouts, Hurricanes and tropical cyclones in addition to cloud and precipitation. It has become very important tool for Nowcasting of very severe weather.

Principle of Radar

A radar contains a transmitter, a receiver, an antenna and an indicator. The block diagram of a radar is given in Fig. 15.1.

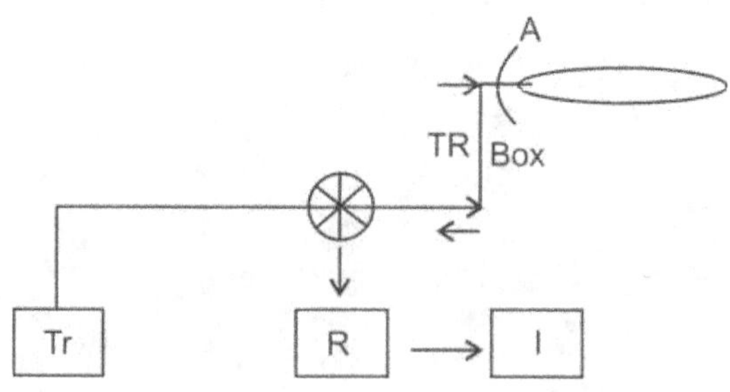

Fig. 15.1 Block diagram of a radar.

where   Tr = Transmitter

         R   = Receiver

         A   = Antenna

         TR = Transmitter Receiver

         I   = Indicator.

A radar sends out a pulse of energy (Pt) in the form of an electromagnetic beam (radio wave beam) which hits the airborne object (airplane, cloud water molecules, aerosols or any reflecting surface) called target and a part of the energy is reflected back and is detected by the radar. The reflected energy received by the radar (Pr) is presented as illustration of target on the radar scope (I). These are called echoes. The pulse of electromagnetic waves transmitted from the radar, for all practical purposes, have velocity of light about 3×10^8 m/s (1.86×10^5 miles/s). The time taken by the radar beam to hit the target and return back to the radar antenna is measured (by error voltage technique).This gives the range of the target (distance between the radar antenna and the target). The radar scope is in fact a form of Television picture tube, on which beam of electrons impinge and cause it to glow. Echo on the scope is an indicator of the object tracked. For scanning the target the radar antenna can be moved horizontally (0 – 360° bearing angle *reckoned* from true north) and vertically (0 – 90° tilt angle horizon to zenith). The horizontal, vertical angles and the range of the target can be read directly on the scopes. A radar scope provides more information than the bearing of the target, which depends on the returned signal called the strength of echo (of the target) or intensity.

Types of Scopes

1. **PPI :** (Plan Position Indicator) : This provides a plan view of the echoes, namely range and bearing (horizontal angle).

2. **RHI (Range Height Indicator):** This provides a view of the range and height of the target echoes.

3. **REI (Range Elevation Indicator)** : This provides the range and elevation of the target echoes.

Besides these, there are arrangements for detecting intensity, gain, isoecho contouring etc in the radar.

The functioning of the radar depends on various factors such as wavelength (λ), Transmitting power (Tr), size and shape of antenna and the number of scopes. If a radar antenna radiates pulse energy 500 kilowatts power, its receiver can receive as small energy as 10^{-14} watts. The minimum detectable signal strength is about 10^{-13} watts. According to the principle of Rayleigh scattering, shorter wavelengths scatter more readily than longer ones. The power which is scattered directly back toward the radar antenna is called back scatter power and it is this power alone is considered to be reflected back by the target. Back scatter is proportional to the cross-sectional area of the target. However in case of very small target areas, such as cloud drops, for a fixed wave length the back scatter power is proportional to the sixth power of

the drop diameter i.e., $P_r \propto d^6$. or $\dfrac{P}{P_1} = \dfrac{d^6}{d_1^6}$. Consider a 3 cm radar is used. If d = 0.1 cm and d_1 = 0.2 cm then $2^6 P = P_1$ or if d = 0.1 cm, d_1 = 0.3 cm then $3^6 P = P_1$. i.e., the back scatter power of diameter is 3^6 times. On the contrary the reflected power depends on the cross sectional area of the target's diameter 0.1, 0.2, 0.3 cm would be P, 4P, 9P respectively. It is evident from this illustration that the back scatter by a drop is dependent on the diameter of the drop.

Back scatter is inversely proportional to the fourth power of the wave length (λ) i.e., $P \propto \dfrac{1}{\lambda^4}$ or $\dfrac{P}{P_1} = \dfrac{\lambda_1^4}{\lambda^4}$.

For example the back scatter of λ_1 = 1 cm radar is 81 times greater than that of λ = 3 cm radar $\dfrac{P}{P_1} = \dfrac{1^4}{3^4}$ i.e., $P_1 = 81P$.

Attenuation

The weakening of the outgoing (or reflected) energy by the intervening clouds, raindrops and suspended particulate matter is called attenuation. It has been found that the attenuation is more for shorter wavelengths as compared to the longer. It is therefore useful to use shorter wavelength radars (1 or 3) for detection of small rain drops (drizzle, rain etc) and longer wavelength radars for heavy precipitation (cyclones, hurricanes) to reduce attenuation. Because of this 3 cm radars are used for cloud and storm detection purposes and 10 cm radars are used for cyclone detection (Hurricane). Depending on the purpose of detection 1, 3, 5, 10 and 20 cm radars are used.

Radar Equation

With the notation of block diagram of radar, the power received per unit area of the airborne object (target) located at a range of 'r' is given by $\dfrac{P_t}{4\pi r^2}$. If G the gain of the antenna of radar (which depends on the shape and size of the antenna), A_t the cross sectional area of the target, then the power intercepted by the target $= \dfrac{P_t}{4\pi r^2} G A_t$.

The reflected power of the object received back by the antenna is given by

$$P_r = \frac{P_t\, G\, A_t}{4\pi\, r^2} \times \frac{A_e}{4\pi\, r^2}$$

Where A_e = effective area aperture of transmission by the target A_t which is now acts as a source of power transmission.

Pr is of the order 10^{-12} to 10^{-13} watts.

For practical purposes the value of $A_e = \dfrac{G^2\, \lambda^2}{4\pi}$.

For parabolaidal reflector $G = \dfrac{8\pi}{3}\dfrac{A_P}{\lambda^2}$

where A_p = aperture area of the antenna.

$$\therefore \qquad Pr = \frac{P_t\, A_t}{(4\pi r^2)} \cdot \frac{G^2 \lambda^2}{4\pi} = \frac{P_t\, A_t}{(16\pi^2 r^4)} \times \frac{64\pi^2 A_p^2}{9\lambda^4} \times \frac{\lambda^2}{4\pi}$$

$$Pr = \frac{P_t A_t A_p^2}{9\pi r^4 \lambda^2}$$

If σ_i is back scatter coefficient, then (A_t is replaced by σ)

$$Pr = \frac{P_t A_p^2 \sigma_i}{9\pi r^4 \lambda^2}$$

$$\therefore \text{ Average} = \overline{Pr} = \frac{Pt\, A_p^2}{9\pi r^4 \lambda^2}\left(\sum_{i=1}^{\eta} \sigma_i \right) \qquad\qquad ......(15.1)$$

If V_m = Volume of cone = $\dfrac{\pi r^2\, \theta\, \phi}{4} \times$ effective depth

$$\text{Area of ellipse} = \pi ab$$

$$V_m = \pi \frac{r\theta}{2}\cdot\frac{r\phi}{2} \times \frac{h}{2}$$

where $\quad \theta$ = horizontal beam width

$\qquad\quad \phi$ = vertical beam width

$\qquad\quad r$ = range

$\dfrac{h}{2}$ = effective depth

$$V_m = \frac{\pi r^2 \theta \phi}{4} \times \frac{h}{2} \qquad\qquad \dots(15.2)$$

where   h   =   pulse width

Finally the radar equation is given by [(15.1) and (15.2)]

$$\overline{Pr} = \frac{P_t\, A_p^2\, \theta\, \phi\, h}{72\, \lambda^2\, r^2} \left(\sum_{vol} \sigma_i \right) \qquad\qquad \dots(15.3)$$

where $\sum \sigma_i$ = radar reflectivity

It follows from eq. (15.3), that

$$\overline{p}_r \;\alpha\; p_t,\;\; \overline{p}_r \;\alpha\; A_p^2\;,\;\; \overline{p}_r\, \alpha\, h,\;\; \overline{p}_r\, \alpha\, \frac{1}{\lambda^2}\;,\;\; \overline{p}_r\; \alpha\; \frac{1}{r^2}$$

According to Mie and Rayleighs theory

$$\sigma i = \frac{64\pi^5}{\lambda^4}\, /K/^2\; a_i^6 \qquad\qquad \dots(15.4)$$

Where $/K/$ = depends on the refractive index of droplet and eletric and magnetic dipole

$$a_i = \text{radius the } i^{th}\ \text{drop}$$

It follows from eq. 15.3 and 15.4 that

1. $\overline{p}_r\; \alpha\; a_i^6$

2. $\overline{p}_r$ is more for water droplets than for ice particles (density of water > density of ice)

3. It has been found that water spheres reflect 5 times that of ice sphere

4. For the same echo strength, for larger droplet wavelength is effective.

Characteristics of Radar Echoes

Weather echoes are classified into three groups (i) feature type, (ii) synoptic type and (iii) seasonal type.

(i) Feature type

Shape, size on various radar scopes 1. Convective, 2. Stratiform 3. Bright Band, 4. Dry holes, 5. Wind shear, 6. Protubarence, 7. Spiral band.

1. Canvective Type

On PPI scope they look circular or oval shaped. Relatively small in size with clear cut and sharply divided edges. Each echo is a single cell or group of closely packed cells. On RHI scope, convective cells appear as tall, narrow columns. Intensity is quite bright and can be detected even on long range. They do not exhibit bright band except when the convective cell is in decay stage. They show isoecho contours. The brightness of echo is due to the high water content, updrafts and downdrafts in the cell. Bright portion of echo denotes severe turbulence. Extraordinary brightness of echo indicates the presence of hail in the cloud. Blurring of edges of convective cell echo is the first symptom of decay of cell. On reduction of receiver gain, the reduction in size of convective echo is comparatively less than that of stratiform cloud echo.

2. Stratiform

On PPI scope stratiform cloud echoes are large in size without clear cut boundaries and they are diffused and less bright. Height-width ratio is small and does not show any vertical development. Uniform intensity. Distant echoes may not be detected. Intervening precipitation may mask the stratiform echoes beyond precipitation. Stratiform echoes exhibit Bright Band and do not display isoecho contour. In decay stage of a convective form echo degenerate into stratiform echo.

3. Bright Band

A Bright Band or bright line is associated with stratiform echoes which can be seen on RHI scope on reduction of receiver gain. This bright band is found in the vicinity of 0°C isotherm and differentiates the cloud particles above and below the freezing level. It helps to decide the height of freezing level, stability of the atmosphere or turbulence free cloud. Bright Band appears in the decay stage of Cb but not in the developing stage of Cb.

4. Dry holes

In a bright large convective cloud echo, sometimes a small area embedded in it is found without any echo. This is called Dryhole. This is different from isoecho contour. The cause of this is not fully undetstood. These are associated with severe thunderstorms with heavy precipitation, strong surface wind. It is also an indication of a tornado or hurricane. In association with cyclonic storm dryhole is observed besides the eye of the cyclone.

5. Wind Shear

A shower associated with a thunderstorm can be seen on RHI scope as separate columns of falling rain. These vertical column often slant out. This represents vertical wind shear.

6. Protuberance and Hook's

On PPI scope the protuberance or 'comma' or six '6' figure shaped echoes indicate very intense convective cloud which is associated with hail or tornado. These can be seen even with higher antenna elevations and low receiver gain.

7. Spiral Bands

On PPI scope spiral band echoes indicate tropical revolving storm. These bands persist for a long period with slight changes.

(ii) Synoptic Type Cloud Echoes

On PPI squall line echoes appear with length and breadth ratio of (10:1) ten to one or more. Airmass front echoes appear like squall line but they are thick and more uniform.

(iii) Seasonal Type Cloud Echoes

These include rainy weather echoes which appear diffused with illdefined edges. Due to attenuation, echoes beyond precipitation may not be seen. Others are globular, Hurricane, Tornadoes, water spouts and Tropical cyclone.

In a tropical cyclone five echo regions will be seen. At the periphery convective cellular echoes will be observed. These are associated with squally weather. Inward spiraling echoes with slight movement is seen. The outer bands towards radar side will be bright while the banding portion farther from the radar will be weak. The inward spiraling echoes will be 500 to 600 km away from the storm centre. The gap between the banding will be about 50 to 80 km which is free from echoes. After banding a large area of intense echo will be observed. Next to this intense echo a cellular structure near the eye which has thick band of echoes arranged in a definite spiral band. The eye of storm is a cloud free zone which is either circular or oval shaped. With the help of these patterns the centre of the storm can be located exactly. In a decay stage of a storm, the spiral bands appear diffused. With the continuous monitoring of a cyclone on radar a running commentary of cyclone intensification, decay, speed and movement can be given accurately as long as it is on the sea. For tracking a cyclone a 10 cm radar is most suitable with less attenuation. Generally the range of the radar is 400 km. It is the best tool to track a cyclone over sea when it is in the range of the radar. A tornado is specified by a strong hook echo (Fig. 15.2). On RHI the vertical extent will be about 16 to 18 km higher. Some small echoes move at right angles to the general motion of the wind towards the right side. A water spout gives a thunderstorm type echo.

INSAT Picture

Tornado

Fig. 15.2 *Contd...*

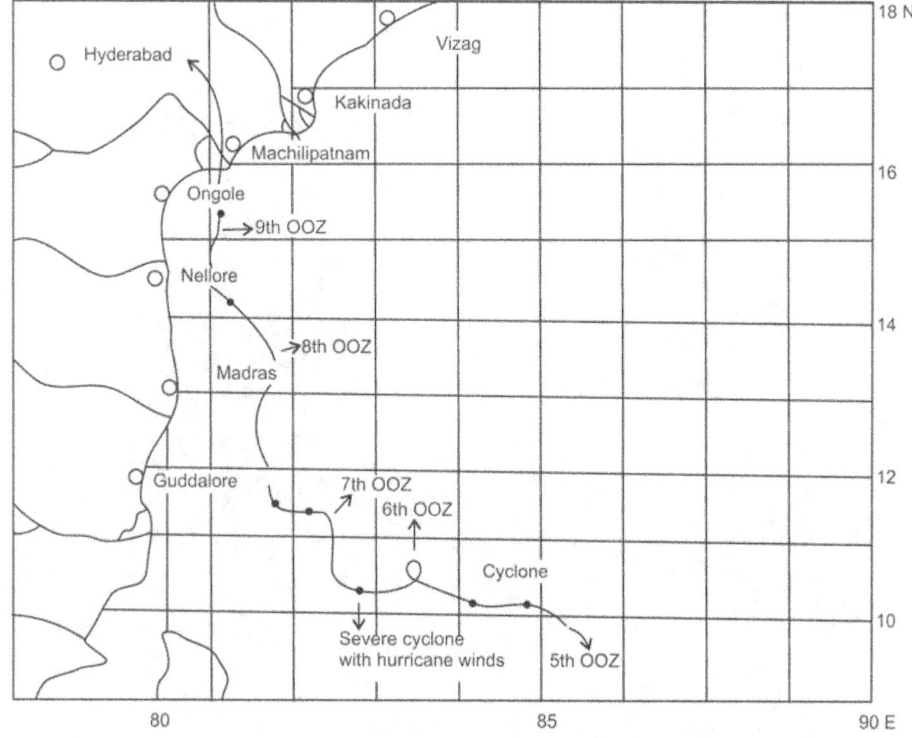

Fig. 15.2 Tracks of cyclonic storms using radar.

Note : Lightning cannot be directly seen on PPI. But with the help of negative and positive of the transparency of the film if we take the photography at that instant by imposing these two we get lightning.

Anomalous Propagation

Non-precipitation echoes, ground clutter. This is due to permanent objects such as hills, tall buildings, lofty trees etc. These are permanent echoes.

Normal Super Refraction Echoes

By bending of the radar beam we get some echoes. This occurs mainly in severe winter due to inversion of temperature.

Abnormal Refraction Echoes

These are multiple refraction echoes also called Radio duct.

Questions

1. What is Radar? State briefly its principle and the type of scopes and their use.

2. Write the Radar equation, state the conclusions?

3. Classify the types of Radar echoes and briefly describe them?

4. Describe the use of radar in cyclone detection and tracking?

CHAPTER 16

Hailstorms

Hail is one of the byproducts of a severe thunderstorm. Thunderstorm that is associated with hail is called hailstorm. Hail is formed in a well developed cumulonimbus cloud, which is characterized by severe updrafts. A typical hailstone when cut shows layer structure, like that of an onion. Hail is a climatic element, varies with time and space. The frequency occurrence of hailstorm is more in middle latitudes and it abruptly reduces to zero towards poles. Over the world, the western coasts of continents have maximum frequency during spring season. Hailstorms are not confined to any part of the world. Most hailstorms occur around lat 10 °N in April to September period and about lat 8 °S in October to March. However the whole of the northern hemisphere between lat 20 °N and 50 °N is prone to incidence of hailstorm in summer. Generally the size of hail is small (pea size, diameter less than 10 mm) and occurs over an area less than one square kilometer. But they cause heavy damage to fruit trees. It is estimated that on an average about 1% of the world's crop is damaged by hail. The mass of hail in general does not exceed 10% of the mass of that thunderstorm.

Hail has several shapes and is classified by its size and compared with the familiar objects. If the diameter is less than 5 mm, it is called shot, 5-10 mm is called pea size, 10-20 mm is called grape size, 20-30 mm is called walnut size, 30-50 mm is called golf ball size, 50-60 mm is called hen egg size and if it exceeds 60 mm it is called tennis ball size. However in general the solid precipitation from cloud is classified into four types. (i) Grauple or soft hail, (ii) small hail, (iii) ice pellets and (iv) hail.

Soft Hail

They are white, opaque and conical in shape. Diameter is less than 5 mm, density 800 kg/m^3 .

Small Hail

They are partly transparent, round or conical in shape. Diameter upto 5 mm, denisity 800-900 kg/m^3.

Ice Pellets

They are transparent, spherical or irregular in shape. Diameter less than 5 mm, density the same as that of ice. They form by the freezing of rain drops.

Hail

They are lumps of ice or ice and water with air inclusion. Diameter more than 5 mm (up to 5 cm). They are partly transparent or opaque. They have alternating layers of ice and air babbles. Generally they are spherical or conical in shape.

Largest Hail

As mentioned earlier, the size of hail generally varies from 5 to 50 mm in diameter. But there are reports of very large size hail. On 30 April 1888, Moradabad in Bihar, India reported large hailstones of cricket ball size, which took the life of 246 human beings and 1600 goats and sheep. In Germany, in 1925, Talman measured an ellipsoidal hail 26 × 14 × 12 cm in size and weighed 2.04 kg. On 11 March 1957 Begumpet Airport (Hyderabad) India reported average hailstones of size 5-8 cm in diameter. On 27 May 1959, in Delhi, India, hail size of 200 mm diameter caused holes in aircraft frames and there were reports of hail size 250-375 mm diameter. On 3 September 1970, at Coffey Ville, Kansas, USA, a hailstone was weighed 758 gm, diameter 190 mm and circumfrence 444 mm. The substantiated reports of hail of 1.9 kg fell in Kazakisthan. 970 gm hail fell near Strasbourg, France in August 1958. The other reported hail weights are 4.5 kg recorded in China, 4 kg Hungary and 3.4 kg India. These reports probably were groups of hail stones clustered or frozen together. There were also many reports of hail stones lying more than one meter deep on the ground. These consists mostly pea to walnut size and piled up drifting by wind from hill sides. During 11-13 March 1981, there were widespread hailstorm over Telangana (A.P), India which caused damage to 33000 acres standing crop, damaged 85000 houses and taken a toll of 18 human lives and 13000 live stock. The size of hail ranged 100 to 1200 gm.

Hail Formation in the Cloud

It is believed that some kind of ice nuclei when falls in the supercooled water, hail embryos are formed. It then grows by collision and accretion process. Hailstones have concentric accretions of clear and opaque ice. When embryo is carried aloft by updraft it is frozen. By repeated circular trips within updrafts of thunderstorm cell, that carry hailstone above and below freezing level, hailstone is formed. Depending on the time spent above freezing level milky hail layer forms by rapid freezing, while the clear ice layer is formed below the freezing level. The number of trips above and below freezing level gives the number of layers within the hailstone. In one hailstone twenty five separate layers have been counted. Calculations show that an updraft of 95 kmph is required to support the hail of diameter 25 mm. A hailstone of 80 mm diameter requires an updraft of 200 kmph to support and a 130 mm diameter hailstone requires an updraft of 377 kmph to support.

Favourable Synoptic Conditions

Various studies indicate the following favourable conditions for occurrence of hailstorm. (i) wind shear in middle troposphere (850 to 200 h Pa). (ii) surface high humidity and also in the lower levels. (iii) low level convergence of moist air masses of contrasting characteristics (such as continental and maritime air masses). (iv) In tropics, movement of upper air trough in zonal westerlies Jet stream (200 h Pa), preferable location of hailstorm is along the axis of trough. (v) North-South trough on sea level chart with associated wind discontinuity/ cyclonic circulation in the lower troposphere. (vi) Steep lapse rate of temperature in the lower levels (surface to 500 h Pa). A number of factors must be considered simultaneously and it is possible that a non-linear combination of parameters will be required. For example, wind shear plays an important role in the organization of the most severe hail storm, but it is likely that if shear is too strong, storm development will be inhibited. A major source of non-linearity in the forecasting problem is the existence of several distinct types of hailstorms. Some principal types are: (i) an airmass or pulsating multiple bubble storm, (ii) a multicell strom and (iii) a large severe super cell storm.

The first type requires great thermal instability and occurs in the absence of wind shear or strong winds. The second type requires moderate to strong instability and a proper vertical wind profile. The third type may occur without great instability, requires strong winds with a properly organized vertical wind profile and is associated with a strong organized downdraft fed by rain cooled air from the middle troposphere. It can produce very large stones, great crop damage and severe wind damage. It seems to be a type of storm associated with tornadoes.

Divergence exists from a trough to downstream crest in upper tropospheric (westerly) wave. In a sinusoidal wave, maximum upper level divergence is located about midway between trough and ridge. Its intensity is greatest for waves with short lengths and large amplitude and large wind speed in the jet stream. It seems that jet stream is a major factor in the development of severe thunderstorm.

A study revealed that wind speed from surface to 500 h Pa was very different from that compared to the wind speed from 500 h Pa to 250 h Pa. The difference between the two was related to the size of the hailstones produced. A wind shear of 65 kmph (35 kt) between the two layers correspond to heavy hail (3-5 cm diameter), 61 kmph (33 kt) to moderate hail (1-2 cm diameter) and 48 kmph (26 kt) to light hail (diameter less than 1 cm), and less than 44 kmph (24 kt) indicate no hail formation.

The terminal velocity of hail is given by an approximate formula

$$V = \sqrt{2W/C_d\,\rho_a\,A}$$

where W = weight of the hail

           A = cross-section area

           C_d = drag coefficient

           ρ_a = density of air, which varies with height.

The fall speed of large hail stones is of the order 40-100 mps.

Note :

Low Level Wind Shear

Low level wind shear is associated with the following meteorological situations. (i) Ground layer inversion during winter months, which results in strong jet like winds aloft. (ii) At the boundaries of air masses. (iii) Sea breeze circulation. (iv) Sharp changes due to topographical features over aerodromes. (v) In the vicinity of thunderstorm.

Microburst

Downdrafts in thunderstorms encircling horizontal gusts of large magnitude is called downburst or micro-bursts. A downburst may be defined as a localised intense down draft with vertical currents exceeding a downward speed of 3.6 mps (12 ft/sec) at an height of 91 m (300 ft) above ground. This speed is comparable to the normal descent rate of a jet aircraft during landing. This threshold speed of downburst tends to double the sinking speed of such an aircraft to adjust balance (or trimmed) for normal approach near touch down at usual 3° glide slope below 150 m (500 ft). A microburst is a smaller but rather more intense version of the downburst. The horizontal extent of the downburst is 4-10 km and the microburst is 1-4 km. When downburst/ microbursts reach the surface they spread out horizontally and causes strong gusty surface winds with average maximal gusts 50 to 60 mps. In such an encounter aircraft loses balance and crashes.

Questions

1. What are the byproducts of a severe thunderstorm? Which are the favourable global locations and seasons occurrence of hail storms?
2. Describe a hail and classify them. State the possible largest hail?
3. Describe the formation of hail in cloud and state the favourable synoptic conditions for hail formation in cloud?
4. What is downburst and micro-burst? Explain?

CHAPTER 17

Tornadoes

Tornadoes are the most violent storms on earth, but they miss detection on synoptic chart. A Tornado is a violent rotating column (spiral motion) of air. It appears as pendent cloud extending from cumulonimbus cloud base to the ground. The column (or funnel) does not always extend to the ground and may be masked by the dust. A funnel cloud is similar to a tornado, except that the funnel does not reach the ground. The diameter of the column is about 100 m. In general tornadoes are more frequent in extra-tropics than in tropics.

The word Tornado is derived from the Spanish word Tronada, meaning thunderstorm. In northern and western parts of Africa tornado still refers to a thunderstorm. In Latin Tornare means "to turn", thus tornadoes are also called twisters. Tornadoes have different names in different parts of the world. In France and Germany they are called Trombe, in Spain and Italy Tromba, in Russia Symerch, in Japan Tatsumaki, in India Hatishnura (meaning Elephants trunk)

Life Cycle of a Tornado

Tornado funnel may be seen as thin rope, conical shape, cylindrical shape or thick dense cloud mass touching the ground. Generally tornadoes rotate in an anti clockwise direction but there may be some that rotate in clockwise direction. During the life span of a tornado the funnel undergoes many changes. The life cycle of a tornado may be divided into five stages.

1. Dust Whirl Stage

In this stage whirling of dust is observed below cumulonimbus cloud but it does not touch the ground.

2. Organising Stage

In this stage the funnel from the base of Cb cloud touches the ground and intensity increases.

3. Mature Stage

In this stage funnel attains maximum diameter (100-250 m). Central pressure in the eye drops (25-200 h Pa) and attains maximum wind speed. Circulation usually stays in continuous contact with the ground through stages 3, 4 and 5. The funnel creates havoc, destroys buildings, poles, trees and sucking debris and dust raises to great heights into the air. Some times motor cars, animal and heavy objects sucks in, lifts aloft and thrown away at considerable distance. The destruction of buildings is also caused by explosive effect due to sudden fall of pressure (in less than a minute) by over 50 h Pa or more. The large pressure defect between inside a closed building and the outside atmosphere leads to an explosion, which bursts the walls and the ceiling outwards. According to some empirical theory, a pressure drop of 100 h Pa causes wind speed 600 kmph around the vortex.

4. Shrinking Stage

In this stage the width of the funnel decreases and tilt of vertical axis increases, fury drastically decreases.

5. Decay Stage

In this stage the shape of the funnel spreads like a spiral of rope, decreases in height and ultimately disappears.

Major tornadoes pass through all five stages while minor Tornadoes abort in stages one – two – five.

Size of a Tornado

The diameter of the funnel generally varies 100-250 m, rarely reaches to 1000 m. The vertical depth of tornado circulation extends to the middle of Cb cloud (about 10 km). The average path length is 5-10 km but it may range 30 m to 500 km. Life span varies 2-3 minutes to 3 hours. Short lived tornadoes have wind speed 50 mps, path width 100 m and travel length 2 km, whereas long life period tornadoes have wind speed 100 mps, path width 200-600 m and travel length 200 km. Life period varies 15 seconds to 8 minutes at a point. An extreme duration of 7 hours along the ground observed in Illinois on 26 May 1917. They generally move in a straight line path over a flat country for a long distance. Movement becomes zig-zag when it passes through hills, tall buildings and limits speed. On some occasions dies down. The average energy of tornado is 10^{-11} times the solar energy received by the earth (3.67×10^7 cal/min)

Central Pressure Drop in a Tornado

The central pressure drop (in the eye) may 100 to 200 h Pa. On 20 Aug 1904, at Minnesota, aneroid barometer recorded a pressure fall of 200 h Pa. The eye of a tornado behaves like a vacuum, consequently strange things happen near the vortex such as, trees are stripped of their barks, sheep lose their wool, chickens lose their plumes, corks of bottles fly off, chests explode and splinters fall in all directions and bursting of closed buildings etc. People who had the experience narrated that they had the bursting experience in their chest and ears.

Synoptic Situations Favourable

The first signs of a Tornado development is a large active cumulonimbus cloud with mammatus. The cloud may acquire green and yellow colour. Green colour lightning or ball lightning also be observed. On some occasions a Tornado may occur without any thunderstorm activity but there will be heavy rain, showers and hail. Cb is associated with every tornado. It is probable a supercell type convective cloud is the genesis of a tornado. The synoptic conditions favourable for the development of a supercell type convective cloud are: (i) Low level convergence with associated upper air divergence, (ii) lot of moisture in low levels, (iii) super adiabatic lapse rate in the lower - mid levels, (iv) strong vertical shear in the horizontal wind (upto 300-250 h Pa or more).

Radar Echoes

As mentioned earlier in thunderstorms hook (6) type echo is associated with a tornado, which is an indication of meso-cyclone. It may be noted here that every hook type echo is not a tornado, it only indicates severe thunderstorm. The echo reflectivity is to the order of 4 or more.

Tornado Vortex Signature (TVS)

A pulsed Doppler Radar when directed to a stationary cyclonic vortex, it will show particles moving away from the radar to the right (positive velocities) of the line joining the radar and the centre of vortex and the left of this line particles will appear to be moving towards the radar (negative velocities). Doppler radar will indicate positive velocities to the right of the centre of the vortex and negative velocities to the left of the vortex centre and zero along the line of the vortex centre. The velocity distribution configuration detected by Doppler Radar is called Tornado Vortex Signature (TVS). TVS is detected in the middle region of Cb. A tornado takes about 15 to 20 minutes to touch the ground after TVS is detected.

Electrical Phenomena

Increased lightning activity will be observed about half an hour before touch down of a tornado.

Acoustic Phenomena

A peculiar whining sound like buzzing of an army of bees is heard when the funnel cloud is high up in the air. When the funnel touches the ground a terrific roar, like the sound of a cannon fire, is heard for a few minutes. Snake hissing sounds heard before tornado touches the ground. This is attributed to the vibrations of the air masses rotating around the funnel.

Precipitation

Heavy rain sometimes associated with hail precede and follow a tornado. The eye of the tornado, like the cyclone eye, is generally free from precipitation.

Tornado Hazards

Prediction of a tornado is still in infant stage as its mechanism of formation is not clearly understood. However with the help of radar nowcasting is done (forecasting its movement and severity in a very short period) to alert people to save their lives. Property damage due to tornadoes seems at present inevitable. The average death toll in United States of America is about 150-250 per annum, but the injury seems many folds of this. This is caused by the flying objects hitting the people and the subsequent fire. The greatest loss of life due to a single day tornado was 689 people recorded on 18 March 1925, while the greatest loss on a single day was about 1200 people recorded on 19 February 1884.

Some Recorded Tornado Incidence

On 30 May 1879 an iron bridge weighing 108 tons was lifted over the River Big Blue in Irving (USA). This incident was recorded in great detail in a book by Finley (1881). A tornado in Minnesota lifted a passenger car with 117 passengers, total weight about 72 tons. In 1879, in Missouri, a horse was lifted and carried a distance of few hundred meters and deposited without any injury. In New Delhi (India) on 17 March, 1978 a tornado in about three minutes took a toll of 28 people, caused injuries to about 700 people and damaged property worth more than one crore rupees. The same tornado lifted a passenger bus with 70 passengers and deposited twenty meters away in a canal. A tempo vehicle was also lifted from a petrol pump and thrown 100 meters away. In a Germany city market a man who was standing under the vortex was lifted high in the air, pelted with hailstones and thrown. When he came back to senses he found himself under two men, one woman and a horse. Another man who was standing close to a post in front of an emergency ward of a hospital in Lefortova was lifted five meters high in the air, carried a distance of 100 m in the garden and deposited on the lawn.

Development of Tornadoes Associated with Fire and Volcanic Eruptions

Under favourable circumstances wide spread fire may develop a tornado. On 1 September 1923, a fire broke out in Tokyo after an earthquake. Within 24 hours about 120 tornadoes and smoke devils developed in Tokyo after the great earthquake. Some tornado acquired intensity of wind 50-70 m/s, picked up cars, people outside the fire line. About one lakh forty two thousand (142000) people died in Tokyo, more by fire than by earthquake itself.

Tornadoes/waterspouts may develop after volcanic eruptions over land/ water surface. In 1963, there was volcanic eruption of Surtsey in the middle of the sea, off the coast of Iceland. Each intense volcanic eruption gave rise to an enormous dense cloud. From these clouds tornadoes of different sizes developed on the leeward side.

Safety Measures

People can adopt safety measures to protect themselves by listening to tornado advisory bulletins broadcast on national TV channel. The following measures to be adopted as in case of lightning strokes and cyclones. Shut off immediately, after alert, electric power, gas supplies and extinguish all fires. Seek shelter quickly in a tornado cell or in sturdy reinforced pucca concrete building. Do not drive a car which may be hit by missile debris or it may even be lifted and thrown away by strong winds. If caught in an open place, run to a nearby culvert or ditch, lie down and hold on to a fixed object to protect from being blown off. Protect your head from the injuries of flying objects.

Fujita-Pearson (FPP) Tornado Intensity Wind Scale

This was proposed by Fujita in 1971. This wind scale designed to link the Beaufort scale of Force 12 with Mach No.1 in twelve steps. The scale is derived from the formula $V = 6.30 (F + 2)^{3/2}$

where V = wind speed in mps. The scale specifications are as follows.

Tornado Statistics

Tornadoes can occur in any latitudes, but they are more frequent in extra-tropics than in tropics, and rare in equatorial belt. The annual frequency of occurrence of tornadoes in USA is about 200. The Mississippi valley has maximum occurrence in the world. Generally tornadoes move from south westerly direction (in northern hemisphere) at the average speed of 15 mps. The maximum wind speed does not exceed 142 mps. The average travel length is about 2 km (range 100 m to 600 km), in USA 7 km. The average width of track is 50 m to 250 m. There may be erratic movement as well. Average life span varies from 2 minutes 2 hours.

Fig. 17.1 The tornado (upper photo) which struck Edmonton, Canada on 31 July 1987 left 26 dead and cause damage amounting to US$350 million (*lower photo*) (**Photos :** *Atmospheric Environment Service Canada*).

Table 17.1

Scale	Wind	Damage description
F0	18-32 mps	Light damage: such as breaking of tree branches, falling of sign boards.
F1	33-49 mps	Moderate damage: Beginning of hurricane wind speed. Moving autos pushed off the roads. Peel surface off roofs. Mobile homes overturned
F2	50-69 mps	Considerable damage: Box cars pushed over; Mobile homes demolished. Large trees uprooted or snapped. Light object missiles generated.
F3	70-92 mps	Severe damage: Roofs and walls tornoff well constructed homes. Trains overturned, Heavy cars lifted off ground and thrown. Most forest trees uprooted.
F4	93-116 mps	Devastating damage: Well constructed houses levelled; cars thrown and large missiles generated, structures with weak foundation blown off some distance.
F5	117-142 mps	Incredible damage: Automobile sized missiles fly through the air beyond 100 meters; strong frame houses lifted off foundations and disintegrated at considerable distance. Trees debarked; incredible phenomena will occur.
F6-F12	142 mps to Mach-1 (=speed of sound)	The maximum wind speeds of tornadoes not likely to reach F6 wind speeds.

Waterspouts

Waterspouts are of two types. Type one that develops downward from a cumulonimbus cloud and it is also called fair-weather waterspout. Type two is simply a tornado over the water. Type two develops upwards from the surface of water and is not directly associated with a cloud. In both of these types water gets sucked up into air along with aquatic creatures such as fish, frog etc. These aquatic creatures may be found in rain associated with waterspout. In the world over Florida keys and Palm beach Florida have the highest frequency of incidence of waterspouts about five per annum per 100 sq km.

Waterspouts undergo a life cycle similar to that of tornado on land. It has five stages of life cycle which are briefly described below.

1. Initial or Dark Spot Development Stage

First visible sign of a vortex is a dark spot on the sea surface. In this initial stage a short funnel pendant may develop from the super cell (Cb cloud).

The dark spots may occur in bunches of two or more, of which one may be dominating the others. The dominant one becomes waterspout and the others decay. The life span in this stage may be about 1-20 minutes. Many may dissipate after this stage. From the droppings of tracers it is gathered that the dark spots are developed by the rotation imposed from above.

2. Spiral Stage

In the second stage, formation of spiral occurs on the sea. Only one dark band (150-1000 m) comes out from a nearby shower band. Spiral bands indicate the lines of confluence or difluence on sea surface while the regions of flow away from the vortex seen along the surface.

3. Ring Spray Stage

In this stage wind strength increases beyond a critical value of about 22 mps and throws up a ring of spray from the surface. In this stage funnel increases in size and the axis of funnel tilts. It begins to move rapidly along the surface as it comes under the influence of the wind shift line. The wind shift line is believed to be associated with the cool breeze from close by shower.

4. Mature Stage

Mature stage lasts about 2 to 18 minutes. This stage is characterized by the strongest wind, funnel tilt and associated with forward wind speed 4-8 mps. In this stage the funnel may acquire double eye wall like that of cyclone and generally moves along a gentle curved path. The spray ring becomes spray vortex. Waterspout may not be visible all through from cloud to water surface. On radarscope hook echo may be seen.

5. Decay Stage

The last stage is decay stage. This stage generally lasts about 2 to 3 minutes. Spiral and funnel disappears, and rain cooled air takes over waterspout. However in decay stage spiral rain may still be seen which lasts about 5 minutes.

Waterspout Statistics

The maximum tangential wind speed estimated by photogrammetric technique is 85 mps (in the lowest levels of 10-15 m amsl). Waterspouts carry objects weighing 5 tones to a height of about 40 m over coastal waters out at sea. The helical funnel of waterspout may go upto about 300 m in air. Waterspout vortex may generate waves of moderate to high amplitude over the sea surface. According to schroder (1977) statistics Hawaiian island has maximum incidence of water spouts about 7 per annum. Photogrammetrically estimated wind speed at an altitude of 38 m is about 55 mps and vertical velocity 25 mps. According to aircraft penetration of waterspouts observation the principal funnel features are: (i) A warm central core region, (ii) Funnel core deficit 1 to 10 h Pa depending on waterspout intensity, (iii) In the core upward vertical velocities 5-10 mps, (iv) Tangential velocities at penetration altitude of 400 m is about 30 mps, (v) Both cyclonic and anti-cyclonic vortices encountered.

In general waterspouts are more common than tornadoes and the average frequency is about 400 per annum per 10^4 sq km in favorable locations. Compared to tornadoes central pressure of waterspouts have higher pressure and lower pressure defect and relatively less ferocious violent.

Dust-Devils

Dust-devils are a frequent phenomena in tropics and sub-tropics, however they can be seen at any latitude under suitable environmental conditions. Generally dust-devils are observed during scorching hot sun in deserts and semi deserts. They are generally much smaller in size and less violent than tornadoes. The funnel core is warm with temperature anomaly of 1 to 10 °C. The core has low pressure with pressure defect 1 to 4 h Pa. Dust and sand whirls in a funnel shape but no pendant cloud is involved. The rotation of wind may be cyclonic or anticyclonic. The horizontal rotation and upward flow of wind in dust-devils exceed 30 kmph w ith average height of column being 200 m. Duration of life varies from few minutes to 7 hours with horizontal path distance of about 50 km.

Favourable Condition for Formation of Dust-Devils

(i) Hot or intense insolation (ii) Dry dusty ground (iii) Steep lapse rate of temperature at the ground (iv) calm or light surface wind.

Questions

1. Describe a tornado and its life cycle?
2. Describe the size, travelling length, life period, central pressure of a tornado?
3. Write the favourable synoptic situations for the development of a tornado. What is its shape an radar?
4. Write few lines on Tornado vortex signature, acoustic phenomena and hazards of tornado?
5. Write the safety measures against tornado hazards?
6. Write the salient features of Fujitha-person (FPP) Tornado intensity wind scale?
7. Write an essay an waterspouts?

CHAPTER 18

Tropical Meteorology

The region between Tropic of Cancer (lat 23½ °N) to Tropic of Capricorn (lat 23½ °S) is called Tropics. The regions between Tropic of Cancer to Arctic circle (lat 66½ °N) and Tropic Capricorn to Antarctic circle (lat 66½ °S) are called sub-tropics or extra-tropics. However the weather systems are not restricted to geometrical limits of these latitudes. Weather mainly depends on the migration of the sun from Tropic of Cancer (summer solstice) to Tropic of Capricorn (winter solstice). It may be noted here that the Hadley cell which is located in lower latitudes of each hemisphere with descending air reaching the earth's surface at about lat 30°. For all purpose we consider the region between the 30th parallels of latitude which are the centres of sub-tropical high pressure belts. In many respects the weather systems of tropics and sub-tropics differ greatly, which are given below.

1. In the lower and middle troposphere the tropical atmosphere is convectively unstable while extra-tropical atmosphere is convectively stable.

2. In tropics, in lower atmosphere, weather systems move from east to west while in extra-tropics they move from west to east.

3. The tropical regions are affected by both westerly and easterly waves, while extra-tropics are affected by only westerly waves.

4. In tropics the mean meridional (north-south) circulation (Hadley cell) is direct while in extra–tropics the mean meridional circulation (Ferrel cell) is indirect.

5. In tropics atmosphere gains westerly momentum (that it exports momentum to extra–topical regions) while in extra-tropics westerly momentum is lost.

6. In tropics cyclones are of warm core, while in extra-tropics the cyclones are of cold core.

7. In tropical stratosphere QBO (Quasi Biennial Oscillations) are prominent while in extra tropics QBO is not marked.

8. In tropics the relative vorticity is equivalent to coriolice parameter (f), while in extra-tropics the relative vorticity is smaller than coriolice parameter.

We shall now consider some important tropical systems.

Equatorial Trough

The trade winds of each hemisphere tend to converge from sub-tropical High pressure belt towards relatively low pressure area near equator . This area is called equatorial trough, commonly known as the doldrums. [The equatorial trough is centred at about lat 5°S in January and lat 12 to 15°N in July. It migrates through 20° latitudes between seasons. This migration of trough influences the seasonal march of cloudiness and rainfall and also the position of formation of tropical cyclones. The annual mean position of the equatorial trough lies at about lat 5°N rather than at the geographical equator. This mean position of lat 5°N is called "meteorological equator".]

The idealized average mean sea level pressure and wind circulation in January and July are given in the Fig. 18.1 and Fig. 18.2 respectively.

The average pressure distribution as given in above Fig. 18.1 and Fig. 18.2 indicates that: (a) In sub-tropics each hemisphere has a pressure maxima of about 1015 h Pa and 1020 h Pa in summer and winter respectively. (b) The axis of equatorial trough is along lat 5°S in January and 12 to 15°N in July. The annual mean position is called the meteorological equator which lies along lat 5 °N. (c) The equatorial trough migrates about 20 latitudes (5°S to 15°N) annually. This influences the seasonal march cloudiness, rainfall and formation of tropical cyclonic storms (d) The sub-tropical high pressure belts are asymmetric with respect to the geographical equator (e) In January the ridges are centred near lat 30 °N and 35 °S while in July they are centred near 35 °N and lat 30 °S. (f) The area between equatorial trough and subtropical ridge is broader in winter hemisphere as compared to summer hemisphere. (g) In each hemisphere the mass of air in the tropical belt decreases from winter to summer. As a consequence of this mass does not escape fully to the polar regions, which implies that a net export of mass takes place from summer hemisphere to winter hemisphere across the equator.

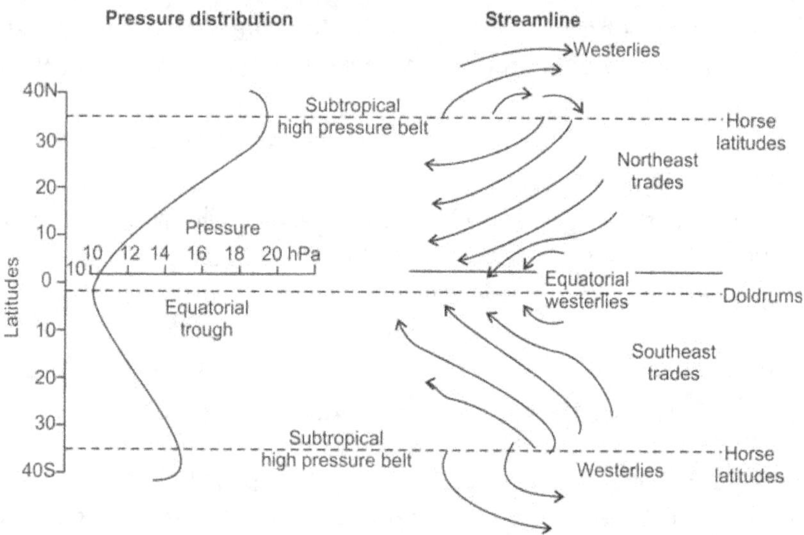

Fig. 18.1 January mean sea level pressure and streamlines.

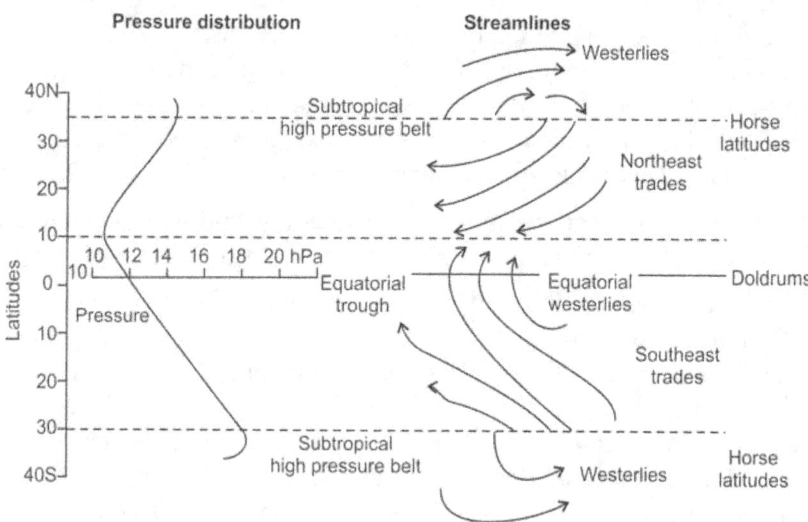

Fig. 18.2 July mean sea level pressure and streamlines.

Some Special Features of Tropics

Weather systems in tropics are different in many ways from extra–tropics. The quasi-geostrophic approximation model works satisfactorily in extra–tropics but fails in tropics, where primitive equation models work well. The theory of geostrophic adjustment of extra–tropical region, namely in a relatively small period of time, adjusting of wind field to pressure field results in quasi geostrophic balance, but this fails in tropics, where pressure field adjusts itself to the wind field over a longer period of time and thereby acquires quasi-geostrophic balance. Geostrophic approximation which suites extra–tropics is not valid in tropics. Similarly Frontal models which suites extra–tropics are not applicable to tropics. In tropics meridional temperature gradient is of the order 1 °C/100km, while in extra–tropics it is of the order 10 °C/100 km. Similarly 24 – hour temperature changes in tropics and extra–tropics are of the order 1°C and 5 °C respectively. Pressure gradients (24 hour pressure changes) in tropics are weak (1 h Pa) while in extra-tropics they are large (10 h Pa). In tropics weather is seasonal in extra-tropics it is continuous. The tropical atmosphere consists of persistent wind systems, with their associated composition and thermodynamic properties, the trades (Trade winds) and monsoon (seasonal winds). Diurnal variations of meteorological elements in tropics are approximately governed by the perturbations of these systems attributable to thermal and orographic disturbances of known origin.

Note : Persistency is defined as the ratio of the mean vectorial wind speed to the mean speed without regard to direction.

Trade Wind Inversion

Trade winds are generally persistent and steady. Over ocean areas cumulus clouds may develop with base about 1km and tops about 2 km. However along coastlines facing trade winds the cloud development is higher and associated with showers. On the lee side of the mountains the region is free from clouds. The restricted cloud height development and fair weather in trade wind zone is attributed to trade wind inversion. The inversion of temperature in the lower altitudes of trade winds is caused by the subsidence of air from the subtropical high pressure belt (STHPB). This inversion persists in trade winds which move towards equator and puts the lid on cloud development over ocean areas. However as the air moves towards equatorial trough, the inversion weakens and cloud height development increases due to instability. This results in heavy, frequent precipitation in equatorial trough zone.

Trade wind inversion is prominent over western coasts of North Africa, South Africa, and over North America and South America. In these areas the base of cloud comes down to about 500 m and the rise of temperature in inversion goes up to 10 °C. From the base of temperature inversion RH decreases to the top of inversion by about 50%. Below the inversion layer air mass is moist and cool and above the inversion layer the air mass is dry and warm. The maximum intensity of inversion occurs in the subtropical belt region; over northeast Atlantic, northeast Pacific and over southeast Atlantic and southeast Pacific and decrease as it moves closer to equatorial trough.

Inter Tropical Convergence Zone (ITCZ)

The northeast and southeast trade winds of the two hemispheres converge (meet) in a narrow zone on a large scale in the lower levels. This is called ITCZ. Generally these trade winds are separated by a wide doldrum area (or equatorial trough), but in some places they meet, where surface wind discontinuity with horizontal convergence in an east-west direction takes place. ITCZ area is associated with cloudiness and thunderstorms, squalls, overcast skies and heavy rains. ITCZ meanders with the movement of the sun. Wind discontinuity exists between westerlies (SE trades deflected after crossing the equator) in the equatorial region and the easterlies (NE Trades) on the other side. ITCZ extends from western Africa to southeast Asia across Indian ocean and another across Atlantic and the Pacific oceans, the wind discontinuity occurs between NE-Trades and SE-Trades. This was earlier termed as Inter-tropical front and Inter-tropical Discontinuity.

Some Features of ITCZ

ITCZ is the equator ward limit of upward limb of Hadley cell. On sea level, averaged over the globe, ITCZ lies north of the equator in the northern hemisphere summer and south of the equator in the southern hemisphere summer. The extreme positions of ITCZ during July passes over south Asian continent and in January penetrates deep into the African continent, in Australia it lies on land, in south America it lies south of the equator, in ocean region it lies north of equator. On land surface ITCZ coincides with highest temperature zone, on sea it coincides with the region of highest sea surfacewater temperature (SST). For this reason ITCZ is regarded as thermal equator. The position of ITCZ on surface differs from the lower levels at 850, 700 and 500 h Pa levels. The north south migration of ITCZ is maximum over south Asia, it goes up to 28 °N over India, in July, and over east Africa it goes up to 18°S in January. In such extreme positions, the air masses on either side of the ITCZ may not belong to different hemispheres. There is cyclonic wind shear across ITCZ with westerly component on the equatorial side and easterly component on the pole ward of ITCZ. Generally no temperature difference exists on two sides of ITCZ. Relatively cold moist maritime air lies on equator side of the ITCZ and dry (less moist) air on pole ward side of the ITCZ. For this reason heat transport takes place from ITCZ to middle latitudes. ITCZ produces bad weather conditions over an extended wide area, with cloud tops penetrating tropopause altitudes (up to 18 km), cloud base may be less than 1 km or even may come down to the earth's surface.

Quasi–Biennial Oscillation (QBO)

Repeating a phenomena in every two years is called Biennial oscillation. Quasi means as if or as it were. Quasi–Biennial Oscillation thus means repetition of more or less the same phenomena in about two years. QBO is observed in many meteorological parameters through great depth of the atmosphere throughout the whole world. However zonal wind Oscillations in equatorial stratosphere is most conspicuous. The relationship between QBO in stratospheric winds over the equator and rain fall over mid-latitudes and other meteorological phenomena over the world is not yet clearly understood. And also the very cause of QBO is still under investigation.

During 1884-85 Clayton discovered 25 months Oscillation in surface pressure and precipitation. However this discovery did not attract the attention of meteorologist for a long time. In late 1950s, the following features were discovered in zonal stratospheric wind flow. Krakatoa volcanic eruption in August 1883 penetrated into the lower stratosphere up to 32 km amsl. The cloud observations indicated that there were easterlies of strength 63 kt in the lower stratosphere. Von Berson, in 1908-09, found westerlies of sufficient strength in lower stratosphere over Africa. It was then believed that the zonal circulation contained Berson westerlies in the lower stratosphere and Krakatoa easterlies above it. In 1954 Palmer found that Berson westerlies were present at about 20 km level in narrow zone at about lat 2° N extending from lat 9° N to lat 5° S. Krakatoa easterlies prevailed above it upto 40 km height over equatorial region, extending from lat 15° N to lat 15° S. Transition height between westerlies to easterlies is found (21 to 27 km) to vary from month to month and year-to-year. In 1956-57 contrary to the belief prevailing then, Mc Creary found easterlies below and westerlies above in the equatorial stratosphere. Subsequently it was found westerlies steadily moved down and easterlies replaced above it, bringing back the situation to normal flow of Berson westerlies in the lower stratosphere and Krakatoa easterlies above it. In 1959, Graystone found year-to-year reversal of winds in equatorial stratosphere. In 1960 Reed found the period of oscillation of stratospheric wind is about 26 month, which is now called 26-month or Quasi-Biennial Oscillation.

According to recent findings, westerlies (or easterlies) flow starts at higher levels (at about 35 km level) and then descends to tropopause (about 16 km). The descent roughly takes place at 1km/month. Maximum wind speed of westerlies (or easterlies) occur at 25 km level.

Westerlies as well as easterlies descent from 35 km to 25 km level with almost the same wind speed and then the speed attenuates (weakens) and disappears at tropopause level. The transition from easterlies to westerlies is more marked as compared to the transition westerlies to easterlies. At about 50 km level and above there is semi-annual (6-monthly) Oscillation of easterlies and westerlies. This flow gets attenuated between 40 to 30 km and below 30 km to tropopause level it is interlocked with QBO. Between 30 to 35 km levels there exists simultaneous half-yearly and two yearly oscillation.

The variation of amplitude with respect to latitude and height particularly between 16 to 30 km (tropopause to 30 km) is described below. Oscillation is symmetric with respect to equator. Maximum amplitude observed at about

24 km over the equator, with small variation above 24 km and strong attenuation below 22 km. At any level there exists strong attenuation of amplitude as we move away from equator, and it reduces to about one-eighth of the (central) value at lat 20°.

The following meteorological phenomena also exhibits QBO (i) Zonally averaged upper air temperature, (ii) Surface pressure, (iii) Rainfall, (iv) Total ozone, (v) tropopause height, (vi) winter and spring stratospheric warming, (vii). surface temperature all over the world with opposite phases in tropical and extra-tropical latitudes. (viii) Depth of equatorial westerlies over Indian ocean, (ix) Indian monsoon activity and (x) Annual rainfall over India as a whole and sub-division wise, (xii) Number of cyclonic disturbances during summer mansoon.

Some Theoretical Aspects of QBO

The phenomena of QBO was first ascribed to the solar heat energy that is received by the earth. The variations of solar energy is dependent on : (i) sunspot cycle, (ii) solar activity, (iii) the heterogeneous rotation of the sun etc. Subsequently it was attributed to the internal forcings in the atmosphere. The recent thinking is, the effect of Kelvin waves and Mixed Rossby-gravity waves. Some important features of Kelvin waves and Mixed Rossby-gravity waves are given in Table 18.1.

Table 18.1

Sl.no	Feature	Kelvin waves	Mixed Rossby-Gravity waves
1.	Period	15 day	4 to 5 days
2.	Horizontal wave length	40000 km	10000 km
3.	Speed and direction	30 mps, east wards	30 mps, west wards
4.	Vertical wave length	8 km	6 km
5.	Amplitude of wave perturbations		
	(a) zonal wind	8 mps	3 mps
	(b) Meridional wind	0	3 mps
	(c) Temperature	3°C	1°C
	(d) Geopotential height	4 meters	30 meters
	(e) Vertical velocity	$4 \times 10^{-3} \, m^2/sec^{-2}$	$4 \times 10^{-3} \, m^2/sec^{-2}$
6.	Momentum transport	Westerly momentum carried upwards	Easterly momentum carried upwards
7.	Penetration in stratosphere	Penetrates the strato-spheric easterties, get absorbed in the zone of westerlies	penetrate the stratospheric westerlies and get absorbed in the zone of easterlies.

Kelvin waves transport westerly momentum upwards, penetrates the zonal easterlies in the lower stratosphere and gets absorbed in the westerlies above and strengthens it. Thus it brings down the upper westerlies to lower levels. In contrast to this Mixed Rossby-gravity waves transports easterly momentum upwards. It penetrates zonal westerlies in the lower stratosphere and gets absorbed in the easterlies above and strengthens it. Thus it brings down the upper easterlies to lower levels. The alternating thin layer of upper westerlies and easterlies in the middle stratosphere are provided by semi-annual (six monthly) temperature and zonal wind oscillations around the equator. In this zone continuous supply of wave energy is provided at the bottom by the gravitational waves of the troposphere. Thus the four agencies, tropospheric gravitational waves, semi-annual mid-tropospheric zonal wind oscillations, Kelvin waves, Mixed Rossby-gravity waves, together create the QBO in the lower tropospheric zonal winds in the equatorial region.

Note: The longitudinally averaged zonal wind circulation in tropical stratosphere presents a semi-annual (half yearly) oscillation at upper levels (between 45-50 km) and an irregular quasi-biennial (about 26 months) oscillation at lower levels (between 25-35 km).

Stratospheric Sudden Warming (SSW)

SSW is dependent on the influence of lower and middle troposphere. To understand this consider first the winter stratospheric circulation. In polar and middle latitudes absorption of UV-radiation by zone considerably decreases after autumnal equinox. Consequently temperature decreases rapidly in winter stratosphere, particularly in polar latitudes as compared to equatorial zone. This creates and strengthens the westerlies in winter stratosphere. Due to temperature distribution a strong westerly jet (speed 80 mps) is located between lat 60 to 70 ° in the vicinity of stratopause in winter hemisphere and easterly jet (speed 60 mps) between lat 60 to 70° in the vicinity of summer hemisphere stratopause. Winter westerly jet stream is called polar night jet stream. In addition to these, there exists a semi-annual oscillations of temperature and winds in tropical upper troposphere (described in QBO) having maximum amplitude near the equator. It has been found in winter very large amplitude of planetary waves of troposphere penetrate into stratosphere and even mesosphere in association with blocking situation in troposphere. In about 7 to 10 days of these planetary waves penetration, the temperature in the stratosphere shoots up in polar regions and zonal westerlies may become easterlies. This warming may even be spread to equatorial region. It may be noted here that stratospheric warmings are accompanied by tropospheric blocking, but all tropospheric blockings are not accompanied by stratospheric warmings.

SSW is classified into two categories. Major warmings and Minor warmings. When westerlies are replaced by easterlies at about 10 h Pa (or 30 km) level it is called Major warmings, in which sudden temperature rise (Δ T) is of the order of 50 °C. When westerlies are not replaced by easterlies at about 10 h Pa level it is termed as Minor warmings, in which temperature rise of the order of 20 °C. The average difference between Major and Minor warmings is about 30 °C at 10 h Pa level. Each year will have either a Major or Minor warming or a series of Minor warmings, but the two types do not occur together on any one year. On an average Major warmings follow QBO. The process of evolution is not yet clearly understood. SSW propagates downward from about 45 km level to lower stratosphere. In middle latitudes warmings at about 40 km level is accompanied by cooling in Mesophere at about 65 km level.

Southern Oscillation

Sir Gilbert walker (1923-24) first found that rainfall over India increases/ decreases with the fall/rise of pressure over southern hemisphere. It was later found that several meteorological elements bear this Seesaw relation with temperature, wind, moisture, sea surface temperature (SST), ocean currents etc. However, for historical reasons, the term southern oscillation (SO) refers to see-saw like pressure oscillation over Tropical Indian ocean and Tropical Pacific ocean. The oscillation period is about 3.8 years.

Surface atmospheric pressure system over eastern Pacific ocean (off south America) and western Pacific (Indonesia) ocean breaks down every few years. If pressure rises over the western Pacific (becomes above normal), simultaneously pressure falls over eastern Pacific (becomes below normal) and vice versa. Because of this, trade winds weaken and during strong pressure reversals the surface easterly winds (west blowing winds) are replaced by westerly winds. Sea surface water over a broad equatorial Pacific area warms up and moves towards south America as a surge, called Kelvin waves. This warming continues for several months. At the end of warming period, the pressures over eastern Pacific rises (becomes above normal), simultaneously pressure falls over western Pacific (becomes below normal). As a consequence the equatorial Pacific now will have easterly wind instead of westerly. This Seesaw of surface atmosphere pressure (reversal) at the opposite ends of Pacific Ocean is called the Southern Oscillation. SO is related to inter-annual variations in SST, sea level pressure, 200 h Pa geopotential heights in northern hemisphere, tropospheric mean temperature in tropics and precipitation at selected stations in equatorial Pacific. Most of them are inter-related. SO period generally varies between 2.5 to 3.5 years. SO phenomena is of great importance in long range forecasting. The major anomalies, of weather round the globe in equatorial belt is useful in forecasting one year in advance. SO phenomena is now recognized as ocean-atmospheric interaction.

Indices of Southern Oscillation (SOI)

The most common indices of the SO are based on the atmospheric surface pressure difference (Δp) at two stations located at the opposite phases of SO region. For example

$$\frac{\Sigma \Delta p}{n} = \frac{\Sigma (P_t - P_{co})}{n} = \frac{\Sigma (P_{sa} - P_{pd})}{n}$$

$$= \frac{\Sigma (P_t - P_{pd})}{n}$$

where P_t = Surface pressure at Tahiti, (18 °S, 149 °W)

P_{co} = Surface pressure at Cocos Island

P_{sa} = Surface pressure at Santiago

P_{pd} = Surface pressure at Port Darwin (12 °S, 130 °E)

Averages are generally taken for December - January – February. The time series of Δp at these places during the above months indicate SO.

A negative value of SOI indicates higher pressure over north Indian ocean or west Pacific while comparatively low pressure over eastern/south east Pacific. This implies poor Indian monsoon.

SO Darwin (Broken line) and Tahiti (continuous line) five months average of sea level pressure departure. This clearly shows when the pressure at Darwin were above normal, the pressure at Tahiti were below normal and vice versa. See Fig. 18.3.

SO has irregular period 2 to 5 years averaging about 3 years. What controls 2-5 year period of SO are not known. The interaction of SO with other parts of global atmospheric pressure is not clear.

EL Nino

Along and off the west coast of south America, generally southerly cool Peru current blows. This cool current promotes upwelling, which brings up nutrient rich sea weeds. Because of this a large fish population particularly anchovies, is found in this area. Due to the presence of large fish population, a large number of sea birds visit this area and whose droppings produces phosphate rich deposits. This supports the local fertilizer industry. Every year around December month a warm south moving current replaces the cold north moving current. This generally lasts for few weeks and then the normal cold current replaces it. Since the warm current appears around Christmas, it is called EL Nino by the local people. In Spanish EL Nino means child, it refers to the

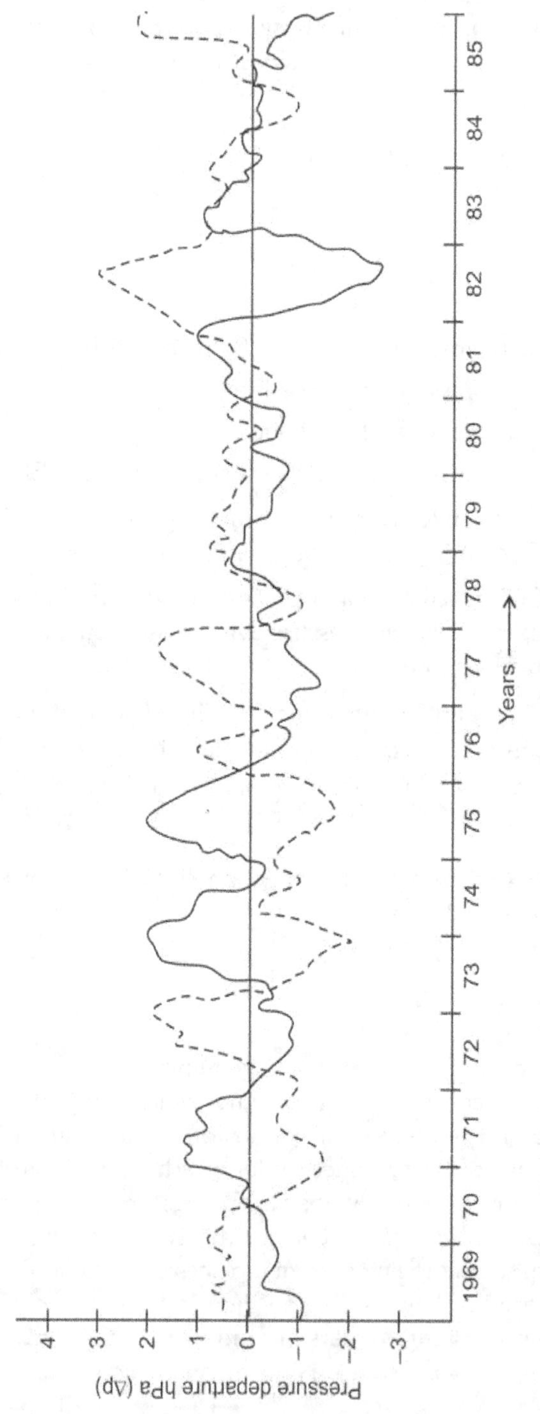

Five month average of sea level pressure departure at Drawin (broken line), Tahiti (continuous line).

Fig. 18.3 Southern oscillation.

Christ child. In some years the warm episode persists for several months and causes death to large fish population and marine plants. Scientists now call these warm persisting episodes EL Nino. Dead fish and birds litter the sea water in beaches of Peru, cause depletion of oxygen from water and hydrogen sulphide engulfs the area. This results in heavy fall in fish catch and effects the phosphate industries and thus effects the economy of Peru. For example, the EL Nino of 1972-73 reduced the anchovy fish catch to little more than one third of 1971-72. This greatly effected the world's fish meal production in 1972, which was used for poultry and live-stock feeding. As a consequence, alternative animal feed (Soyabeans) prices shot up by 40% in United States. The strongest EL Nino occurred during 1982-83. During this period equatorial east blowing (westerly) winds were stronger, which pushed the sea surface water. This caused elevation of the sea level in east Pacific and fall in sea level in the west Pacific. Eastward moving water warmed up as much as 6 °C in the equatorial Pacific than the normal. This warm water invaded the cold north moving water off Ecuador, Peru and extended thousands of kilometers westwards along the equator. It also spread northwards up to west coast of North America and south Canada. This episode added moisture into the atmosphere and increased the thunderstorm activity and caused flooding in Ecuador and Peru. The global influence of EL Nino and SO is not clearly understood but it is air-sea interaction. During 1982-83 EL Nino period, a severe drought prevailed over Indonesia, south Africa and Australia.

La Nina

EL Nino and La Nina are the two complementary phases of the SO. La Nina is associated with cooler SST in the equatorial Pacific with isolated atmospheric convergence zones. La Nina is also a phenomena of air-sea interaction in which atmospheric heating decreases, compression of convection reduces to very small areas of instability. This instability strengthens the trade winds and ocean currents.

EL Nino Index

It is generally defined as an average of anomalies of SST along the west coast of south America.

$$\frac{\Sigma\Delta T}{n} = \frac{\Sigma(SST - SSTa)}{n}$$

where   SST = Actual sea surface temperature

SSTa = Average of sea surface temperature

Averages are taken for the months of Dec, Jan, Feb. Annual time series for these months are compared for assessing severity.

Note : EL Nino index is not clearly defined like that of Southern Oscillation Index.

ENSO (EL Nino + Southern Oscillation)

The El Nino and SO phenomena studied separately for decades but recently discovered that they are linked phenomena of air-sea interaction. In the beginning SO is thought of as seesaw of pressure at two ends of Pacific and ELNino as SST off western coasts of South America. It is now treated both these phenomena as single and called ENSO. ENSO had both favorable and adverse effects on Indian monsoon. During twentieth century it showed more favorable effects as compared to adverse effects. ENSO phenomena has pronounced effects on inter-annual climate variability. Major ENSO episodes became the cause of temporary migration of rainfall regions of tropics leading to droughts over large area and floods/torrential rains over regions where normal rainfall is very low. ENSO cycle period seems to be about 12 to 18 months.

Studies indicate that ENSO events present a standing wave pattern in equatorial pressure field and SST. Some cases indicate standing wave characteristics. Sea level pressure anomalies indicate migratory type quasi-biennial period while stationary type shows 3 to 7 year period oscillation. Thus ENSO presents dual character of stationary and migratory waves. The general circulation model (developed by Barnett et al 1991) suggests the migratory nature of anomalies along equator from western Pacific to eastern Pacific.

Questions

1. Define Tropics. In what respects the weather systems are tropics differ from Extra--tropics?

2. What are the main features of pressure distribution in Tropics in the months of January and July? Write short notes on Equatorial Trough?

3. What is Meteorological Equator? Write some special features of Tropics?

4. Write briefly on (i) Trade wind inversion (ii) ITCZ (iii) QBO (iv) SSW (v) SO.

5. State Southern Oscillation Index and its effects?

6. Write briefly (i) ELNino (ii) LaNina (iii) ENSO.

CHAPTER 19

Jet Streams

Definition

According WMO, "A jet stream is a strong narrow current, concentrated along a quasi horizontal axis, in the upper troposphere or in stratosphere, characterized by strong vertical and lateral wind shears and featuring one or more velocity maxima. The speed of the wind must be greater than 30 mps or 60 kt ". Rossby gave the name "Jet stream" on the analogy of "gulf stream".

Generally a jet stream is thousands of kilometer (km) in length, hundreds of km in width and a few km in thickness. Vertical wind shear is of the order of 5-10 mps/km, lateral wind shear is of the order 5 mps/100 km. The path of the maximum speed is called axis and the tubular volume around it is called core.

In aviation jet streams are important weather phenomena. Jet stream models are used in flight planning and air traffic control. During second world war jet streams were encountered by army pilots and were used by them in navigation for bombing operation in far east.

Jet streams are embedded in the atmospheric wind flow in extremely shallow layers with strong wind speed. Horizontal dimension is measured normal to the jet axis. Wind speed reduces to (V/2) half its value at jet core (V) at about 200 km left of the jet core and 300 km right of the jet core (that is, cyclonic shear is more than anti cyclonic shear about the jet core). In vertical, the half value (V/2) speed occurs at about 5 km above or below the jet core (Maximum wind). Fig. 19.1 Turbulence (or perturbation) components are superimposed both in horizontal and vertical directions. The basic current varies gradually with time. The necessary and sufficient conditions for the existence of jet stream are: (i) a meridional temperature gradient (about 3 °C per degree latitude), (ii) angular momentum imparted by the rotation of the underlying surface. These conditions are verified by the geophysical model experiment. In a rotating dishpan, temperature gradient is maintained from the rim to the centre. Atmospheric jet stream may be simulated by the rate of rotation and the temperature difference between pole and equator.

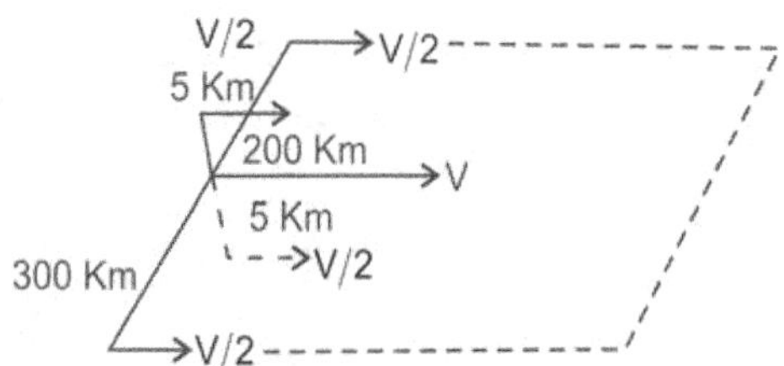

Fig. 19.1 Wind field around jet core.

In a well developed jet stream, positive vorticity maxima exists at the cyclonic side of jet core, which is caused by the horizontal wind shear. The vorticity value is further added up if the jet contains elongated trough. Jet maxima is considered ageostrophic because it is attributed to the effects of geostrophic perturbation motions, which are superimposed on ageostrophic basic current (that is a current which obliges the gradient wind equation).

Jet-Stream Types

Jet streams essentially are of two types 1. Polar Front Jet stream (PFJ), 2. Sub tropical Jet Stream (STJ). In addition to these, there are jet streams called 3. Tropical Easterly Jet (TEJ), 4. Low Level Jet (LLJ), 5. Polar Night Stratospheric Jet (PNSJ) and 6. QBO jet in lower equatorial stratosphere.

1. Polar Front Jet Stream

The sub-tropical westerlies and the polar easterlies which have contrasting air mass characteristics meet at sub-polar lows. These are called polar Fronts. Polar front jet stream is located near polar front (Lat 25–35 °N) at an altitude of about 9 km (300 h Pa), attains maximum intensity of more than 100 kt in winter. In summer it is less intense and moves north wards to about Lat 35-45 °N. The region below and above the PFJ extending to ground and stratosphere is a baroclinic zone. The zone above PFJ is of negative baroclinicity. The strength of PFJ is closely related to the frontal zone strength and position in the troposphere. During summer the baroclinic zone is less developed and moves north. It may be noted here that the temperature gradient is not evenly distributed over all latitudes from thermal equator to the pole, but it is concentrated around the frontal zone. PFJ is irregular or discontinuous zonally but the main jet stream core is associated with the Rossby long waves. In the vicinity of PFJ core, cyclonic shear (left of the core) is more than anticyclonic shear (right side of the core), where absolute vorticity will have smaller positive values. The region of maximum cyclonic shear is observed slightly below the jet core level and maximum anticyclonic shear slightly above the core level. Jet core maximum generally does not coincide with the tropopause. Jet core is observed to be right of tropopause break.

Note :
Definition

Tropopause is the lowest level at which the lapse rate decreases to less than or equal to 2 °C/km, provided that the average lapse rate of the 2 km layer above does not exceed 2 °C/km.

The height of the tropopause is higher in summer than in winter. The height of the tropopause near equator is about 16 km which decreases to 8 km near the poles.

Sub-tropical Jet Stream (STJ)

STJ is similar in structure to PFJ, but occurs at higher potential temperature than PFJ. It is generally located above sea level sub-tropical ridge line (lat 25-30 °N) at an altitude of 12-15 km (200-150 h Pa) and attains the maximum speed about 250 kt (over Japan and south Indian ocean). Due to latitudinal migration STJ enters the zone of PFJ and as a result it loses its separate identity of structure. Infact STJ is generally attached to the remnants of PFJ baroclinic zone. STJ is caused by the import of angular momentum from equator side (south to north). According to Palmen STJ would form on the poleward edge of the meridional circulation cell (shown in Fig. 19.2). It implies that STJ formation is greatly influenced by meridional and planetary waves.

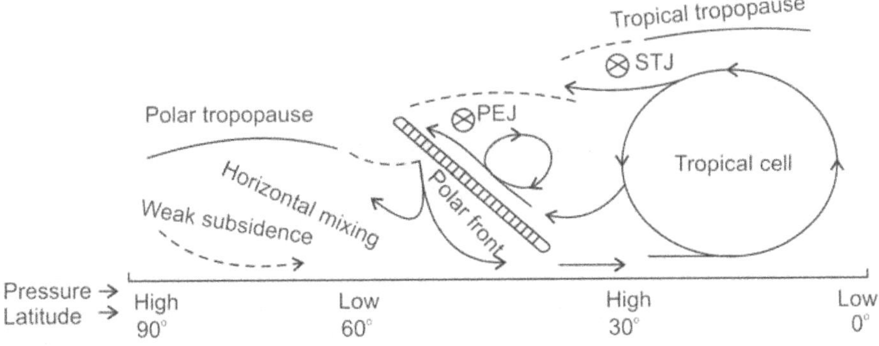

Fig. 19.2 Northern hemisphere mean merodional circulation.

STJ is more intense in winter than in summer and it is persistent in direction and location. STJ also occurs between tropopauses breaks, lies below the higher tropical tropopause about 150 km to the right of the break line. STJ is dependent on temperature gradient of upper troposphere while PFJ is dependent on steep temperature gradient caused by the interaction of polar and tropical air masses. Compared to PFJ, the ratio of vertical to horizontal dimension is smaller and horizontal wind shear increases with the increasing speed at jet core. The strongest cyclonic shear (left side of jet core) is observed below the jet core and anti-cyclonic shear (right side of the jet core) slightly above the core. In the lower altitudes PFJ transforms to STJ when the baroclinicity below 7.5 km (400 h Pa) is completely removed.

Tropical Easterly Jet stream (TEJ)

TEJ is found over India and north Africa during summer at an altitude of 14 to 16 km (150-100 h Pa), with speed 60-100 kt between latitudes equator to 15 °N and longitudes 50 – 80° E. TEJ is not observed over tropical Atlantic and Pacific. The average easterly wind speed at 200 h Pa is about 50 kt. During summer easterlies are found in tropics around the hemisphere but they do not reach the jet speed. During winter the easterlies are weak and found about maximum speed of 20 kt on average over Indonesia and over central Africa. They can not, however, be called jet streams. TEJ is related to the steep lateral temperature gradient, with the upper air becoming progressively cooler to the south. The most important characteristic of TEJ is the intense summer monsoon rainfall which is confined to the right entrance (RE_t) and left exist (LEx) of the jet core Fig. 19.3. Monsoon rainfall in general increases with the intensification of TEJ and decreases with the weakening. In general RE_t and LEx of a jet core (be Easterly or Westerly jet) are regions prone to disturbed weather in the northern hemisphere. In the southern hemisphere the reverse is true.

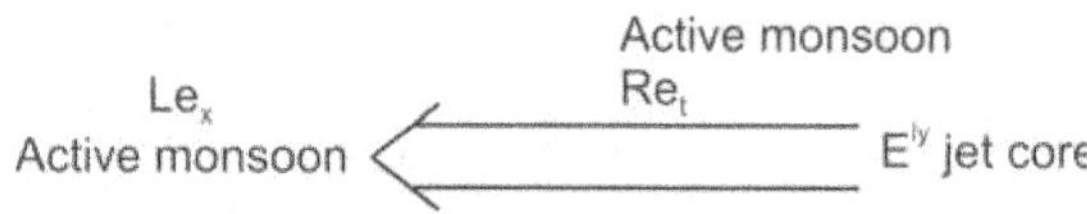

Fig. 19.3 TEJ and monsoon activity.

Low Level Jet Stream (LLJ)

There are strong winds that occur between 1 to 3 km asl but they do not mostly satisfy WMO definition of Jet stream. These are strong winds as compared to monthly and seasonal average winds. These strong winds are referred to as Low Level Jet streams. LLJ are effectively influenced by orography, friction, diurnal heating. They are limited in horizontal and vertical extents and attain average speed about 50 kt. LLJ are prominent over east African coast and over Peninsular India during summer. The other locations are west central America (USA), Nambian coast in south Africa. Favorable locations are mountain slopes, continental coasts and narrow mountain gaps (example Palghat gap in south India).

East African coast LLJ appears in summer months. This originates near Mauritius, moves in a southeasterly direction, crosses the equator and moves northwards along the east coast of Africa (Kenya, Somali). It then moves in a south-westerly direction in Arabian sea to west coast of India. It acquires maximum speed in the late night or early morning. According to observational evidence, LLJ along Kenya and Somali acquires the speed of 60 Kt or more (satisfies WMO definition of jet stream) on many occasions. The strongest current is found in July and August at an altitude of 1 to 1.5 km asl. Mean Maximum wind speed is about 30 kt near north Malagasy and 35 kt near Somali. It has steady southerly direction (190°).

The LLJ over West Central USA

This occurs in summer. It runs parallel to the Rocky mountains (direction SSW) over the eastern slopes at an altitude of about 1 km, with a mean maximum speed about 25 mps. It also acquires maximum speed during late night or early morning.

The LLJ along Peru coast runs parallel to the western slopes of Andes from southerly direction at a very low altitude of 0.3 km asl. It acquires a maximum speed of about 12 mps for most part of the year. Because of its very low altitude it touches the sea water, cools it and aids upwelling along the Peru coast. This subsidence of air is presumed to be the cause of desertification along the coast. The LLJ along the Namibian coast is similar to the Peru coast LLJ. It moves along the west coast of south Africa in south-east Atlantic, from southerly direction. It also causes upwelling (off the coast) in the cool Bengula sea current.

Marsabit (Africa), and Palghat gap (South India) winds are examples of wind channeling effect. In both cases wind acquires jet steam winds speed when wind blows perpendicular to the barriers and passes through the mountain gaps. Such orographic wind channeling effect may be found in many places. These gaps are very useful places for deriving wind energy.

Jet Streams over India

The principal jet streams over India are (i) STJ and (ii) TEJ. However during monsoon period on some occasions LLJ is observed over Peninsular India which is the extension of LLJ found off Somalia coast in Africa.

(i) STJ

It is a westerly jet stream. Generally occurs during non-monsoon months (October to May) over north India. It is strongest in winter months. Its mean position is about Lat 27 °N, altitude 12 km (200 h Pa) and mean wind speed at core level is 100-120 kt. The southern most position of STJ over India is Lat 22 °N, occurs during February with wind speed 150–200 kt, direction mainly westerly. During monsoon months STJ migrates to north of India, weakens and disappears. The mean speeds at jet core are the lowest 60 to 75 kt in October and May and altitude about 12-14 km.

STJ over India is confined to troposphere only. It appears there are two maxima cores. One is located over heat low area of West India, Jodhpur (27 °N, 72 °E), at an altitude of 12 km and mean speed about 100 kt. Another is located over north east India, Gauhati (26 °N, 92 °E) at an altitude of 10.5 km, mean wind speed about 100 kt. In both cases direction is westerly. Mean wind speed of 60 kt or more is found north of lat 20 °N at an altitude of 12 km. North of lat 23 °N, this speed begins at 300 h Pa level over west and in Assam (over east) at 400 h Pa. In these latitudes wind speed of 60 kt or more extends up to 150 h Pa level. Horizontal wind shear is about 5 kmph/100 km south of the jet core. Vertical wind shear is about 20 kmph/100 km at 3 km below the jet core maxima. Below the jet core the lapse rate of temperature is more to the south than north. Above the jet core the lapse rate is reverse and at the jet core it is slack. Reversal of meridional temperature gradient exists near the jet core level.

(ii) TEJ

It is very prominent during monsoon period (June–Aug) over India. It extends from south China sea to Africa through Malaysia, Indian penisula. It covers more or less global monsoonal area. TEJ occurs between 14-16 km close to the tropical tropopause, along Lat 15 °N, with mean wind speed of about 70 kt. TEJ is not found over tropical Pacific and Atlantic Oceans. It weakens to less than 50 kt over India in September (at an altitude 14-16 km). When TEJ appears over Indian area, the STJ over north India disappears. This occurs more or less simultaneously. During break monsoon conditions TEJ moves north ward up to lat 20 °N. See Fig. 19.4.

In July winds are easterly 10-20 kt at 300 h Pa level south of 27 °N and westerly north of 32 °N with transition in between. Over the peninsular India easterlies strengthen with height up to 150 hPa/100 hPa where it attains maximum wind speed. At Chennai mean wind at 150 h Pa is 085°/60kt, at 100 h Pa 085°/80 kt and at 070 h Pa 080°/

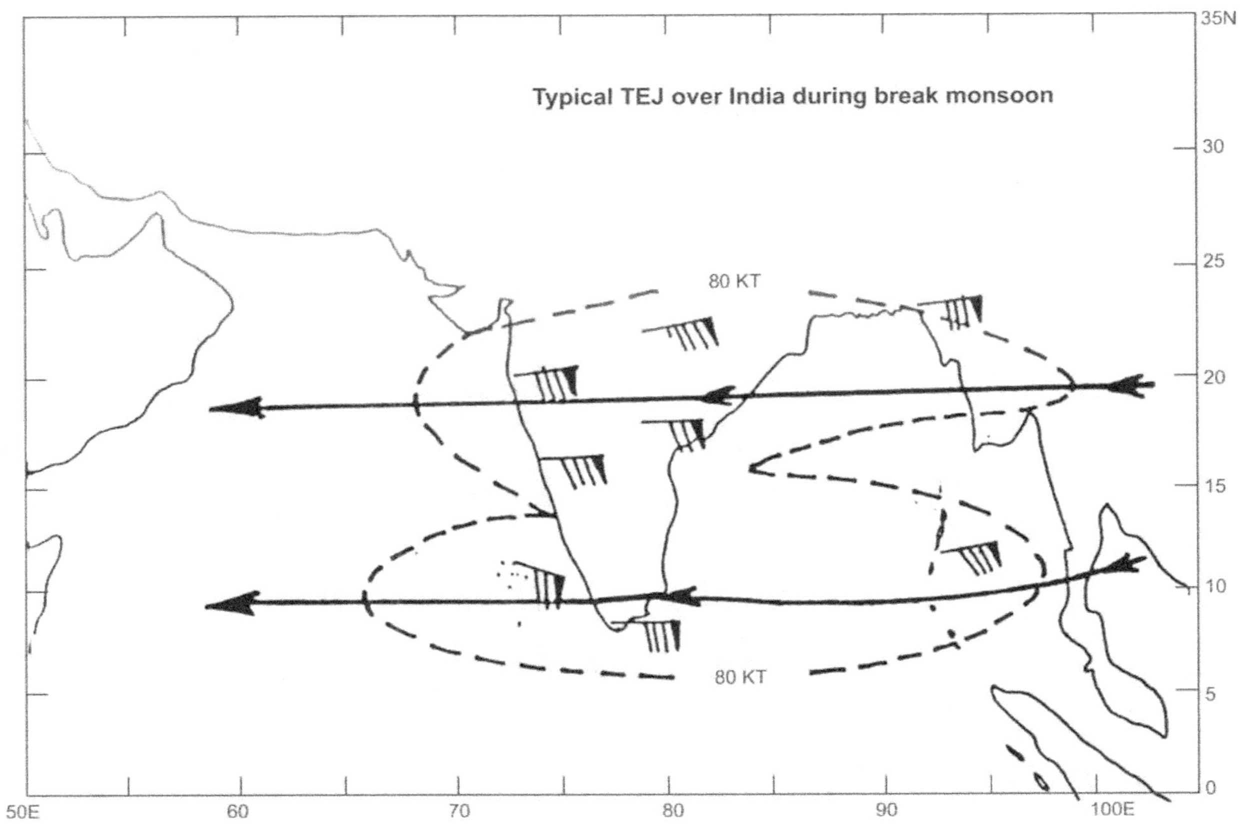

Fig. 19.4 Maximum wind 0000 UTC of 14.8.88.

25kt. The easterly jet core occurs over India at lat 13 °N and height 100 h Pa. Easterly regime at 150 h Pa level extends south wards up to 5°S and north wards up to lat 25–30 °N. AT Delhi (28 ° 30' N) it is easterly 10 kt, while at Gan island (1°S) 55 kt. Easterlies do not extend north of 30 °N. Ambala (30° 30' N) shows westerlies between 150 h Pa-70 h Pa.

The cause of easterly Jet stream over Peninsular India is the effect of south ward decrease of temperature over the region in the entire troposphere. In fact it is the position of the Sun 20 °N at this time of the year. Horizontal wind shear north of the jet core (100 h Pa level) is about 7 kmph/100 km while south of the core 9 kmph/100 km. Vertical shear between 300 h Pa and 100 h Pa varies between 15–20 kmph/100 km and the maximum shear occurs 5 km below the core. Above the core the shear between 100 h Pa and 70 h Pa is about 22 kmph/100 km. In September strong easterlies are found south of lat 20 °N. At Chennai a maximum speed of 50 kt is found at 150-100 h Pa.

The temperature distribution is characterized by (i) large temperature gradients exist below jet level (ii) a little or no gradient of temperature at the jet level, and (iii) a reversal of temperature gradient above jet level. Temperature anomaly charts indicate that Jet is located in the Col region.

PNSJ

The Polar night Stratospheric Jet stream is located in the stratosphere around Arctic and Antarctic circles. It is westerly jet located at altitude of 35 km and lat 65 °N/S and speed 150-250 kt. PNSJ in southern hemisphere is comparatively stronger than the counterpart in the northern hemisphere. During SSW this weakens greatly and may be replaced by weak easterlies. (SSW : Stratospheric sudden warming)

QBO Jet in Lower Equatorial Stratosphere

During July-Aug easterlies in the upper troposphere are strong. These penetrate into lower stratosphere. If the QBO in July-August coincides in pahse with easterly regime, then a jet appears in the lower stratosphere. Similarly a westerly jet appears in lower stratosphere during winter season.

Tropopause

Definition

According WMO recommendation, "The first tropopause is defined as the lowest level at which the lapse rate falls to 2 °C/km or less, provided that the average lapse rate of the 2 km layer above does not exceed 2 °C /km".

Above the first tropopause if the average lapse rate between any level (and at all higher levels) within 1 km exceeds 3 °C/km, then a "second tropopause" is defined by the same criteria as for the first tropopause. This second tropopause may be either within or above the 1 km layer. The tropopause formation seems governed by water vapour and ozone. The radiative process of troposphere is governed by water vapour while the stratosphere is governed by ozone. Water vapour splits ozone into oxygen. The lower tropical troposphere holds more moisture (water vapour) than extra tropical and polar regions due to convective activity. Because of this more water vapour is pumped into tropical troposphere. Hence the tropical ozone regime of stratosphere has higher base. Tropopause is a surface discontinuity for moisture in the atmosphere (above the tropopause RH is very low about 3%). Because the presence of water vapour the tropopause height varies from equator (16 or 17 km) to pole (8 km).

There are three types of tropopauses (i) Polar tropopause, (ii) Extra-tropical or middle latitude tropopause and (iii) Tropical tropopause. Polar tropopause occurs north of lat 70 °N and located north of the polar front jet stream. Average height is 8 km. Polar tropopause is diffused, in which inversion is less marked or absent. Middle latitude tropopause occurs between lat 30–70 °N (that is between sub-tropical and polar jet streams). Average height is 11 km. In this tropopause there is sudden change of lapse rate, at times becomes isothermal but it has no inversions.

Tropical Tropopause occurs between equator and the subtropical Jet stream. Its average height varies between 12 to 17 km . It is well defined by the inversion temperature above it. It is found throughout the year south of lat 30 °N. In North of Lat 30 °N, tropical type tropopause is found in monsoon months. In winter when the subtropical Jet stream appears over India, the extra-tropical tropopause is found over north India. In such situations tropical tropopause is found at higher altitudes (16 km) with marked inversion and extra tropical tropopause below it (10-11 km). Such double tropopause is common over north west India during Jan, Feb months.

Note :

1. Fronts, inversions and tropopauses may be treated as discontinuities. The sloping surfaces of tropopauses, inversions and fronts may be assessed for forecasting. According to Sutcliff: "For prediction as distinct from analysis, fronts are valuable mainly for two reasons: their motions may be extrapolated with some confidence for short periods and when well defined they tend to give similar weather in similar synoptic situations".

2. Weather in tropics tend to be concentrated around air mass discontinuities (where convergence takes place).

Westerly Waves

We know that Rossby (or long) waves move from west to east in middle and higher latitudes. They are sinusoidal, 4-5 waves cover a hemisphere and move slower than average wind speed in middle and upper troposphere. Westerly waves are superimposed on these long waves and have shorter wave length and amplitude. Westerly waves move faster and are completely guided by the movement of Rossby waves. Westerly waves are not seen on surface charts but observed in upper levels as waves. On surface chart it may be seen as trough and some times we may get a closed isobar if drawn at 1 h Pa interval, when the wave is well marked.

Weather disturbances originate in these westerly waves or these waves act as breeding places for extra tropical depression/cyclones. The area of convergence lies ahead of the trough line where convective clouds, broken to overcast sky with heavy precipitation is seen. In the rear of the trough line generally fair weather cumulus is observed.

Easterly Waves

In tropical zonal easterlies sinusoidal oscillations often develop, particularly during monsoon period. Troughs in mid-tropospheric easterlies may form closed systems, however they are not seen as a closed low on sea level pressure chart. The main characteristics of these easterly wave troughs are: they form in ITCZ, have wave length 200-2500 km, period 3-4 days, speed of propagation 8 mps (or 6-7 longitudes per day). These waves generally originate between lat $15°$ and equator, move east to west between altitudes 600 h Pa–200 h Pa. These waves have a tilt of northeast to southwest in the northern hemisphere and northwest to southeast in southern hemisphere. They are located south of the easterly jet core. These waves are steered by tropical easterly current. The waves become weak, stagnate or even dissipate over land area.

Easterly waves are found in tropics on both sides of the equator throughout the year. They are marked during monsoon months June-September. Tropical disturbances originate in these waves. Easterly waves seems to originate in west central Pacific as weak low pressure areas and propagate. These disturbances develop into depressions/cyclones in the vicinity of ITCZ. The remnants of west Pacific systems enter Bay of Bengal after crossing Myanmar. They rejuvenate and under favorable conditions they transform into monsoon depressions over Head Bay of Bengal. Easterly waves may cross Peninsular India, enter Arabian sea as weak systems. From there they appear over north Africa where they gain intensity. They cross Africa enter into central Atlantic, move to west Atlantic (Caribbean). Due to long travel over Atlantic, these easterly waves intensify and develop into Hurricanes. Some cross central America and enter into east central Pacific as weak systems. From there they move to west Pacific and develop into typhoons in south China sea. During its global journey these easterly waves undergo a lot of changes in strength, size, horizontal and vertical tilt, cloud structure etc. In general southern hemispheric easterly waves are similar to northern hemispheric easterly waves but they are comparatively less marked except in its summer (Dec-March).

It is generally accepted that the energy for easterly waves is provided by both barotropic and baroclinic instability. A model of easterly wave is shown in Fig. 19.5.

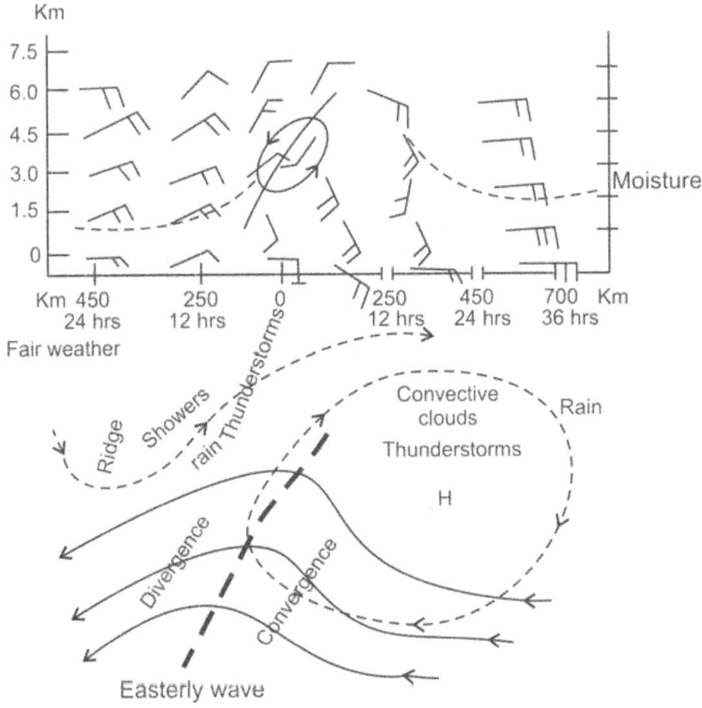

Fig. 19.5

Cloud and weather sequence over sea during the passage of typical easterly wave, are given below.

1. **In Ridge :** Trade Cu average height, no precipitation.

2. **Ahead of Trough :** Cumulus humilis, strong haze, no pptn.

3. **Close to the Trough Line :** Large cumulus, Ci, Ac, scattered showers, improvement of visibility

4. **At Trough Line :** Cumulus congestus, broken to overcast Ci, Ac, frequent rain or showers

5. **Rear of the Trough Line** : Winds veer, decrease of temperature, Cumulus congestus, Cb with layers of Sc, As, Ac and Ci. Frequent moderate to heavy showers with light rain between showers. Thundershowers.

6. **Eastern Outskirts** : large Cu, Cu-congesturs, isolated Cb cells, SCT, Sc, Ac, Ci. Moderate showers-decreasing.

   (SCT = Scattered)

Questions

1. Define a Jet Stream and Describe salient features about its structure.

2. What are the types of Jet Streams. Describe briefly about PFJ, STJ and TEJ, LLJ.

3. Describe the salient features of STJ and TEJ over India.

4. Define Tropopause. Describe briefly about tropopause types.

5. Write briefly on (i) westerly waves, (ii) easterly waves.

6. Draw a typical easterly wave and write about the cloud and weather sequence ahead, at and in the rear of easterly wave over sea.

CHAPTER 20

Monsoons

The word "Monsoon" is derived form the Arabic word "Mausam" meaning season. It connotes seasonal reversal of winds and rainfall. The economy of tropical countries dependent on agriculture and which largely dependent on seasonal rains from ancient times. In ancient times the Phoenicians, Greeks, Romans and Arabs had maritime trade with Kerala. They depended on wind-driven sail ships. Hippalus (45 AD) a Greek mariner, who started around summer solstice (June) took about 40 days to reach Kerala from Red sea across Arabian sea. In the return voyage, which started about winter solstice (December) he took about the same period from Kerala to Red sea port. In both side voyages he found winds were favorable. These favorable winds were called "kala varsha" in Kerala and "Hippalus winds" by Greeks. However these seasonal reversal of winds with rains are now called "monsoons" in the scientific literature. The monsoons are now defined based on the following criteria.

1. Shift in prevailing wind direction by more than $120°$ between January to July.
2. Season : June to September.
3. Monsoon region extends from lat 35 °N to 25 °S and long 30 °W to 170 °E.
4. Mean wind speed in the month : 3 mps or more.
5. In every two years, in each month there is less than one cyclone or anticyclone in a 5 square latitude and longitude.

The characteristic feature of monsoon air mass is moist maritime and extends to great depths (3 to 9.5 km or 700 h Pa to 300 h Pa). From ancient times the economy of monsoon region is dependent on agriculture, which is heavily dependent on monsoon rainfall. More than 80% of the annual rain fall occurs in this region during monsoon period. As a consequence, the onset of monsoon, its onward progress, variations in intensity of rainfall, duration of monsoon, breaks in monsoon have a tremendous impact on the economy/business of the region. The greatest boon of this region is the cyclic behavior of monsoon over a very large area. The study and prognosis of monsoon rainfall and variability has acquired paramount of importance in these regions. The normal monsoon forecast has a magic spell on business community as a whole. In fact the occurrence of normal monsoon both in space and time is a rare phenomena.

The time and space variation of rainfall during monsoon is an inherent phenomena but the forecast particularly below normal rainfall has adverse effects on business and agriculture but this seems a routine phenomena considered over large area and time. Keeping this in view a number of scientists tried to unravel the causes of variations in monsoon rainfall. First we shall consider all these aspects of monsoons over Indian sub-continent.

Monsoons Over India

The description of monsoon over land and sea differs. Over sea areas (Bay of Bengal and Arabian sea) the word monsoon indicates prevailing surface wind while on land it indicates rainfall. According to IMD the strength of monsoon over sea/land areas is described below.

Over sea area

Description	Wind speed (in kt) reported or inferred to be existing
Weak monsoon	Upto 12 kt (23 kmph)
Moderate monsoon	13 to 22 kt (24 – 42 kmph)
Strong monsoon	23 to 32 kt (43 – 60 kmph)
Vigorous monsoon	33 kt or more (61 kmph or more)

Over land area

Description	Rainfall compared with long period average called normal at a place or sub–division.
Weak monsoon	R.F less than half the normal
Normal monsoon	R.F. ½ to less than 1 ½ times the normal
Active/strong monsoon	(i) R.F. 1 ½ to 4 times the normal (ii) R.F. in at least two stations should have 5 cm if the sub–division is along west coast of India and 3 cm if it is elsewhere. (iii) R.F. in that sub-division should be FWD or WD.
Vigorous monsoon	(i) R.F. more than 4 times the normal (ii) R.F. in at least two stations should be 8 cm if the sub-division is along the west coast and 5 cm if it elsewhere. (iii) R.F. in that sub-division should be FWD or WD.

Normal Rain Fall (R.F)

It is the average R.F. of long period of a station.

Isolated (one or two places) R.F. over a sub-division means less than 25% of the stations of the sub-division reporting rainfall.

Scattered RF (or at a few places) means 25 to 50% of the stations of the sub–division reportng rainfall.

Fairly wide spread (FWD or at many places) stands for 50 to 75% of the stations reporting R.F in the sub-division

Wide spread (WD or at most places) stands for 75 to 100% of the stations reporting R.F in the sub-division.

Note

(i) In any sub-division where the percentage of hill stations is high, then the hill stations must also be taken into account for describing the activity of monsoon. In other sub-division the hill stations will be excluded.

(i) The monsoon activity will be described in all the sub-divisions of north east India as is done for sub-division of other regions.

(ii) The monsoon activity need not be described over Bay islands and the Arabian sea.

Monsoon dynamics is viewed (like land and sea breeze circulation) as the cause of differential heating of vast land (Asian continent) on the north and the sea (Indian Ocean) in the south. The Coriolis force in the tropics cannot balance large pressure gradient. Except in disturbed condition, the pressure distribution is of little consequence in monsoon forecasting as compared to wind circulation. We shall consider onset of monsoon over Kerala, then its steady march over rest of India, and withdrawal.

On regional scale, the onset of summer monsoon (or southwest monsoon) is closely related to the cross equatorial wind flow (or inter hemispheric flow) and the transition of winter to summer. Onset of monsoon means commencement of summer rains, however a precise definition is remained as a debatable point. Monsoon is basically a moist air stream. Its arrival and further advancement is a gradual process which builds up moisture in the lower troposphere. Similarly, withdrawal of monsoon is also a gradual process but it is much faster than onset. In a nut shell, there is a great rhythm or cyclicity of onset and withdrawal of monsoon phenomena. Broadly speaking monsoon is an annual oscillation of atmosphere with great regularity.

IMD has fixed the normal dates of onset of monsoon based on the sharp increase in pentad (five days) rainfall Fig. 20.1 and changes in wind circulation,

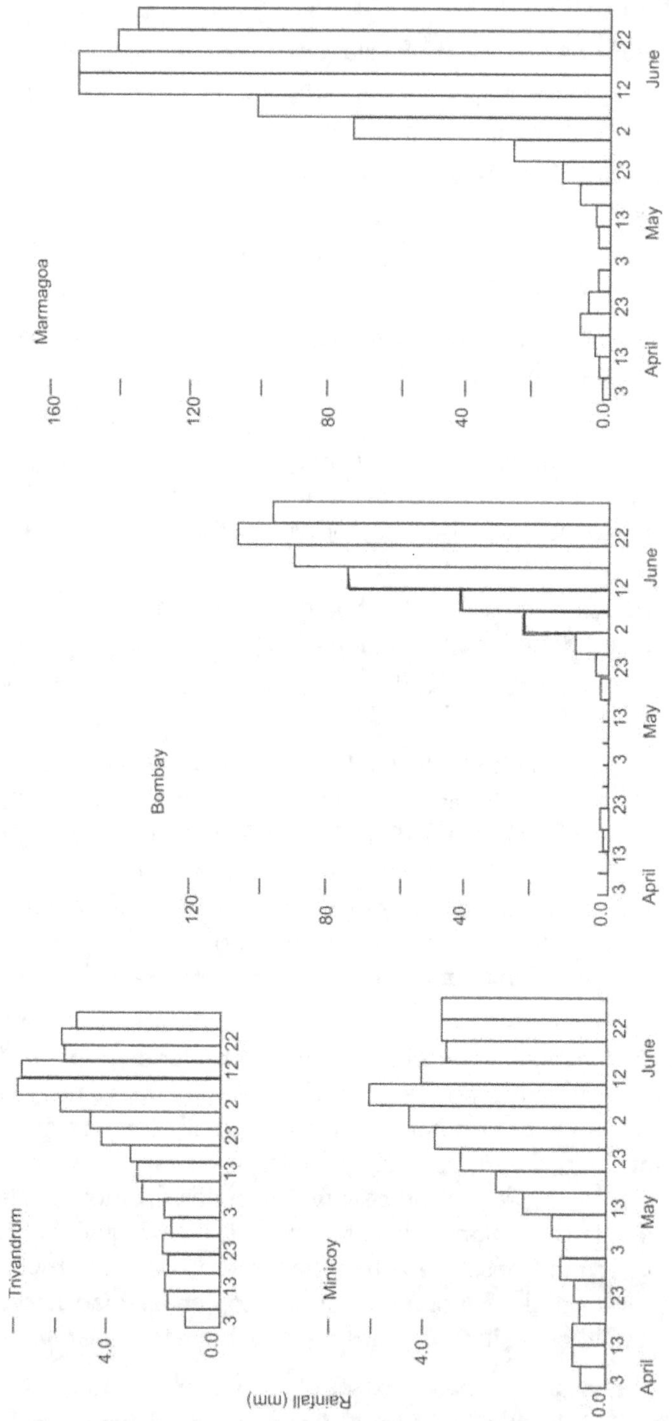

Fig. 20.1 Normal pentad rainfall.

particularly wind directions and moisture. The normal date of onset of monsoon in Kerala is 1st June and the standard deviation is 8 days. During individual years onset varies. It means that if the sharp rise in pentad rainfall occurs between 24 May and 9 June, it is considered normal . But if it occurs before 24 May it is called early onset and if it occurs after 9 June it is called delayed (or late) onset. The earliest onset of monsoon was recorded on 11 May in 1918 and the most delayed onset was 18 June in 1972. The normal dates of onset and withdrawal of monsoon is shown in Figs. 20.2, 20.3 and 203(a).

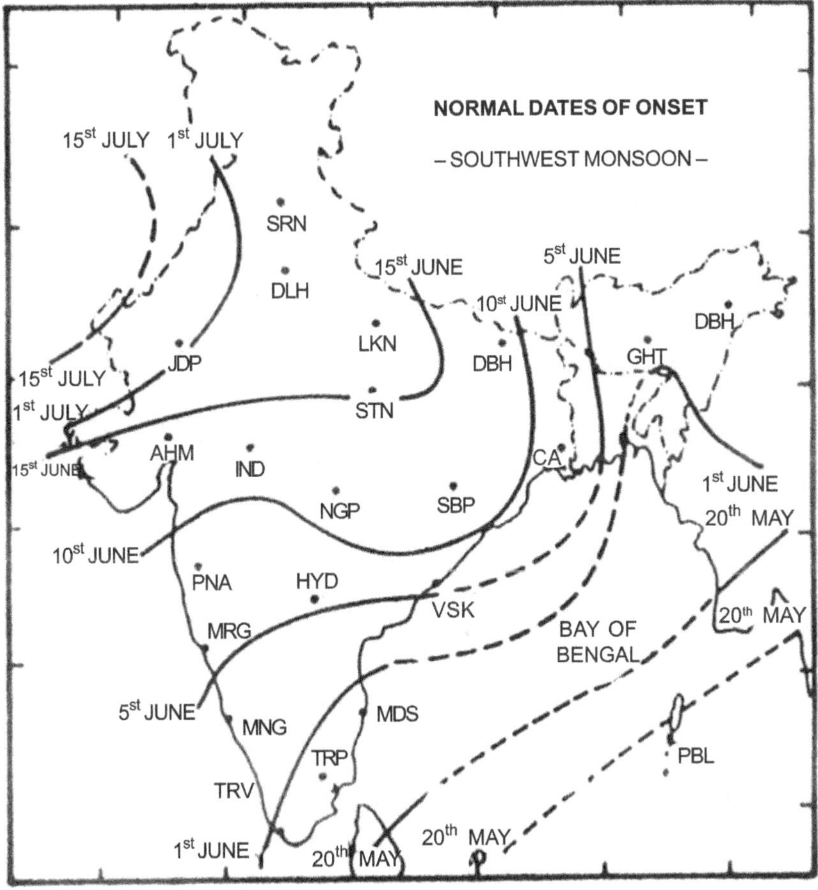

Fig. 20.2 Normal dates of onset of southwest monsoon over India (Rao Y.P 1976).

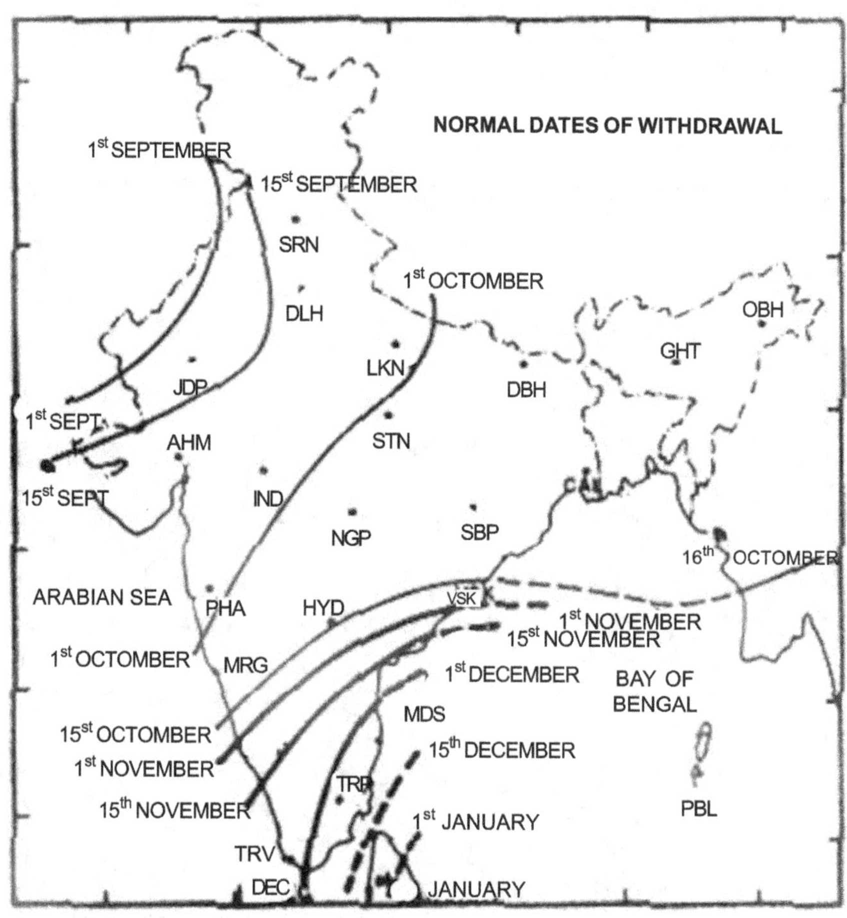

Fig. 20.3 Normal dates of withdrawal of southwest monsoon over India (Rao Y.P. 1976).

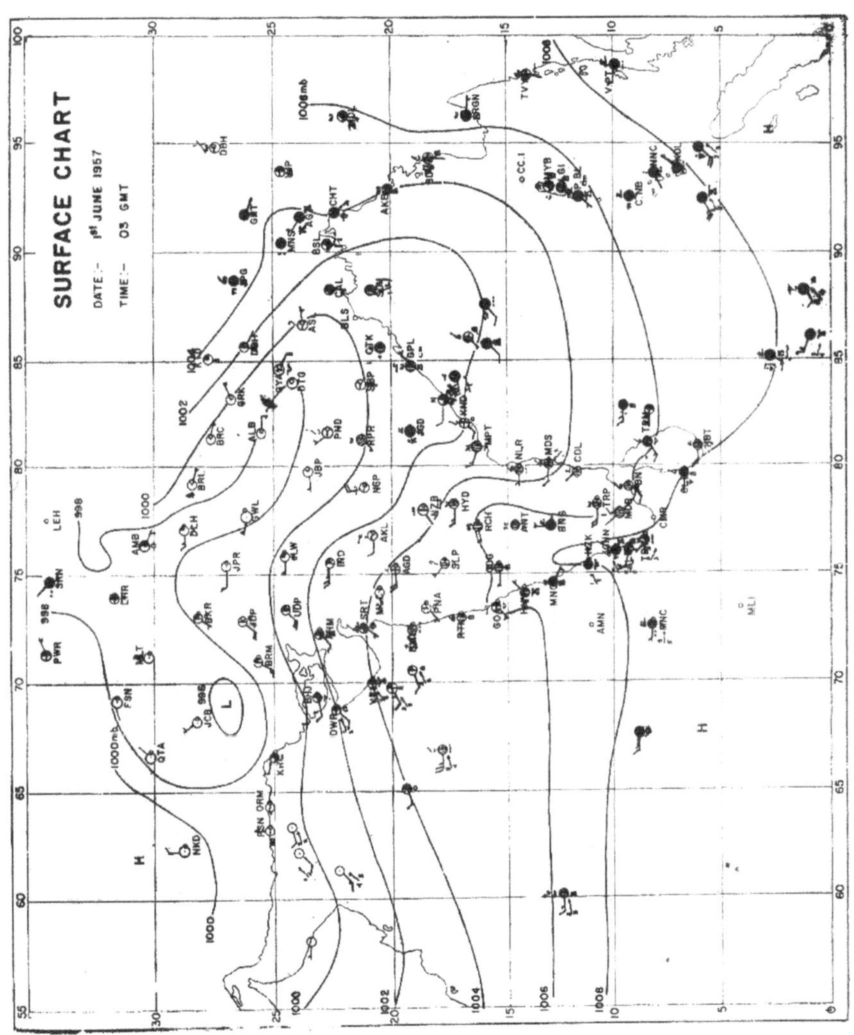

Fig. 20.3(a) Onset of monsoon typical chart.

Declaration of Onset of Monsoon Over Kerala

If five stations out of seven stations get rainfall (RF) 1 mm or more for two consecutive days, on second day monsoon will be declared advanced over Kerala. If monsoon is confined to south of lat 13°N, monsoon Recession can be declared if 3 or more stations out of seven report no rainfall for three consecutive days. In 1987 the onset of monsoon was delayed by five days over peninsula south of lat 20 °N and east of long 85 °E. But over majority of rest India it was between 5-25 days and in some pockets 25-35 days. This delay and subsequent monsoon activity proved 1987 to be a year of severe drought.

The phenomena of onset of monsoon over India is one of the spectacular events of nature. It has a pulsatory and fluctuating nature during onset and advancement over India. There is large variation of rainfall from year to year and place to place. The definition of onset of monsoon based on pentad rainfall increase has good reasons for an acceptance. Considering the monsoon season period as a whole one day's rainfall comparsion seems too small, while decadal (10 days) period is large. Consequently five day has been chosen. It is difficult to discern the pentad pressure changes at different places but it is easily recognizable in pentad changes in rainfall. This yard stick of measure of onset of monsoon however is difficult in northeast India, where considerable premonsoon rain fall occurs due to thunderstorm activity in May. In this area, the onset is guided by the notion of steady march of monsoon in one direction. There may be periods of stagnation of northern limit of monsoon for a number of days. Further advancement is generally associated with a fresh small scale or transient disturbance. Each disturbance pushes the northern limit of monsoon further than the previous position. The rainfall belt also moves forward in the northerly direction. However in case of Gangetic valley, the rainfall belt rather moves from east to west along the axis of monsoon trough following the movements of low pressure systems that form over Head Bay of Bengal.

Satellite pictures called cloud imageries, show that there is a systematic movement of equatorial cloud bands northward during onset and advance of monsoon and confirms the cross equatorial flow. Weekly mean cloudiness data clearly shows the northward movement of cloud bands. It is found, when the cloud band reaches about lat 10 °N, the onset of monsoon occurs over Kerala. At about the time of onset of monsoon over Kerala, the subtropical westerly jet stream moves away from north India and simultaneously an easterly jet stream is observed along lat 12-13 °N in mid and upper troposphere. The onset and advancement of monsoon is mostly (more than 75% of the occasions) is associated with synoptic situations such as sea level troughs, formation and movement of low pressure systems (lows, monsoon depressions etc), upper air cyclonic circulations over Arabian sea, Bay of Bengal and over sub-continent. These systems generally move westwards.

Withdrawl of Monsoon

Monsoon loses its activity over remote north India (Jammu and Kashmir) by the end of August. ITCZ recedes back, and sub-tropical ridge appears and establishes over northwest India indicating the process of withdrawal of monsoon. The normal withdrawal of monsoon dates are shown in Fig. 20.3 which commences from 1st September and by 15 October withdrawal completes over India baring Tamilnadu adjoining few districts of Andhra Pradesh, where onset of northeast monsoon commences (we shall discuss this separately). The following Fig. 20.4 and 20.5 show the schematic representation of summer and winter monsoon [schematic representation. Solid line indicate lower tropospheric circulation, dotted line represent upper tropospheric circulation].

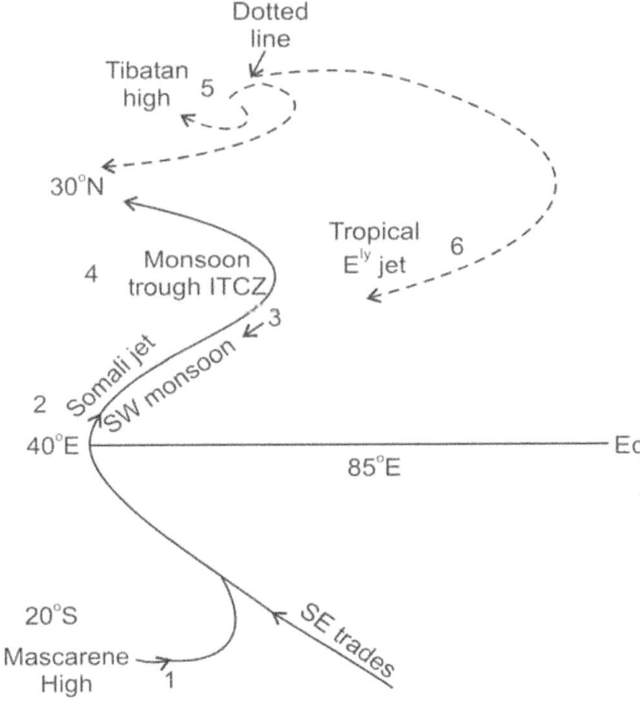

Fig. 20.4 Schematic representation of summer monsoon.

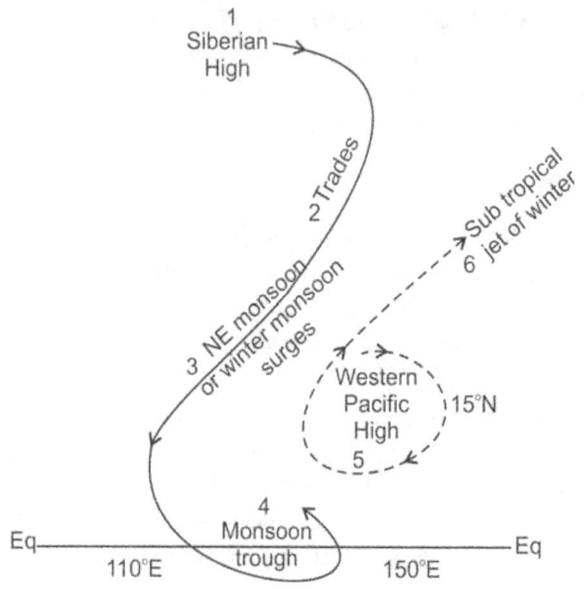

Fig. 20.5 Schematic representation of winter monsoon (not to any scale).

Pressure Distribution During Summer Monsoon

During June to September intense monsoon low (heat low) lies over north Africa and south Asia. A trough of low joins these two systems from west Sahara to China passing through Arabia, Iran, Pakistan, north India, Bangladesh. An intense heat low forms over Pakistan and adjoining Rajasthan (Thar desert) where the sea level pressure drops to about 995 h Pa. The lowest pressure in the monsoon trough may go down to about 985 h Pa. The equatorial trough moves to lat 25-35 °N over Africa, south Asia but it remains close to the equator over western Pacific. During monsoon (June-Sept) a semi-permanent anticyclone cell lies over south Atlantic (1025 hPa), Mascerene High (1025 h Pa) over Indian ocean, Australia (1020 h Pa) and southwest Pacific (1010 h Pa).

During May the pressure gradient over Arabian sea and Bay of Bengal is very slack with isobars practically running along west coast and east coast.

The pressure difference between Mumbai and Tiruvananthapuram is nearly zero. Similarly between Bhubneswar (BBW), Kakinada (KND) and Pamban (PBN) is nearly zero. Weak high pressure cells still persists over Arabian sea and Bay of Bengal. Weak low pressure cells lie over Rajasthan, Bihar and Assam. The pressure difference between Low and High pressure is about 5 h Pa (1006 – 1011 h Pa). This configuration changes after the onset of monsoon (by first week of July). The variation of pressure at Jodhpur, Mumbai, Tiruvananthapuram during April to June indicates that pressure near equator

(TRV) falls gradually from 3 April to middle of May (about 2 h Pa) and then rises gradually (about 1 h Pa). In the middle (MBI) it falls continuously till June end (about 5 h Pa) while in the north near heat low (JDP) the pressure fall is steep till the end of May (about 8 h Pa) then falls slowly (about 2 h Pa).

Temperature Distribution During Summer Monsoon

Before the onset of monsoon generally moderate to severe heat wave conditions prevail in many parts of India. With the onset of monsoon the sweltering heat is abated and surface temperature over many parts of India cools down acquiring nearly uniform temperature except over Rajasthan and adjoining Pakistan. Temperature decreases from Baluchistan (in Pakistan) to Bangladesh across India with a range of temperature of about 10 °C. During break monsoon conditions temperature rises and sweltering conditions prevail over India. The lowest day temperature of about 26 °C prevail along west coast of India, and also along Mynmar coast. Over Deccan plateau (Andhra Pradesh, Tamilnadu, Karnataka) temperature of the order 30–31 °C prevail in July. The mean daily temperature in July is shown in Fig. 28.1

Mean night temperatures generally uniform throughout India of the order 26 ± 2 °C. They are maximum about 33 °C over Baluchistan and minimum about 24 °C over west coast of India and Mynamar coast. The diurnal variation of temperature of about 8 °C prevails over north west India and Pakistan and along east coast of south India, while it is about 6 °C over rest of India. Near Cherrapunji, the heaviest rainfall zone in India, the diurnal range of temperature is about 4 °C.

Monsoon Rainfall Over India and its Variability

Fig. 20.6 shows long term average monsoon rainfall over India. We can say Indian monsoon rainfall in general is a function of orography. From the Fig. 20.6 it is obvious Western Ghats and Khasi-Jaintia hills receive maximum rainfall. This is because the prevailing westerly winds blow perpendicular to the Ghats. Deccan plateau being on leeward side of prevailing winds is in a shadow zone gets less rainfall. Similarly southward of the Khasi-Jaintia hills get more rainfall while northward side of Khasi-Jaintia hills also gets less rainfall as it is being on the leeward side of prevailing winds. Rajasthan and part of Gujarat gets the least rainfall. On the wind ward side of western ghats rainfall increases from west coast along the slopes of Ghats. Cherrapunji which is located on the southern side (wind ward side) of Khasi-Jaintia hills receives worlds maximum annual rainfall of 1142 cm (monsoon season rainfall 840 cm) while Brahmputra valley, north of KJ hills receives 120 cm annual rainfall. Wind ward side of western ghats receives more than 500 cm rainfall during monsoon while on the leeward side (Deccan plateau) gets about 80 cm during monsoon season. Major part of India receives bulk of annual rainfall (75-80%) during monsoon season. Minimum rainfall belt extends from northwest Rajasthen to central Bengal which coincides with the axis of monsoon trough. Kanyakumari district

Rainfall cm

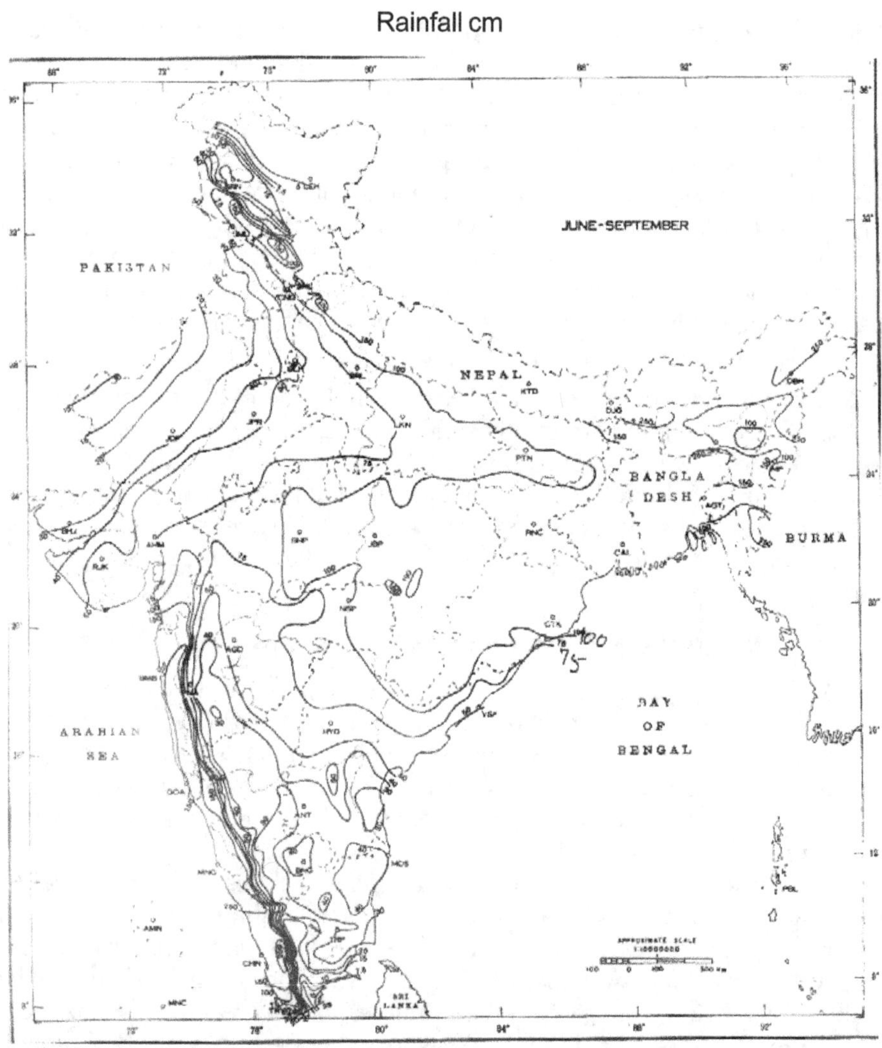

Fig. 20.6

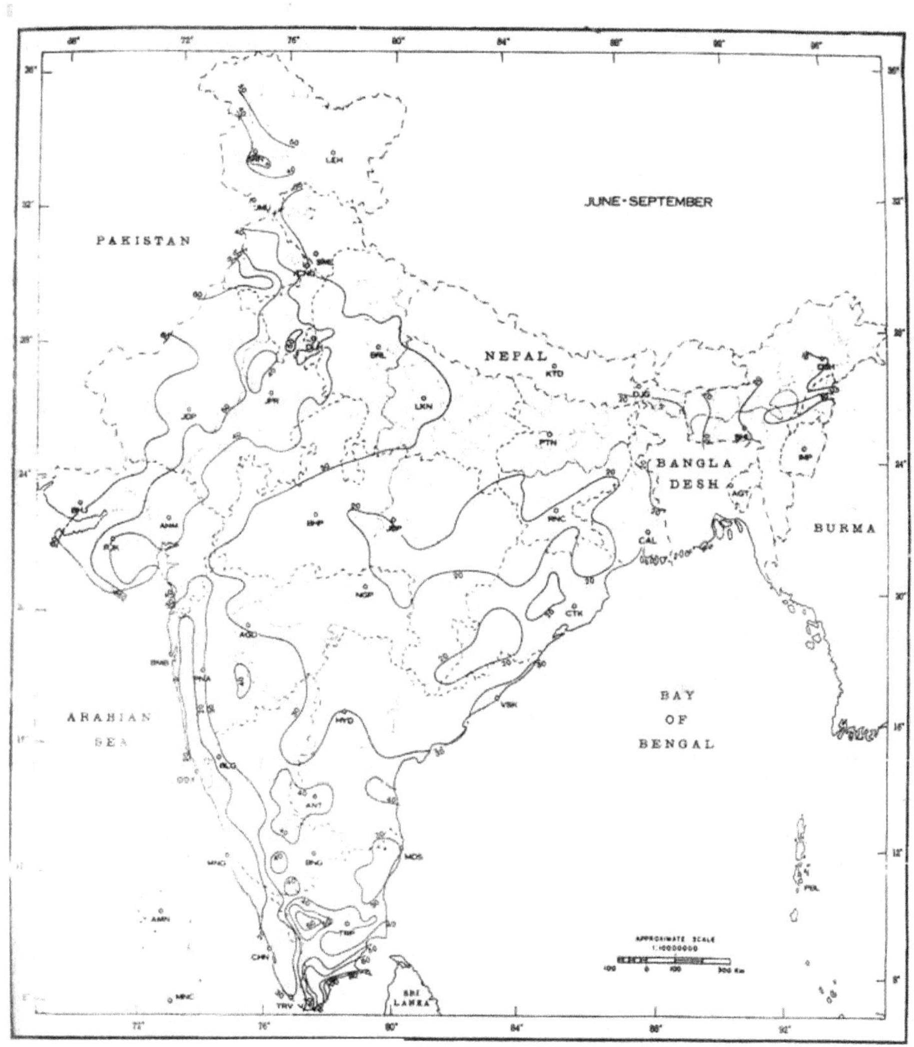

Fig. 20.6(a) Coefficient of variation of rainfall.

in Tamilnadu gets seasonal rainfall of 10-20 cm and the Thar desert of Rajasthan and adjoining Sindh also gets 10-20 cm rainfall. Eastern Himalayas gets more rainfall than the western Himalayas.

Monsoon rainfall over India varies from more than 800 cm to less than 40 cm. The coefficient of variation is 20% in north-east India and Konkan to 60% (maximum 80%) in Rajasthan and south Tamilnadu. See Fig. 20.6(a). There is large variation in year to year and there seems no trend in rainfall pattern. However some meteorologists are of strong view that there is some trend in consonance with sunspot cycle of 11 years.

According to IMD convention, a meteorological sub-division (part of the country) is said to be affected by drought, if it receives seasonal (monsoon) rainfall less than 75% of the normal (long period average is called normal). It is further classified as moderate drought if the seasonal rainfall departure from normal is –25 to –50% and severe drought if the departure is more than 50% (deficiency). The year is considered to be a drought year if 20% of the area or more is affected by the drought. There are some other definitions of drought which have only theoretical consideration. One such is 10% below normal is a drought year. Study of 100 years RF data shows that, there were no occasions of two consecutive years of drought as defined by this definition 10% deficiency criteria. There were 3 occasion in these 100 years when two consecutive years had drought namely 1904-1905, 1965-1966 and 1986-1987. The period 1920 to 1960 was the best period so far in regard to monsoon rainfall. During 1899, 1918, 1972, 1979 rainfall deficiency was more than 20%, while in 1917 rainfall was in excess more than 20%. Events of extreme summer monsoon rainfall (1881-1980) is given below.

Deficiency of Rainfall

S.No.	Year	Area affected	Departure
1	1899	68.4%	–21%
2	1904	34.4%	–11%
	1905	37.2%	–14%
3	1918	70%	–21%
4	1920	38%	–16%
5	1951	35.1%	–16%
6	1965	38.3%	–19%
7	1966	35.4%	–14%
8	1972	40.4%	-22%
9	1974	34%	-14%
10	1979	34.8%	-20%
11	1986	19.7%	-13%
12	1987	47.7%	-17%

Excess Rainfall

S.No.	Year	Departure
1	1916	+13%
2	1917	+20%
3	1933	+13%
4	1942	+15%
5	1961	+17%
6	1975	+14%

The decadal averages of rainfall was below normal during 1891-1900, 1901-10, 11-20, 61-70, 71-80 while they were above normal during 1921-30, 31-40, 41-50, 51-60. It is also observed that the annual mean temperature anomalies were also mostly positive (Actual-normal) during 1920-60 while they were negative (Actual-normal) during 1880-1920. This indicates the intense relation between temperature anomalies and rainfall. Further it indicates 40 year epoch of cool temperatures followed by warm. However during 1881-90 decadal average rainfall was above normal (actual 89.6 cm, normal 88.75 cm) while the decadal annual temperature anomaly was negative (–0.57) that is in cooler epoch.

Inter Monsoon (Seasonal) Rainfall Variability

Inspite of regular annual recurrence (or periodic nature) of summer monsoon there are large variation of rainfall within the season. These variations are attributed to synoptic scale disturbances, oscillation of monsoon trough, mid-latitude effects and quasi periodic oscillation of 30-40 days. Synoptic scale disturbances, such as lows, depressions, cyclonic storms, mid-troposphere cyclones, off-shore vortices etc., cause rainfall variations in periods of 5-7 days. The intensity, frequency of these disturbances have dominant role in floods. The oscillation or meandering of monsoon trough (on sea level chart) from its normal position cause rainfall variability in periods of 10-15 days. Shifting of eastern end of monsoon trough to the foot hills of Himalayas generally causes floods in the river Brahmaputra, while most of the country gets scanty rainfall. On the contrary if the monsoon trough shifts south wards of its normal positions wide spread rainfall occurs over most part of the country. Interaction of extra-tropical weather systems (extra tropical cyclones, western disturbances, low index cycles etc) or southern hemispheric systems effects the monsoon circulation. It appears a large cloud mass cycles develop over south Indian ocean and moves north ward gradually in bursts and reaches north India in about 40 days. EL Nino and southern oscillation together called ENSO. There is teleconnection between ENSO and monsoon activity.

We shall now discuss five semi permanent features of the monsoon.

Heat Low

The formation of heat low over Pakistan and adjoining Rajasthan is a semi permanent feature of monsoon over India. The heat low is very shallow and extends up to 850 h Pa (1.5 km) level. Above the heat low, there exists a well marked ridge which is a part of Sub-Tropical High. Rainfall in association with this low over the area is comparatively very small, which resulted in the formation of Thar desert in Rajasthan. The surface day temperature in the area is high, which evaporates the moisture. Since the associated upper air cyclonic circulation (UACYCIR) extends only to about 1.5 km and marked subsidence aloft, inspite occurrence of cloudiness the precipitation is very small. However the intense heat low (pressure departure is below normal) acts as suction devise for moist air along the monsoon trough and to some extent related to good monsoon over India. During weak heat low (pressure departure above normal) monsoon rainfall over India is greatly affected and results in deficient or even scanty rainfall over vast area of the country. In 1987 the central pressure over heat low area was mostly above normal, which proved to be a drought year. Satellite measured estimates of long wave radiation (or earths radiation) indicates that tropical / subtropical deserts are heat sinks. It was pointed out in the earths heat budget that tropical areas are heat sources (where insolation exceeds outgoing long wave radiation of the earth) and transports heat to polar regions. Incase of deserts it is found to be the reverse.

Monsoon Trough

The monsoon trough plays an important role in the activity of monsoon over India. During summer monsoon period an elongated low pressure area extends from heat low over Pakistan to Head Bay of Bengal. This is called monsoon trough. This is a semi permanent feature of monsoon circulation and is attributed to the east-west orientation of Himalayan ranges and north-south orientation of Khasi-Jaintia Hills. Generally the eastern side of the monsoon trough oscillates. It migrates sometimes southwards and some times northwards. South ward migration results in active/vigorous monsoon over major part of India. In contrast to this the northward migration of trough leads to break monsoon condition over major part of India and heavy rains along the foot hills of the Himalayas and many a times results in floods in Brahmaputra river. South of the axis of monsoon trough winds are westerlies or south westerlies while north of the axis they are easterlies. This wind circulation extends up to 500 h Pa level and on some occasions extends up to 400 h Pa. The upper air axis of the trough slopes south-wards with increasing height. The sloping is more prominent on west side (heat low side). The air temperature north of the trough axis observed to be slightly warmer than the south. The normal position of the axis of monsoon trough at different height level is shown in Fig. 20.7. Earlier it was believed that monsoon trough is formed by mechanical effect associated with the orographic alignment of the Himalayas and the mountains of Myanmar. However recent numerical models indicates that the orientation of monsoon trough is related with the radiation balance of the lower troposphere. The monsoon trough is regarded as the equatorial trough (thermal equator) or Inter Tropical convergence zone (ITCZ) in Indian latitudes. It migrates from lat 8 °N in July to lat 18 °S during January with cyclonic wind shear.

Mascarene High

It is a high pressure area that is found around Mascarene Islands (in south Indian ocean) during monsoon period. This is responsible for cross equatorial flow through south Arabian sea and it acts as a southern hemispheric linkage. The variation in the intensity of High pressure causes monsoon surges across equatorial flow. These surges are responsible for heavy rains along the west coast of India (Fig. 20.7(a)).

Tibetan High

It is a large warm anticyclone located over Tibetan plateau (centre lat at 28 °N long 98 °E) in middle/upper troposphere during June-September period. It is marked at 300 h Pa level with centre 30 °N, 90 °E and extends 70-110 E. Shifting its position east or west causes variation in monsoon activity over India. West movement causes monsoon transients from Bay to northwards. It must be noted here that Tibetan High is different from Pacific High. The mechanism of formation of Tibetan High is not yet known.

Somali Jet

Low level (L.L. 1 to 1.5 km asl) inter hemispheric cross equatorial flow of air attains Jet speed at the west end of monsoon regime along the east coast of Africa. The Jet originates near Mauritius and northern part of Madagascar in the southern hemisphere. The jet reaches Kenya coast (at about 3 ° S) covers the plains of Kenya, Ethiopia and to Somali coast at about 9 ° N. During May moves further north into eastern Africa, then into Arabian sea and reaches west coast of India in June. It attains maximum strength in July. The cross equatorial low level current attains jet speed during summer monsoon. Short period (8-10 days) fluctuations are observed in LL jet stream are attributed to the northward propagation of surges upto Mozambique channel and or due to the variations of the strength of monsoon–Hadley cell. The strengthening of LL Jet gives rise to strong monsoon over peninsular India.

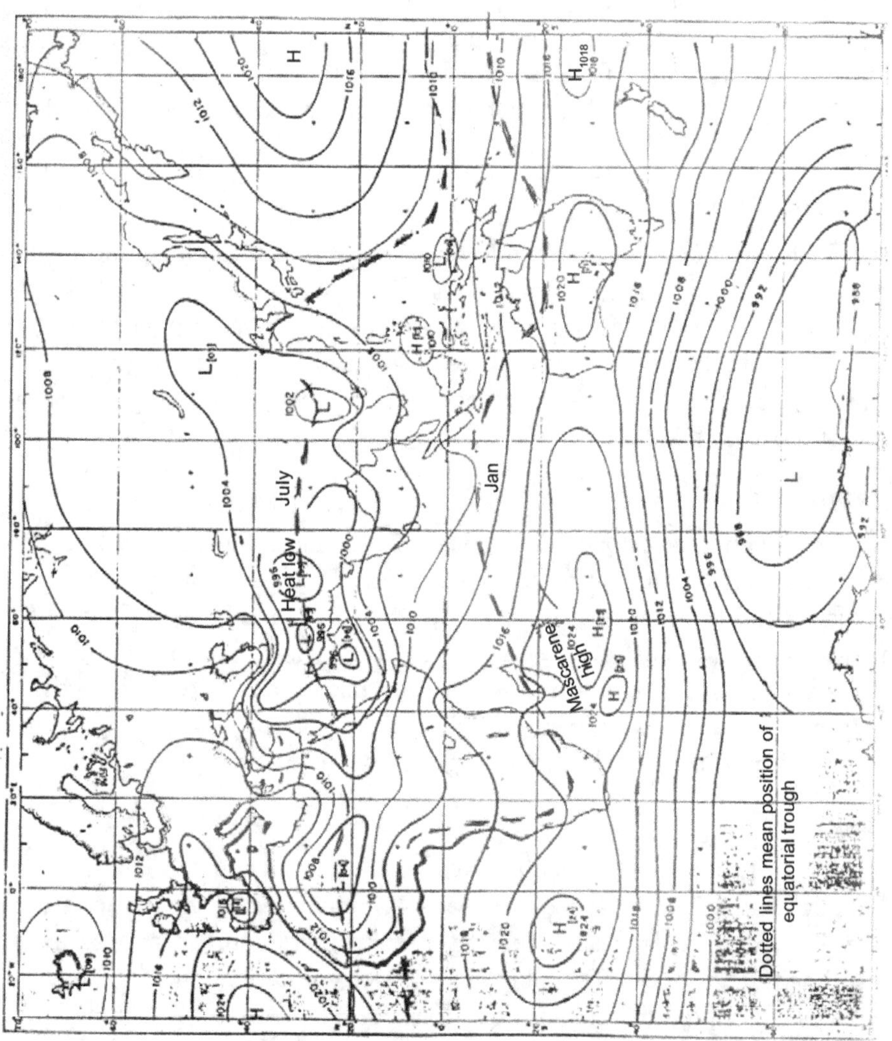

Fig. 20.7 Sea level pressure (mb) : July.

MEAN AXIS OF MONSOON TROUGH JULY
900 hPa
850 hPa
900 hPa
850 hPa
700 hPa
700 hPa
600 hPa
600 hPa
500 hPa
500 hPa
30N
20N
10N
70 E
80 E
90 E

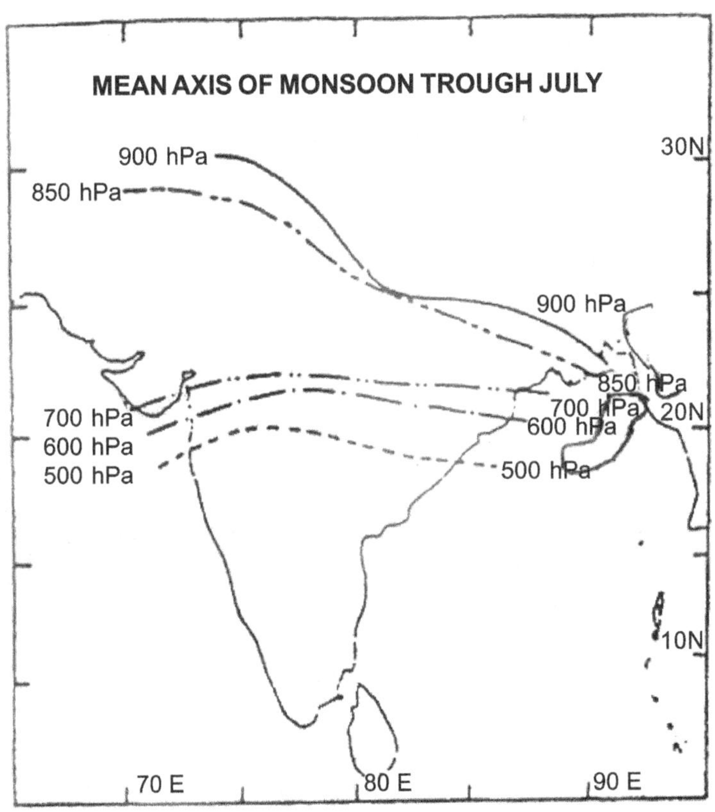

Fig. 20.7(a)

Troughs in Monsoon Circulation Over India

During monsoon period different troughs are observed which have different effects on rainfall. The following are important.

(i) Sea level monsoon trough (already discussed) (ii) Sea level trough along and off west coast (Gujarat coast to Kerala coast) (iii) trough along east coast (Orissa to Tamilnadu coast) in lower troposphere, (iv) East-west trough in lower and mid troposphere over central and southern parts of India, (v) Mid-tropospheric north-south trough over south peninsula, (vi) Trough in upper tropospheric easterlies and westerlies, (vii) ITCZ (or equatorial trough).

(ii) During summer monsoon trough of low pressure observed along the west coast of India throughout the coast or in parts. Strengthening of this trough causes active monsoon over peninsula. The frequency of occurrence of this trough is more during June and July (three times) as compared to August and September. These troughs are generally very shallow, extend up to 1.5 km asl in association with low/depression, or it may extend upto 5.8 km asl in association with MTC which is described subsequently.

(iii) N-S troughs are observed over central parts (which move from west to east) and along the east coast of the country (which move east to west) particularly during break monsoon conditions, when the westerly wind regime prevails over entire country in lower and mid-troposphere. Weather associated with these troughs are mostly convective nature (thunderstorms). Generally these troughs do not cause active/vigorous monsoon conditions.

(iv) E-W oriented troughs are observed during break monsoon conditions along lat 10-12 °N. Any formation of low or depression in this trough moves east to west and sometimes north and cause revival of monsoon.

(v) Troughs in upper easterlies move west wards. This aids formation of intense low or depression.

(vi) Troughs in upper tropospheric westerlies in middle-latitudes on some occasions extend south wards and interact with monsoon circulation over India. These move from west to east and cause delay in onset of monsoon during June. In some cases they help formation of depression over Head Bay of Bengal. In July, August months these troughs pull the monsoon trough to foot hills of the Himalayas and cause floods in Brahmputra.

(vii) ITCZ/Equatorial trough. We have already discussed about ITCZ in general. Most of the tropical disturbances (depressions/cyclones) over the globe develop in the ITCZ area. The frequency of disturbances vary with the activity of ITCZ and as a consequence rainfall over Tropical areas. Over Indian sub-continent a cloud band which moves north wards during onset phase of monsoon is associated with equatorial trough/ITCZ. During monsoon period (June-Sept) trough of low pressure area (equatorial trough) which migrates with considerable rainfall activity is identified as ITCZ or near equatorial trough. ITCZ is warm at the surface, cold in the mid-troposphere and again warm in the upper troposphere.

Mid-tropospheric Cyclone (MTC)

During summer monsoon period cyclonic vortices that are confined to mid-troposphere are called Mid-tropospheric cyclones. MTCs are found only in mid-troposphere (700-500 h Pa) and nothing is seen on surface weather chart. These systems cause heavy rains (as much as 30 cm in 24 hrs) over northeast Arabian sea and along west coast of India. Incase of intense system cycir lies between 850-500 h Pa. These systems generally remain stationary or have very little movement for many days. The horizontal dimension of the MTC is 500-1000 km. and vertical extension 3 to 6 km. The maximum vertical velocity is found to be at 600 h Pa. The maximum vorticity (15 to $20 \times 10^{-5}\, sec^{-1}$) is observed at 600 h Pa, while at surface the cyclonic vorticity is very weak and cycir is not seen in the lower levels (up to 1.5 km) where only trough is seen. The thermal structure of MTC shows that it is of cold core between 700-500 h Pa and warm core between 500-300 h Pa. MTCs are mostly found during June and July with average frequency of 1-4 per season. The life period varies 3-7 days. In the vicinity of MTC at 500 h Pa, RH is found to be exceeding 50%. Large vertical wind shear is also found. MTC_s are found to accentuate the upper westerly troughs and cause heavy rain over NW-India. (north east Rajasthan, Hariyana, Himachal Pradesh and even J and K).

Off Shore Vertex

During summer monsoon a shallow trough of low pressure is observed (on sea level chart) along west coast of India. In this trough vortices develop and move along the coast (south to north) at a speed of about 100 km/day and cause heavy rains along west coast over a small area. These shallow vortices are called 'off shore vortices'. They are prominent in July during active/vigorous monsoon condition and are absent during weak or break monsoon conditions. The size of these vortices are small and cannot be detected on synoptic weather charts. They are mostly detected on the basis of surface winds of coastal stations which show easterly implying the eastern part of the vortex. On some occasions lower level winds, say at 0.9 and 1.5 km asl, provide clues of the presence of vortices. The diameter of the vortex is inferred to be about 150 km and vertical extension about 1 km above the surface. It is also inferred that these vortices are formed due to wind blowing at right angles to the western ghats which are spread in a north-south direction with altitude varying 1 to 1.5 km.

Orographic Effect

Orography plays an important role in precipitation. Rainfall increases with altitude on wind-ward side and decreases on lee-ward side. A study shows that in tropical region rainfall has maximum at about 1 to 1.5 km height, while in equatorial region the maxima lies close to the sea level and decreases upward. In mid-latitudes/sub-tropics precipitation increases with height. In Indian sub-continent orientation of mountain barriers play major role in monsoon rainfall.

The amount of precipitation depends on several factors, of which orography, airmass characteristics, low pressure system are important. Strong orographic influence is observed in monsoon rainfall in westernghats and in the Khasi-Jaintia hills. Rainfall (amount) increases with height on wind side. Convection also occurs off western ghats of India. Similar convection is also observed off western coast of the mountains of Myanmer, Thailand and Phillipines.

Monsoon Disturbances

During monsoon period (June-Sept) a series of low pressure systems form over Head Bay of Bengal and generally move in a west-north-westerly direction along the monsoon trough and weaken over west Madhya Pradesh or Rajasthan. On their way these systems give copious rain. Some of these low pressure systems may intensify into depressions or deep depressions and rarely become cyclonic storms. The study of monsoon depressions is of great importance because they give abundant rain. According to convention a monsoon disturbance is called a depression if the associated wind speed is 18-33 kt (33-60 kmph) over sea or may have two to three closed isobars on sea level chart. It is called a cyclonic storm if the surface wind speed over sea exceeds 33kt (60 kmph). The study of 100 years data reveals that the average cyclonic disturbances during monsoon months is about 2-3 per month or about 10 per season. In the early June and late September these disturbances first appear in mid-troposphere (4.5 km) as cycir and descend to sea level, during other period they appear first on sea level as trough or low and then concentrate into depression. The frequency of occurrence of these disturbances in Bay of Bengal is much more as compared to over land and Arabian Sea. The number of cyclonic storms that form in these months are much less as compared to depressions and deep depressions. The disturbances that form over Arabian sea are far less than over land.

During monsoon season a marked seasonal trough of low pressure lies over north Bay of Bengal. A series of low pressure waves/easterly waves travel to north Bay across Myanmar. These waves are generally the remnants of Typhoons/Depressions of Pacific ocean. When these waves reach the Head

Bay of Bengal they rejuvenate and concentrate into depressions. However some may form insitu without the arrival of any remanent system from Pacific. The meteorological conditions that prevail at the time of formation of depression are: surface low pressure area/trough dips into Bay of Bengal. Associated with this low level cycir descends to surface. When a remanent of a Typhoon emerges into Arakan coast, westerlies strengthen over central and peninsular India. Pressure falls around the Head Bay (which is detected by 24 hr changes and departure from normals). Sea surface temperature (SST) is found to be warmer over the depression field compared to the surrounding coastal waters. Satellite cloud imagery indicates persistent bright, organized cloud mass with T- number 1 or 1.5. Easterly jet axis moves away from the field. Vertical wind shear decreases over the field.

Movement of Monsoon Depressions and Associated Rainfall

Tracks of past monsoon depressions indicate that they generally move in a west-north-westerly direction. The movement may be inferred by 24 hr maximum pressure falls or direction of steering current just above the level of non-divergence (normally 400 h Pa wind direction) over the field.

Wind Field

In the lower levels westerlies prevail south of the monsoon trough axis and easterlies north of it. UACYCIR (upper air cyclonic circulation) associated with a depression extends 600-500 h Pa (and in some cases upto 400-300 h Pa) level. Upper winds above CYCIR will be easterlies. When upper easterly winds are not strengthened, particularly in June and in break monsoon period, it is found that the depression initially moves in a north westerly or northerly direction and then recurve and move eastwards (Fig. 20.8(a) and (b)). This may cause floods in sub-montane regions of Himalayas and in Brahmaputra valley.

The field of monsoon depression horizontally extends to about 1000-1500 km and vertically 500-400 h Pa. In the lower troposphere maximum convergence is found just ahead of the depression centre and relatively weak field in the rear. The core region of the depression is found to be comparatively cooler in the lower levels (up to 3.1 km) and warmer aloft. Associated with a depression the UACYCIR extends to 6 km tilting southwest wards with height and anticyclonic circulation aloft. In a west moving system, comparatively more moist air mass is found in southwest side of the depression centre and also more rain. The average life period of the depression is about 6 days. These depressions are responsible for rainfall variation both in space and time (about a week). On rare occasions monsoon depressions concentrate into cyclonic storms and cause havoc with torrential rain and winds.

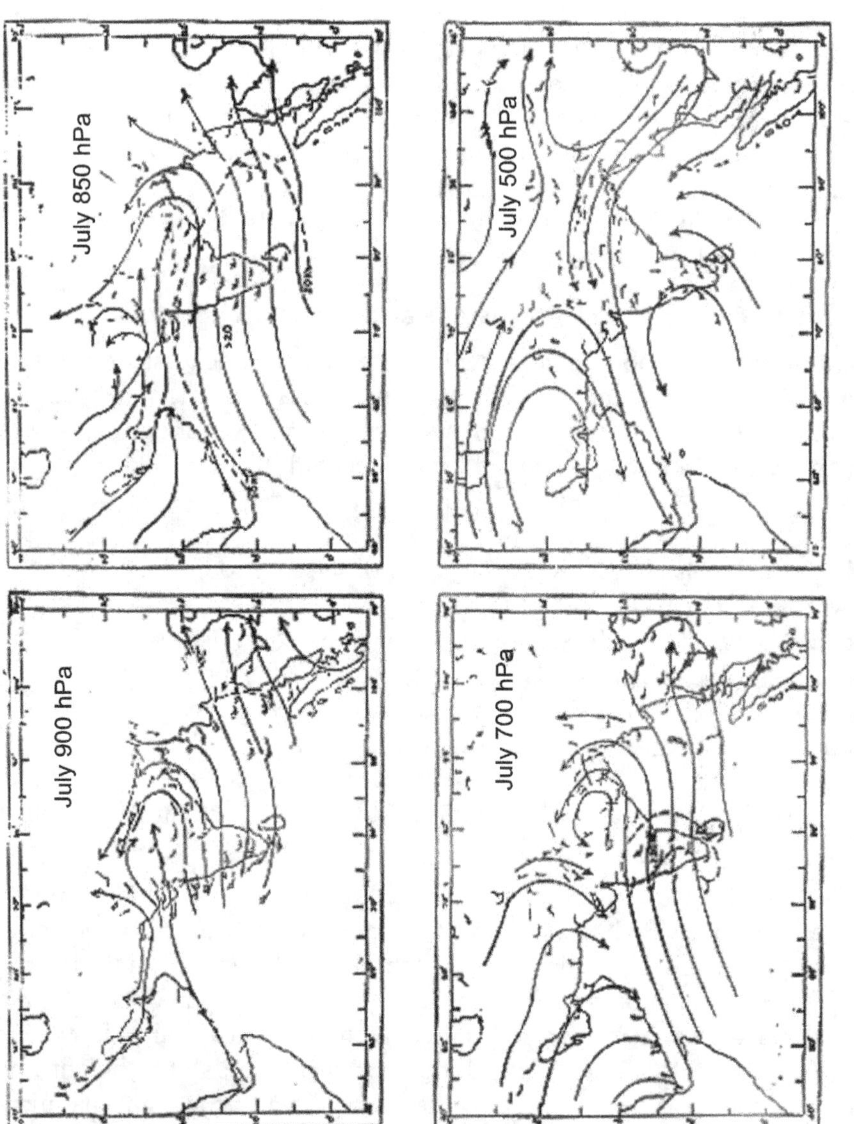

Fig. 20.8(a) Upper winds.

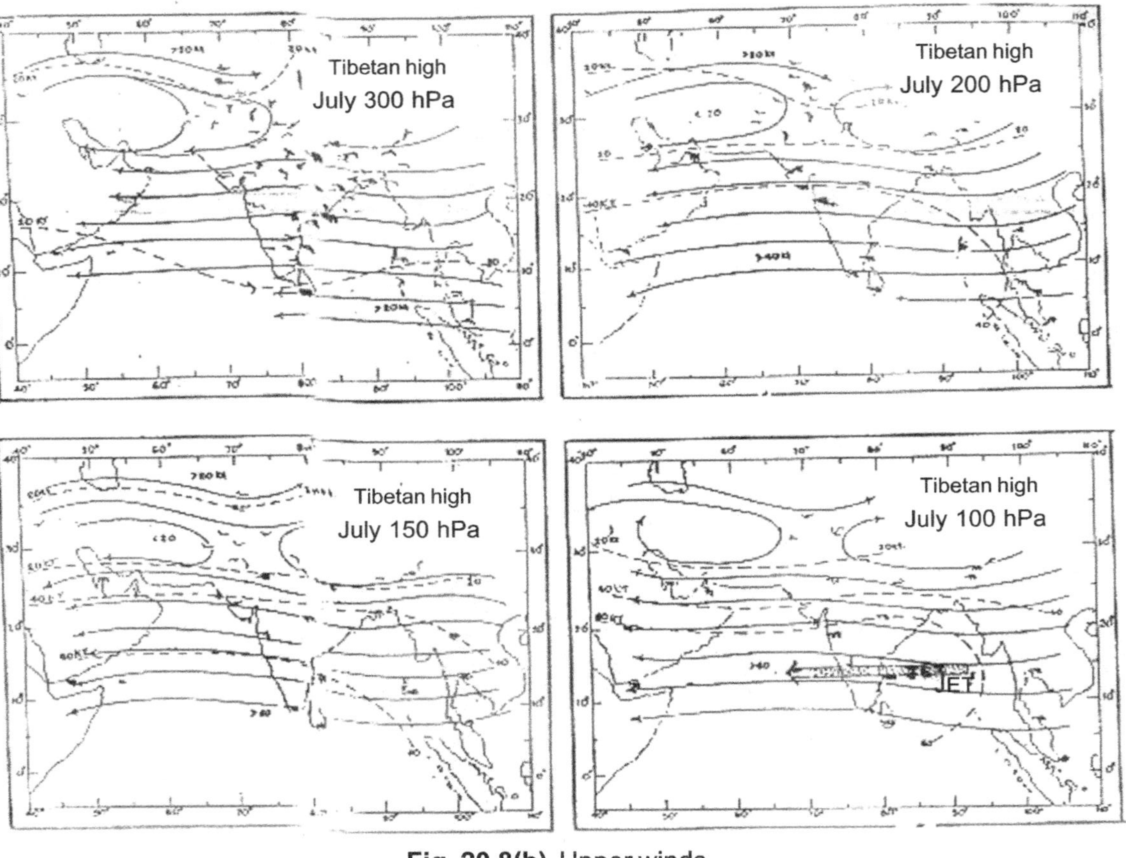

Fig. 20.8(b) Upper winds.

Breaks in the Monsoon

Monsoon fully establishes over India by the end of June and thereafter sustained monsoon conditions prevail over whole of India during July and August. Withdrawal begins in September middle. However during July, August months there were occasions of short period, say about a week, when rainfall over most part of the country is absent. This low rainfall activity is termed "Breaks in monsoon" which is associated with change in established monsoon circulation. A convention is followed that if the monsoon trough is absent on sea level chart and at 850 h Pa level for more than two days then it is called break in monsoon. According to this convention the longest break of 20 days occurred in 1906 (29 July to 18 Aug). A statistical study of 80 years data (1887-1967) shows that the average period of breaks in July 5.8 days and August 6.5 days and most frequent duration in these months 4 and 3 days respectively and the longest period in these months being 17 and 20 days respectively. Standard deviation is about 3 days. There was no break in 12 years out of 80 years indicating the probability of occurrence of break is 0.85.

Synoptic Situations During Breaks

Shifting of Monsoon trough to the foot hills of the Himalayas causes break. This may be caused by (i) the eastward movement of active western disturbance north of the country, (ii) pronounced low index westerly circulation in middle latitudes, (iii) Eastward movement of large amplitude trough in westerlies (between 500-300 h Pa) over north of India across Tibetan Plateau, (iv) movement of Bay depression north wards to foot hills of the Himalayas, (v) movement of a western Pacific Typhoon to north of 30 °N over China or its recurvature northeast wards.

Break in monsoon may be caused by the formation of a low in lower latitudes which moves west wards along lat 10 °N over south peninsula or by the formation of High pressure ridge at 500 h Pa level over Central India.

During break monsoon conditions mid-latitude upper tropospheric ridge (at 300 h Pa) shifts southwards to about 26 °N, mean sea level pressure falls to below normal over south peninsula (south of 13 °N) where surface temperature becomes above normal. This warming is observed in the entire troposphere. Over central parts of India sea level pressure rises to the order of 4 h Pa.

During break monsoon, heavy rains occur along foot hills of the Himalayas and over southern parts of India (Tamilnadu, south Andhra Pradesh), while rainfall activity drastically reduces over western and central parts of India. In upper troposphere easterly Jet shifts northwards as far as 23 °N (normal position is 13 °N). Two Jet core maxima are located, one at about 12 °N (south of its normal position) and another at about 22 °N. Two westerly jet cores are also observed, one at about 32 °N and another at about 40 °N.

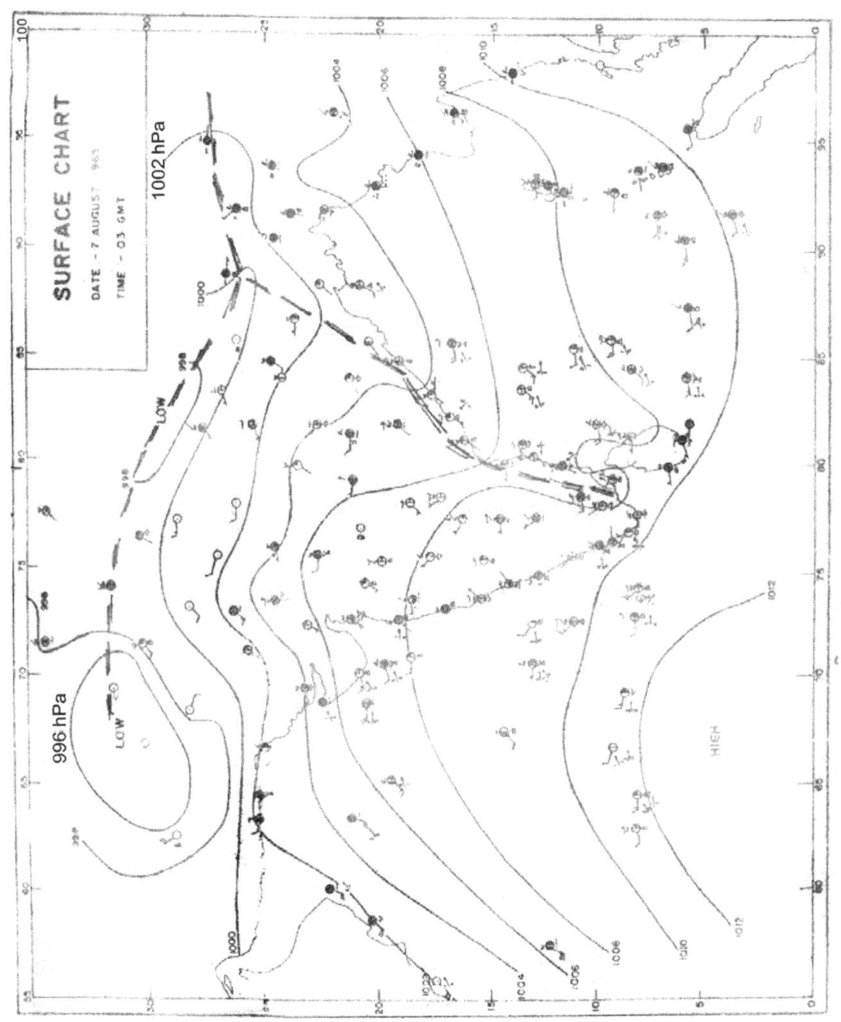

Fig. 20.9(a) Break monsoon conditions typical surface chart.

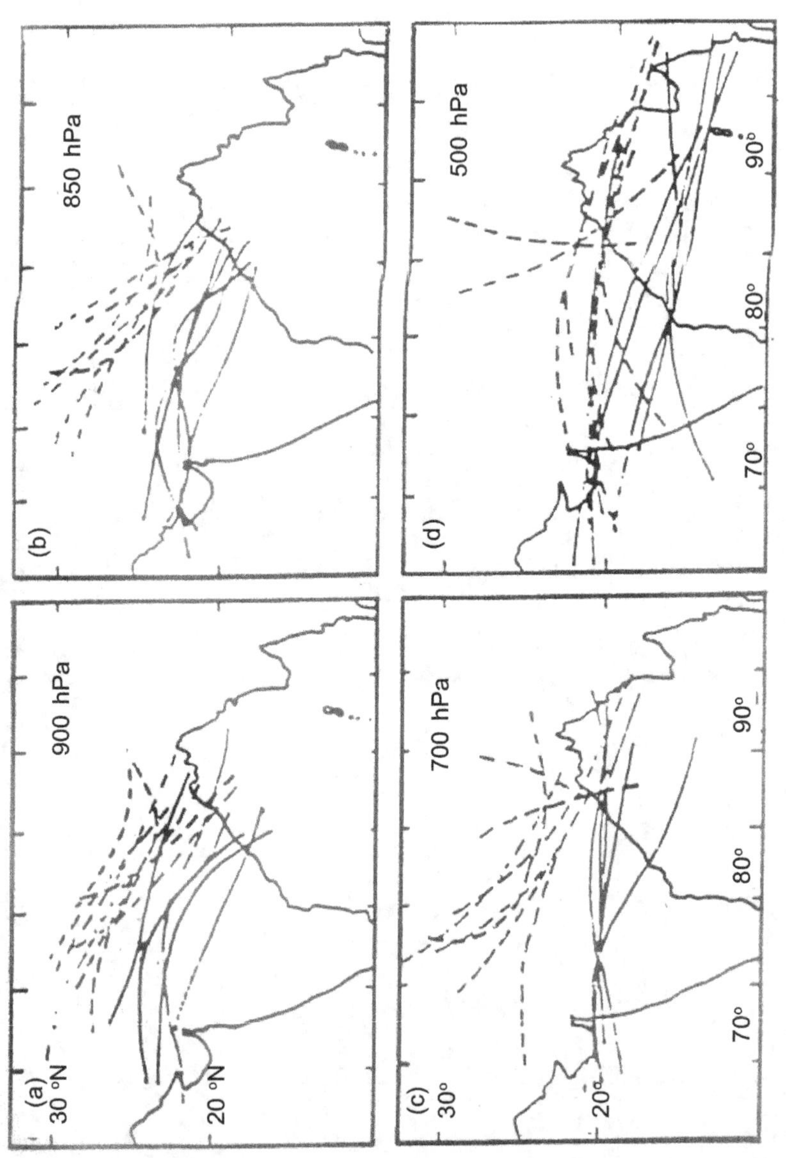

Fig. 20.9(b) Mean position of the monsson trough during strong (normal) and weak (break) monsoon; (a) at 900 hPa, (b) at 850 hPa (c) at 700 hPa (d) 500 hPa.

Long Range Forecast of Monsoon Rainfall

Correlation and regression methods have been used by IMD in issuing monsoon rainfall forecast since 1909. Statistical method of Time series analysis is used for studying monsoon rainfall. One study found that the annual rainfall is normally distributed over India except in west Rajasthan, Jammu and Kashmir and interior parts of peninsula. Another study found that the frequency distribution of monsoon rainfall over India does not differ significantly from normal distribution except in Kutch, west Rajasthan, south Assam, certain pockets of Himalayas, south Tamilnadu and adjoining areas of south Kerala. Various studies brought to the light that monsoon rainfall is related to certain parameters which are related temperature, wind, pressure and snow cover. Some of these parameters are positively (+) correlated and others negatively (–). Using sixteen (16) parameters a multiple regression equation was developed and used for forecasting is given below. The model has shown good encouraging results for a decade.

The model equation is $R = c_o + \sum_{i=1}^{16} c_i x_i$

Where c_o, c_i are model constants derived by least square method and

R $\quad$ = monsoon rainfall

X_1 $\quad$ = EL Nino in current year (–)

X_2 $\quad$ = EL Nino in previous year (+)

X_3 $\quad$ = Average north Indian temperature in March (+)

X_4 $\quad$ = Average temperature of east coast of India in March (+)

X_5 $\quad$ = Average temperature of central India in May (+)

X_6 $\quad$ = Average temperature of northern hemisphere in Jan and Feb (+)

X_7 $\quad$ = Latitudinal position of ridge at 500 h Pa along longitude 75 $^\circ$ E in April (+)

X_8 $\quad$ = East-west extent of trough-ridge at 50 h Pa over northern hemisphere (–)

X_9 $\quad$ = Westerly wind strength at 10 h Pa in January (+)

X_{10} = Pressure anomaly (SOI) between Tahiti-Darwin during spring (+)

X_{11} = Darwin pressure during spring (–)

X_{12} = Pressure anomaly of Argentina (south America) in April (–)

X_{13} = Pressure anomaly of equatorial Indian Ocean during Jan-May (–)

X_{14} = Himalayan snow cover during January-March (–)

X_{15} = Snow cover in Eurasia during Dec. of previous year (–)

X_{16} = Pressure anomaly of northern hemisphere Jan-April (–)

Regression technique is extensively used in USA for forecasting seasonal temperature and rainfall. Similarly in Europe seasonal weather forecast being tried fitting regression equation using sea surface temperature as one parameter.

Note : The 16 – parameter regression equation study indicates that the probability occurrence of favorable parameters during below normal (less than 90% of normal) or deficient monsoon rainfall is about 0.3, while the probability occurrence of favorable parameters during normal monsoon (more than 90% of normal) is about 0.6. The probability occurrence of failure of forecast, that is when predicted normal rain, actual would be below normal and when predicted below normal rain would result in normal is about 0.1. For example in 1957 there was only one parameter favorable (ridge position at 500 h Pa in April) while other parameters were unfavorable but resulted in normal rainfall. In 1966, there were 8 favorable parameters out of 14 still it resulted in deficient rainfall.

Monsoon QPF : For quantitative precipitation forecast, a non-linear multiple parametric regression equation has been employed is given below.

$$R = 621 + \sum_{i=1}^{16} c_i$$

Where R (mm) = monsoon rainfall over India

Xi = Sixteen parameters

ci and pi are model constants which are determined by the method of least square fit.

Northeast Monsoon

In India the period October to December is called northeast monsoon. In this period south coastal Andhra Pradesh, Rayalaseema, Tamilnadu and neighbour-hood gets bulk of the annual rain. For Tamilnadu (TN) and neighborhood it is main rainy season. Coastal districts of TN get 60% of annual rainfall while interior districts get 40-50% of annual rain. Rayalaseema gets about 36% of annual rain during September, October months, which are the rainiest months. According to IMD classification of seasons, Oct-Dec period forms post monsoon or retreating south–west monsoon because it follows the SW-monsoon (June-Sept). SW– Monsoon begins to retreat from north India from the middle (15th) of September and completes withdrawal from whole of India by 15th October. Simultaneously monsoon circulation pattern also changes. Instead of low pressure systems forming over north Bay (during SW monsoon) a low pressure area forms over central and south Bay of Bengal, which further shifts to south Bay during November and south of Lat 5°N during December. An east-west trough extends from this low to south Arabian sea with associated upper air trough extending to 3.0 km asl. In the mid and upper troposphere easterlies prevail over south peninsula but the easterly jet which was present during SW-monsoon disintegrates.

The following are the IMD criteria for declaration of NE-Monsoon over Tamilnadu:

(i)   Withdrawal of SW-monsoon upto lat 15°N.

(ii)   Onset of persistent surface easterly winds along Tamilnadu coast.

(iii)   Depth of easterlies extend upto 1.5 km

(iv)   At least fairly widespread rainfall over coastal Tamilnadu and south coastal Andhra Pradesh and neighbourhood.

(v)   Onset not to be declared before 10th October even if all above conditions exist.

Rainfall Distribution Over Coastal Andhra Pradesh and Rayalseema

These areas get rainfall both in summer monsoon and NE monsoon but receives more in SW-monsoon. As in case of SW-monsoon there is no abrupt increase of rainfall (September to October) with the onset of NE-monsoon. Rainfall in October remains practically same as in September. It decreases rapidly from October to November and by middle of December it becomes insignificant. See Table 20.1 below

Table 20.1 Comparison of southwest and northeast monsoon.

SW-monsoon	NE-monsoon
1. Rainfall increases abruptly onset of SW-monsoon.	There is no significant change in rainfall on onset of NE-monsoon.
2. Rainfall increases from the coast to western ghats.	Rainfall decreases from the coast to interior Tamilnadu.
3. Variation of rainfall is high in regions of poor rainfall.	Variation rainfall is high in the regions of highest rainfall.
4. Presence of tropical easterly jet in the upper troposphere.	Tropical easterly jet completely disintegrates.
5. Tropical disturbances assume the form of depression and rarely become cyclones.	Tropical disturbances frequently concentrate into cyclonic storms.
6. Maximum rainfall occurs south of the track of depression.	Heavy rains occur both north and south of the cyclone / depression track.
7. Active / vigorous monsoon conditions occur along west coast frequently (about 25% of occasions).	Active / vigorous monsoon conditions are less frequent along east coast.

Rainfall Distribution in Tamil Nadu

Tamilnadu gets more rainfall in October-December as compared to SW-monsoon. Rainfall considerably increases from September to October and continues in November. Coastal districts of Tamilnadu get more rain in November than in October, where as the interior districts get more rain in October than in November. By December rainfall decreases substantially. Thus NE-monsoon rain mainly occurs in October and November.

In Kerala October is the rainset month of NE-monsoon. Here rainfall decreases in November.

Time Series Variations of Rainfall

Study of point rainfall of individual stations all over India over a period of 100 years or more indicate the following.

1. Many stations indicate significant increase or decrease of rainfall amount from a particular year onwards.
2. Significant Oscillations of different magnitude were observed.
3. 8-15 year cycle (solar cycle) observed at many stations.
4. A number of stations showed quasi-biennial oscillations.
5. The mean annual rainfall of India as a whole (based on 60 years data 1901-1960) was found to be 119 cm with standard deviation of 9.5 cm. Country as whole received deficient rainfall during the year 1905, 1918, 1920 and 1941.

Questions

1. Define the modern criteria of monsoons and IMD description of strength of monsoons over sea and land areas.

2. Describe the onset of monsoon over Kerala and its march over rest of India. Draw a schematic representation of SW monsoon.

3. Describe the withdrawal of monsoon over India and onset of NE monsoon over Tamilnadu. Draw a schematic onset of NE monsoon and Temperature field.

4. Describe the pressure field of summer monsoon over Indian sub-continent?

5. Write briefly about the monsoon depressions and rainfall variability over of India.

6. What are the semi permanent feature of Indian summer monsoons, describe them briefly?

7. Write briefly on (i) Troughs in monsoon circulation over India (ii) Mid-tropospheric cyclones, MTC (iii) Off shore vortices (iv) Orographic effect on monsoon rainfall.

8. Write briefly on disturbances in windfield in Indian sub-continent.

9. What are break monsoons? Write the synoptic features associated with it.

10. Write the salient features of long range forecast of Indian monsoon rainfall.

11. Write briefly on NE monsoons. Compare the features of SW monsoon and NE monsoon.

CHAPTER 21

A Bird's Eye View of Global Monsoon

In the equatorial regions, where ITCZ migrates and seasonal reversal of winds takes place is called monsoon region. Further, regionally periods of during which 75% or more of annual rainfall occurs are termed monsoonal areas. Regions with less than 250 mm annual rainfall categorised as deserts. These conditions are observed over Western Africa, Eastern Africa, Indian sub-continent, Tibetan Plateau, China, Southeast Asia (comprising Southeast Bay of Bengal, Malaysia, Indochina, South China sea and Phillipines), north Australia. Salient features of monsoon over these areas are described below.

Monsoon Over West Africa

Over west coast of Africa, the ITCZ remains north of the equator throughout the year. In July it occupies the northernmost position and in January the southernmost position. ITCZ does not move steadily southwards or northwards but oscillates with seasonal march northwards or seasonal retreat southwards. Rainfall extends to the northernmost latitude in August and migrates with the sun southward latitude in January. The advance of rains to the north is gradual while the retreat (withdrawal) of rains is quick. In the region north of lat 5 °N the maximum rainfall occurs during August- September. The region south of this experiences two peaks with light or no rain during July.

Westerlies prevail in the lower troposphere and easterlies aloft. Westerlies decrease in intensity while easterlies increase in intensity with height in the lower troposphere. Westerlies are cool and moist. Easterlies have two maxima, one in the lower troposphere (at 600 h Pa) called African easterly jet, and another in the upper troposphere (at 150 h Pa) called Tropical easterly Jet. Speed decreases 600 to 400 h Pa and then increases. Both the jets conform to the WMO definition.

Easterly Waves Over Africa

During northern summer easterly waves are prominently seen in the neighbourhood of ITCZ. Associated with easterly wave disturbances, squall lines are common over west Africa.

Monsoon Over East Africa

During October, the Near Equatorial Trough (NET) south of the equator acts as ITCZ. Along east coast of Africa it moves 2 °S (in October) to 12 °S (end of December) and where it stays till end of January. From this position this trough moves back (from January to end of April) to about 2 °S and gives up its role as ITCZ. The southward movement of ITCZ causes rain in Kenya and Uganda from mid-October to mid–December. This period is locally called 'short rains' season. These rains are associated with southeasterly winds which are caused by Anticyclone over Arabian Sea. During northward journey of ITCZ (November), the north NET acts as ITCZ (March to June), these regions again get rain, which are locally called 'Long rains' season. Tanzania receives rain during November to March.

Monsoon Over Tibetan Plateau

Tibet is the highest plateau in the world whose average height is about 5 km. It is also called the roof of the world. Along with the Himalayas, Tibetan Plateau was probably formed by the plate tectonic movement, which was originally located in sea. Tibet has extreme climate. Near the ground it has lowest moisture (absolute humidity), lowest temperature, lowest pressure but have strong winds as compared to other parts of the world in that latitude belt. On the contrary its planetary boundary layer (which is highest in the world) has highest temperature in summer with lowest pressure (Thermal Low) and highest pressure (Cold High) in winter. The annual rainfall varies 5 cm (in the west) to 60 cm (in the east) and classified as arid or semi-arid region. Rain occurs from April to October, but bulk of the rain is received during June to August. The annual frequency of thunderstorm days is about 90 with about 55 hail days (which is again the highest in that latitudinal belt).

During summer in the upper troposphere, the sub-tropical westerly jet (STWJ) lies north of the plateau but relatively weak, while easterlies prevail to the south of the plateau. In winter STWJ shifts southwards and strengthens. It has two branches. The stronger branch runs south of the plateau while the weaker (comparatively) branch runs north of the plateau. Cold High prevails from November to March with peak in December. Warm low prevails from June to September with maximum intensity in July. Because of this reversal of pressure, wind systems this is called Tibetan plateau monsoon. The plateau experiences cold dry weather in winter and warm, humid and rainy weather during summer. Spring and autumn are transition periods.

There appears a see-saw like monsoon activity over central India and Tibet. When monsoon is active over central India, Tibet experiences break monsoon condition. Similarly during break monsoon epochs in central India, Tibet experiences active/vigorous monsoon conditions.

Monsoon Over China

China experiences both summer and winter monsoons.

Summer Monsoon

Summer monsoon begins over south China and progresses in steps to central and then to north China. Withdrawal of summer monsoon also takes place in steps, starting from north China to south. The step movement of monsoon is closely related to the migratory movements of sub-tropical ridge (between 125 °E to 140 °E)

Summer monsoon current consists of three component currents :

1. Southwesterly current from central Bay of Bengal,

2. Southerly current of cross-equatorial flow from the vicinity of Australia, and

3. Southeasterly current from the southern flank of subtropical High located over Pacific.

During mid-April to first week of June, the sub-tropical ridge migrates 15 to 18 ° N. Over south China moist, warm southwesterly winds blow in the lower troposphere and experiences showers/rains. Locally this period is called "sub-tropical southwest monsoon". During first fortnight of June, the subtropical ridge migrates to 20 °N. In this period all the three currents referred above invade over south China. The current moves over to central parts of China and causes rains up to Yangtze River valley (Mei rains, Plum rains) and up to over south China it is called "tropical south west monsoon". By the end of first week of July the sub-tropical ridge migrates to 25 °N and summer monsoon rains commence over Yellow river valley. By the end of July the sub-tropical ridge further migrates to farther most 30 °N. Now summer monsoon rains commence over north China whereas the south China receives southeasterly current rains and hence locally called "southeast summer monsoon". By the end of August and first week of September the sub-tropical ridge migrates southwards to 27 °N. This completes withdrawal of monsoon from north China and rainfall is confined up to Yellow river valley. It may be noted here that monsoon over north China stays for about one to one and half months only. This is the main rainy season in that area. On some occasions summer monsoon may reach north China very late or remains there for very short period. This leads to drought in that area. On the contrary when monsoon is vigorous, this area experiences floods. By the last week of September the subtropical ridge comes back to 25 °N which completes withdrawal of monsoon over Yangtze valley (central China). By the middle of October Sub-tropical ridge comes back to the position of 15-18 °N which completes the withdrawal of summer monsoon over South China.

Winter Monsoon

During (October to March) winter the Siberian Anticyclone (High) gets intensified by the passages of migratory extratropical cyclone waves. As a result at the eastern end of the High north–easterlies below over east China and cause winter precipitation. The strengthened Siberian High periodically propagates the cold surges with north easterly bursts. These north easterlies propagate over south China and then to near equatorial trough (NET). Easterly waves move in the neighborhood of NET, which trigger for development of depressions/cyclones. These systems give copious rain in the equatorial region and Malaysia, north Indonesia and south China. This is winter monsoon over China.

Monsoon Over Southeast Asia

Monsoon has a pulsatory character and advances in steps. By the second week of May monsoon establishes over southeast Bay of Bengal, Malaysia, Indochina, South China Sea extending up to 22 °N. By the end of June it advances over Bangladesh, northernmost India. By the first week of July it advances over southeast china up to lat 33 °N and reaches north china 40-45 °N by the end of July. In these regions summer monsoon is reflected by tropical cyclones, locally called Typhoons. Typhoons are the most furious/violent and devastating systems over the globe. One such system took a human toll of 3 lakh people at Haiphong in 1881. The frequency of Typhoons is the highest in Phillipines with average 20-22 per annum. Summer monsoon commences withdrawal from northernmost region in the last week of August and completes withdrawal by November end.

Phillipines experiences both summer and winter monsoons. Summer monsoon extends from May to September with peak in August. Easterly waves travel close to ITCZ and help formation of tropical depressions/cyclonic storms (locally called Bagiuos) during June to December. Winter monsoon extends from November to February.

Monsoon Over North Australia

About 90% of the annual rainfall over north Australia occurs during summer November to April and experiences dry season during winter (May to October).

During onset and withdrawal phase of monsoon rainfall amount suddenly increases and drops. Like the Indian monsoon, tropical disturbances such as Lows, monsoon depressions and cyclones (locally called Willy-Willys) are experienced in Australia. The ITCZ zone in this region is characterized by horizontal low level convergence of moist air and divergence in the upper troposphere.

According to one local definition, after 1 November if at least four out of six stations around Darwin experiences rain, and if the average rain fall over p-days exceeds 19 (p + 1) mm, then the pth day is declared as onset day of monsoon. In addition to rainfall, wind circulation upto 0.9 km should be westerly with at least moderate speed and that on pth day the cumulative west component speed should exceed 5.15 (p + 1) m/s.

Based on satellite imagery, the onset of monsoon is taken when the large scale blow up of tropical convection imagery acquires width–length ratio about 1:3 (width 10° latitude and length 30° longitudes).

Australian monsoon wind circulation shows that westerlies in the lower troposphere and easterlies in the upper troposphere. The monsoon triggering mechanism is believed to be cold surges of south China sea and low level southerlies along the western coast of Australia.

The Planetary Scale Aspects of Monsoon

Monsoon circulation may be viewed as a planetary scale differential heating like the land and sea breeze circulation. Differential heating causes potential temperature and isobaric surfaces to intersect (like in the case of land and sea breeze). This intersection of surfaces generates APE (available potential energy) both in equatorial plane (x – p plane, Walker cell circulation) and meridional plane (y-p plane, Hadley cell circulation) [APE = Total potential energy (TPE) (which is sum of internal and gravitational PE) of a closed system – (minus) the minimum TPE that develops due to hypothetical redistribution of mass adiabatically]. APE is converted into KE of monsoon flow in the lower and upper stratosphere through overturnings in the zonal (Walker cell) and meridional circulation (Hadley cell). A steady monsoon circulation is maintained by the diabetic heat sources and sinks (diabetic meaning non-adiabetic). EL Nino-Southern Oscillation (ENSO) phenomena affects planetary scale monsoon. About 77% of the equatorial belt (between 10 ° N to 10 °S) around the globe is occupied by the oceans. On long term average the western Pacific is the warmest. This generates climatic Walker circulation in the x–p or equatorial plane. However this changes once in a few years when the eastern Pacific anomalously becomes warmer and disturbs the Walker circulation and in other ways also affects the planetary scale monsoon circulation.

Questions

1. Explain what is monsoon region. Describe briefly (i) monsoon over West Africa, (ii) monsoon over east Africa, (iii) Easterly waves over Africa.

2. Write short notes on monsoon over Tibetan plateau.

3. Write short notes on : (i) summer monsoon over China. (ii) winter monsoon over China, (iii) monsoon over southeast Asia, (iv) monsoon over north Australia (v) The planetary scale aspects of monsoons.

CHAPTER 22

Drought

According to US weather Bureau, drought is defined as "Lack of rainfall so great and long continued as to affect injuriously the plant and animal life of a place to deplete water supplies both for domestic purpose and for the operation of power plants, especially in those regions where rainfall is normally sufficient for such a purpose".

The term drought has different connotations in different parts of the globe. In Egypt, any year the River Nile does not flood is a drought year irrespective of rainfall. In Bali, a period of six days without rain is a drought. In parts of Lybia, droughts are recognized only after two years without rain. According to Henry (1906), 21 days or more when rainfall is 30% or less of average for the time and place is a drought. Extreme drought when rainfall fails to reach 10% of normal for 21 days or more. According to Hoyt (1936), any amount of rainfall less than 85% of normal is a drought.

The definition of drought varies from the use of water and its scarcity. Based on this a few standard verities of droughts are: Meteorological, Hydrological, Agricultural and Economical. Of all these, meteorological drought is important which directly or indirectly influences or affects other droughts.

Definition of Aridity and Drought

Aridity is both low precipitation (pptn) and low effective precipitation, where

$$\text{effective precipitation} = \text{pptn} - \text{Evaporation}$$

$$\text{Index precipitation effectiveness} = \frac{r}{t}, \text{ where } r = \text{annual rainfall in mm,}$$

t = mean annual temperature in °C.

If $\frac{r}{t}$ <40 arid, if $\frac{r}{t}$ > 160 per humid.

Aridity is a permanent climatic feature of a region of low rainfall and high temperature. While drought is a temporary feature related to variability of rainfall, when the rainfall is appreciably below normal.

Drought may occur in any rainfall or temperature regime. It may be confined to a single area or river basin or may be widespread invading over many states. Even incase of widespread it may be severe only on a small area compared to the total area affected. An American study shows that the occurrence of drought is frequent over an area where the coefficient of annual rainfall variation is more than 35% and is less where the variation is less than 25%. Further the drought occurrence is frequent where annual rainfall is small and coefficient of variation is high and they tend to go together.

If we consider a few days without rain it is of little consequence but if continuous for weeks or months, particularly at a place in rainy season it greatly affects the agriculture. If there is a succession of two or more seasons (or years) with deficient rain it may lead to disaster. In Sahel region (Africa) on the southern part of the Sahara desert, there were scanty rains between 1968-73, which lead to severe drought in 1972/73. Such deficient rain also frequently occurred in other parts of Africa, in northeast Brazil and over a large area of Australia, Indonesia, Russia and adjoining sates. This phenomena will certainly recur in future and in all parts of the world. Such prolonged droughts caused total crop failures, to decimation of livestock and widespread starvation and deaths. It has been estimated that droughts and famines in India during 1965-67 took a toll of about fifteen lakh people and caused economic losses over US $ 100 millions.

Drought essentially a meteorological phenomena, consequently meteorologists are deeply involved in national and international efforts to mitigate its damaging effects. The disasters caused by droughts are also influenced by diverse factors such as agricultural practices, change in population density and the ability of a country to provide alternative supplies of food, water and employment. As yet the occurrence, cessation, continuation and recurrence of drought cannot be predicted reliably or stopped. However it can be reasonably inferred using statistics of past data which may be helpful to governments and local authorities to combat it. From the past data of any locality it is possible to indicate probability occurrence of drought of various intensities at different times of a year presuming that the meteorological conditions recorded in the past will be repeated. This information is useful for planning land use and agricultural development by using drought-resistant crops, design of irrigation projects etc., emergency plans for the next drought when it occurs (which is of course certain).

The causes of drought are many and complex and it is not yet fully understood. Research, including simulations using general circulation models reveal linkages between ENSO events and world-wide pressure variations. It is evident that droughts in Australia and parts of south America are closely related with strong ENSOs. Monsoon rain over India, east Africa and parts of Asia are also linked with ENSOs. The severity of droughts may be measured by using the following parameters.

1. Deficiency in rainfall
2. Deficiency in river run off,
3. Soil moisture reduction,
4. Fall in ground water levels and
5. The storage required to meet the normal (prescribed) demands or drafts.

IMD Definition of Drought

A meteorological sub-division (India is divided into 35 Met. Sub-divisions 1981 on wards and 31 sub-division 1875-1980) which receives total seasonal (monsoon seasonal) rainfall less than 75% of the normal (long term average) value is considered affected by drought. If the seasonal rainfall received is 50 to 74% of the normal it is classified as moderate drought and if it is received less than 50% of the normal then it is classified as severe. If in any year the areas affected by drought of the above two types exceeds 20% of the total area of the country (about 7 sub-divisions or more) that year is considered as a drought year for the whole of the country.

Drought statistics of India for the years 1875-1987 is given in Table 22.1 and its analysis follows.

Table 22.1 Drought statistics.

Year	No.of sub-divisions affected by drought	No.of sub-divisions affected by severe drought	Percentage of Area affected	Rank of drought for whole of country
1875	2			
1876	3			
1877	16	(7)	59.5	3
1878	3			
1879	1	(1)		
1880	0			
1881	4			
1882	0			
1883	6	(1)		
1884	6	(1)		
1885	4	(1)		
1886	1			

Table 22.1 Contd...

Year	No.of sub-divisions affected by drought	No.of sub-divisions affected by severe drought	Percentage of Area affected	Rank of drought for whole of country
1887	2			
1888	3			
1889	2	(1)		
1890	2			
1891	8	(1)	22.7	21
1892	1			
1893	0			
1894	1			
1895	2			
1896	5			
1897	0			
1898	3			
1899	20	(7)	68.4	2
1900	1			
1901	7	(2)	30.0	16
1902	4			
1903	1			
1904	10	(4)	34.4	13
1905	13	(2)	37.2	8
1906	0			
1907	8	(1)	29.1	17 A
1908	2			
1909	0			
1910	0			
1911	9	(4)	28.4	19
1912	1			
1913	6		24.5	20
1914	0			
1915	6	4	22.2	22
1916	0			
1917	0			
1918	22	(7)	70.0	1
1919	0			
1920	12	(1)	38.0	7
1921	2			
1922	2	(1)		

Table 22.1 *Contd...*

Year	No.of sub-divisions affected by drought	No.of sub-divisions affected by severe drought	Percentage of Area affected	Rank of drought for whole of country
1923	3	(1)		
1924	1			
1925	5		21.1	24
1926	0			
1927	0			
1928	6			
1929	3			
1930	0			
1931	1			
1932	2			
1933	0			
1934	2			
1935	0			
1936	1			
1937	2			
1938	3			
1939	8	(2)	28.5	18
1940	0			
1941	9		35.5	9
1942	0			
1943	0			
1944	1			
1945	0			
1946	0			
1947	0			
1948	3	(2)		
1949	1			
1950	1			
1951	10		35.1	11
1952	6			
1953	0			
1954	0			
1955	0			
1956	0			
1957	1			
1958	0			

Table 22.1 *Contd...*

Year	No.of sub-divisions affected by drought	No.of sub-divisions affected by severe drought	Percentage of Area affected	Rank of drought for whole of country
1959	2			
1960	1			
1961	0			
1962	1			
1963	1			
1964	1			
1965	11		38.3	6
1966	8		35.4	10
1967	0			
1968	7		21.9	23
1969	4	(1)		
1970	0			
1971	5			
1972	16	(1)	40.4	5
1973	1			
1974	8	(2)	34.0	14
1975	0			
1976	2			
1977	1			
1978	0			
1979	10	(1)	34.8	12
1980	2			
1981	3			
1982	10		29.1	17 B
1983	1			
1984	4			
1985	9	(1)	32.3	15
1986	7	(1)	19.7	25
1987	15	(6)	47.7	4
1988	0			
1989	1			

Analysis

N = No. of years of data = 115, total number of sub-diversions affected by drought = 401.

$$\text{Average} = \frac{401}{115}$$

Mean no. of sub-divisions = 3.5

Standard deviation = 4.47 sub-divisions

Of the 115 years, all sub-divisions were free of drought in 33 years.

5 sub-divisions were effected by drought in 3 years,

6 sub-divisions were effected by drought in 6 years,

7 sub-division were effected by drought in 3 years etc.

Table 22.2

No.of sub-divison effected by drought	0	1	2	3	4	5	6	7	8	9	10	11	12	13	14	15
No. of years or frequency	33	22	14	9	5	3	6	3	5	3	4	1	1	1	0	1

No. of sub-division effected by drought	16	17	18	19	20	21	22	23-35
No. of years or frequency	2	0	0	0	1	0	1	0

P(D) = Probability of a drought year

F(D) = Frequency of drought years

F(DD) = Frequency of drought year preceded by a drought year.

P(DD) = Probability of a drought year preceded by a drought year.

From the above data we infer that the probability of occurrence of drought free year in all subdivisions of India is $P(D) = \dfrac{33}{115} (= \dfrac{F(D)}{N})$

The probability occurrence of severe drought free year in all sub-divisions of India where $P(D) = \dfrac{88}{115}$

The probability of getting drought in 7 or more sub-divisions of India $= \dfrac{23}{115}$

The probability of getting severe drought in 7 or more sub-divisions of

India $= \dfrac{3}{115}$

Case I

Assume the whole country is affected by drought (moderate/severe) when 7 or more sub-divisions of India are affected by drought. Under this assumption, the probability occurrence of drought over whole of India is

$$\frac{F(D)}{N} = \frac{23}{115} = 0.20$$

$$F(DD) = 4, \ N = 115, \ F(D) = 23,$$

$$P(D) = \frac{F(D)}{N},$$

$$P(DD) = \frac{F(DD)}{F(D)}$$

The probability occurrence of drought preceded by drought year over

India $= \dfrac{F(DD)}{F(D)} = \dfrac{4}{23} = 0.17$

Case II

If we consider India as a whole is affected by drought when 6 sub-divisions or more affected by drought, then the probability of getting a drought year is :

$$F(D) = 29, \ F(DD) = 4, \ P(D) = \frac{F(D)}{N} = \frac{29}{115} = 0.25,$$

$$\text{Two consecutive years of drought} = \frac{F(DD)}{F(D)} = \frac{6}{29} = 0.21$$

The probability occurrence of drought $P(D) = \dfrac{F(D)}{N}$ and occurrence of

drought preceded by years of drought $P(DD) = \dfrac{F(DD)}{F(D)}$ in the various Met

sub-divisions of India is given in Table 22.3.

 The Science of Weather and Environment

Table 22.3

Met sub-division	F(D)	F(DD)	P(D)	P(DD)
Coastal Andhra Pradesh	12	1	0.104	0.083
Telangana	21	2	0.18	0.1
Maratwada	16	1	0.14	0.06
Viderbha	15	1	0.13	0.07
Madhya Maharashtra	9	1	0.08	0.11
Konkan and Goa	9	0	0.08	0
North Interior Karnataka	8	1	0.07	0.13
South Interior Karnataka	7	0	0.06	0
Coastal Karnataka	4	0	0.03	0
Kerala	9	2	0.08	0.22
Tamil Nadu and Pandichery	10	0	0.09	0
Gujarat	26	2	0.23	0.08
Saurastra and Kutch	26	5	0.23	0.19
West MP	12	1	0.10	0.08
East MP	8	1	0.07	0.13
West Rajasthan	30	7	0.26	0.23
East Rajasthan	19	2	0.17	0.11
Jammu and Kashmir	25	7	0.22	0.28
Punjab	20	5	0.17	0.25
Haryana, Delhi and chandighar	21	4	0.18	0.18
Himachal Pradesh	20	4	0.17	0.2
West UP	13	0	0.11	0
East UP	13	1	0.11	0.08
Bihar Plateau	4	0	0.03	0
Bihar Plains	10	0	0.09	0
Orisa	4	0	0.035	0
Sub-Himalayan west Bengal	6	1	0.05	0.17
Gangetic west Bengal	2	0	0.02	0

The Table 22.4 gives the probability levels of drought [m = rank. n= no. of years of record, probability level $F\,a(m) = \dfrac{100m}{n+1}$] in ranking order, n = 115.

Table 22.4

Year	Rank (m)	% area affected (No.of sub-divisions affected)	Probability level (in percent)
1918	1	70 (22)	0.86
1899	2	68.4 (20)	1.72
1877	3	59.5 (16)	2.59
1987	4	47.7 (15)	3.45
1972	5	40.4 (16)	4.31
1965	6	38.3 (11)	5.17
1920	7	38.0 (12)	6.03
1905	8	37.2 (13)	6.90
1941	9	35.5 (9)	7.76
1966	10	35.4 (8)	8.62
1951	11	35.1 (10)	9.48
1979	12	34.8 (10)	10.34
1904	13	34.4 (10)	11.21
1974	14	34.0(8)	12.07
1985	15	32.3 (9)	12.93
1901	16	30.0 (7)	13.79
1907	17 (A)	29.1(8)	14.66
1982	17 (B)	29.1(10)	14.66
1939	18	28.5 (8)	15.52
1911	19	28.4 (9)	16.39
1913	20	24.5 (6)	17.24
1891	21	22.7 (8)	18.10
1915	22	22.2 (6)	18.97
1968	23	21.9 (7)	19.83
1925	24	21.1 (5)	20.69
1986	25	19.7 (7)	21.55

If follows from Table 22.4, that the probability 50% area of the country affected by drought is 3.28% i.e. in 100 years 50% area of the country may be affected by drought 3 to 4 times. The probability 25% area of the country • •• •• •• ••• cted by drought is 17.13% i.e. in 100 years 25% of the country may be affected by drought is 17 to 18 times or in 10 years about 2 times.

To sum up droughts may start at anytime and may last for a long time, attain many degrees of severity. They are caused by general climatic fluctuations associated with persistent large scale aberrations of atmospheric circulation which favored subsidence over the region. The likely factors are: air sea interaction, injunction of large scale ash and dust into the atmosphere, changes in the composition of atmosphere, particularly water vapour, carbondioxide and ozone (which are selective absorbers of radiation that could modify heat balance of the earth).

Questions

1. Define drought. Distinguish between drought and Aridity. State different connotations of drought over different parts of globe.

2. What are the meteorological features or causes of drought and its effects? How to counter act? Summarize them.

3. State IMD definition of drought. Bring out the salient features of Indian drought from past studies. Is it possible to state the probability occurrence of drought from the past historical data?

CHAPTER 23

Natural Disasters

Extreme natural events become disasters when they affect human settlements, economic and social activities. Natural hazards have no political or geographical boundaries and no one can stop them. They are associated with geological, atmospheric, ecological and biological events. Natural hazards can be categorized as meteorological (including hydrological) and other types. The most frequent and common natural hazards that are associated meteorological events are: Tropical cyclones, Hurricanes, Typhoons, Tornadoes, waterspouts, thunderstorms (including hailstorms), Nor westers, Sand and dust storms, heat and cold waves, widespread snowfall. The hydrological hazards are floods and droughts. The other types of disasters are volcanic eruptions, earthquakes, atmosphere and water pollution, forest fires infestations etc. The main three disasters that occur suddenly are earthquakes, tropical cyclones and floods. Of these the later two may be forecasted nearly precise and more practicable. It may be noted here that non-atmosphere hazards prediction is not yet reached the stage of practical use while atmospheric hazards predication achieved tremendous progress. In advanced countries (USA, India, European countries and Japan) loss of life due to cyclones and floods have been drastically reduced through excellent weather forecasts, warnings and preparedness. The devastation of atmospheric natural hazards further can be reduced to a great extent through early warning system, good emergency measures, cyclone shelters, protection barriers against floods and storms surges and prompt delivery of relief. Telecommunication plays an important role in hazard reduction through prompt delivery of forecasts and relief. Public education and awareness helps to reduce the impact of any disaster. It may stressed here that natural disasters will remain the same in future as at present but a great deal could be done to reduce the death, distruction that they cause.

According to World Bank estimates global natural disaster damages between 1950-1990 rose from $ 400 lakh to $ 6520 lakh. Number of natural disasters between 1975-2005 rose 100 to 400 (four times) and it is said that "Nature creates hazards but man makes it disasters". According to one study (emergency data base) of the world natural disasters the most frequent are in India, which ranks fourth among the top ten considering human casualties. If we consider number of people affected, India stands third after China, Bangladesh.

It is worth noting that in 2004 hurricane a poor state Greneda suffered 200% of its GDP while rich Bermuda faced even stronger hurricane (category 5) suffered moderately.

The study of natural disasters in India from 1900-2005 indicates that there had been 160 floods, 21 droughts and 24 earthquakes.

Statistics of ten major types of natural disasters that occurred world over during 1947 to 1980 given in Table 23.1 to enlist their ferocity by ranking, (Bindi V. shah 1983).

Table 23.1

S.No	Type of disaster	No.of deaths nearest thousand	Percentage of total deaths about
1.	Tropical cyclones, Hurricanes, Typhoons	499000	41%
2.	Earthquakes	450000	37%
3.	Floods (other than associated with cyclones)	194000	16%
4.	Thunderstorms and Tornadoes	29000	2.4%
5.	Snowstorms	10000	About 1%
6.	Valcanoes	9000	0.7%
7.	Heat waves	7000	0.6%
8.	Avalanches	5000	0.4%
9.	Landslides	5000	0.4%
10.	Tidal waves including Tsunamis	5000	0.4%

Some of the worlds worst natural disasters are given in Table 23.2.

Table 23.2

Year	Type of disaster	Place	Approximate no.of deaths
526 BC	Earthquake	Syria	250000
1556	Earthquake	Shanxi, China	830000
1703	Earthquake	Japan	200000
1737	Earthquake and Cyclone	Calcutta – India	300000
1976	Earthquake	Tianjin, China	242000
1943-44	Famine	Bengal, India	1500000
1969-71	Famine	Northern China	20000000
1642	Flood	Huanghe River, China	300000
1887	Flood	Henan, China	900000
1939	Flood	Henan, China	1000000
1450	Bubonic plague (Black Death)	Europe and Asia	75000000
1918	Influenza	World wide	22000000
1881	Typhoon	Vietnam, Haiphorg	300000
1970	Cyclone	Bangladesh	200000
2004	Tsunami	Sumatra, Indian Ocean	300000
1876	Cyclone	Backerganj, India	300000

Landslides and Avalanches

Any disturbance in natural balance triggers landslides and avalanches in high mountains like the Himalayas. These cause disruption of road communication, loss of property and life in hilly terrain. Landslide and avalanches are of great concern to road design and geotechnical engineers, road users, Government administration, geologists and defence services besides local inhabitants.

Landslides

A landslide or mud flow is a downward movement (in hilly region) of rock, soil or debris flow under gravity. Landslides occur when the ground is stressed beyond its frictional strength. The following factors contribute to the landslides.

1. Erosion by rivers, glaciers or ocean waves creating over steepened slopes (soil erosion is the wearing away of land surface by natural agencies of wind and water).

2. Strength of the rock or soil.

3. Seismic zones

4. Topography, ground, water and vegetation.

5. Accumulation or gathering of excess weight by rain, snow, flood and waste piles.

6. Man made causes like mining, terrain cutting and filling. Of all these the steepness of slope and amount of water have the greatest correlation with landslides.

In 1920 an earthquake induced landslide in China (Kansu province) took a huge toll of human life (about two lakh people). In Columbia (in 1985) a devasting landslide wiped out town and villages. It took a human toll of about (20000) people. In India Jammu Kashmir, HP, Uttatranchal, Sikkim, Siachen, Kulu, Manali, (southern peripheries of Himalalyas) the incidence of landslide, avalanches are a regular phenomena every year. February 2005 witnessed very intense snowfall (15 to 45 ft depth) in Kashmir, and neighbourhood. This caused landslides and avalanches for a full one week and completely cut off the valley from rest of India, several hundred people lost their lives despite army and airforce best assistance on war footing to the local administration.

Similarly in December 1995 a landslide near Kulu buried about 100 people.

Preventive Measures

Preventive measures differ from place to place. Earth retaining structures, both cement masonery type and flexible type are strongly recommended. Afforestation and turfing of slopes with suitable trees/plants is a long term measure against landslides.

Avalanches

An Avalanche consists of snow, ice, air, water and soil impurities. An avalanche is the downward slide or descent of a large mass of snow/ice on slope having tremendous velocity and force. Avalanche is categorised as powerful and destructive natural force. A sizeable part of our country in the foot of Himalayan region remains cut off from the rest of the country due to avalanches. Snow avalanches occur in winter months and take a heavy toll of life, a few highways remain blocked for several months (5-7) in a year. The frequency of avalanches are more where the slope is 35° to 45°.

Snow avalanches and landslides are grave problems in northwest and Central Himalayan regions which are inhabited or frequently used by local population, for communication, winter sports, mountaineering and defence.

Avalanche Hazard Mitigation in India

In India there is no civil organisation neither at the state level nor at the national level to mitigate the avalanche hazards. Snow and Avalanche Study Establishment (SASE) of DRDO laboratory provides snow avalanche information and forecasting to armed forces. The SASE aims at to provide nowcasting and also forecasting about avalanche hazards to the population residing in landslide/avalanche prone Himalayan region. For this purpose SASE established snow measuring meteorological observatories in Jammu and Kashmir (J & K), Himachal Pradesh (HP), Uttaranchal (UR). These observatories communicate data to Avalanche Forecasting and Mountain Meteorological Centres at: (i) Srinagar-for Kashmir valley, (ii) Sasoma-for Siachen and Nubra valley and (iii) Manali-for Himachal Pradesh. These centres collect data from automatic weather stations, satellite remote sensing data pertaining to latest terrain, qualitative and quantitative snowfall information, precipitation and snow cover. Satellite based sensors provide high resolution data from visible and IR, thermal wave lengths. This data facilitates identifying terrain conditions, avalanche sites, snow albedo, surface temperature and snow cover. GIS (geographical information system) together with satellite data helps SASE forecasters to prepare Avalanche Bulletin. These bulletins are broadcast on AIR and DD regularly, which are proved to be of immense help to the local inhabitants, winter tourists, traffic regulation authority, winter sports, mountain- earing department. It created great awareness among the users and thus helping in mitigating the avalanche/landslide hazard impact.

Preventive Measures

Soil stabilization by terracing geofabric, grouting etc., should be resorted. Afforestation on large scale in areas prone to landslide help arrest slides.

Disaster Preparedness

The aim of disaster preparedness is safety of life and lessening the damage. After the occurrence of a natural disaster people often talk about the phenomena of disaster and conclude that nothing could have been done to prevent its occurrance. Though we may not be able to change weather itself/ disaster phenomena, but we can certainly prepare ourselves to withstand its damaging effects. This is precisely the objective of preparedness. Help can be provided before, during and after a disaster. Meteorological services render help that is relevant whether the disaster is of meteorological nature or non-meteorological. It is an established fact that weather information contributes a lot to effective responses during emergencies. To achieve optimum level of preparedness, education about natural disasters, disaster management, environmental studies must begin at school level and should continue throughout the life and activities. The investment in weather and environmental science education fetches immense benefit in the long run of a nation and an individual in society.

The Role of a Weather Forecaster in Disaster Preparedness

Prediction of a weather phenomena at a certain place and time is a tough job, but over a large area and time the prediction can be accurate and precise. For example it is very difficult to predict the passage of thunderstorms, tornado or cyclone at a particular village/town/city but certainly be accurate over an area say 500-1000 sq Km and over a period of next 12 hours. In western countries nowcasting (running commentary of weather passage) has been introduced to avoid uncertainty.

Early warning of any severe weather helps the management to take necessary precautions to avoid disaster. Weather satellites have been proved to be a boon to meteorologist in collecting of space based weather data from inacessible remote areas of the globe and in dissemination of warnings to those areas.

Using computers and general circulation weather models it is now possible reasonably to predict weather 5 to 10 days in advance (this is termed Medium Range Forecast). It is now an established fact that there is a drastic reduction in cyclone related deaths throughout the world. This is achieved through improved timely warnings to the concerned people and Government officials and awareness among the people, advanced communication system like satellites, TV etc. For example in case of November 1977 Andhra cyclone about 10000 people lost their lives, but in May 1990 Andhra cyclone of the same intensity, the loss of life camedown to less than 1000, although property loss in the both cases was very heavy. It is emphasized that even the 99% correct forecast by itself is of little use if it does not reach in time to the concerned people, the decision makers in an unambiguous local language and the gravity of the situation is not understood by them, and the executive machinery not equipped to combat it.

Disaster Preparedness for Floods and Droughts

We shall now consider about the preparedness for floods and droughts.

Floods are assessed by guage heights, peak discharge and volume of flow. Of these gauge height at fixed locations along the river are more realistic as other parameters of flood are derived from it. Flood forecasting includes mapping of river system, its drainage basin, assessment about the run off of rainfall or snow melt and the capacity of river to drain water from its basin area and time. Flood preparedness includes flood frequency analysis of various magnitudes with the help of past data, return period analysis of floods, probable maximum precipitation . Flood forecasting models are based on anticident rainfall, real time rainfall and gauge heights and quantitative precipitation forecasts.

Watershed management is necessary for hazard reduction in vulnerable areas prone to the incidence of cyclones, floods and droughts. Construction of contour bunds across the slopes, micro level check dams, percolation tanks, afforestation, pasture development, soil and moisture conservation are preparedness measures.

In India about 40 million hectares of area is prone to floods according to National Flood Commission report. About 20% of this area is affected annually. Of this 3.5 million hectares is a cropped area. Floods take a toll of 1400-1450 human lives every year. Indo-Gangetic-Brahmaputra plains susceptible to floods every year. Generally lower parts of all river basins in India are prone to floods. Uttaranchal, UP, Bihar and West Bengal and Assam are chronic flood prone areas. Lower reaches of Mahanadi, Godavari, Krishna are also flood prone areas. Between 1954-1989 about Rs. 2500 crore were spent for the construction of new embankments, drainage, channels and afforestation.

Flood plain management is the first requirement in flood damage mitigation. Flood plain management takes into account of flood–process, population density and property situated in it. Flood frequency studies provide the basis for land use planning with objective of maximising long term productivity and minimising the risks to life and property in case of floods. It helps in design flood control structures and water use projects. The flood disaster scope and risk changes with the land development and increasing population. In general deforestation leads to soil erosion and rise in flood frequency. Similarly flash flooding increases with the expansion of urban areas. To minimise such potential risks land use pattern has to be reviewed periodically. This method also helps in damage control due to droughts which is the reverse of floods. Drought studies help in assessing severity of water shortage that is expected.

Advanced technologies can be successfully employed for flood monitoring. Multiple satellite data from IRS series, Landsat, ERS and Radarsat are useful in flood mapping, flood damage assessment, river configuration mapping, flood hazard zone mapping. Optical (visible image) data is not possible when it is cloudy. Microwave data has all weather capability.

Preparedness for droughts is a complex subject because its occurrence, continuation, cessation and recurrence can neither be predicted reliably nor can be stopped. However using past data the probability occurrence of various intensity of droughts at different periods (seasons) of a year can be inferred reasonably. This information has to be used in planning land and water use, agricultural development by taking up drought resistant crops, design irrigation projects and emergency plans for the next drought when it occurs.

Agricultural drought mitigation can be achieved in four ways :

1. Soil and moisture conservation, contouring across the slope, scooping land, opening ridges and furrows, rain water harvesting.

2. Reducing transpiration (using hydroxy sulphorates), growth retardants (using cycocel), radiation reflectant (using hydrated lime), plant hardening (using calcium chloride).

3. Uses of drought resistant plants–short duration crop, deep root system, dwarf plants, thick stems and disease resistant plants.

4. Reorientation of cropping pattern, inter cropping, life saving irrigation and contingent crop planing in case of erratic monsoon.

Preparedness to Mitigate Climate Change Disaster

Climate change disaster mitigation, preparedness seems an impossible task by any Country. "Mother nature is fortunately a lady of infinite variety. She cannot be constricted into a straight jacket tailored by modelers". We know that many ancient civilization and animal species have been wiped out because of climate change. The climate of the earth has been changing very slowly right from the beginning and it will continue to do so. However during the last century after industrial revolution, there has been lot of damage to the climate. Human activities causing severe pollution of atmosphere, water and soil by toxic pollutants. More land has been brought under cultivation during last 100 years than in all preceding human history. Global water use increased by five folds between 1940-2000 and also fossil fuel consumption. As a result of these, global warming is taking place. There had been green house gases concentration, depletion of stratospheric ozone layer (Antarctic ozone hole), contamination of vital food chains over land and sea by toxic chemicals and acidification of lakes, dying of forest due to acidic deposition and regional smog formation. Radioactive fallout from nuclear bomb tests/explosions. Long-range transport of air borne pollutants observed from Chernobyl power plant accident in USSR.

According to WMO report, global warming of 1.5 to 4.5 °C may result in sea level rise of 20 to 140 cm, increase in frequency and severity of floods in coastal regions. United Nations Environment programme UNEP and Koyoto convention aimed taking actions at : (i) lowering the rate of change in the composition of the atmosphere, (ii) entire elimination of CFCs (which are green-house gases and destroyer of stratospheric ozone layer). (iii) low rate of building up of CO_2 in the atmosphere, (iv) low down in the destruction of tropical forests and reforestation in all parts of the world.

Tsunamis

In Japanies language Tsunamis meaning harbour waves. In Pacific ocean and Hawaian Islands Tsunamis are prominent. They cause great damage to life and property in coastal areas. Tsunamis are very long waves, caused by submarine earthquakes, explosions/volcanoes and great landslides into oceans. Tsunami waves travel round the globe. Like Rossby waves, they have great wavelength about 600-1000 km, wave height 5-50 m and travel with a speed of about 500-1000 Kmph in a circular wave form, the centre of the circle being the position of earthquake/explosion/landslide. These waves show their fury in shallow waters, but not so much in deep water. Though they travel round the world but their destruction diminishes radially as they move farther and farther from the position of earthquake/explosion in sea bed. The wave height ranges 15-50 m depending on the intensity of the earthquake/explosion. Tsunami wave heights are magnified when they coincide with the lunar tide. Even the earthquakes also trigger, more violently when they coincide with solar/lunar tides (of full moon and new moon). Any sub-marine earthquake intensity above six on Richter scale likely to develop Tsunami waves. Tsunami waves become more and more catastrophic with the increasing intensity of the earthquake.

In case of Sumatra Tsunami (26 December 2004) it was noticed that the submarine earthquake intensity was 8.9 (subsequently corrected as 9.3) on Richter scale, created wave heights ranging 15-45 m. The wave height decreased as it travelled a radial distance of 20° lat or longer.

It caused devasting effects in and around a radial distance of 15° lat/long. It virtually swallowed everything within a radial distance of 5-10° lat/long. It took a death toll of more than 3 lakh human beings, of which about 2.3 lakh people were within a radial distance of 5° lat/long. The Tsunami wave fury was more on western and northern direction as compared to eastern and southern direction. This may be due to earth's rotation and coriolis effect. It has to be made clear that Tsunami waves are entirely different from storm surge (tidal) waves, the later are dependent on the central pressure defect (in the eye) and movement of the cyclone.

Earthquakes

Introduction

An earthquake is a detonation of a system underground that develops under suitable conditions of nature of fault and time. ChangHeng, a Chinese philosopher invented the first earthquake recording instrument in 132 AD, which detected an earthquake that occurred at a distance of about 640 Km (400 miles) away and it was not felt by the people at the site of the instrument. The mechanism of the instrument, however was not known to others. Nowadays earthquakes are recorded by a seismographic network of instruments.

There are many types of earthquakes, the main are : tectonic, volcanic and plutonic earthquakes. During an earthquake the slip of one plate (very huge block of rock) over another releases energy that causes the ground to vibrate and this process continues. In this way earthquake energy travels in a wave form.

Different aspects of an earthquake are measured by differents ways, of this magnitude is the most common. It measures the size of the fault at the source of the earthquake and commonly measured by Richter scale.

Richter scale measures the wriggle (twist and turn) on the recording, but other magnitude scales measure the different parts of the earthquake.

Intensity is a measure of the shaking and the damage caused. Logarithmic magnitude scale measures the size of the earthquake. This magnitude scale is called ML, where L stands for local. This is eventually became the Richter scale. This scale is valid for certain frequency and distance range.

On similar lines of Richter scale ML, body wave magnitude–MB, and surface wave magnitude–MS are recently developed by Japanese. MB and MS scales are used beyond 6.5 and 8.3 respectively.

It is observed, during the formation of an earthquake, the temperature near the epicentre region increases. The susceptibility of matter is inversely proportional to the absolute temperature and the permeability of the matter varies with temperature. The variations of susceptibility and permeability changes the earth's horizontal magnetic field, dip and declination. Hourly changes of these elements provide us a clue for forecasting the occurrence of an earthquake.

Interior Structure of the Earth

Seismology is the science of earth quakes. In order to understand the nature of seismic wave propagation and detection a quick survey of the interior structure of the earth is essential. The salient features of the interior of the earth consists of : (i) Lithosphere or earth crust, (ii) Mantle and (iii) Core.

The earth's outer shell or layer is called the earth's crust or Lithosphere. Litho-means stone. Earth's crust is not monolithic but divided into strata. The top stratum is made of sediments or sedimentary rocks, second layer is made of granite rocks and the third layer is made of basaltic rocks. The density of basaltic > granite > sedimentary rocks. The three crust layers are noticed everywhere in continents. The boundary between the sedimentary and granite rocks has not been given any name, the boundary between granite and basaltic rocks is called Konard discontinuity. The boundary under basaltic (between crust and mantel) is called Mohorovicic or simply Moho discontinuity.

The average thickness of the crust is about 35 km which is very thin as compared to the radius of the earth (about 6400 km). The average thickness of the earths crust under oceans is 5-10 km, while its thickness is about 50 km below continental mountain ranges. The average density of the earth is 5.5×10^3 kg/m^3, Lithosphere 2.8×10^3 kg/m^3, Mantle 3.2 to 5.7×10^3 kg/m^3, and in Core 9 to 12×10^3 kg/m^3. The mass of the earth is 6.00×10^{24} kg, lithosphere 5×10^{22} kg, mantle 4.05×10^{24} kg (68% of the earth), and Core 1.88×10^{24} kg (31% of the earth).

Note :   Volume of the earth $= \dfrac{4}{3} \pi R^3$

$$V = \dfrac{4}{3} \times \dfrac{22}{7} \times (6370)^3 \approx 1.08 \times 10^{21} \text{ m}^3$$

$$\text{Average density of the earth} = \frac{M}{V}$$

$$= \frac{6 \times 10^{24}}{1.08 \times 10^{21}}$$

$$\cong 5.55 \times 10^3 \text{ kg/m}^3$$

The materials below the earth crust have higher densities. The depth of mantel is about 3000 km, while that of core is 3000 km.

A discontinuity means an interface and indicates marked change of material properties. Below the earth crust lies Moho discontinuity (at an average depth of 35 km, where density is 3.3×10^3 kg/m^3), which separates the earths crust from mantel (the interior of the earth). The central core (below the mantel) probably consists of liquid outer core (thickness about 2000 km) and solid inner core (thickness about 1000 km). Gutenberg discontinuity separates the mantle from the central core and is located at a depth of about 3000 km, where the density is about 5.7×10^3 kg/m^3. There is still uncertainty about the composition of materials in the interior of the earth. It is generally viewed that the principal constituents of Lithosphere are : Oxygen (93.88%), Silica, Aluminium, Iron, Calcium, Sodium, Potassium and Magnesium (granite and basaltic rocks) all together they makeup 98.5% of the earth's crust by weight. Earth's crust predominantly contains Sial (Silica, Aluminium compound) and sima (Silica, Magnesium compound). Mantle consists of Oxygen, Silica, Magnesium and Iron (Iron, Magnesium silicates). The common silicate in the mantle is probably olivine, (formula $(Mg \, Fe_2) \, Sio_4$). These silicates are similar to those found in stony meteorites. The core probably consists of Iron, Sulphur in the combined form as FeS and Nickel. (Fe, FeS, Ni). Core contains nife (Nickle, Iron compound).

The building blocks of lithosphere are rocks. The generic classification of rocks are : Igneous, Sedimentary and Metamorphic rocks. The rock cycle gives the relationship between three types of rocks. Igneous rocks are formed by melting of earth materials. Sedimentary rocks are formed by sediment deposits on the seafloor or continental deposits of fragmentary materials. Metamorphic rocks are the transformed rocks of Igneous and Sedimentary rocks deep down the earths crust. In Greek, Meta means change, Morpha means form. The contour of the continental crust of the earth is abruptly cut off. It gives way to ocean zones where the granitic layer is missing and the basaltic layer comes close to the surface as shown in Fig. 23.1. Sedimentary layer is very thick below the continents and becomes thin in ocean area, where as the granitic layer almost absent (in ocean area).

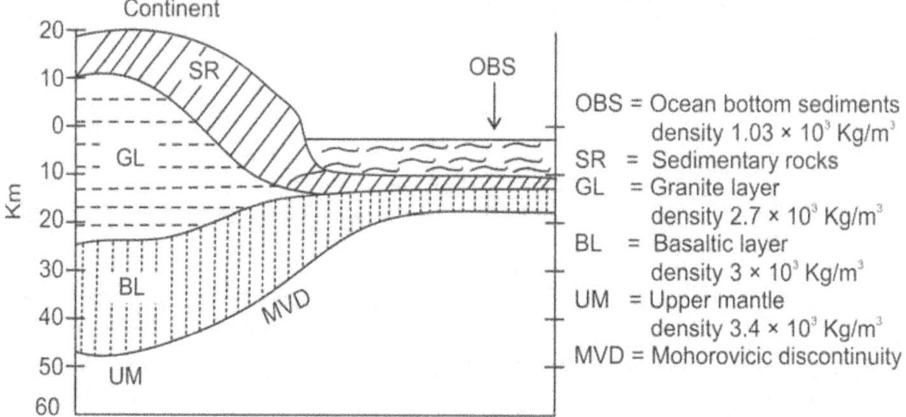

Fig. 23.1 Constituents of earth's crust.

The temperature of the interior of the earth increases with depth. In lithosphere it increases about 1 °C for every 30 meters. However this increase will not continue to the centre of the earth. Observations indicate that the average increase in temperature is about 1 °C / km. The cause of increase in temperature with depth is suggested to be the three sources of heat, (i) generated by the pressure of overlying rocks, (ii) primordial heat (original heat from the time of the earth's formation) and (iii) by the radioactive mineral disintegration.

The interior of the earth behaves like solid elastic in respect of earthquakes and earth-tides. It is known that ocean rises and fall twice in 24 hours due to gravitational attraction of the moon on earth. In a similar way an earth-tide occurs in the earth crust and mantle of the earth which causes rise and fall of earth crust twice a day. It is interesting to know that the buildings rise and fall about 30 cm under the action of these earth tides. Geneva based synchrotron works only during complete rest period and it will not work even if there is slighest change in earths crust due to ebbs and flows occurring far away from the unit. It is found, the Geneva Synchrotron operates about 30 hrs a week and the rest period the surface of Switzerland is vibrating. It has been scientifically proved that the entire territory of Moscow daily raises or falls about 50 cm from a certain average level. These rise and fall are caused by the gravitation of the moon and the sun.

Minerals of the Earth Crust

The material of the earth's crust contains various elements and compounds but they are not uniformly distributed. Minerals are naturally occuring solids composed of one or more elements in a definite chemical composition and atomic structure. Examples Gold, Salt, Calcite, Quartz, Topaz, Graphite, Diamond etc. The properties of minerals depend on : Form (crystalline, amorphus, crystallised), structure (fibrous, granular, columnar etc), cleavage (break along certain planes), fracture or appearance of broken surface of a mineral (splintery, uneven etc), tenacity (brittle, malleable, elastic, ductile etc), colour, lustre, transparency and hardness. Moks scale hardness of certain minerals are given below.

1. Talc, 2. Gypsum, 3. Calcite, 4. Fluorite, 5. Apatite, 6. Orthoclase, 7. Quartz, 8. Topaz, 9. Corundum, 10. Diamond.

Plate Tectonics

According to Indian cosmology (the science of the origin of the universe) the earth consisted at one time of seven continents joined together. They separated like the lotus petals (or leaves) from Mount Meru, the centre of the universe. Afterwards the continents floated and drifted away from the centre and were separated by seven oceans. An amateur Russian astronomer Y.V. Bykhanov observed a remarkable coincidence of the outlines of the American and Euro-African coastlines that they fit well without a crack if these are moved together. By this observation, in 1877, he conjectured that once a uniform continent split into parts and ever since they have been moving away. In 1910, Alfred Wegener (1880-1930) a German geophysicist propounded a theory of continental drift. According to him the original single continent Panghela (all earth) was broken up into pieces. Like the lotus leaves, they separated from one another and floated away giving rise to the modern continents divided by oceans. In 1960's this hypothesis was further modified by the evidence of ocean floor structure and named tectonics of plates or global tectonics. According to this hypothesis instead of continental move, plates of large areas of the earth's crust containing both continents and adjoining sections of ocean floor moved. According to this plate tectonics hypothesis there are six major plates. (i) Euro-Asiatic plate, (ii) African plate, (iii) Antarctic plate (iv) Indo-Australian plate, (v) American plate and (vi) Pacific plate. In addition to these major plates there are several minor plates located between them and they move to some extent independently. (See Fig. 23.2).

According to Staub, another German scientist, the earth consists of constant shifts of the continent Gondawana Southern continent (occupied by the present India, Australia, Africa and South America together with large portions of Indian and Atlantic oceans) and the Northern continent Laurasia (which was located in the present position of Eurasia and north America). Laurasia and Gondawana kept colliding with each other under the action of centrifugal

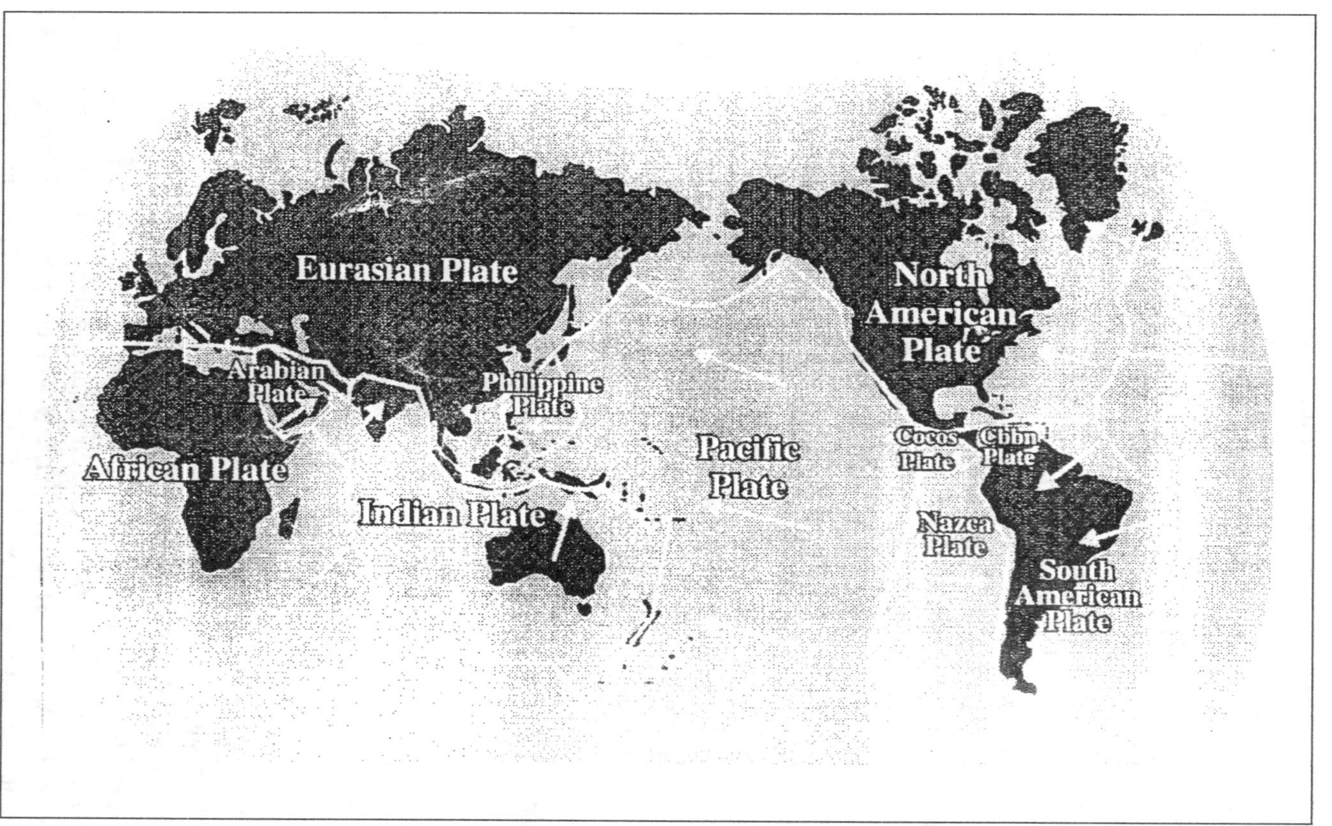

Fig. 23.2 Major global tectonic plates and average direction of motion. Subduction occurs along colliding plates. After toksoz (1975).

forces and drifting away along the meridians. Recent thinking is that Gondawana ended by vertical shifts, various sections of Gondawana displaced sometime upward and sometimes downward along the lines of faults. Further it is believed that a part of Gondawana sank in the present position of Indian ocean. The islands in Indian ocean are the peaks of high mountains in that part of Gondawana. It is also believed that parts of the Pacific ocean and the region of Atlantic ocean between south America and Africa are also a sunken section of the integral continent of Gondawana.

> **Note :** Gonds is a tribe in Madhya Pradesh, Wana a country in India. These two words together make up the name Gondwana.

There are three types of plate boundaries (i) Extension or divergent boundaries (ii) Compression or convergent boundaries and (iii) Transform faults.

Extension boundaries are formed when two adjacent plates move apart. Material from below swells up and a new crust is produced at the crests of the oceanic ridges. Thus the plates on both sides are added up. Compression boundaries are formed when two adjacent plates approach each other. In this case surface is destroyed. The line along which plate destruction takes place is called trench. A third type of boundary forms when the plates move laterally relative to each other. These are called transform faults. In this case neither crust in formed nor destroyed. It has been found that Eurasia and American plates are converging at a rate of 2-4 cm per year, while Amercian, Eurasian and African plates are increasing in size and that of Indian and Antarctican plates are not changing greatly in size.

Note :

Fault : A fracture in a rock mass or rock layer whose opposite faces move independently is called fault.

Tectonics

According to one hypothesis earth's crust has large plates. These plates move due to convective forces that emanate from beneath the crust and create rifts. It is assumed that the upper mantle consists of Newtonian viscous fluid and the convection currents are generated due to heating from below or insitu. The lower mantle consists of very dense fluid which inhibit convection.

Folds

During tectonic movements stratified rocks develop bending. These are called folds.

A new crust is being formed at some boundaries of these plates. All such boundaries are located in the oceans. The earth's crust building up towards American plate on one side and towards African and Euroasiatic plates on the other side. The plates collide at the boundaries which leads to the submergence or subduction of one under the other plate. Such submergence taking place in case of Pacific plate under Euroascatic plate. It is theorised that where old crusts are buried they provoke for earthquakes and volcanic erruptions. The present scientific thinking is that earthquakes are caused by the friction on the boundaries of the plates moving together. Powerful tremors are caused by the accumulated shear stresses which periodically exceed the rock strength. From the heating of the sedimentary layers of the submerging plate volcanic erruptions take place.

Theory of plate tectonics provide a simple and clear explanation of the cause of earthquakes. According to this theory massive ocean ridges and deep ocean trenches are the active zones of earthquakes. Volcanic action is frequent along ridges. Trenches indicate breaks in the earths crust between the plates. A third kind of plate edge is called transformed fault. This edge is a long tear in the lithosphere where horizontal movement takes place. Along ocean trenches crustal rock is destroyed as one plate is pushed beneath another. This is called subduction.

Earthquake Waves

A disturbance which progresses from one point to another point in a medium with transfer of energy but without the transfer of matter is called a wave motion. Elastic waves are mechanical disturbances propagated in an elastic medium. A wave is called longitudinal or compressional if the particles of the medium vibrate in the direction of wave propagation. Longitudinal waves travel through solids, liquids and gases.

A wave is called transverse or shear if the particles of the medium vibrate at right angles to the direction of propagation. Transverse waves travel through solids but not through liquids and gases.

Earthquakes are the vibrations or tremors induced in the earths crust that agitate a part of the crust with all structures and things lying on it. These waves are generally caused by abrupt movement of earth along the line of fracture or break in the rock structure.

An earthquake wave generates different kinds of waves. Primary or P-waves, which are compressional and travel through solids, liquids and gases. Secondary or S-waves, which are shear waves and hence travel through solids only.

P and S- waves may not result in actual displacement of a land mass on the surface of crust. When a part of land mass in a region displaces from its original place and occupies a new position at a lower level it is called land slide.

L-wave or long surface waves travel through surface layers of the earth.

The velocity of P-waves is given by $\upsilon = \sqrt{\dfrac{\lambda + 2\mu}{\rho}}$

Where ρ = density of the medium λ and μ are elastic constants depend on the rigidity of the medium.

The velocity of S-waves is given by $v_1 = \sqrt{\dfrac{\mu}{\rho}}$

The study of seismic waves produced by earthquakes provide valuable information about the nature of matter inside the earth in its path. Seismic waves travel deep down into the earth from earthquake site and return to the earth's surface at some distant point. Seismographs are used to detect the seismic waves on their arrival at the surface of the earth. They also provide the information of type of the wave, its intensity and time of arrival. The speed of seismic waves partly depend on the density of the material through which they pass.

Earthquake Parameters

(i) Time of origin is the time at which the earthquake has occurred (ii) Focus or Hypocentre is the point inside the earth where from the earthquake originated. (iii) Duration of an earthquake generally less than one minute. (iv) Epicentre: The surface point vertically above the focus is called epicentre. It is expressed in latitude and longitude of the point. (v) Focal depth is the depth of the focus from the surface of the earth. The hypocentre of an earthquake is the combination of epicentre and focal depth.

Classification of Earthquakes

(i)   Shallow eathquakes have focal depth $\leq$ 70 km

(ii)   Intermediate earthquakes have focal depth between > 70 km, $\leq$ 300 km.

(iii) Deep earthquakes have focal depth > 300 km

Earth quakes have not recorded focal depth exceeding 720 Km.

Magnitude (M)

The magnitude of an earthquake is a kind of instrumental measure of its size or energy (E), and is given by a linear equation

$$\log_{10} E = a + b\,M \qquad\qquad .....(23.1)$$

Or $\qquad\qquad E = 10^{(a + bM)}$

where a and b are constants, E = Energy, M = Magnitude

The values of constants are generally found to be a = 11.8, b = 1.5

If M = 4 we have $E_1 = 10^{\,a + 4b}$

M = 6 we have $E_2 = 10^{\,a + 6b}$

$$\frac{E_2}{E_1} = \frac{10^{a+6b}}{10^{a+4b}} = 10^{2b}$$

or $\qquad E_2 = E_1 \ 10^{2b} = E_1.10^3$

Thus the energy radiated by earthquake of magnitude 6 (E_2) is 10^3 times the energy radiated by the earthquake of magnitude 4 (E_1).

If E_3 is the magnitude of earthquake 8, then

$$E_3 = E_1 \ 10^{4b} = E_1 \times 10^6$$

i.e., $\qquad E_3$ is 10^6 times E_1.

$\log E_s = 11.8 + 1.5 \ M$, where E_s in ergs, M is Richterscale magnitude.

If $M = 8.25$, E_s = Energy = 3.68×10^{10} kwh or 13.248×10^{23} ergs.

$M = 7.5$ $E_s = 2.86 \times 10^9$ kwh or 10.296×10^{22} ergs.

Seismic energy (E_s) yield for different magnitudes (M) given in Table 23.3.

Table 23.3

M	E_s in TNT	M	E_s in TNT
5.0	32000 tons	8.5	5 billion tons
5.5	80000 tons	9.0	32 billion tons
6.0	1 million tons	10.0	1 trillion tons
6.5	5 million tons	12.0	160 trillion tons
7.0	32 million tons		
7.5	160 million tons		
8	1 billion tons		

Intensity

It is based on effects of the earthquake on buildings, topography, land slide etc., i.e, on macroseismic effects.

The waves spreading from an earthquake centre pass through different rocks at different velocities. Seismic wave have a maximum velocity of 12.5 km per second. On passing through the Mohoviricic boundary the primary seismic waves speed up from 6.5 to 8 kmps and secondary waves from 3.7 to 4.5 kmps. However on passing from mantle into core the primary waves speed drops 12.5 to 8.5 kmps and the secondary waves from 7.5 to 5 kmps. The spreading waves from an earthquake centre pass through different rocks at different velocities. Since the waves are reflected, and refracted on their way, they reach observatories at different times. On an average more than one lakh (10^5) earthquakes of varying intensities are registered over the globe every year by the seismological observatories. The data is evaluated by using 12 grade modofied Mercalli (M.M) scale. The scale can be divided into several groups

1. Slight Internsity or First Group

It consists of first three grades which are weak and not imperceptible earth tremors. These are sensed by some animals. Most of domestic animals become restless, birds fly away from the place of earthquake. Cats fur stands on end. It is also said that second and third grade tremors are sensed by nervous people.

2. Moderate Intensity or Group Two

It consists of grades 4, 5 and 6. These shocks are felt by every one. In this group objects hanging on walls move, hanging lamps/bulbs, chandliers swing to and fro. Cracks develop in some houses. Tall factory chimneys may fall down.

3. Severe Intensity or Group Three

It consists of 7, 8 and 9 grades. The strength of this group is distructive and devasting. Tall buildings may fall down, cracks may occur in the ground and occasional human casualities are noted.

4. Catostrophic or Very Severe Intensity or Group Four

This last group consists of 10, 11 and 12 grades. The strength is described as catastrophic. Many buildings will collapse except structures on monolithic rocks which may not be affected. The shocks and yawning cracks will be very severe. These may cause eletrical fires.

Causes of Earthquakes

The most common cause of earthquake is tectonic activity. This mechanisms is shown in Fig. 23.3. The vertical dashed line (PQ) shows fissure or fault in the solid earth crust. By slow prolonged tectonic movement in the lithosphere one side X of the fault (Fig. 23.3 A) is displaced in relation to the other Y. This is shown in (Fig. 23.3 B), by a deformation of straight lines drawn across the fault (P' Q'). This process continues until the stresses thus generated in the fault zone overcomes the friction between the two sides X and Y. Then a rupture (sudden displacement) occurs. After this the configuration is shown in Fig. 23.3 C. It is this sudden rupture constitutes an earthquake. The slow process is repeated and a new shock occurs at some time later. This mechanism of earthquake is called elastic rebound theory of tectonic earthquakes. Almost all earthquakes occur by this mechanism. However there may be some tremors by volcanic activity. Large earthquakes may occur with a volcanic explosion. The collapse of cavities can be the origin of minor tremors.

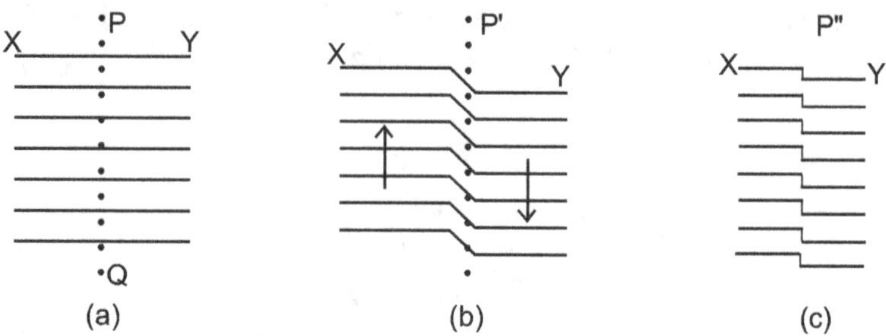

Fig. 23.3 Elastic rebound theory of tectonic earthquakes.

Long ago the super continent Pangea broken and drifted like the lotus leaves. A plate containing India broke away from it and drifted towards Asian land mass. When this plate collided with Euro-Asian plate sediments of ancient sea bed sqeezed together in huge folds and slowly rose to form the Himlayas.

Indian plate consisting of India and parts of Indian ocean is moving at an average speed of 5 cm / year in the north-northeast direction and colliding with Eurasian plate along the Himalayas. This resulted in faults and fractures in the Himalayas. These are responsible for some great earthquakes in the Himalayan region in the past.

Categorisation of earthquakes	**Richter scale**
Tremor or microearthquake	Magnitude < 3.0
Slight earthquake	Magnitude ≥ 3 to < 5.0
Moderate earthquake	Magnitude ≥ 5 to < 7.0
Great earthquake	Magnitude ≥ 7.0

Important Seismic Belts

1. First belt, the most important Seismic belt runs along the Pacific and includes western coast of South America, North America, eastern coast of Asia, the Island of sourth coast Pacific and New Zealand.

2. Second belt runs from south Pacific Islands through Jawa, Sumatra and Central Asia mountains, further passing through Caucasus mountains to Greece, Italy and Spain.

3. Third belt which is not so important runs from north to south in the middle of the Atlantic ocean.

Seismcity of India

IMD maintains a catalogue on earthquakes in India and neighbourhood from available historical records and also instrumental data. See (Fig. 23.4 & 23.5). This catalogue is continuously updated. From this records it is observed that moderate to great earthquakes have occurred all along the Himalayan region,

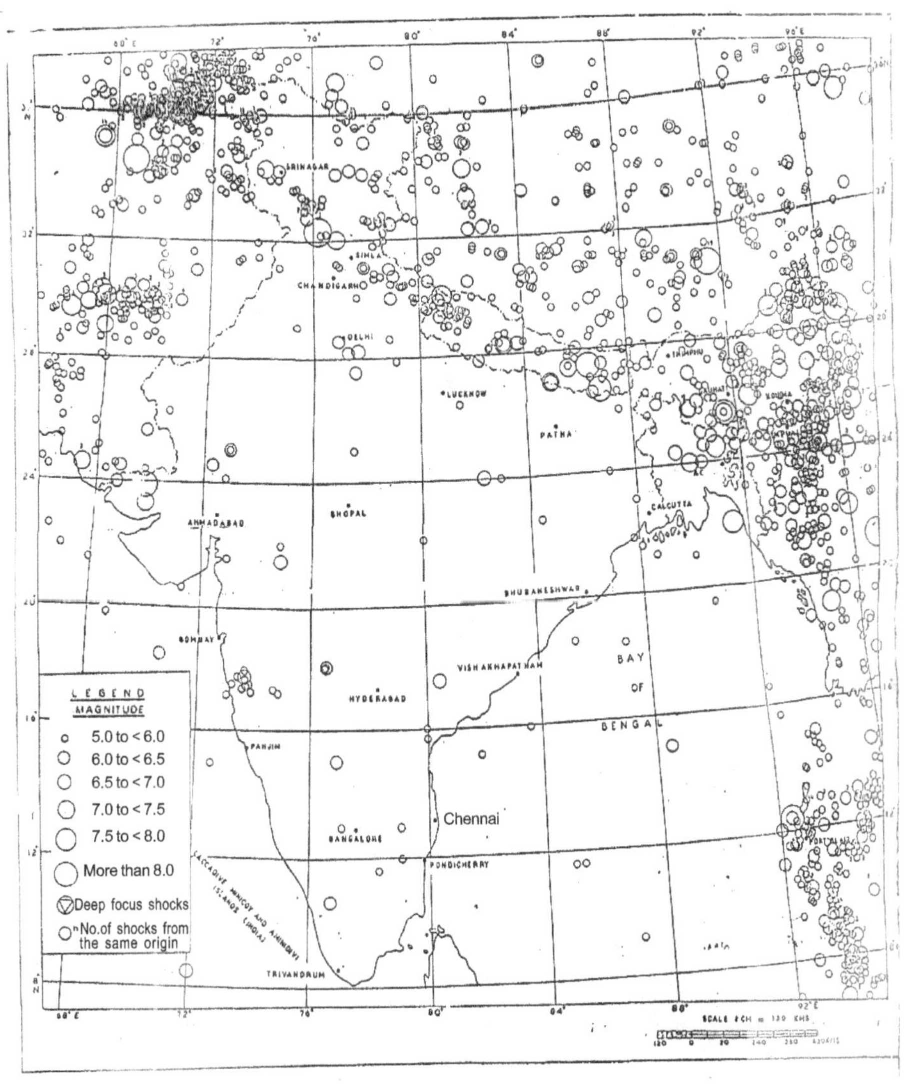

Fig. 23.4 Map of India showing epicenters (upto 1994).

Source : India Met Dept.

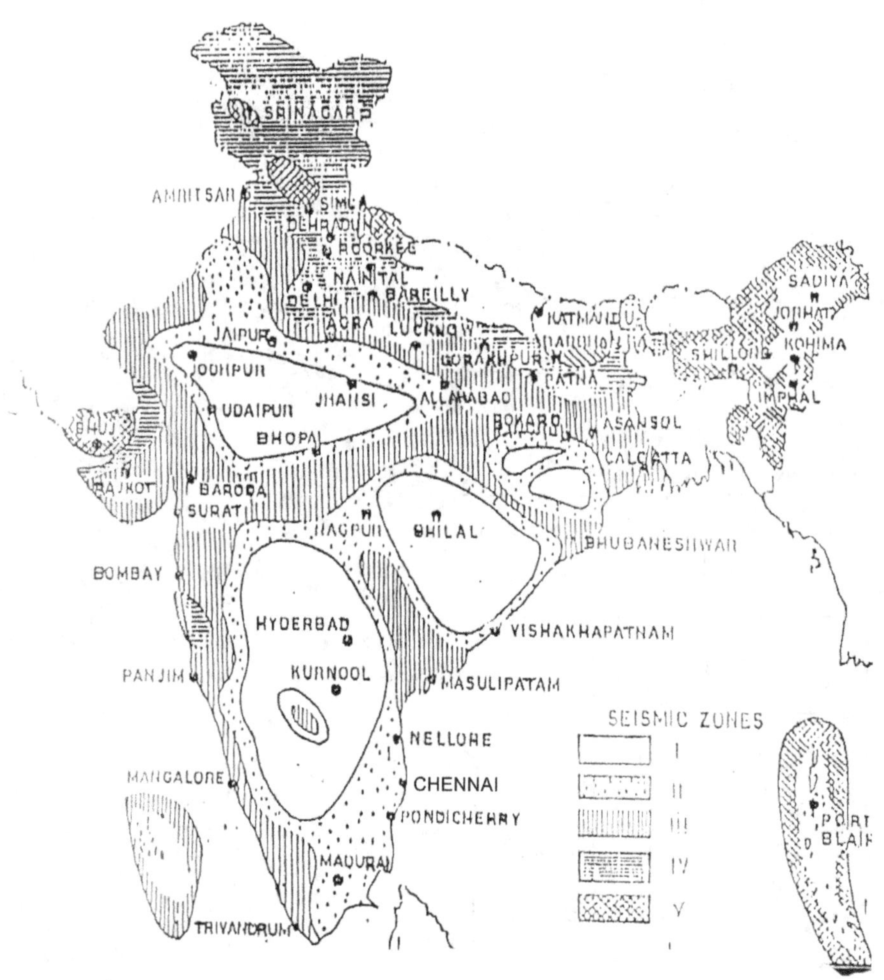

Fig. 23.5 Sketch map of India showing seismic zones (source IMD).

the Rann of Kutch, Manipur, Mynmar (Burma) belt and further continuation to Andaman and Nicobar islands. Scattered earthquakes also occurred but less frequent in peninsular India with magnitude less than 6.5. Latur (Osmanabad district) earthquake had magnitude 6.3. The catostrophic earthquakes of magnitude ≥ 8 is given in Table 23.4.

Table 23.4

Date	Place	Magnitude
12.06.1997	Assam	8.7
15.08.1950	Arunachal Pradesh	8.5
15.01.1934	Bihar-Nepal border	8.3
26.06.1941	Andaman islands	8.1
04.04.1905	Himachal Pradesh	8.0
16.06.1819	Rann of Kutch	8.0

Seismic Zoning of India

According to Bureau of Indian Standards (vide : Is . 1893: 1984, published in 1986) India is divided into five seismic zones based on a number parameters (see Fig. 23.4). Zone V is the most active while zone I is the least. The zoning MM. intensites are given in Table 23.5.

Table 23.5

Zone	MM Intensity grades	Areas
I	5 or less	Remaining parts of India covered by zones III, IV, and V
II	6	
III	7	Kerala, parts of Maharastra, Gujarat, Goa, MP, Punjab, UP and parts of WB.
IV	8	Remaining parts of J & K, H.P, Bihar, northern parts of U.P, W.B, parts of Gujarat, Western parts of Maharastra coast.
V	9 or more	Entire NE-India, parts of J & K, H.P, hills of W-U.P, Rann of Kutch, north Bihar, Andaman and Nicobar Islands.

Questions

1. What are the natural hazards associated with meteorological (hydrological) events and others? What is the difference between these two?

2. What are the overall preparedness to mitigate the natural disasters.

3. Name the first five global disasters and write briefly about them?

4. Describe briefly about Landslides.

5. Write briefly about the Avalanches and its mitigation in India.

6. Write salient points of disasters preparedness and the role of meteorologist.

7. Write short notes on Tsunamis.

8. What are floods. Name the measures of the flood disaster mitigation.

9. What are features of Koyoto convention on global warming and its likely adverse effects.

10. Write an essay about earthquakes?

11. Write briefly the structure of the interior of the earth with discontinuties, densities and temperature variations?

12. Write short notes on : (i) Minerals of the earth's crust (ii) Plate tectonics, (iii) Folds.

13. Write different types of earthquake waves their propagation velocities.

14. Name the earthquake parameters, classification earthquakes.

15. Write the formula of earthquake magnitude and energy of different Richter scale magnitudes.

16. Write the earthquake intensity and its classification.

17. Explain the earthquake elastic rebound theory of tectonics.

18. Write short notes on the (i) Global seismic belts (ii) Seismicity of India.

CHAPTER 24

Tropical Cyclones

Tropical cyclones are the most destructive phenomena of atmospheric nature. They are (unsurpassed) foremost in their violence, destruction and duration. They strike the coastal areas extending thousands of square kilometers with their ferocity of wind, torrential rain and inundate the area with storm tides. Compare with the other natural disasters, an earthquake causes much more devastation of life and property but it is short lived, only one or two minutes. Tornadoes are the most violent storms on earth but they are small and cause destruction in their narrow path for a short period. The extra-tropical cyclones of winter are the largest phenomena of atmosphere but they are much milder as compared to their tropical cousins. Besides the loss of life and property, the trail left behind by the tropical cyclones even after their passing is faced with lot of difficulties in reconstruction. It is estimated that a normal cyclone precipitates water about 2 Giga (10^9) tons in an hour and dissipates about 36×10^{10} kilowatt-hours of kinetic energy. A moderate cyclone releases large amount of energy, of which about 3% is in the form of wind and wave energy. This energy is sufficient for 100-150 million people for a year. With all its ferocity and destruction, the blessings of cyclone is that it gives large quantities of water which fills lakes, ponds and reservoirs without which many places would have been arid.

Henry Piddington, curator of the Calcutta (India) museum (in 1855), coined the word cyclone from the Greek words Kuklos (meaning circle) and Kykloma or Kukloma (meaning wheel or coil of snake). Thus cyclone meaning coils of a snake

Tropical storms are known by different names over the globe. Over Indian seas they are called cyclones. In western Pacific they are called Typhoons (when wind speed exceeds 118 kmph or 64 kt), Hurricanes in Atlantic and eastern Pacific. In Australia, they are called willy-willies, in Philippines as Baguios and in Mexico as Cordonazo. The word Typhoon is derived from Chinese-taai means great and fung means wind. Thus it designates intense storm. Hurricane is derived from Hurakan a French word, which is borrowed from Caribbean Indian to denote tropical storms in Gulf of Mexico and Atlantic.

Our knowledge of cyclones is gradually built up as and when more and more sophisticated instruments are invented. Initially our information of cyclones was based on ship reports and coastal observations when they approached the coast. Subsequently air craft reconnaissance reports, radar observations were added up. Since 1960 weather satellite observations immensely contributed to enhance our knowledge and now with these observations none of the cyclones escape detection.

What are cyclones? Tropical storms are intense low pressure (or T.C) systems that form mostly over sea areas. They germinate as tropical disturbances.

The classification of these disturbances by IMD using wind speed over sea areas is given Table 24.1.

Table 24.1 Classification of tropical disturbances.

Pressure system	Beaufort scale	Wind Speed		Satellite picture current Intensity	Pressure deffect h Pa	Pressure gradient/ 250 km h Pa
		Knots	Kmph			
Low pressure area	4	lessthan 17 (10-17)	19-28	–	–	5
Depression	5, 6	17-27	29-49	1.5	4.0	5-7.3
Deep depression	7	28-33	50-61	2.0	4.5	7.3-12
Cyclonic storm	8, 9	34-47	62-88 (17-23 mps)	2.5-3.0	6.1-10	13-18
Severe Cyclonic storm	10,11	48-63	89-117 (24-32 mps)	3.5-4.0	15-21	18
Very severe cyclone	12 or more	64-119	118-220 (33-61 mps)	4.5-6.5	29-80	18 or more
Super cyclone	more than 12	more than 120	more than 220 Kmph (more than 61 mps)	more than 6.5	more than 80	18 or more

The Table 24.2 gives the International Hurricane Scale (IHS).

Table 24.2

IHS number n	Wind speed Vn			
	mps	kmph	knots	mph
1.0	33	118	64	73
1.5	40	144	78	90
2.0	46	167	90	104
2.5	52	186	100	116
3.0	57	204	110	127
3.5	61	220	119	137
4.0	65	235	127	147
4.5	69	250	135	156
5.0	73	263	142	164
5.5	77	276	149	172
6.0	80	288	156	180
6.5	83	300	162	187
7.0	87	311	168	194
7.5	90	322	174	201
8.0	93	333	180	207
8.5	95	343	185	213
9.0	98	353	191	220
9.5	101	363	196	226
10.0	103	372	201	232

Relationship $V_n = 32.7 \sqrt{n}$ when expressed in mps.

$V_n = 63.5 \sqrt{n}$ when expressed in Knots

$V_n = 117.7 \sqrt{n}$ when expressed in kmph

$V_n = 73.2 \sqrt{n}$ when expressed in mph

Where n is IHS number, Vn wind speed.

Tropical cyclone resembles a giant heat engine see Fig. 24.1 that derives its energy mainly from the transfer of sensible and latent heat from sea to air. The main energy input is water vapour (a form of latent heat). A suitable wind arrangement acts as the starting mechanism. The kinetic energy of these winds provide the initial independent energy. It acquires spin (vorticity) from the coriolis effect. Condensation of water vapour during vertical ascent transforms latent heat into sensible heat. Upper level outflow (divergence) together with low level inflow (convergence) develop radial motion. This circulation in the vertical plane must stand the influences that tend to destroy it. High tropospheric currents provide the cooling system which carries the excess heat to other regions of the globe. Consequently, tropical cyclones form over warm sea waters, where sea surface water temperature must be 26 °C or more. These systems dissipate (die) if they enter over cold waters. Equatorial regions between latitudes 5 °N and 5 °S are not favorable for the formation of cyclones due to the weak or absence of coriolis force. The average life span of a cyclone is about six days until they land or recurve into temperatare latitudes. Some storms last for only a few hours, while others last as along as two weeks.

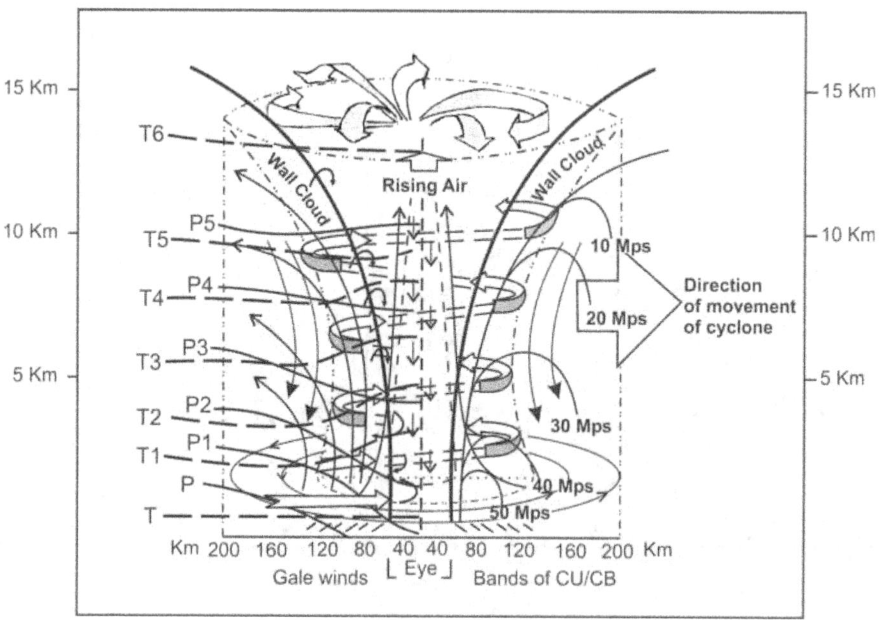

Fig. 24.1 Schematic diagram of a mature cyclone,
with cloud, temperature, pressure and wind distribution.

Tropical cyclones have four stages of life cycle: Formative, immature, mature and decaying stage. There is no set duration that a storm may be in one stage. The storm may skip any one stage or go through in such a short period that it is not possible to distinguish with the available data. On some occasions it is difficult to say in any one stage of development.

Formative Stage

This stage spans the period from the genesis of cyclonic circulation as a low pressure area to the stage when the system intensifies into a severe cyclonic storm. In this stage closed vortex emerges, winds are confined to a small area. Pressure fall is slow. Central pressure reaches 10 h Pa below normal. Formative stage is a very slow process and may spread over several days.

Immature Stage

In this stage central pressure falls rapidly, winds strengthen and these two achieve their maximum limits. This may occur in 24 hrs or sometimes develop from Deep Depression to severe cyclonic storm in few hours. Cloud bands get organized together with rain. The strong winds will be confined to a small area.

Mature Stage

Circulation expands outwards but central pressure remains steady (constant). The area of hurricane winds (more than or equal to BF12) increases from the centre. This stage may last for several days. Symmetry may be lost.

Decaying Stage

In this stage system weakens which may be due to storm crossing the coast or entering over cold water or recurve and move over to extra tropical latitudes. Decay stage may also be rapid.

It may be noted here that on most of the occasions (85%) depressions may intensify into cyclonic storms within 48 hrs of their formation, and within 12 to 24 hrs on about 40% of the occasions. Cyclonic storms attain severe intensity within 36 hrs on 75% occasions while depressions intensify into cyclonic storms within 12 hrs.

More than 50% of the cyclonic disturbances that form in the Bay of Bengal in March, April, May, November and December intensify into cyclonic storms.

A fully developed cyclonic storm can be divided horizontally at surface level into four parts: Eye, Wall-cloud region, Belt of strong winds, outer region (see Fig. 24.1).

The Eye

The centre of the cyclone has calm or light winds, no rain and clear sky. It is called the eye of the cyclone. It is generally circular and has diameter of the order 20-50 km. Because of this there is saying 'uneasy calm before a storm". The lowest pressure in the storm is recorded in the eye.

Wall Cloud Region

A ring shaped inner region surrounding the eye with towering Cb cells (or wall-cloud) with heavy rain and maximum wind is called wallcloud region. The width of the ring varies 10 to 20 km. According to some eye witness reports it looks like a burning inferno due to intense sheet lightning.

A Belt of Hurricane Winds

A belt of hurricane winds (speed 118 Kmph or more) surround the wall cloud region, radius extending to 80 Km from the centre. This region contains hot towers (Cb clouds with heavy convection) about 100 to 200, which produce squall lines, torrential rain, thunder and lightining. Wind in squalls exceed 300 Kmph

Outer Region

In the outer region (extending 200–500 km from the centre) of the cyclone winds decrease radially outwards. This region is also characterized by convection but to a limited extent. The diameter of a tropical cyclone is of the order 100 to 1000 km or more. Some have diameter as small as 30 km.

Between the belt of strong winds and outer region, there is relatively clear or Moat area with little convection.

The greatest damage caused by a cyclone is due to the hurricane winds and tides. The area of hurricane wind is always small. No one knows precisely how high wind speeds can rise in tropical storm. At sea, storms produce a distinctive heavy swell, that effects ocean shipping. As the storm approaches a coast the piling up of water by strong winds produce a disastrous storm surge (tidal wave). A hurricane wind speed of 100 kt creates a roughly a wave height of about 30 m and the KE of about 10^{19} J. The liberation of this much energy and its conversion into heat however does not warm the ocean surface water but cools it. It mixes the ocean water to a depth of about 200 to 400 m in an area about 50–100 km of radius.

Tropical Storm statistics

The global annual average of Tropical cyclones (based on about 100 years data) is 80, and its annual variation is about 8 (i.e., 10%) see Fig. 24.2. Year to year variation of cyclonic disturbances and cyclonic storms are large. The average life period of global Tropical cyclones is 6 days. In Indian seas the annual average of cyclonic disturbances is about 16 (Bay of Bengal 13) and standard deviation is 3.1. Its variation is 7 in 1984 and 23 in 1927. The annual average of Tropical cyclones is about 6 (or 7% of the global total), standard deviation is 1.85. Its variation is 1 in 1949 and 10 in each of the year 1893, 1926, 1930 and 1976. The average life period of cyclones in Indian seas is 1.5 days. In northern Indian ocean 65% of the cyclonic disturbances do not recurve, while the remaining 35% recurve.

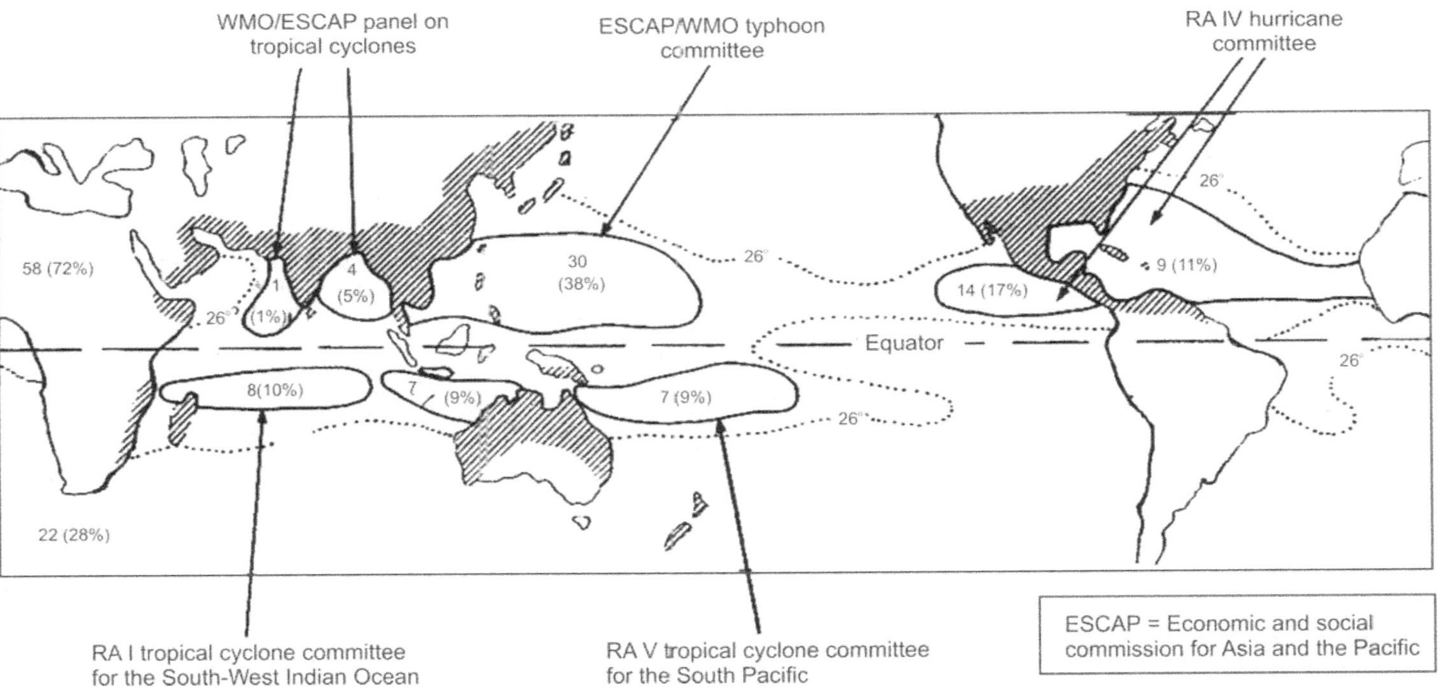

Fig. 24.2 The global occurrence (numbers and percentages) of tropical cyclones. Cyclones form only where sea temperatures exceed 26°C (dotted line); the land areas affected are shown hatched (after W.M. Gray. 1975). The names of tropical cyclone bodies and the basins covered by their programmes are also indicated..

The number of Tropical cyclones (cyclonic storms and severe cyclonic storms) that approach or strike the east coasts of Srilanka, Tamilnadu, Andhra Pradesh and Bangladesh is highest during October to December, while that affects the Orissa coast, west Bengal coast and Arakan coast is highest during June to September and March to May (Post monsoon and monsoon period).

The worlds longest record life period of TC is 31 days, in case of Hurricane Ginger, formed in Atlantic ocean in 1972, while in Indian seas 14 days (2-15 Nov 1886, 16-29 Nov 1964). The radius of global Tropical cyclone (TC) vary 50-100 to 2000 km. A large number of the storms in north Indian ocean had diameter less than 100 km.

Maximum TCs form in western parts of north Pacific (about 33% of the global TC may form through the year). Southern parts of eastern Pacific and south Atlantic are free fromTC.

In Indian seas most of the cyclonic storms form in the months of April, May and in October, November and December. The cyclonic disturbances over Bay of Bengal are more frequent than over the Arabian sea. Tropical disturbances of land origin are confined to southwest monsoon period (June to September). During January, February and March the tropical storms do not form in the Arabian sea. In the Bay of Bengal they are a few and far between. They usually originate between latitudes 5 °N and 8 °N, move in a westerly or north–westerly direction and strike the North Tamilnadu coast or the east coast of Srilanka. During April, May they form between latitudes 8 °N and 15 °N, move in a north–westerly or northerly direction initially and then recurve. In Bay of Bengal the whole of east coast of India. Coastal areas of Bangladesh and Arakan coast of Myanmar are likely to be hit by the storms, particularly in May. In the Arabian sea these storms move towards the coast of Arabia. A few move in a northerly direction towards Maharashtra, Gujarat coast. In June there is a striking change from May. Storms in Bay of Bengal originate between latitudes 16 °N and 21 °N and west of longitude 92 °E. They move in a north–westerly direction and weaken after crossing the coast. In Arabian sea most of the stroms confined to the area north of lattitude 15 °N and east of longitude 65 °E. During July and August stroms in the Bay of Bengal form between lattitudes 16 °N and 21 °N and west of longitude 92 °E. They move in a north–westerly or west–north–westerly direction between latitudinal belt of 20 °N and 25 °N. There is an abrupt fall in the frequency of storms in the Arabian sea from June to almost nil in July, August and September. In September Bay storms originate north of latitude 15 °N and west of longitude 90 °E, move initially in a westerly or North westerly direction and later recurve towards north-north-east. During October, November the frequency of storms increases rapidly. Bay storms originate between latitudes 8 °N and 14 °N, move initially in a northwesterly or west-north-westerly direction. Most of them later recurve to north-east wards. During these months north coastal Tamilnadu, Andhra coasts and Bangladesh coast are relatively vulenerable to the incidence of storms. In Arabian sea initially they move in westerly or north-westerly direction up to lat 15 °N, then recurve north-east wards and strike the Maharashtra – Gujarat coasts. During December the frequency of disturbances fall. Most of the Bay storms originate between latitude 5 °N and 10 °N, move initially in a northwesterly direction and strike Tamilnadu coast or northeast coast of Srilanka. A few of these cross over peninsula and enter into Arabian sea. It may be noted here that the position formation of storms migrates with the movement of the sun both north wards and south wards.

Some Special Features of Tropical Cyclone

Eye

Formation of an eye indicates the stability of vortex. Surrounding the eye has wall cloud region which is characterized by strong vertical upward motion. In contrast the eye has downward vertical motion or subsidence. In general hurricane eye develops when surface wind speed exceeds 65 kt. Wind temperature inside the eye is comparatively warmer than the neighborhood. During intensification eye shrinks and development of eye is a sign of intensification of cyclone. As said earlier the lowest pressure is recorded in the eye. This is also called pressure eye. Barographic trace shows a flat region with little bumping caused by the fluctuations see Fig. 24.3. In mature storms, in Indian seas (Bay of Bengal), the diameter of the eye is about 10–25 km, while in case of Typhoons, it is about 50–100 km. Radar observations indicate that eye does not remain constant but undergoes constant transformation. This is called Radar eye. Satellite eye features will be discussed separately.

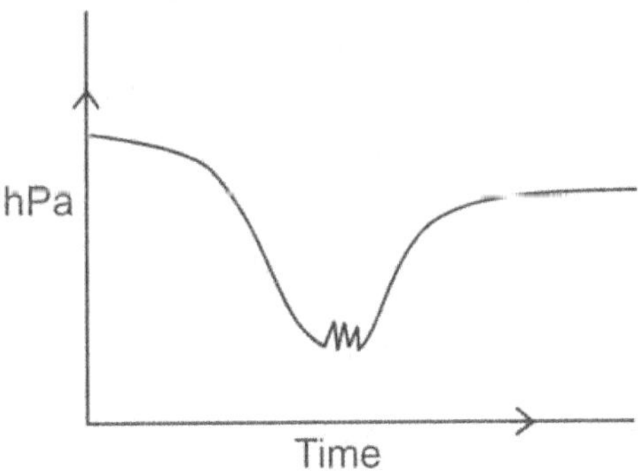

Fig. 24.3 Barograph trace of cyclone eye.

Double Eye

In a very severe cyclone there would be two cloudless zones in the central region and two associated cloud walls. These are called inner and outer eye or in generally called double eye and double cloud wall. The two concentric eye walls with different radii from the centre of a cyclone is an indication of a very severe nature of a cyclone. Double *eye* wall is noticed in continuous radar observation. The *eye* wall looks like a funnel with narrow section at sea level and broder section at the top of the cyclone vortex (compare with the shapes and structures of dust devils, Tornadoes). In case of weakening of cyclone radar observations indicate that the outer eye wall (radius about 30-35 km) gets contracted to about half its original size while the inner eye wall weakens and disappears. Rainfall in the cloud wall will be very heavy (may be more than 5 cm/hr) but decrease rapidly and becomes nil at the centre of the eye. Rainfall also decreases outward from the wall cloud region but gradually.

Spiral Bands

Radar and satellite pictures clearly indicate axial asymmetry of cyclone structure. This fact is also observed in wind, cloud and rainfall. It is observed that spiral bands (some times called rain bands or feeder bands) converging from the outer periphery which gradually contract inward to eye wall and there is strong inflow into the eye wall. Heavy to very heavy precipitation occurs in these spiral bands (3-5 cm/hr). The spiral bands appear to have a life period of several days but the convective cells in these bands have short life period of about half an hour.

Pressure Field

The lowest pressure in a cyclone occurs in its eye region. In the cyclone field the constant pressure surfaces dip downward towards the centre from sea level to about 200 h Pa and there after it reverses to form a hump at the eye region Fig. 23.4. This shows horizontal convergence of winds from surface to about 200 h Pa and divergence above it. Pressure field is not uniform and is not symmetric about the centre. Isobars rarely appear circular. In case of intense cyclone, during mature stage the inner most isobars may appear circular otherwise elliptical. The pressure gradient to the right side of the direction of motion is more (about twice as compared to the left). The intensity of a cyclone depends on the pressure deficiency at the eye and generally the defect is 5-10% from that of normal. The global lowest pressure of 870 h Pa was recorded in Typhoon "Tip" in Pacific on 12-10-1979. In this case the pressure defect was about 13% from normal sea level pressure (1000 h Pa) and the sustained wind speed was 165 kt. The next lowest pressure of 899 h Pa was recorded in Hurricane Allen over northwest Atlantic on 7 Aug 1980 which had maximum sustained wind speed of 165 kt.

The most intense cyclone on record in Indian seas was False point, Orissa cyclone, in which the central pressure of 918.9 h Pa was recorded on 22 September 1885. In this cyclone the pressure drop was 91.1 h Pa and maximum sustained wind speed was 136 kt. However in case of Diviseema, Andhra cyclone of 19 November 1977, the estimated central pressure was 911 h Pa based on Satellite NO AA-5 with intensity T-7. In this case the pressure defect was 99 h Pa and maximum sustained wind speed (MSWS) was 140 kt. (see Fig. 24.5).

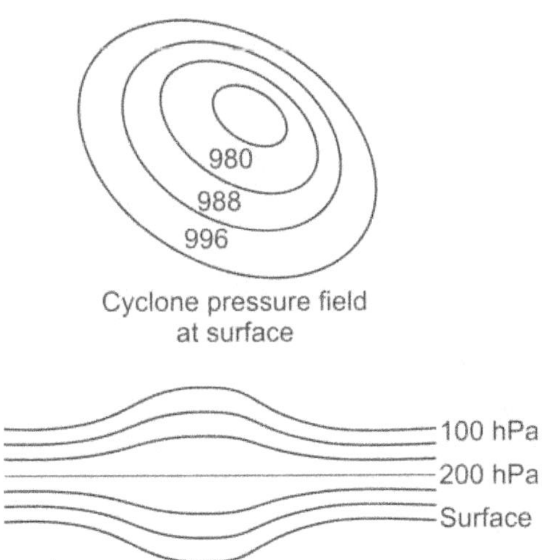

Fig. 24.4 Constant pressure surfaces near the cyclone eye.

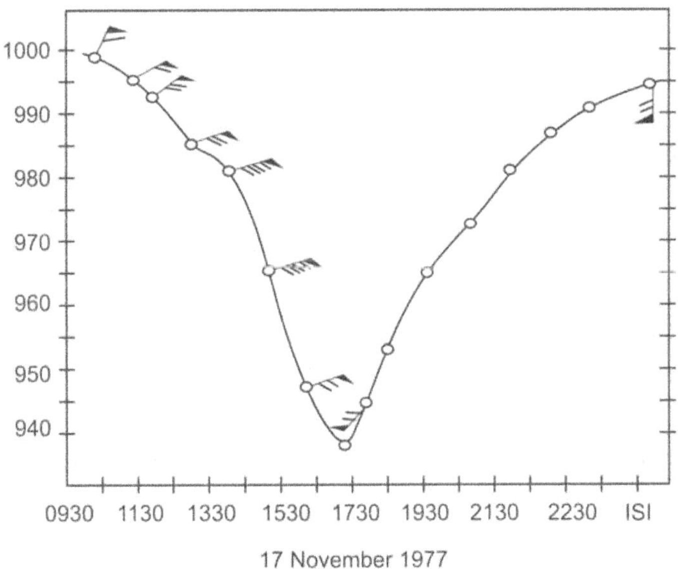

Fig. 24.5

Wind Field

There exists a distinct low level convergence and high level divergence in the storm field. However, along the vertical axis of the cyclone, wind field may be divided into three layers, namely, lower, middle and upper tropospheric level. In the lower troposphere, particularly in PBL, pronounced in-ward radial cyclonic windflow (convergence) occurs. In the mid-troposphere radial flow drastically decreases while in the upper troposphere extending to the top of cyclonic storm marked anticyclonic out flow (divergence) occurs. Maximum vertical velocity of 1.5 to 1.9 m/s occurs at about 900 h Pa level and maximum outflow occurs at about 200 h Pa level. Air that enters the cyclonic vortex in the lower troposphere rises along the wall cloud and rain bands. Outflow takes place from the top of cyclone vortex, which sinks at some distance and acts as a transporter of heat. A small part of the outflow also take place in the eye of the cyclone. At surface wind speed in the eye is very light or calm. Wind speed rapidly increases from the eye wall and attains maxima having cyclonic shear in the wall-cloud region and thereafter anti-cyclonic shear exists till the outer boundary of the cyclone. The cyclonic wind vortex has a maximum diameter at the surface level which gradually decreases with height and anti-cyclonic circulation strengthens above it. The band of strong wind would be 20-30 km from the centre of the storm. In this band of strong wind speed decreases rapidly towards the eye region. While outwards wind speed decreases slowly and irregularly.

Wind speed at surface level depends on the pressure drop at the centre. For Indian seas, a mathematical relation was found to estimate the maximum sustained wind using pressure drop (difference at the centre in relation to the peripheral pressure) which is given below

$$V_{max} = 14.2 \sqrt{Pn - Po}$$

where V_{max} = Maximum sustained wind speed in kt
       Pn = Peripheral pressure in h Pa
       Po = Lowest pressure at the centre in h Pa

Fletchers empirical formula which is used in typhoons and hurricanes is given below.

$$V_{max} = 16 \sqrt{Pn - Po}$$

In addition to these two formulae there are some similar relations have been in use in different parts of the world.

In a cyclonic storm surface wind is one of the most destructive phenomena of nature. It was inferred that several wind speed maxima occur in wall cloud region and surrounding spiral bands. Since in many cases wind instruments were blown off, it is not clearly known how strong winds could be in a severe cyclonic storm. With the available data from cyclone field, both while crossing the coast and on high seas and aircraft reconnaissance flight data, the maximum wind speed estimating empirical formulae (given above) derived. The Table 24.3 gives the empirical estimate of maximum wind speed for Indian seas.

Table 24.3 Data from Indian seas.

Date	Maximum sustained estimated / recorded wind speed (kt)	Peripheral Pressure (h Pa)	Lowest msl pressure estimated/ recorded in the eye hPa	Δp = pressure difference hPa
24 May 1963	104 (RFF)	1006	947	59
24 May 1963	70 ship	1006	988	17.7
12 June 1964	80	1002	969.5	32.5
21 Nov 1964	80 ship	1010	983	27
22 Dec 1964	85	1010	978	32
3 Nov 1966	115	1010	961	49
10 Nov 1966	70	1008	965.1	42.9
6 Nov 1969	75	1012	975.2	36.8
7 Nov 1969	80	1010	975.2	34.8
12 Nov 1970	80	1008	982.6	25.4
29 Oct 1971	100	1006	966	40
9 Apr 1972	80	1008	983	25
10 Sept 1972	110	1004	969	35
21 Sept 1972	85	1002	978	24
24 Oct 1972	65	1008	995.7	12.3
21 Nov 1972	80	1008	983.0	25
5 Dec 1972	75	1006	984.0	22
5 Nov 1973	74	1008	982.5	25.5
22 Oct 1975	91	1006	972.0	34.0
11 Sept 1976	80	1006	971.5	34.5
31 Dec 1976	75	1009	989.0	20
17 Nov 1977	104	1008	941.0	67
8 Nov 1978	100	1012	959.0	53
12 May 1979	70	995	963.0	32
22 Sept 1979	70	1006	986.0	20
3 June 1982	100	996	952	44
Total	2198			8804
Mean	84.54			33.86

Using first formula Vmax = 14.2 $\sqrt{33.86}$ ≅ 82.4

Observed average Vmax = 84.54, calculated Vmax = 82.4, error about 2.5%

Considering only those cases which have MSWS more than 100 kt, we have

								sum	mean
MSWS (kt)	104	115	100	110	104	100	100	733	105
Δp (hPa)	59	49	40	35	67	53	44	347	49.57 ≅ 50.0

Vmax = 14.2 $\sqrt{49.57}$ ≅ 99.7 ≅ 100, error is about 5%

Table 24.4 gives relationship between T-number and minimum sea level pressure (MSLP) or pressure drop $\Delta p = Pn - Po$ (Adopted from D vorak 1975 and extended for Indian seas by Mishra and Gupta, 1976)

Table 24.4

T-number	MSLP for Atlantic h Pa	Atlantic and Pacific ΔP	MSLP NW – Pacific h Pa	Maximum wind speed Kt. $V_{max} = 14.2\sqrt{\Delta P}$	Pressure drop $\Delta p = Pn\text{-}Po$ (Arabian sea and Bay of Bengal
2	1009	7	1003	30	4.5
2.5	1005	11	999	35	6.1
3	1000	16	994	45	10.0
3.5	994	22	988	55	15.0
4	987	29	981	65	20.9
4.5	979	37	973	77	29.4
5	970	46	964	90	40.2
5.5	960	56	954	102	51.6
6	948	68	942	115	65.6
6.5	935	81	929	127	80.0
7	921	95	915	140	97.2
7.5	906	110	900	155	119.1
8	890	126	884	170	143.3

$$V_{max} = 14.2 \sqrt{\Delta p}$$

Storm Surges

The word storm surge means the rapid rise in the sea level water due to the winds, pressure changes, ocean waves and ocean currents associated with a single storm at the time of crossing the coast (or landfall). Storm surges inundate coastal areas and cause devasting damage to life and property. More than 75% of human loss of life and property is caused by storm surges. Following are the principal factors that contribute to the piling of sea level water in storm surge.

Pressure Fall

Sea surface water rises at a place where horizontal atmospheric pressure is low. This effect is called Inverted Barometer. The rise in the sea level is about one meter for every 100 h Pa pressure defect as compared to normal pressure of neighbouring region (Fig. 24.5(a)).

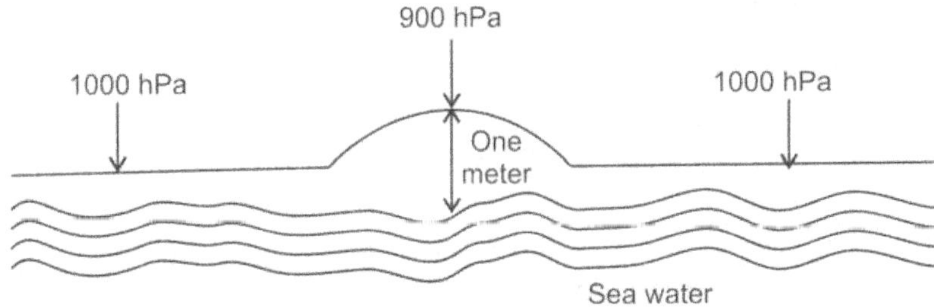

Fig. 24.5(a) Inverted Barometer effect.

Wind Stress

The frictional force (due to the blowing of strong winds) acting on the sea surface is known as wind stress. This pushes the sea water (fetch of sea water) towards the coast in the sector in which the winds are onshore, and away from the coast in the sector in which the winds are offshore. In northern hemisphere the onshore wind will be on the right of the track of storm while in the southern hemisphere on the left.

Coastal Shelf

In general the depth of sea increases as we move away from the coast into the sea (normal to the coast). As the cyclone moves towards the coast the piling up of water takes place to normal to the coast, the right of track in the northern hemisphere and to the left in the southern hemisphere.

Astronomical Tide

Russian scientist established that the entire territory of Moscow daily rises or falls about half a meter from a certain average level. These rises and falls are due to daily ebbs and flows caused by the gravitation of the Moon and the Sun. The gravitational pull of the Moon and the Sun on the earth including ocean waters, there is a rise and fall in the sea level. This is called Astronomical tide. This depends on the phases of the Moon and the Sun and is independent of the other terrestrial features. This rise of sea adds up to the storm surge.

Local Topography

Local topography such as mouth of a river, esturies play a role in cyclone storm surge flooding , in the event of storm crossing the coast.

Storm Surges in Bay of Bengal

The studies of Indian Institute of Tropical Meteorology, Pune revealed the following characteristics of the storm surges in Indian seas.

All the storm surges are associated with severe cyclonic storms, but all severe cyclonic storms, crossing the coast are not accompanied by storm surges. The Hurricanes, Typhoons are much more severe than the Bay cyclonic storms. But the storm-surges developed by Hurricanes and Typhoons in general are not so high as in the Bay cyclonic storms. Bay of Bengal seems to be the worst tropical cyclone region of the globe in regard to storm surges. This may be due to coastal orography, bathymetry etc. All storm surges occur either along the storm track or right of it. On some occasions recession of sea occurs on left of the track. Storm surges may occur before or after the storm crossed the coast. In general slow moving storms are not accompanied by storm surges, in otherwords   fast moving severe storms are likely to be accompanied by storm surges. There seems no fool-proof method of predicting storm surges. Storm surges in the Bay of Bengal are quite active and have maximum surge height in see Fig. 24.9 north easterly track upto 2.5 meters above astronomical tides for a pressure defect of 50 hPa, 2.0 m height in northerly track and 1.5 m in north westerly/westerly track for the same pressure defect. Most of the severe storm surges in the Bay occur in the months of October and November while in the premonsoon period of April, May are less severe.

Some important cyclones with associated low pressure and tidal (surge) heights are given Table 24.5.

Table 24.5

S.No.	Cyclone	Date of crossing coast	Lowest Pressure in the eye	Tidal height (m)	T. number
1.	Orissa Paradeep	28 october 1999 cyclone	924	3 - 5	6.5
2.	Andhra cyclone Machilipatnam	5-10 May 1990	921	4 - 5	7.0
3.	Andhra cyclone Deeviseema	14-20 Nov 1977	911	5	7.0
4.	Falsepoint Orissa	Sept. 1885	919	22 ft (6-7m)	
5.	Backerganz (Bangladesh)	1 Nov 1876	–	10-40 ft 3-12 m	
6.	Bangladesh	11-12 Nov 1970	–	40ft/12 m	
7.	Hoogli river mouth Calcutta West Bengal	7 October 1737	–	40ft/12 m	

Mathematical Model of Storm Surge

It is quite complex and beyond the scope of this book. However an empirical model (by Ghosh 1977) has been described below which is successfully used as a first approximation. He prepared nomograms to estimate the peak surge for Bay of Bengal north of Lat 10 ^{0}N. These nomograms used :

    (i)   pressure defect Δp, (at the eye as compared to the cyclone outer boundary pressure,

    (ii) radius of maximum wind (wall cloud region),

    (iii) speed of cyclone and

    (iv) the gradient of the coastal shelf.

In general two sets of nomograms are used to estimate the peak surge in Bay of Bengal. One set is used for the coast between lat 10 °N. to 20.5 °N and the second set north of lat 20 °N.

The empirical model for predicting storm surge for east coast of India is given by

$$\eta = \alpha_o\, \Delta p + \alpha_1\, \Delta p^2 + \alpha_2 C$$

where η = storm surge height above astronomical tide

        Δp = Pressure difference between the outer peripheri of the storm and the lowest central pressure in the eye

        C = Speed of the storm

α_0 α_1, α_2 are constants which vary according to the directions of movement of the storm (α_o : for northeast, α_1 : for north and α_2 : for northwest), determined by the method of least square fit. given in the Table 25.5.

Table 25.5

Track	α_o ($\times 10^2$)	α_1 ($\times 10^4$)	α_2 ($\times 10^2$)
North east	9.59	$-$ 0.91	$-$ 4.60
North	2.88	3.08	$-$ 1.20
North west	8.24	$-$ 1.60	$-$ 5.15

Numerical values of α_o, α_1 and α_2 (Das et. al., 1974)

Based on the above equation the surge prediction nomograms are prepared for NE, N and NW movement shown in Fig. 24.6 (a) (b) & (c).

Sp: Preliminary Surge Peak Estimate

Using Δp (from satellite T- Classification), R = distance between the center of eye and the center of wall-cloud region (either from radar observation or from satellite picture) preliminary Sp is estimated using nomograms.

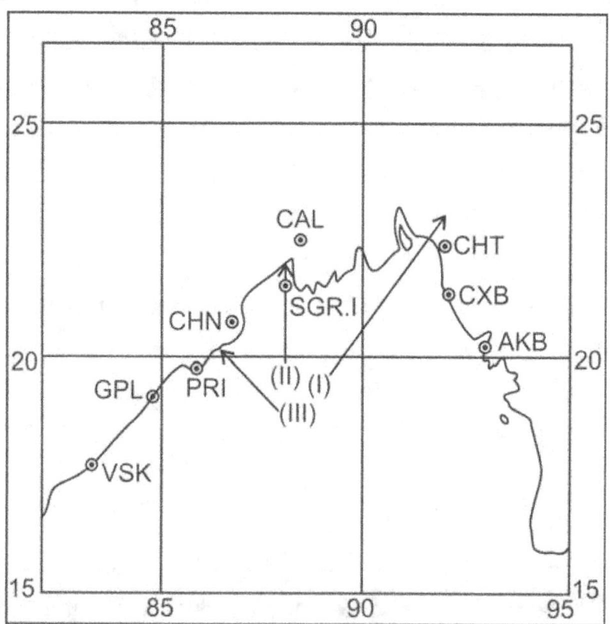

Idealized storm tracks (I), (II) a (III)
for surge prediction diagram

Fig. 24.6

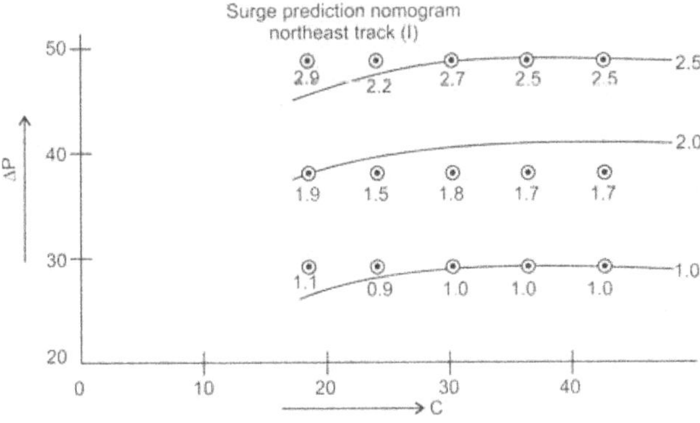

Fig. 24.6(a)

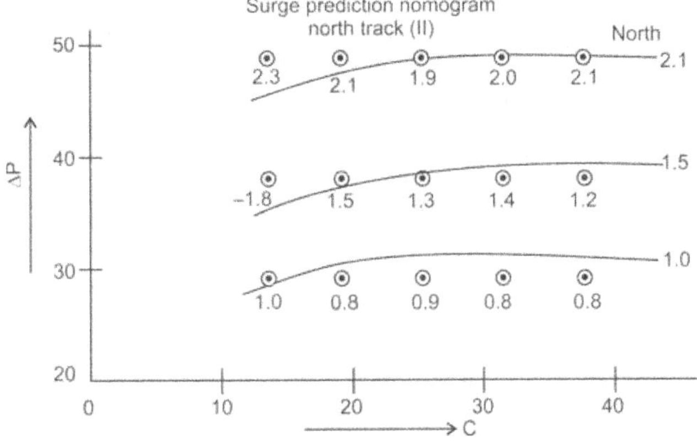

Fig. 24.6(b)

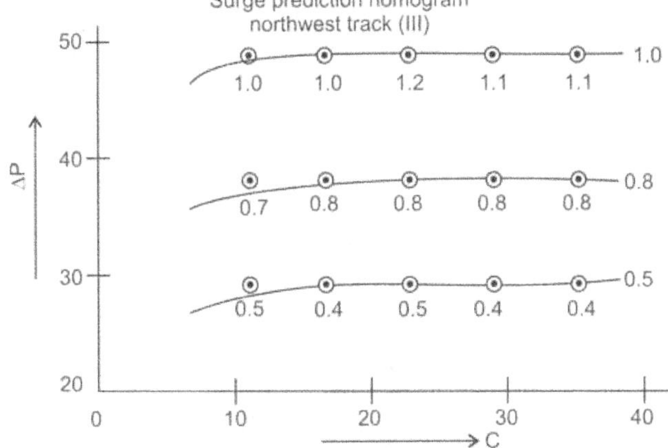

Fig. 24.6(c)

F_{SE} : Shoaling Factor or Correction for the Bathymetry

The nomogram Fig. 24.7 gives the correction term for bathymetry. In this nomogram vertical lines are spaced 100 km apart on the coast. Important towns/cities are shown along x-axis. The y-axis gives the shoaling factor. To use this nomogram, cyclone land fall location has to be predicted (using synoptic charts, satellite or radar track). Facing the coast from sea to the landfall point, measure the distance from the point to the nearest station in kilometers. s = distance which is to be reckond positive if the point is on the right and negative on the left. Add radius of maximum winds to this (R + s). R + s gives the location of the peak surge relative to the station. Find shoaling factor using nomogram.

F_m : Correction for Storm Motion Vector

A third set of nomogram gives the correction factor for the storm motion vector. The nomogram Fig. 24.8 & 24.8(a) contains radii of storm speeds (kmph), rays are crossing angles (θ degrees) of storm track to the coast. Also contains radius of maximum wind 50 km. Speed of storm and angle of attack to the coast are estimated on the basis of synoptic charts.

Final Peak Surge Sea Level Elevation (η)

This is the product of preliminary estimate Sp, bathymetry correction F_{SE} and correction for vector storm motion F_m.

$$\eta = \text{peak surge} = Sp \times F_{SE} \times F_M$$

Total sea level elevation

$$= \eta + \text{astronomical tide height at the time of storm crossing the coast.}$$

Effect of Wind on Sea Surface Waters

When wind blows over the ocean water, wind energy is transferred to the surface layers. Some of this energy is used up in generation of surface gravity waves – which results in small movement of water in the direction of wave propagation and some energy is spent for driving currents.

The frictional force acting on the ocean surface water due to wind blowing over it is called the wind stress. It is generally denoted by τ and is given by the formula

$$\tau = C\omega^2$$

where ω = Wind speed

C = is not constant but a function of prevailing atmospheric conditions and increases with wind speed.

2.6
2.5
2.4
2.3
2.2
2.1
2.0
1.9
1.8
1.7
1.6
1.5
1.4
1.3
1.2
1.1
1.0
0.9
0.8
0.7
0.6
0.5
0.4
0.3
0.2
0.1
0.0
Paradip
Puri
Gopalpur
Calingapatnam
Vishakhapatnam
Kakinada
Masulipatnam
Nellore
Madras
Pondicherry
Nagapatnam
Kilometres × 10²
Shoaling factor, F_{SE}

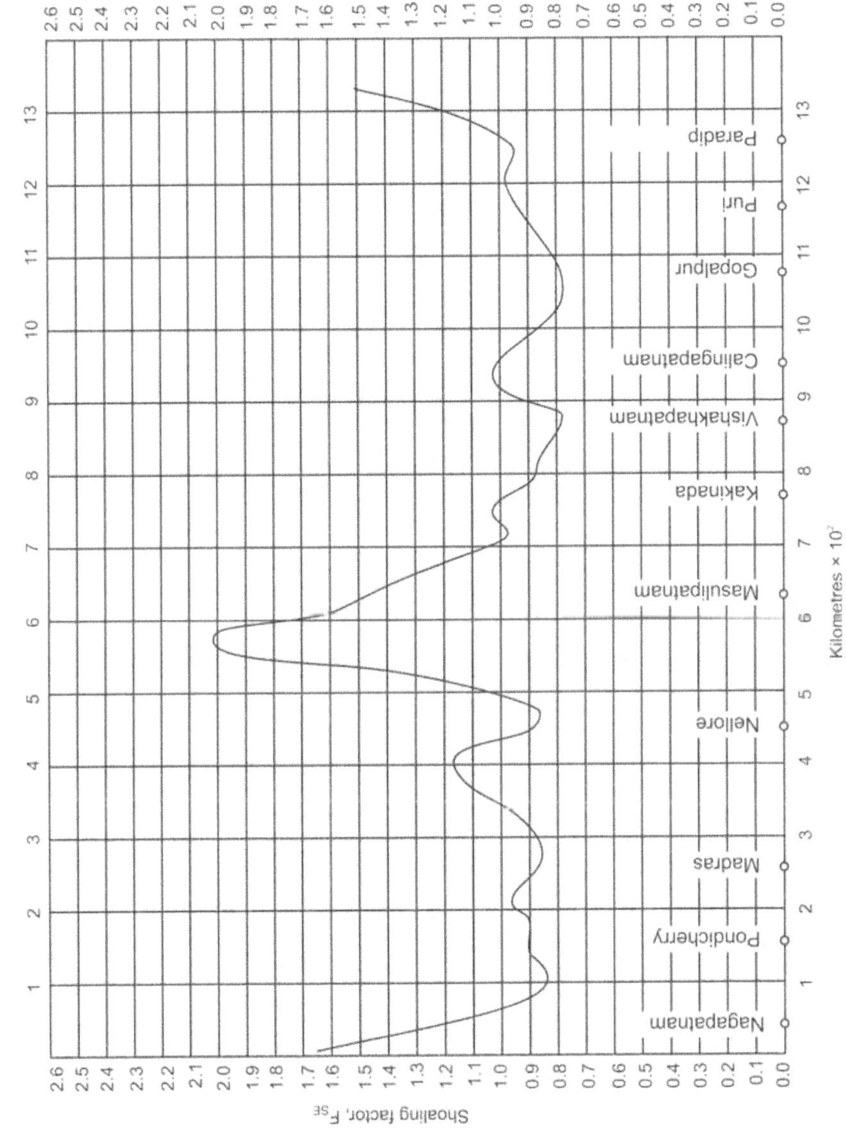

Fig. 24.7

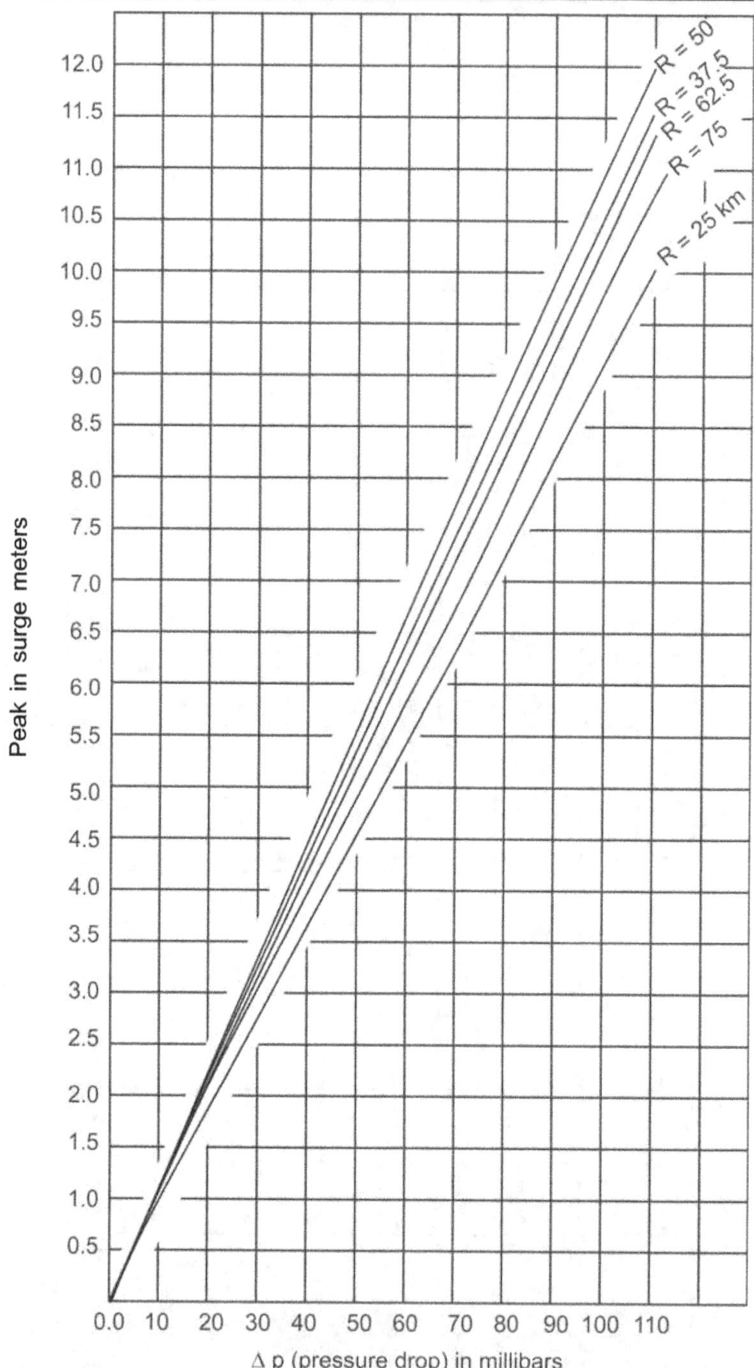

Fig. 24.8 Nomogram of peak surge as a function of pressure drop and radius of maximum winds for the northeast coast of India.

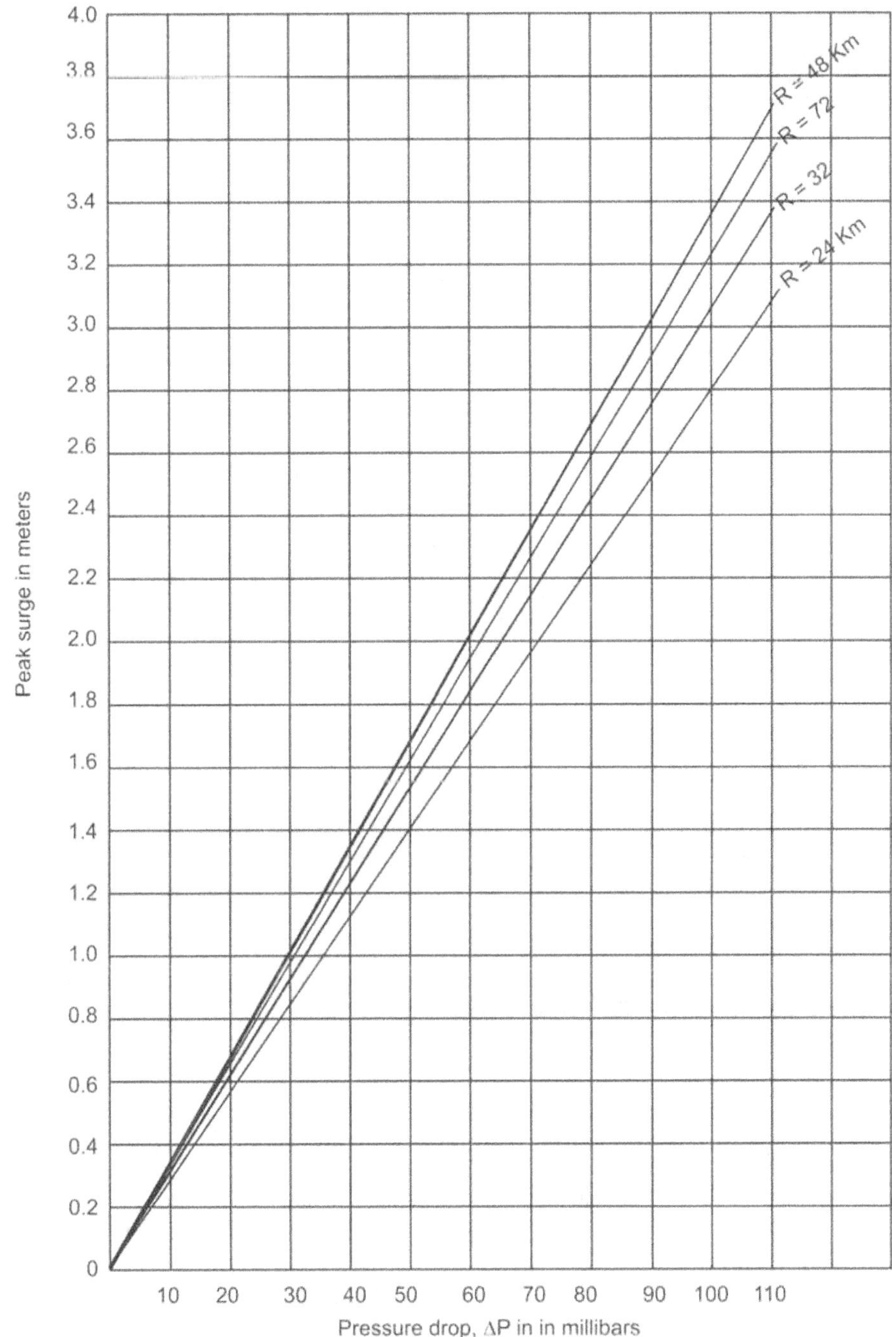

Fig. 24.8(a) Nomogram of peak surge on the open coast as a function of pressure drop and radius of maximum winds (R in Km) for the southeast coast of India.

As C increases (i) the turbulent convection increases in the over laying atmosphere and (ii) increases the roughness of the surface.

An empirical relation is that sea surface current is about 3% of the wind speed. Thus 20 mps (40 kt) wind may create sea surface current of about 0.6 mps.

$$\text{Coriolis force} \; \alpha \; \sin\phi,$$
$$\text{Coriolis force} = m \times 2 \; \Omega \; \sin\phi \times u$$

where m = mass of the particle

u = speed, ϕ = Latitude

Ω = Angular velocity of the earth about its axis.

f = Coriolis parameter = $2 \; \Omega \; \sin\phi$

If $\phi = 40, \Omega = 7.29 \times 10^{-5} \, s^{-1}$

$f = 2 \times (7.29 \times 10^{-5}) \times 0.643$

$= 9.4 \times 10^{-5} \, s^{-1}$

Coriolis force = $m \, f \, u$

For homogeneous infinite ocean, the speed of the sea surface water current (u_0) is given by the formula (derived by Ekman)

$$u_o = \frac{\tau}{\sqrt{v\rho f}}$$

Where

τ = wind stress

v = Coefficient of eddy viscosity for vertical mixing

ρ = density for ocean water

f = Coriolis parameter.

Theories of Tropical Cyclone Formation

The following are the necessary qualitative conditions favourable for the formation of a Tropical Cyclone.

SST : Sea Surface Temperature

Sea surface water temperature up to a depth of 60 m must be 26 °C or more (since the underneath water upto 60 m depth is found to be affected by a tropical cyclone).

Ocean thermal energy potential per unit area of 60 m

= Thermal energy above 26 °C and down to a depth of 60 m

$$= \int_0^z \rho s (T - 26) dz \rho s$$

where sea water temperature is 26 °C or more,

$-Z$ = depth of sea water upto which temperature is 26 ^{0}C or more ($-Z$ is found to be -60 m or more depth)

ρ = Sea water density

s = Specific heat of sea water.

T = Sea water temperature.

If a tropical cyclone moves over an area in the rear of another cyclone which produced upwelling and lowered SST, then the cyclone will weaken or dissipate over that water.

Presence of High Humidity in the Mid–Troposphere

Regions of high relative humidity in mid-troposphere as compared to the seasonal normal are favourable places for the genesis of tropical cyclone. In this region convective clouds, that develop in the planetary boundary layer, are further triggered. This is because of moist entrainment of mass and subsequent latent heat of release. However such high relative humidity in mid-troposphere does not differ much in case of widespread convective activity. The regions of low relative humidity in mid-troposphere is *inhibitive for tropical cyclone genesis*.

Vertical Gradient of Equivalent Potential Temperature

$$\left(\frac{\partial \theta_e}{\partial p_o} \right)$$ **between Surface and 500 h pa**

This should be (15 to 20 °K) large. This gives the conditional instability. Cumulus convective instability is presumed to be the primary mechanism for the genesis of tropical cyclone.

Low Level Relative vorticity (ζ)

ζ should be positive. Larger value indicates more possibility of tropical cyclone genesis. Low level vorticity variation shows strong correlation for intensification of a tropical cyclone.

Coriolis Parameter ($f = 2\,\Omega \sin \phi$)

The magnitude of the Coriolis force or deflective force is given by fv or 2Ω V sin ϕ, where Ω = angular velocity of the earth, V = horizontal wind speed, ϕ = latitude. f is small at the equator (ϕ = 0, Sin ϕ = 0) and large at the poles (ϕ = 90, Sin ϕ = 1). Because of this, tropical cyclones do not form near the equator 5 degrees either side. Of the global total 65% of the tropical cyclones have genesis in the latitude belt of 10° to 20°. In northern hemisphere cyclone genesis extend upto 35° N and about 13% of tropical cyclones form beyond lat 20 °N, while in the southern hemisphere cyclogenesis do not occur beyond lat 22 °S.

Vertical Shear of Horizontal Wind

Cyclones generally do not form when the shear of zonal wind flow between 950 h Pa and 200 h Pa is more than 10 mps. Tropical cyclones form in a region where the vertical shear of horizontal wind is minimum between the lower and upper atmosphere. In such case the advection of heat and moisture is small and as compared to the surroundings temperature and moisture increases throughout the disturbance. This leads to surface pressure fall and tropical cyclogenesis.

There are several theories for breeding of tropical cyclones and equal controversis. The following are a few noteworthy theories of tropical cyclone formations.

Convectional Hypothesis

It is the oldest hypothesis. According to this hypothesis, for some reasons unusually large number of thunderstoms, squalls and heavy rains occur over tropical oceans. This activity leads to surface level convergence and the formation of cyclonic circulation.

Inter Tropical Convergence Zone (ITCZ)

Most of the researchers view this theory credibly. ITCZ shifts to 12 to 15 °N during northern hemisphere summer and south of 5 °S during southern hemisphere summer. The convergence of trades/air masses from the two hemispheres breed cyclogenesis.

Easterly Waves

Waves in baroclinic easterlies are the origins of Tropical Cyclones. Easterly waves are accompanied with succession of Highs and Lows. Some times these Lows and Highs loose their stability which gives rise to the formation of a Tropical Cyclone. Invasion of the tropics by cool air masses from the polar and temperature latitudes gives rise to the formation of tropical cyclone.

Conditional Instability of Second Kind (CISK)

In tropics, convectively unstable atmosphere (in cumulus convection – a mesoscale) vertical upward motion releases convective instability of static type. This is called convective instability of first kind (CIFK). Release of CIFK in a big way (on synoptic scale) in horizontal direction affects the dynamic stability if the atmosphere, that is, CIFK gives rise to a dynamic instability, which is termed – conditional instability of second kind (CISK). Studies indicate under favourable conditions of moisture supply CISK mechanism converts a low pressure system into a hurricane (when the atmosphere is conditionally unstable).

Questions

1. Explain, Tropical cyclones are the most destructive phenomena of atmospheric nature.

2. In what way a TC resembles a giant heat engine. What are the favourable the conditions of TC genesis. Describe briefly life cycle of a TC.

3. Describe briefly the horizontal divisions of a well developed T.C.

4. Write briefly the formation, movement of cyclones in Bay of Bengal and in Arabian sea during pre and post monsoon periods.

5. Describe some special features of T.C.

6. Write the empirical formula to calculate maximum sustanced wind speed in T.C. in Indian seas and that of Fletchers empirical formula.

7. What is a storm surge ? What parameters are effective in deciding storm surge ?

8. Write briefly about storm surges in Bay of Bengal.

9. Write empirical formula for estimating assessting storm surge height in Bay of Bengal, write its dependent parameters.

10. Write short notes on : (i) Wind speed effect on sea surface waters, (ii) Sea surface water temperature for cyclogeneses (iii) CISK

CHAPTER 25

Cyclone Assessment Through Satellite images

With the launching of artificial satellite Sputnik–I by Soviet Union on 4 October 1957, the space age has began. The Sputnik–I orbited the earth once in every 96 mintes in an elliptical orbital plane inclined at an angle of about 65° with the equatorial plane, having perigee of 227 km and Apogee 947 km. Observing weather from space has began on 1 April 1960, with the launching of TIROS-1 (Television Infra Red Observational Satallite) by USA. This all fools day is an important milestone in the history of Meteorology, which was clearly shown the global weather observations from space. INSAT (Indian National Satellite) series is a multipurpose operational satellite, used for domestic long distance telecommunication, weather observations and nation wide direct TV broadcasting.

Meteorological satellites are of two kinds: (i) Sun synchronus or near polar orbiting and (ii) Geostationary

Sun syschronus satellites orbit (elliptical orbit or circular orbit) at various altitudes, very close to the poles and always pass over the earths equator at the same sun time on each of its orbits but at some other longitude which is west of its earlier (preceding) crossing position, depending on the orbital time t (if t is in hours then 15 t° long west of its earlier crossing position). Polar orbiting satellites or a space vehicle cover the whole of the earth and see every location of the earth twice a day at the same approximate time. It can provide (i) atmospheric vertical temperature profile of the globe, (ii) High resolution picture transmission (HRPT), (iii) Automatic picture transmission (APT), (iv) collection of data from automatic weather stations and transmission.

Geostationary satellites move west to east in a circular orbit in the equatorial plane at an altitude of about 36000 km and have velocity (464 m/s) same as that of the earth. Consequently they appear stationary with respect to a fixed ground object on equator. They provide continuous weather data over a fixed area round the clock, vertical temperature profile, cloud motion vectors (wind) and sea surface temperature (SST). It acts as a platform for reception of data from automatic weather stations and transmit to the ground station.

The photographs taken by the satellites are called imageries. They are basically two types. (1) Visual, which are obtained by T.V. cameras, and (2) Infrared (IR), which are obtained by scanning radiometer. The visual photographs taken by T.V. cameras look black and white and represent what we could see with eyes from a satellite. This imageries is a measure of earth's albedo. IR images represent temperatures and temperature contrasts (warm or cold, clouds at different heights).

TV cameras on board satellites measure radiation in visible portion of electromagnetic spectrum about 0.4–0.7 µm. This is possible only from sunlit portion. The IR scanning radiometer measures long wave radiation emitted by the earth, atmosphere (cloud, fog etc) and water bodies in the wave lengths 8 to 13 µm.

Some Characteristics of Satellite Data

Day visual

The brightness of TV images is a measure of earth's albedo (reflectivity). White patches indicate the areas of high albedo and black or gray patch shows the areas of lowest albedo. Different gray shades correspond to scenic information. White may represent clouds, snow, ice or sand (as seen by naked human eye). Vegetation, soil and clouds of different heights (high, low, medium clouds) and thickness appear as shades of gray to black to white. Some satellites (SMS/GOES) take visual photographs of the earth about every half an hour and have resolution 1 to 8 km at the satellite sub-point (a point directly below the satellite at the time of photograph). The resolution 2 km simply means that any ground element/cloud as small as 2 km size may be clearly identified from the satellite photographs. The visible channel thus capable of providing a wealth of data such as thunderstorms, cumulus clouds or convective clouds and low level/high level wind flow. With the help of every half hourly data, meteorologist can identify whether the convective cloud is developing or dissipating, cloud motion wind vectors provide wind speed.

Night Visual

Besides day time visual images there is provision in some satellite based instrumentation to take night time visual images with high resolution. It is a low light sensor, able to detect even in moon light. In the absense of moonlight, city lights and aurora phenomena, volcanoes, oil and gas fields, forest fires etc. can be detected.

IR-Images

In this imagery high clouds look white, low/convective clouds as gray and hot land/water surface as black. This scheme may be reversed or colour films may be used to detect various features such as ocean currents. In general black depicts hot or warm and white denotes cool or cold relatively. The thermal IR imagery is viewed in the interval range of wave length 8–13 μm. A VHRR (very high resolution radiometer) provides IR pictures with resolution of 1 km.

The large diurnal variation of land surface produces marked change in appearance of day and night time IR images. During day time the warm land surface appears dark as compared to the cool water bodies which appear gray and mild cold water bodies together with high clouds appear as white. During night time the cold land appears bright shade of gray in the IR images. The land which is very close to the water bodies may acquire the temperature of the water bodies consequently the image contrast between land and water bodies is not marked. This makes coast line and lake shores less obvious. The variation in IR images is also caused due to seasonal and latitudinal temperature variation. In winter, particularly during night, clouds over cold land areas are not easily detectable, However in tropics there is always temperature difference between land surface and middle/high clouds and hence easily detectable.

As pointed out earlier the brightness differences in visible (TV) pictures indicate difference in reflectivity (albedo) while the brightness difference in IR pictures indicate temperature difference of the emitting surfaces.

The Table 25.1 gives the difference of cloud pictures as seen in visible and IR imagery.

Table 25.1

			Visible imagery			
			White	**Gray**	**Dark gray**	**Black**
IR imagery	(a)	Black			Fair weather cumulus and desert	Warm Ocean
	(b)	Dark gray	St			Clod Ocean
	(c)	gray	St, Sc	Thick Ci		High frosted terrain
	(d)	light gray	Ci, As, Cb			
	(e)	white	Frontal cloud and active snow	Dead Cb anvil	Dense Ci	

Note : All surfaces with temperature above absolute zero emit radiation.

Note: Satellite imagery clouds classified as

(i) Convective or cumuliform clouds,(ii) Stratiform clouds and (iii) Cirri form clouds.

Before the advent of weather satellites cyclone detection was based on coastal synoptic observations, Radar observations and a few ship observations (which were very sparse or none from the cyclone field). Lack of observations was a great draw back in detection and forecasting of a cyclone, its intensity, movement (direction and speed) and land fall. There might have been occasions of missing detection when cyclones were formed far away from coast (away from radar range of 400 km) and dissipated over the sea itself. After the advent of geostationary weather satellites there was not even a single occasion of missing detection of cyclone over the sea. The geostationary satellites stationed over equatorial circular orbit are GOES – east (long 15 °W) GOES - west (long 135 °w) of USA, METEOSAT of Europe at zero meridian, INSAT-ID of India at 85 °E, GMS of Japan at 140 °E see Fig. 25.1. These virtually cover the global detection of cyclones upto 60 ° latitude on either side of the equator. Earlier it was believed that the intensity of cyclone was a function of its size but it is now confirmed by satellite images that it is not so. Dvorak, an American meteorologist propounded an empirical cyclone intensity code called Dvorak T-number (T-stands for tropical cyclone). Even now with some minor modifications Dvorak intensity code is followed. (See Fig. 25.2, 25.3, 25.4).

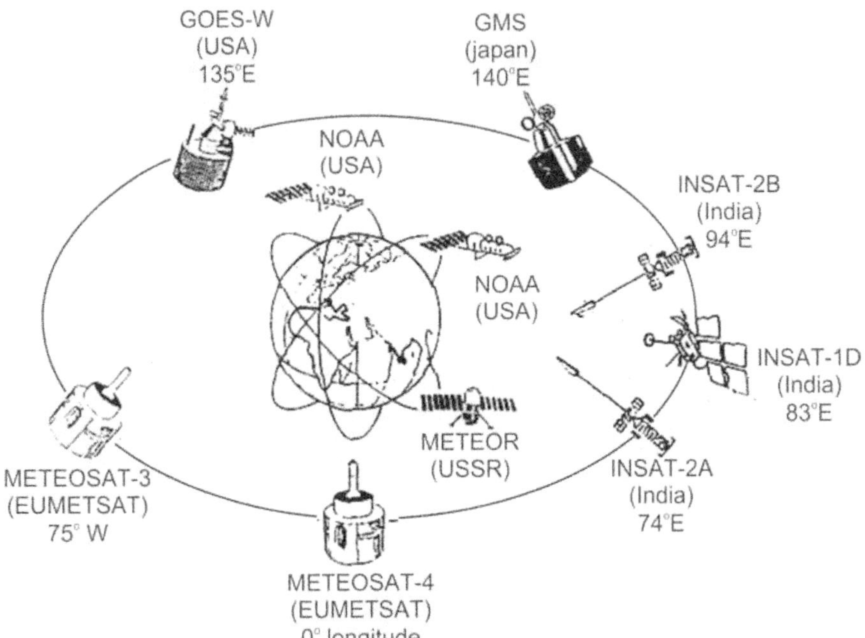

Fig. 25.1 Global system of operaional meteorologocal satellites.

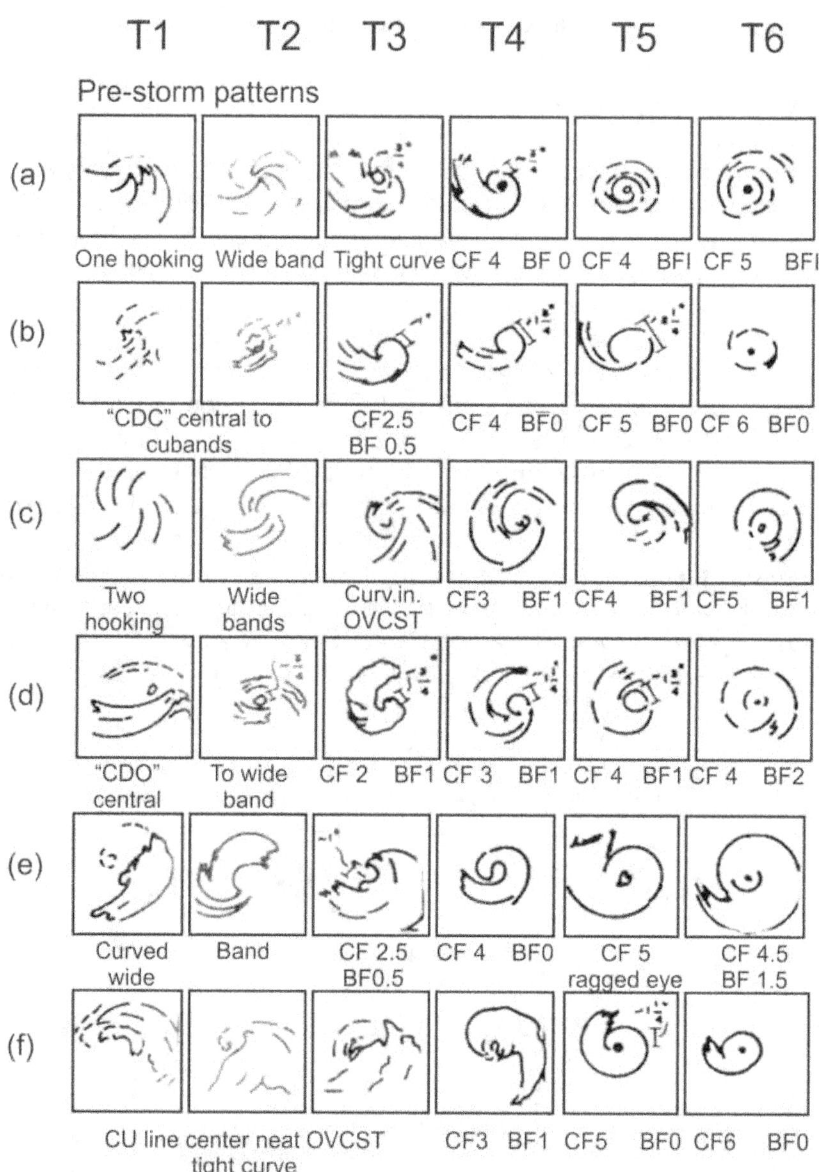

Fig. 25.2 Common tropical cyclone patterns and their corresponding T-numbers. The T-numbr shown must be adjusted for cloud systems displaying unusual size. The patterns may be rotated to fit a particular picture of a cyclone (Dvorak. 1975).

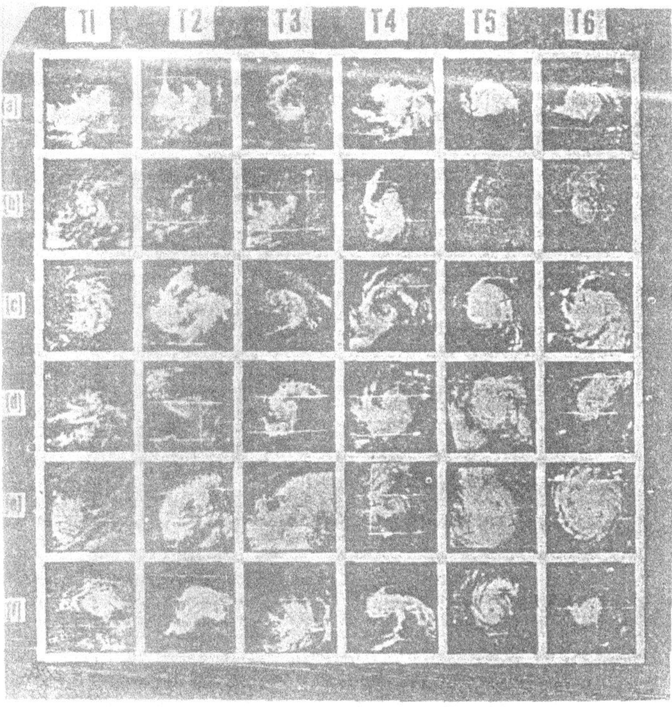

Fig. 25.3 Typical satellite photographs of six stages (T1-T6) of development of tropical vortex from pre-storm stage to intense hurricane stage (Dvorak, 1975).

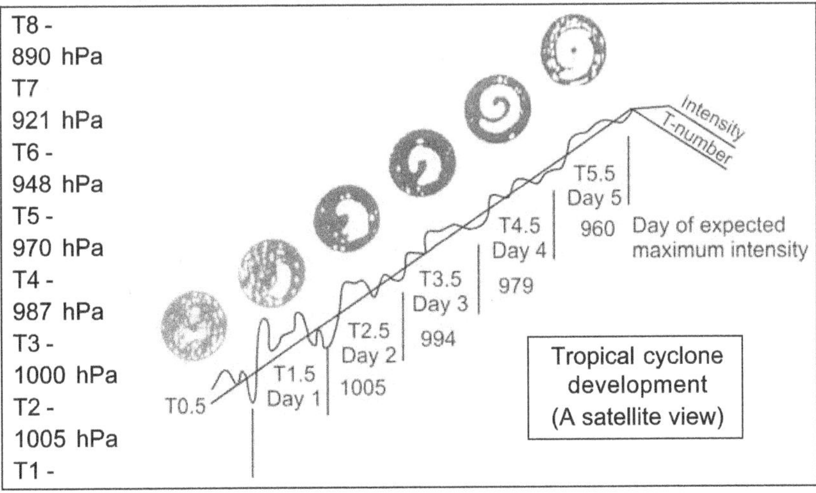

Fig. 25.4 Model of tropical cyclone development used in intensity analysis (curved band pattern type).

T-number is determined on the central features (CF) of Central Dense Over cast (CDO) size, embedded distance of Eye and Banding Features (BF). Digital IR data of cyclone field temperatures help in location of Eye (warmest part) of cyclone and the lowest temperature part indicates the wall cloud region (about 50 km from the centre of Eye).

Note

1. T-number = CF (= size of CDO + embedded Eye distance) + BF (= Amount of banding coil around CDO + vertical depth of wall cloud)

2. Digital IR data of cyclone field provides temperature field of cloud tops. This is determined by gray scale. Warm temperature region (T_w) of central region indicates cyclone Eye and the surrounding coldest (Tc) region indicates wall cloud region. The temperature difference of these two (Tw-Tc) gives an idea of the intensity of the cyclone. The Dvorak scheme categorises the cloud pattern by the CDO, spiral bands and the eye. The CDO features such as round, oval, angular, irregular, ragged edge defines CSC (cyclone system centre). OR cloud line (band features) define CSC.

Note : (Centre of the Eye wall is called CSC which is visible when the cyclone Eye is clearly formed).

The Dvorak's procedure enlists the following steps.

Step 1 : Located cloud system centre (CSC).

Step 2 : Determine relationship between CSC with CDO. This is done by comparing with standard pattern. If the measurement in this step is clear cut, the intensity is final. If not we go to step 3.

Step 3 : Model Expected T-number (MET). In this method current picture is compared with the previous days picture to ascertain whether the storm is intensified or weakened by the past trend.

Step 4 to 6 : MET is further refined comparing with the standard patterns corresponding to the intensity and then upgraded or downgraded the intensity. The intensity thus fixed is subject to certain rules to make final adjustments. The rules are–to hold variation of intensity to one T-number per day in the pre-storm stage and to within one number of the MET during the later stage. As an example, let in the prestorm stage two days earlier the T-number T_{-2}, one day earlier T_{-1}, to day T, the next day T_1 and so on.

The main features of the standard patterns (description of photo chart) may be categorized as (i) curved band, (ii) shear, (iii) Eye, (iv) CDO (v) Embedded centre. In addition to this another rare pattern is (vi) central cold cover (CCC) which indicates the halt of further development of the storm. The CCC pattern analysis is taken as Step 3. When the storm centre is masked by cold overcast clouds, the CCC pattern is recognized in which outer curved lines and curved bands appear very weak or not seen. As a result visible pictures are not useful for determining CCC pattern. The persistence of CCC pattern is an indication that storm development is halted. It may be noted that very cold, commapattern contains cirrus streaks/lines on the peripheri of dense overcast, however such cirrus lines are absent in CCC pattern.

Step 7 : This covers the rules that are followed in determining T-number. When measurements of cloud features are clear the data T- number (D T- number) is fixed and it is final. When the features are obscure or not clear, the satellite pictures are compared with standard pattern and assign pattern T-number (P – T number). When both D-T number and P – T number are not clear it is then MET (Model Expected T- number) is used. The last technique is used when data are received 3-hourly as in the case of INSAT pictures.

Step 8 : This covers assigning final T-number imposing some constraints. The constraints are. "Hold the final T-number close to one number per day" during the pre-storm evolution stages and within one number of the MET during rest of the evolution. It may be noted here that no lowering of T-number is allowed during night in the beginning two days.

Current Intensity (CI)

CI gives directly the cyclone intensity. In development stage the CI is equal to T-number but CI number > T-number when the cyclone is weakening. As a rule CI number is not lowered for about 12 hours after the cloud features show the signs of weakening. At that time CI number = T-number + ½ to 1. If a storm weakens and then intensifies the CI number is not lowered even if the CI number is greater than T-number. The empirical relation between the CI number and storm Maximum wind speed (MWS) is given Table 25.2.

Table 25.2

CI number	MWS (kt)	MSL Pressure Atlantic h Pa	MSL pressure NW- Pacific h Pa	Δ p = pressure drop Indian seas h Pa
1	25			
1.5	25			
2	30	1009	1000	4.5
2.5	35	1005	997	6.1
3	45	1000	991	10.0
3.5	55	994	984	15.0
4	65	987	976	20.9
4.5	77	979	966	29.4
5.0	90	970	954	40.2
5.5	102	960	941	51.6
6	115	948	927	65.6
6.5	127	935	914	80.0
7	140	921	898	97.2
7.5	155	906	879	119.1
8	170	890	858	143.3

MSL : Minimum sea level (pressure)

msl : Mean sea level

(Two charts standard pattern to be inserted).

IR picture colour code No's with temperatures given in Table 25.3.

Table 25.3

Code No	Colour	Temp °C
1	OFF white	9 to -30
2	Dark gray (DG)	-31 to -41
3	Medium gray (MG)	-42 to -53
4	Light gray (LG)	-54 to -63
5	Black	-64 to -69
6	White	-70 to -75
7	Cold light gray (CLG)	-76 to -79
8	Cold Medium gray (CMG)	-80 to -84
9	Cold dark gray (CDG)	-85 to -89

Naming of Cyclones in Bay of Bengal and Neighourhood

In 2005, a panel of 8 member countries (India, pakistan, Srilanka, Bangladesh, Myanmar, Oman, Thailand and Maldevis) was set up by WMO for recommending names, which will be adopted and declared by IMD. Of the 40 names proposed by the panel includes :

1. Payaar – Myanmar
2. Baaz – Oman
3. Fanoos – Pakistan
4. Mala – Srilanka
5. Mukda – Thailand

The India names include :

1. Akash
2. Bijli,
3. Jal,
4. Lahar, and
5. Megh

The naming of cyclones were introduced in 2005.

Questions

1. Write short notes on : (i) Sun synchromus and "Geostationary satellites and the satellite based visual and Infrared images.

2. Write the important geostationary satellite stationed on equatorial orbit. Write Dvorak T-number classification with that of Δp and maximum sustaned wind.

3. Write briefly IR pictures colour code numbers and corresponding temperatures.

4. Write short note on naming of cyclones. Write the latest WMO recommended based cyclones names adopted for Bay of Bengal and Neighbourhood.

CHAPTER 26

Artificial Satellites

We shall study elementary space dynamics which begins with kepler's laws of planetary motion, Newton's laws of motion and Universal gravitational law.

The Law of Orbits (Kepler's First Law)

The orbits of planets are ellipses with the sun being at one of the foci of ellipse.

The Law of Areas (Kepler's Second Law)

The line joining the sun to any planet sweeps out equal areas in equal intervals of time.

From this law, the velocity of a planet or satellite at any instant is derived as

$$V = \sqrt{\mu\left(\frac{2}{r} - \frac{1}{a}\right)}$$

where

μ = constant, depends on the gravitational field

a = semi-major axis of the ellipse (equation of ellipse is $\dfrac{x^2}{a^2} + \dfrac{y^2}{b^2} = 1$,

where a,b are the semi major and minor axes).

r = radial distance of the planet from centre of the sun.

If $a \to \infty$, $V \to V_E = \sqrt{\dfrac{2\mu}{r}}$ which is the escape velocity, and the trajectory becomes a parabola.

If $a = r$, then the trajectory becomes circular.

and the velocity (Vc) is given

$$Vc = \sqrt{\dfrac{\mu}{r}} \text{ or } V_E = \sqrt{2}\ Vc$$

We have $V \lessgtr V_E$, depending on the trajectory ellipse, parabola and hyperbola respectively.

The Law of Periods

The square of planetary time period (T) is proportional to the cube of the major axis (a) of the orbit, viz $T^2 \propto a^3$ or $T^2 = \dfrac{4\pi^2 a^3}{\mu}$

where   T = Orbital period

         a = semi-major axis of the ellipse

         μ = constant, dependent on gravitational field.

The orbits of meteorological satellites are nearly circular (ellipses). The advantage of circular orbit is that the pictures taken from any point of the orbit will have the same size. In case of elliptical orbits the image sizes differ at different point of the orbit.

An artificial satellite can fly only in a circular or an elliptical orbit and the orbital plane passes through the centre of the earth. It cannot travel above any latitudinal or longitudinal parallels except the equator. The flight altitude depends on its traveling speed. If we assume that there is no air resistance, then an artificial satellite launched from the surface of the earth with speed of 7.9 km/sec it will make one complete circle in the sky around the earth in 1 hr 24 min and 25 sec. This period is called sidereal circuit period. The circuit period increases with the increase of altitude. Suppose 'Re' is the radius of the earth. For an altitude of 2 Re, the sidereal time would be 7 hr, 17 min, for altitude 4 Re or 6 Re the sidereal time will be 15 hr, 44 min and 26 hr, 3 min respectively.

$$Ts = \dfrac{2\pi\left(R_e + h\right)}{u},$$

where Re = radius of the earth = 6378 km

        h = altitude of the satellite above earth

        u = circular speed of the satellite.

We know the Newtons law of universal gravitation is given by

$$F = G\,\frac{m_1 m_2}{d^2} \qquad \ldots\ldots(26.1)$$

where F = Force of attraction between two bodies of masses m_1, m_2

        d = distance between the two body masses m_1, m_2

        G = universal gravitation constant = 6.67×10^{-11} Nm^2 kg^{-2}

The attraction g_e, of the earth on unit mass (1kg) placed on the surface of the earth is given by

$$g_e = G.\,\frac{1 \times M_e}{Re^2} \qquad \ldots\ldots(26.2)$$

where M_e = mass of the earth = 5.98×10^{24} kg

        Re = radius of the earth = 6380 km

        G = universal gravitational constant = 6.67×10^{-11} Nm^2 Kg^{-2}

$$g_e = \frac{6.67 \times 10^{-11} \times 5.98 \times 10^{24}}{\left[6380 \times 10^3\right]^2}$$

$$= 9.8 \text{ m s}^{-2} \qquad \ldots\ldots(26.3)$$

Now consider a unit mass (1kg) kept at a point P in space, distance R from the centre of the earth, whose altitude from the surface of the earth is 'h' that is $R = Re + h$

Let g_R be the gravitational force of earth acting on this unit mass.

Then $$g_R = G\,\frac{1 \times Me}{R^2} \qquad \ldots\ldots(26.4)$$

From Eq. (26.2). and (26.4) we have

$$\frac{g_R}{g_e} = \frac{\left[\dfrac{GMe}{R^2}\right]}{\left[\dfrac{GMe}{R_e^2}\right]} = \frac{R_e^2}{R^2}$$

or $$g_R = g_e\,\frac{R_e^2}{R^2} \qquad \ldots\ldots(26.5)$$

Escape Velocity (V_E)

The minimum velocity with which a body be projected into space from the surface of the earth, so that it may travel to infinity is called escape velocity. In other words the body escapes from the earth's gravity.

The work done (dw) by external force in moving a unit mass from the point P in space through a small distance dR in the direction OP (where O is the centre of the earth) is given by

$$dw = g_R \, dR \quad (\text{work} = \text{Force} \times \text{distance})$$

$\therefore$ The work done (w) in moving unit mass from P to infinity (∞) is given by

$$W = \int_R^\infty g_R \, dR$$

$$= \int_R^\infty g_e \frac{R_e^2}{R^2} dR \qquad (\text{using Eq.(26.5)})$$

$$= g_e R_e^2 \int_R^\infty \frac{dR}{R^2} = g_e R_e^2 \left[-\frac{1}{R}\right]_R^\infty$$

$$W = g_e \frac{R_e^2}{R} \qquad \qquad \dots(26.6)$$

$\therefore$ The amount of work done (W) to move a satellite of mass m_s to infinity from the surface of the earth is given by

$$W = m_s \int_{Re}^\infty g_e \frac{R_e^2}{R^2} \, dR$$

$$= m_s \, g_e R_e^2 \left[-\frac{1}{R}\right]_{Re}^\infty$$

$$= m_s \, g_e R_e^2 \left[\frac{1}{R_e}\right]$$

$$W = m_s \, g_e R_e \qquad \qquad \dots(26.7)$$

This work is equal to the KE of the satellite required to escape from the earths gravitational force.

$$KE = \frac{1}{2} m_s V_E^2 \qquad \qquad \dots(26.8)$$

where V_E = escape velocity of the satellite.

Equating (26.7). and (26.8). we have

$$\frac{1}{2} m_s V_E^2 = m_s g_e R_e$$

$$V_E^2 = 2 g_e R_e$$

or $\qquad V_E = \sqrt{2g_e R_e}$ .....(26.9)

Putting $\qquad Re = 6.400 \times 10^6 \text{ m}, \ g_e = 9.81 \text{ m/s}^2$

We get $\qquad V_E = \sqrt{2 \times 9.81 \times 6.4 \times 10^6}$

$$= 11.2 \text{ km/sec}$$

Satellite Orbital velocity

The speed of a satellite in its orbital motion is called orbital velocity.

When a satellite moves in a particular orbit, its centripetal force is equal to the gravitational force of the planet.

$$\text{The centripetal force} = \frac{m_s V_0^2}{R} \qquad \qquad .....(26.10)$$

where $\quad m_s$ = mass of the satellite

V_o = Orbital velocity of the satellite

R = distance between the centre of the earth and the centre of the satellite.

The gravitational force of the planet on satellite located at a distance of R from the centre of the planet

$$= G \frac{Mm_s}{R^2} \qquad \qquad .....(26.11)$$

where $\quad M$ = mass of the planet

Equating (26.10). and (26.11), we have

$$\frac{m_s V_0^2}{R} = G \frac{Mm_s}{R^2}$$

Or $\qquad V_0^2 = G \frac{M}{R}$ .....(26.12)

In case of Earth, $\qquad V_0^2 = G \frac{M_e}{R}$ .....(26.12(a))

Where R = Re + h, Re = radius of the earth, h = altitude of the satellite above earths surface

we have proved
$$g_e = \frac{GM_e}{R_e^2} \quad \text{Eq. (26.2)}$$

or
$$G M_e = g_e R_e^2$$

Substituting this value in Eq. (26.12(a)) we have

$$V_0^2 = \frac{g_e R_e^2}{R}$$

or
$$V_0 = R_e \sqrt{\frac{g_e}{R}} \qquad \qquad \dots\dots(26.13)$$

The orbital velocity of a satellite close to the earth, (that is approximated to $h \ll R_e$ or $R = R_e$)

$$V_{oe} = \sqrt{g_e R_e} \qquad \qquad \dots\dots(26.14)$$

From (26.9). and (26.14). we have

$$V_E = \sqrt{2}\, V_{oe} \qquad \qquad \dots\dots(26.15)$$

that is the escape velocity is $\sqrt{2}$ times orbital velocity of the satellite which is close to the earth.

The time period (T_o) of a satellite located at a distance of R (moving in a circular orbit) from the centre of the earth is given by

$$T_o = \frac{2\pi R}{V_o} = \frac{2\pi R}{R_e \sqrt{\dfrac{g_e}{R}}} \qquad \text{using Eq. (26.13)}$$

$$T_o = \frac{2\pi R}{R_e} \sqrt{\frac{R}{g_e}} \qquad \qquad \dots\dots(26.16)$$

where
$$R = R_e + h$$

The time period (Te) of a satellite close to the earth [putting $R=R_e$ in Eq. 26.16)]

we have

$$T_e = 2\pi \sqrt{\frac{R_e}{g_e}} \qquad \qquad \dots\dots(26.16(a))$$

If $V_o < V_E$, a satellite reaches an altitude 'h' above the earth's surface (KE = work done)

$$\frac{1}{2} m_s V_0^2 = \int_{R_e}^{R} m_s\, g_R\, dR$$

$$\frac{1}{2}V_0^2 = \int_{R_e}^{R} g_e \frac{R_e^2}{R^2}\, dR \quad \text{(using Eq. (26.4). } g_R = g_e \frac{R_e^2}{R^2}\text{)}$$

$$\frac{1}{2}V_0^2 = g_e R_e^2 \int_{R_e}^{R} \frac{dR}{R^2} = g_e R_e^2 [-\frac{1}{R}]_{R_e}^{R} \quad \text{where } R = R_e + h$$

$$= g_e R_e^2 [\frac{1}{R_e} - \frac{1}{R}] = g_e R_e^2 [\frac{R - R_e}{RR_e}]$$

$$V_0^2 R = 2g_e R_e [R - R_e]$$

$$= 2g_e RR_e - 2g_e R_e^2$$

or
$$2g_e R_e^2 = R[2g_e R_e - V_0^2]$$

putting $R = R_e + h$, in the above equation, we get

$$\therefore \qquad R_e + h = \frac{2g_e R_e^2}{[2g_e R_e - V_0^2]}$$

or
$$h = \frac{2g_e R_e^2}{2g_e R_e - V_0^2} - R_e$$

$$h = \frac{V_0^2 R_e}{2g_e R_e - V_0^2} \qquad\qquad .....(26.17)$$

As $h = \to \infty$, $2g_e R_e - V_0^2 \to 0$. This implies V_o becomes escape velocity V_E

Thus we have $V_E^2 = 2g_e R_e$ which is the same as equation 26.9.

Note

(i)   If $V_o < V_E$, the trajectory or the path of the satellite (or rocket) will be ellipse or circular.

(ii)   If $V_o = V_E$, the path is a parabola, and

(iii)   If $V_o > V_E$, the path is a hyperbola.

Example

we have proved that the circular (orbit) velocity of a satellite at an altitude h or at a distance R from the centre of the earth ($R = R_e + h$) is given by

$$V_o = R_e \sqrt{\frac{g_e}{R}} \quad \text{and the orbital period } T_o = \frac{2\pi R}{R_e} \sqrt{\frac{R}{g_e}} \quad \text{(eqs. 26.13, 26.16)}$$

The orbital velocity of a satellite close the earth (i.e., $R \cong R_e$ is given by

$V_{oe} = \sqrt{g_e R_e}$ and the time period $T_e = 2\pi \sqrt{\dfrac{R_e}{g_e}}$

Given $g_e = 9.81$ m/s^2, $R_e = 6.37 \times 10^6$ m, $\Omega = 7.292 \times 10^{-5}$ rad/s

Earth–sun distance $= 1.5 \times 10^{11}$ m

We have orbital velocity near earth

$$V_{oe} = \sqrt{9.81 \times 6.37 \times 10^6} \cong 7.9 \text{ Km/s}$$

Orbital time $T_e = 2 \times 3.143 \sqrt{\dfrac{6.37 \times 10^6}{9.81}} = 5065$ seconds $= 84$ min, 22 sec

Escape velocity $\qquad V_E = \sqrt{2g_e R_e} = \sqrt{2} V_{oe}$

$$= \sqrt{2 \times 9.81 \times 6.371 \times 10^6} = 11.18 \times 10^3 \text{ m/s}$$

$$= 11.2 \text{ km/s}$$

Example

If h= 1000 km, $R_e = 6.371 \times 10^6$ m, $g_e = 9.81$ m/s^2, R = Re + h

Orbital velocity $\qquad V_o = R_e \sqrt{\dfrac{g_e}{R}} = 6.371 \times 10^6 \sqrt{\dfrac{9.81}{7.371 \times 10^6}}$

$$= 7.35 \text{ km/s}$$

orbital time period $\qquad T_o = \dfrac{2\pi R}{V_o} = \dfrac{2 \times 3.143 \times 7.371 \times 10^6}{7350} = 6034 \text{ s}$

$$= 1 \text{ hr } 45\text{m}$$

Example

If h = 35840 km, Re = 6.371×10^6 m, $g_e = 9.81$ m/s^2. Mean orbital speed of the earth on its axis of revolution 2977.0 m/s

Here R = Re + h = $6.371 \times 10^6 + 35.840 \times 10^6 = 42.211 \times 10^6$ m

Orbital velocity of a satellite

$$V_o = R_e \sqrt{\dfrac{g_e}{R}} = 6.371 \times 10^6 \times \sqrt{\dfrac{9.81}{42.211 \times 10^6}}$$

$$\cong 3071 \text{ m/s}$$

Time period

$$\frac{2\pi R}{V_o} = \frac{2\pi R}{Re}\sqrt{\frac{R}{g_e}}$$

$$= \frac{2 \times 3.143 \times 42.211 \times 10^6}{6.371 \times 10^6}\sqrt{\frac{42.211 \times 10^6}{9.81}} \cong$$

86400 s $= 24$ hrs.

So the height of geostationary satellite orbit is about 35840 km.

Note : A circular orbit (in equatorial plane) in which a satellite moves from west to east with the orbital time period of 24 hrs is called Clarke orbit or 24 hour orbit.

Importance of Satellite Based Data

Man is dependent on nature for feeding himself and animal. He extracts food, clothing, habitation, minerals etc. all from nature. From the dawn of civilization man has been observing and studying the natural hazards associated with tropical cyclones, floods, earthquakes, volcanic eruptions and trying to minimize its effects. Meteorology may be regarded as an observational science. Scientists are collecting observational data of these natural hazards from atmosphere, earth and oceans. With all its technical and scientific advances man could only collect crucial observations of weather and sea from about 20% of the global area and the remaining 80% was blank. This lack of observational data from this vast area is a severe draw back in understanding the ocean and atmospheric behaviour for evaluating and predicting the severe natural phenomena. This lack of data are now compensated by satellite based observations.

The principal parameters of atmospheric study consists of temperature, wind, relative humidity, rainfall (precipitation), atmospheric pressure at sea level, cloudiness, composition of atmospheric gases. Ocean study includes sea surface temperature, wind stress on sea surface water, sea surface level, sea surface currents, heat content in the top layer of the sea, ocean eddies, sea-sub-surface circulation and precipitation over the sea. The satellite based instrumentation in single measurement provide sea surface temperature which is equivalent to about thirty thousand research vessels observational data. Because of such voluminous precious data, satellite based observations gained incredible importance. A number of remote sensing instruments have been developed for getting information (observations) of atmosphere, oceanography hydrology, geology, forestry and agriculture. The following points further add to the importance of satellite based data.

Satellite based observations provide a regular supply of data from remote places which are not accessible for conventional observations. Large scale weather systems (cyclones) scanned from high altitudes provides information of the system in a single view. Geostationary satellites provide continuous weather data of a major portion of the globe which is useful in monitoring and warning of short lived weather phenomena such as thunderstorms, Tornadoes.

Voices from the Space

With the advent of communication satellites, all the political and cultural frontiers have been fused into a global market. Different parts of the globe have been linked round the clock with the help of these satellites. They are used to relay large volume of radio and TV broadcasts, telephone exchanges, telex and facsimile services. Live via satellite has become a common feature on occasions of Olympic games, international cricket, football, tennis etc. matches, election results.

The first soviet satellite Sputnik–1 (1957) may be regarded as communication satellite by virtue of radio beacon. The first human voice returned to the earth from space was US president Dwight. D.Eisenhower a tape recorded Christmas message through project SCORE. The first true communication satellite was Telstar–1, launched on 10 July 1962, which linked United States of America and Europe on global T.V. Communication. Molniya series of erstwhile Soviet Union were the world's biggest domestic communication satellite network which linked various Soviet state centres by black and white T.V. programmes, Telephones and Telegraphs. This network had a major social, political and economic impact on Soviet Union states and its allies.

Satellite based communication system immensely used for rapid transmission of data or reception of data from Automatic stations and their transmission to the meteorological centres immediately. Weather data are highly perishable commodities, its reception in time enhances its value. As compared to other modes of transmission satellite communication system is the fastest. Satellite based communication system is the fastest. Satellete based communication system used for broadcast of stored data in orbit to read-out stations. It is also used for real time broadcast of data to world-wide readout stations. This system collects environmental data from buoys, remote stations, ships, aircrafts and transmits to central processing stations and perform facsimile transmission of processed weather data to field stations. Geostationary satellites are used to monitor the condition of magnetic field of the earth, to measure the flux of energetic particles in the neighbourhood of the satellite, to monitor X-ray emission from the sun. These observations are transmitted to central processing facility station. The geostationary satellite derived main products include: quantitative rainfall estimates, cyclone/hurricane detection, classification, cloud motion wind vectors, cloud top heights and temperatures, transmission of storm bulletins.

Summing up, satellite based various kinds of sensors 1. monitor the conditions of the atmosphere and on the surface of the earth, 2. receives data sent to them from the sensors located at other places and 3. transmit the data or information they acquired/received to the ground centres. Monitoring of atmospheric weather conditions or on the surface of the earth is generally carried by (i) vidicon system, (ii) different types of TV cameras and (iii) largely by radiometers.

We have learnt that a radiometer is a device for measuring radiation amount in all spectral bands. In case of meteorological satellites they measure in visible (λ = 0.4 to 0.7 μm) and thermal bands (IR λ : 0.7 to 1000 μm, microwave λ : 1000-100000 μm). Satellite based radiometer devised such that at any instant it receives and measures radiation from a small area (unit) of which the image is required. By scanning mechanism it takes images of very large number of tiny units in rapid succession and thus measures the radiation amount received from each unit. An image is formed at the ground receiving station in more or less the reverse process. Radiation measurements received from the satellite are converted into brightness values, which are presented at their relative positions on photographic or electronic digital display device (such as TV screen). Images are of two kinds ; visible (produced by visible radiation) and infrared (produced by IR radiation) which were discussed earlier.

Resolution of an Instrument

The smallest area on the ground that a given instrument is able to identify (or scan) is called its spatial resolution or simply resolution with respect to a unit angle of view (or field of view) of the scanner.

The scanning swath (wide band on the ground) width is dependent on the altitude of the satellite and the swinging angle of the rotating mirror or the receiving picture tube. Scanning radiometer (SR) works on the principle of mechano–optical scanning.

The resolution of a scanning radiometer is a function of the sensor field of view, altitude of the satellite and the sanning rate of the sensor. The resolution of SR sensor on board NO AA/ITOS satellite is about 7.5 km (along sub-satellite point). In terms of temperature error it varies from $\pm$ 2 °C at the warmest end (300 °K or 27 °C) and $\pm$ 8 °C at the cold end (185 °K or – 88 °C). The resolution of SR sensor of METEOR satellite is 15 km or in terms of temperature error $\pm$ 2 °C at the warmer end (300 °K) and $\pm$ 4 °C at the cold end (220 °K)

Remote Sensing Radiation Measurements

Remote Sounding

We have learnt that Radiometer is a generic term used for any instrument that measures radiation. Radiometer on board satellite measure the mean albedo (or reflectivity) of the earth–atmosphere system and its variation in space and time. The study of the vertical temperature profiles of the atmosphere may be conceived as in adjoining Fig. 26.1. In this figure the atmosphere is stratified into layers L_1, L_2, L_3, L_4 and L_5. The top of the boundary layer L_1 is denoted by T, which is the top of the atmosphere. The lower boundary of the lowest layer L_5 (the surface of the earth) is denoted by S. In the wavelength range A, the terrestrial radiation reaching the satellite is absorbed by the layers L_2 to L_5 except L_1, that is the radiation received by the satellite comes only from the top layer L_1. This corresponds to the average temperature t_1 of the layer L_1. This average temperature t_1 is taken with regard to the density, consequently the measurement is influenced by the lower layers L_2 to L_5. In the wavelength B, the radiation received by the satellite corresponds to the average temperature of the layers L_1 and L_2 and is denoted by t_{12}. The difference t_{12}-t_1 gives the average temperature of the layer L_2. Similarly in the wavelength range C the radiation received by the satellite represents the average temperature t_{13} of the layers $L_1 + L_2 + L_3$. The difference $t_{13} - t_{12}$ gives the average temperature of the layer L_3. Likewise in the wavelength range D the average temperature of the layer L_4 is given by $t_{14} - t_{13}$. The layer L_5 of wavelength range E receives some radiation from the surface of the earth and/or cloud tops. This is the lower limit to the altitude for deducing temperature. Thus the layer L_5 average temperature corresponds to the actual earth's surface or cloud top temperature, which is given by $t_{15} - t_{14}$.

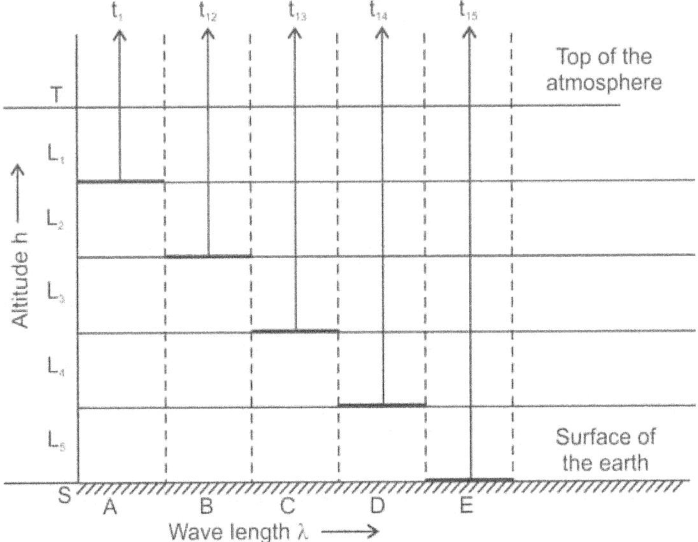

Fig. 26.1

A number of sensory devices are used for radiometric measurements, which operate over a number of wavelength ranges. There are three distinct electromagnetic spectral bands 0.4 to 1.2 μm (visible and near infrared), 3 to 5 μm and 8 to 13 μm (thermal infrared) and 1mm to 10 m (radio). Narrow spectral bands are selected from these to deal with specific problems. Oceanographic application requires about 100 spectral bands. The spectral range 0.85 to 1.1 μm provides maximum information for oceanographic application. For atmospheric studies the following wavelengths are important. 0.57 to 0.62 μm, 0.69 to 0.75 μm, 0.79 to 0.82 μm, 1.1 to 1.2 μm and 1.26 to 1.28 μm. In these spectral regions, terrestrial radiation is absorbed by the atmospheric gases, particularly Oxygen, Water vapour, Ozone, Methane, Nitrous oxide, Carbonmonoxide. The spectral region 0.3 to 0.4 μm is obscured by the atmospheric haze/smog and hence not useful for imaging.

Basic Principle of Remote Soundings

The remote sounding of the atmosphere from space is principally based on the property of selective absorption of the atmospheric gases. Satellite based radiation measurements in any spectral band depends on (i) the concentration, and (ii) temperature of the absorbing component. For all practical purposes all the energy of solar radiation is contained in the wavelength range (λ) of 0.1 to 100 μm with maxima about 0.5 μm. The out going terrestrial radiation is contained in wavelength range $\lambda = 1$ to 100 μm with maxima about 10 μm. For estimation of atmospheric temperature and moisture the various absorption spectral bands of some important gases are given below.

1. Carbondioxide (CO_2) absorbing spectral bands wavelengths (λ) are: 13.5 to 16.5 μm (centre 15 μm), 4.2–4.4 μm (centre 4.3 μm), 10.4 μm, 9.4 μm, 5.2 μm, 2.0 μm, 1.6 μm and 1.4 μm. Weak bands in 0.78–1.24 μm.
2. Molecular oxygen (O_2) absorbing spectral bands (λ) : 1.07 μm, 1.27 μm, series of lines between 50-70 G Hz, one line at 118.75 G Hz.
3. water vapour (H_2O) absorbing spectral bands (λ) : 5.5-7.5 μm : (centre 6.3 μm), 2.6–3.3 μm (several centres), 1.9 μm, 1.4 μm, 1.1 μm, 0.94 μm, 0.81 μm, 0.72 μm, 22.24 and 183.31 G Hz, lines from 300 GHz to 8 μm.
4. Ozone (O_3) absorbing spectral bands (λ) : 14.1 μm, 9.6 μm, 9.0 μm, 5.75 μm, 4.75 μm, 3.59 μm, 3.27 μm, 2.7 μm. Lines between 9.2–43.65 G Hz, lines at 96.23 G Hz, 101.74 G Hz, 118.36 G Hz, 20 lines between 160 and 380 G Hz. 1800-3400 °A (centre at 2600 °A), 3200–3600 °A, 4400–7400 °A.
5. Methane (CH_4) absorbing spectral bands (λ) at 3.3 μm, 7.7 μm
6. Nitrous Oxide (N_2O) absorbing spectral bands (λ) : 4.5 μm, 7.8 μm, 17.0 μm and 25 G Hz, 75 G Hz, 50 G Hz.

7. Carbonmonoxide (CO) absorption spectral bands (λ): 4.67 µm, 2.35 µm, 1.57 µm. 1.19 µm, 2.38 to 25 µm (several centres), 230.77 GHz, 115.27 GHz.

8. Atmospheric windows or least absorption bands (λ) : 8-13 µm, 18 µm, 4.7 µm, 4.0 µm, 3.8 µm, 2.3 µm, 1.65 µm, 1.25 µm, 1.05 µm, and 1.4 to 10.0 GHz, 33-36 G Hz, 80-100 GHz, 125-140 GHz, 210-235 G Hz. Also in visible green.

For temperature soundings, generally oxygen band measurements are used, while for humidity measurements water vapour bands are used. It is assumed that CO_2 and O_2 are homogeneously mixed in the atmosphere and are invariant in time and space. This assumption is not valid in case of ozone and water vapour which are highly variable. Radiation sounding measurements are biased in the presence of clouds. For this reason instead of single field of view, a multiple field of view methods are used. Satellite derived temperature profiles are commonly compared with radiosonde observations at standard constant pressure levels.

The pentachannel radiometer of TIROS satellite described below.

Channel 1

Spectral band width 6.0-6.5 µm. This is the main water vapour absorption band. This band sensor measures temperature near the top of the atmospheric water vapour (about 400 h Pa level). The distribution of water vapour varies both in time and space. The level may be much lower over desert and higher over moist tropics. Lower temperature indicates high altitude (more depth of moisture) and high temperature indicates low altitudes (shallow depth moisture, nearer to the surface).

Channel 2

Spectral band width 8-12 µm. This is atmospheric window (or lowest absorption). This band sensor measures surface temperature in the cloud free area or cloud top temperature (or heights of cloud tops). During night it measures cloud coverage.

Channel 3

Spectral band width 0.2-6 µm. This band sensor measures earth's albedo or reflected solar radiation.

Channel 4

Spectral band width 7-30 µm. This band sensor measures the long wave radiation or IR radiation emitted by the earth–atmosphere system or the exhaust of the atmospheric engine.

Channel 5

Spectral band width 0.55–0.75 μm. This band sensor measures reflected solar radiation in the red part of the visible spectrum or reflected solar radiation in the same visible region as the TV cameras. It provides the cloud cover during day time.

Indian National Satellite (INSAT)

On 18 July 1980 India became the seventh nation (after Soviet union, USA, France, Japan, China and United Kingdom) to launch an artificial satellite by (satellite launch vehicle) SLV–3.2 Rohini rocket from Sriharikota in Andhra-Pradesh. INSAT-1 a multipurpose satellite, used for (i) Long distance telecommunications (telephony, data, facsimile etc), (ii) round-the-clock meteorological earth observation and data relay and (iii) direct TV broad casting.

The first INSAT-1 A, launched on 10 April 1982 was deactivated on 6 september 1982 due to mallfunctioning. INSAT-1B launched on 30 August 1983 and was put into operational use from 15 October 1983. It was geostationary satellite, located at long 74 °E in the equatorial circular orbit, moving west to east at an altitude of 35680 km. Its orbital time period was 24 hrs. INSAT–1 system covers the space-segment and telecommunications, Meteorological ground segments. The telecommunication component of INSAT-1 space segment provides over 8000 two-way long distance telephone circuits potentially accessible from any part of India, even the remotest without regard to intervening terrain and terrestrial distances.

To enhance Met (Meteorological) Data reception and quality, India launched METSAT on 12 September 2002 by PSLV (Polar Satellite Launch Vehicle) C4 (continuous 4) from Sriharikota Andhra Pradesh. This was first time that ISRO used the PSLV to launch a meteorological satellite in the geo-synchronus transfer orbit (GTO). VHRR on board METSAT has the capability of imaging earth in the visible, thermal infrared and water vapour bands. It has data relay transponder which could collect the data from unattended meteorological platforms and relay it to the MDUC (Meteorological Data Utilisation Centre).

Meteorological Utilisation of INSAT–1

It provides : (i) round-the-clock regular, half hourly synoptic images of weather system including severe weather, cyclones, sea-surface and cloud top temperatures, water bodies, snow etc., over entire territory of India and adjoining land and sea areas, (ii) collection and transmission of meteorological, hydrological and oceanographic data from unattended remote platforms, (iii) timely warnings of impending disasters from cyclones, floods, storms etc., and (iv) dissemination of meteorological information including processed images of weather systems to the forecasting offices.

Weather data provided by INSAT–1 system is very useful in Agriculture, Aviation, Ports and shipping, Hydrometeorological services and in flood forecasting. It is very useful in detection and tracking of cyclones and in its early warning.

INSAT–1 Meteorological ground segment consists of Meteorological Data Utilisation Centre (MDUC), 20 Secondary Data Utilisation Centres (SDUCs), 100 land based and 10 ocean based automatic unattended Data collection platforms (DCPs) and Disaster warning system (DWS). MDUC is located in Mausam Bhavan (IMD Head Quarters) New Delhi, where INSAT data is processed and the processed data is disseminated to the users. The meteorological data from INSAT-1B first received in the form of electro-magnetic signals at the Delhi Earth Station (Sikandarabad of Buland Sahar district in UP) from where they are transmitted to MDUCs over microwave link.

INSAT-1 based VHRR (Very High Resolution Radiometer) has two channels to measure radiation in the wave length band 0.55-0.75 μm (visible micrometer) and 10.5-12.5 μm (Infrared micrometer). It has a field of view (FOV) about 2.75 km × 2.75 km in the visible and 11 km × 11 km in IR at the Satellite Sub-point. It has bidirectional scan which takes one second to cross 20° in east-west direction and 0.2 seconds to 30.7 micromeridians in the north-south direction. The cycle then continues in reverse in the west-east direction. It takes about 22 minutes 44 seconds to scan 20° by 20° image of the earth. It has the facility to scan 20° (east-west) by 5° (north-south) image.

MDUC processes the INSAT-1 VHRR data. It converts VHRR data into pictures of earth's cloud cover over India and adjoining land & sea areas. These pictures are called satellite derived cloud imageries. The data are received on a round-the-clock basis, which is useful in 12 hourly day-to-day weather forecasts. Using half hourly imageries upper winds (direction and speed) are derived over India and adjoining oceanic areas. The VHRR data are also used for derivation of sea surface temperature (SST) over the Indian seas.

The cloud imagery data processed at MDUC is transmitted to the SDUCs, about 20, located at different operational forecasting centres of IMD, through P & T communication links.

The Data Relay Transponder (DRT) on board INSAT-1 is used for collection of Meteorological, Hydrological and Oceanographic data from remote, uninhabited locations over the land and sea, where DCPs are installed. The DCPs data are transmitted through DRT to the earth station, from where it is received by MDUC through microwave link.

Disaster Warning System (DWS) Via INSAT

This scheme was implemented initially in coastal Andhra Pradesh (CAP) and coastal Tamilnadu (CTN) in India. In this scheme warnings against disastrous weather conditions such as cyclonic storms, floods etc are directly sent to the areas likely to be affected through INSAT. The warning messages are originated from the Area Cyclone Warning centre Chennai will be transmitted to INSAT-1 from Chennai earth station (P & T). A S-band transponder on board INSAT-1 will receive and relay back these signals for the reception by Disaster warning receivers installed in CAP and CTN. The warning signals are selectively addressed to specific DWS Receivers. Receipt of timely warnings will help the local authorities to take action on cyclone mitigation. DWS scheme is proved to be very effective and fruitful in disseminating timely warnings directly to the effected people and in turn saved life and property.

Some Important Terms used in Satellite Meteorology

Apogee/Perigee : The points in its elliptical orbit at which the satellite is farthest/closest from the centre of the earth (that is the points of maximum/minimum satellite distance from the centre of the earth)

Nodes : The points of intersection of the equatorial plane by the satellite orbit.

Ascending /Descending Node : As the satellite crosses from the southern hemisphere into northern hemisphere the satellite orbit cuts the equatorial plane at the ascending node; if it passes from the northern hemisphere into southern hemisphere its orbit cuts the equatorial plane at the descending node.

Attitude : The orientation of the satellite spin axis (optical axis) in space. It is expressed in terms of right ascension (going up) and declination.

Declination : The angular distance of an object located north (+ positive) or south (– negative) of the celestial equator measured along the hour circle passing through the object.

Oriental Inclination : The angle between the equatorial plane and the orbital plane having its vortex in the ascending node. If the satellite moves due east, the inclination angle ranges $o < i < \dfrac{\pi}{2}$ but if it moves due west it ranges $\dfrac{\pi}{2} < i < \pi$

Nadir Angle (η) : The angle measured at the satellite between a specific axis (or ray) and the local vertical. The nadir angles of the satellite spin axis and a camera optical axis are called the "Satellite nadir angle" and the "tilt" respectively.

CDA (Command and Data Acquisition Station) : A ground based station (on earth) at which various functions to control the satellite operations and to obtain data from the satellite are performed. The CDA transmits programming signals to the satellite, and commands transmission of data to the ground. Processing of data (by electronic machines and by hand) is accomplished at the CDA station. Raw and processed data are disseminated from CDA stations.

Celestial Sphere : An imaginary sphere of infinite radius whose centre is the centre of the earth, upon which appear projected the stars and other astronomical bodies. This sphere is fixed in space, but appears to rotate from east to west, because of the earth rotation from west to east.

Celestial Equator : The great circle in which the plane of the earth's equator intersects the celestial sphere.

Degradation : The diminution of the picture quality due to noise, rotation of satellite etc; or any optical, electronic or mechanical distortion in the image system.

Direct Pictures : Pictures taken while the satellite is within telemetry range of CDA. Direct pictures are transmitted directly to the ground station without being recorded on tape in the satellite.

Distortion–Optical, Electronic : An apparent warping or twisting of the picture image received from a satellite. The distortion may be caused due to electronic or optical. Electronic distortion is caused by (i) imperfections in the circuitry, (ii) the tape recorder, (iii) the vidicon tube structure, (iv) the transmission system or (v) the signal characteristics. Optical distortions are caused by the characteristics of the lens and optical alignment.

Ecliptic : The great circle in which the plane of the earth's orbit intersects the celestial sphere.

Horizon Distance : Distance measured in degrees, along the principal line between image principal point (IPP) and apparent horizon.

Noise : A voltage received through an antenna system or within the ground amplifier and other electronics that does not belong to any intended signal. The noise appears in the satellite pictures in the form of small specks or bands or streaks.

Radiometer : An instrument to measure electromagnetic radiation energy. Satellite based radiometers provide radiation budget of the surface-air system, sea surface temperature (SST), cloud cover distribution and cloud top temperature. It operates in the visible, infrared and microwave wavelength regions.

Spectrometer : A satellite based spectrometer provides chemical composition of the surface layers in the atmosphere and ocean (that is, it provides a clue to look into the physical, chemical process involved there) and measures atmospheric concentrations of water vapour.

Radioaltimeter : An instrument which gauges space craft flight altitudes and enables to measure variations of ocean surface level and wave height.

Spatial Resolution : The unit area on the ground that a given instrument is able to scan is called its spatial resolution or simply resolution, having a corresponding unit angle of view (called field-of-view) of the scanner.

Orbit : The path which a celestial object (satellite) follows in its motion through space, relative to some selected point. The closed orbit may be a circle or ellipse.

Orbit Number : It refers to the particular circuit beginning at the satellite ascending node. The orbit number begins from launch at the first ascending node as zero and there after the number increases by one at each ascending node.

Optical Axis : The ray perpendicular to the image plane, passing through the lens nodal points.

Pass : A single circuit of the earth by meteorological satellite passes begin at the time the satellite crosses the equator in ascending node (from southern hemisphere to northern hemisphere). Pass is used to measure the period of time that the satellite is within the telemetry range of CDA (or data acquisition station)

Path : The projection of the satellite orbital plane on the earths surface (or the locus of the sub-satellite point) is called path. As the earth is moving under the satellite, the path of a single orbit is not a closed curve. The path and track are used interchangeably. On a mercator projection the path is a sinusoidal curve.

Perspective Grid (Incorporating Tilt and Height)

A perspective grid is a network of lines constructed for an oblique photograph, with lines representing corresponding to the lines of an imaginary quadrilateral on the surface of the earth.

Prograde Orbit (Direct Orbit)

The orbit inclination θ lies between $0°$ and $90°$.

Retrograde Orbit

The orbit inclination lies between $90°$ and $180°$ expressed as $180 - \theta$, where θ is prograde inclination.

Inclination of the Satellite Orbit (i)

The angle between the equatorial plane and orbital plane having its vertex at the ascending node. i is positive corresponding to the angle as reckoned anti-clock wise from the eastern direction at the equator. If the inclination varies from 0 to $\dfrac{\pi}{2}$ $(0 < i < \dfrac{\pi}{2})$, the satellite moves due east but if i varies from

$\dfrac{\pi}{2}$ to π $(\dfrac{\pi}{2} < i < \pi)$, the satellite movement is due west.

Questions

1. State Kepler's laws of planetary motions.

2. Define escape velocity from the earth's gravity, derive expression. Find the escape velocity of an object if given radius of earth $\simeq$ 6400 Km, and ge = 9.81 m/s².

3. Define satellite orbital velocity, derive a period equation to find the orbital velocity and Time period of a satellite around the earth. Deduce these if the satellite height (h) is small as compared to the radius of the earth.

4. Find the orbital velocity of a satellite, given the height of satellite is 35840 km, radius of earth 6371 Km, ge = 9.81 m/s², mean orbital speed of the earth about its axis 29770 m/s.

5. Write the importance of satellite based meteorological data.

6. Write the importance of satellite based communication and in particular in cyclone disaster warning.

7. Write satellite remote surrounding of vertical temperature profile of atmosphere. Write the basic principle of remote surroundings.

8. Describe briefly the penta channel radiometer of TIROS satellite.

9. Describe INSAT, write its utilization meteorology (IMD) and in particular in disaster warning system.

10. Explain briefly : (i) CAD, (ii) Noise (iii) Radiometer (iv) Spatial resolution, (v) Pass of a satellite (vi) Prograde orbit.

CHAPTER 27

Climatology

Weather and Climate

The physical state of atmosphere at any particular location (station) seldom exhibits a steady state even during short intervals of time. The ever changing physical state of atmosphere constitutes the weather. It is described in terms of instantaneous values or short period mean values of the meteorological elements on surface of the earth and in atmosphere, such as temperature, pressure wind, humidity, state of sky (cloud), precipitation etc. The average conditions (not only numerical values but available information) of weather at a place over a long period (more than 30 years), together with its extremities constitutes the climate of the place. In Greek climate means inclination, which refers to the inclination of the solar rays which is the function of time and latitude.

The phenomena of weather and climate has been changing right from the origin of atmosphere and life on earth and reached a balance which is suitable for life on earth.

Climatology involves the collection, presentation, analysis and interpretation of weather data over, long periods of time and space. It may be broadly described as statistical meteorology in relation to geography or natural environment. Climate never fits into regid demarcation. It changes from one type to another, one generation to the next, from one century to the next and from one ice age to the next. Climatic normals are average values of weather elements over a place for about 10 days or one month (say 1-10 Jan or Jan month), averaged over long period about 30 to 100 years of record. These normals are used as yardstick for describing variation of weather/climate.

Climatology may be divided into three parts.

    (i)   Physical,

    (ii) Regional and

    (iii) Applied.

Physical climatology deals with the causes of variation in heat exchange, moisture exchange and movement of air over space (latitude/longitude) and time (over months of a year). We know weather of a place is described by the meteorological elements wind, visibility, atmospheric pressure, temperature, cloudiness, fog, precipitation, radiation, duration of sunshine etc. These elements involve in a number of processes, such as heat, moisture momentum exchanges or transfers. These process in turn depend on latitude, altitude, soil type, water bodies, orography (plane or mountain barriers) and topography.

Regional climatology deals with the description of world climates which includes climate classification.

Applied climatology deals with the relationship between climate with other sciences particularly human health (Bioclimatology), Building climatology, Agricultural climatology, urban climatology etc.

Climate Classification

Climate classification is a complex subject. In essence it is a way of effectively presenting average weather information in a simplified and general form. Most of the classifications are based on the relationship between climate and vegetation or soil. Natural vegetation is the best classification of the world's climate.

A systematic division of climate is based on temperature, moisture, general wind circulation with probabilities of weather occurance. Thus it may be classified as :

    (i)   Empirical

    (ii) Genetic and

    (iii) Applied

Empirical

It is based on observable features. Based on temperature we may have hot, warm, cool and cold climates. To these we may associate precipitation criteria or moisture.

Genetic

It is based on latitude and temperature zones, namely Torrid (between latitudes $23\frac{1}{2}\,°N$ to $23\frac{1}{2}\,°S$), Temperate (pole ward in either hemisphere between $23\frac{1}{2}°$ to $66\frac{1}{2}°$) and Frigid (beyond polar circles to poles in either hemispheres).

In addition to these general wind circulation, distribution of land and sea, orography effects are considered. Thus we may have polar, tropical, trade wind littoral, maritime, continental and highland climates.

Applied or Functional Classification

It defines limits (climate indices) in terms of climate effects on other phenomena. Mostly natural vegetation is chosen to integrate climatic factors on vegetation. The common classifications use vegetation terms such as Rainforest, Desert, Steppe, Tundra etc. Human health and comfort may be chosen in relation to clothing, housing, physiology and medicine.

Astronomical Classification

The earliest classification of climate was based on astronomical consideration of insolation. Originally classified into three categories as Torrid zone, Temperate zone and Polar or Frigid zone. Subsequently added two marginal zones, namely, two sub–tropical High pressure dry belts in North and South hemisphere and snow forest or Boreal Tree forest.

Torrid Zone

It extends from Tropic of Cancer to Tropic of Capricorn. This zone is characterised by small variation of temperature in a year and abundant of rain. During a year Sun reaches zenith twice at noon and on any day at noon its height is not less than 43^0.

Temperate Zone

In northern hemisphere it extends from tropic of Cancer to Arctic circle and in southern hemisphere from Tropic of Capricon to Antarctic circle. In this region sun never attains zenith. In summer hemisphere the length of a day increases from 12 hrs to 24 hrs as one moves from Tropic of Cancer (Tropic of Capricorn) to Arctic (Antarctic) circle. In winter sphere the reverse (12 hrs to 24 hrs night).

Polar or Frigid Zones

It extends from polar circles to poles, In summer hemisphere the length of a day increases from 24 hrs to six months while in winter sphere the reverse 24 hrs night to six months night.

Two Sub–tropical High Pressure Dry Belts

In both hemispheres this is a transition zone from Torrid to Temperate zones.

A Zone of Snow Forest or Boreal

It is a transition zone from Temperate to Polar, in northern hemisphere. In southern hemisphere this does not exist, as there is no land mass in this belt.

Koppen's Climate Classification

Wladimir Koppen (1846–1940) was a Russian–born Biologist. He intensely devoted his life for studying climatic problems. He classified the climates based on vegetation, temperature, rainfall and seasonal characteristics. It is an empirical form. Koppens classification extensively used in geographical teachings. The salient points of his classification are given below.

Temperature Criteria

Using mean monthly temperature he classified climate into five categories and denoted these by capital letters.

A : Tropical forest climate. All seasons hot. Coldest month temperature is greater than 18 °C (No cold season),

B : Dry climates,

C : Warm temperate rainy climate with mild winter. Coldest month temperature between – 3 to 18 °C and warmest month temperature greater than 10 °C,

D : Cold boreal or snow forest climate. Severe winter. Coldest month temperature less than – 3 °C (minus three degrees) and warmest month temperature greater than 10 °C,

E : Polar or Tundra climate. Warmest month temperature less than 10 °C (No warm season).

The Significance of Temperature Limits

10 °C summer isotherm is approximately the pole ward limit of tree growth, 18 °C winter isotherm is the critical limit for tropical plants.

– 3 °C is isotherm denotes a few weeks snow cover.

Additional second letter symbols (f, s, w, m) denote rainfall regime third letter (h, k, k', i, g) temperature characteristics and the fourth letter (S, W, T, F) denotes a special feature of the climate.

Aridity Criteria

It is defined as in Table 27.1

Table 27.1

	Steppe (BS)/ Desert (BW) boundary	Forest /Steppe boundary
Winter pptn maximum	$\dfrac{r}{t} = 1$	$\dfrac{r}{t} = 2$
Even pptn throughout the year	$\dfrac{r}{t+7} = 1$	$\dfrac{r}{t+7} = 2$
Summer pptn maximum	$\dfrac{r}{t+14} = 1$	$\dfrac{r}{t+14} = 2$

Significance of Criteria is that

$$\frac{r}{t} < 1 \qquad \text{desert or arid}$$

$$1 < \frac{r}{t} < 2 \qquad \text{Semi–arid}$$

Notation

S : Semiarid or steppy

W: Arid or desert T : Tundra F : Ice Cap (Frost)

t = mean annual temperature °C;

r = annual precipitation in cm.

pptn = Precipitation

Main Sub-Divisions are made as under

(I)   With regard to seasonal pptn. The common notations are

    f   :   no dry season (constant wet)

    m   :   monsoon climate along with short dry season.

    s   :   summer dry season.

    w   :   winter dry season.

(II)   With regard to temperature characteristics.

    For the B – climate notation are:

    h   :   mean annual temperature greater than 18 °C

    k   :   mean annual temperature less than 18 °C (warmest month >18 °C)

    k'   :   mean annual temperature and warmest month <18 °C

The Principal Sub-Classifications are

Af, Aw, BS, BW, cf, cs, cw, Df, Dw, Ef while As, Ds are rare occurrence, ommitted.

Additional Sub groups

    i   :   When the annual range of temperature is < 05 °C

    g   :   Gangetic type. Maximum temperature occurring before summer solstice.

The summary of the Koppen's classification is given in Table 27.2

Table 27.2 Summary of Koppen's classification.

Groups	A	B	C	D	E
Temperature of the coldest month	> 18 °C	–	< 18 °C > – 3 °C	< 18 °C > – 3 °C	< 18 °C
Temperature of the warmest month	–	–	>10 °C	>10 °C	<10 °C
Precipitation (cm)	–	$\dfrac{r}{t+7}, \dfrac{r}{t}, \dfrac{r}{t+14} \leq 2$ r = 2(t + 7) is Bf r = 2t is BS r = 2(t + 14) is BW			

Note : If $\dfrac{r}{t+7}, \dfrac{r}{t+14}, \dfrac{r}{t} > 2$ then it does not belongs to the group B.

Merits and Demerits of Koppens' Classification

It is an empirical classification and is not based on genetics. Since it is based on numerical values of temperature and pptn units, it has universal acceptance. In this classification important meteorological parameters have been used. It takes into account of seasonal trends of temperature and pptn, besides annual means. Instead of long names of climates symbolic abbreviations are used. e.g., Af stands for hot and humid climate. Elaborately constantly hot average with coldest month temperature > 18 °C and constant wet with more than 6 cm rain in all months.

In this classification boundary regions between different climates are not specified. It does not take into account of wind and relative humidity. This classification is not suitable for high elevation stations.

Example

For the given meteorological data of a station classify Koppen's climate type.

Month	J	F	M	A	M	J	J	A	S	O	N	D	Annual
Temp (°C)	26.3	27.0	27.3	27.8	28.2	27.3	27.5	27.0	27.2	27.1	26.7	27.1	27.2 average
R F (cm)	25.0	17.5	19.6	18.5	17.5	17.1	17.0	20.0	17.4	20.5	25.7	25.3	241.1 Total

(i)   Temperature of the coldest month 26.3 is > 18 °C. It is type A.

(ii)   Range of temperature during the year 28.2 – 26.3 = 1.9 is $\angle$ 5 °C. It is – group i.

(iii) It is wet throughout the year. It is – f

(iv) $\dfrac{r}{t+7} = \dfrac{241.1}{27.2+7} > 2$. It is not B type.

∴ the climate classified as A f i

Example

For the following meteorological data of a station classify the Koppen climate.

Month	J	F	M	A	M	J	J	A	S	O	N	D	Annual
Temp (°C)	13.3	13.9	15.5	18.4	21.9	23.7	25.2	25.4	24.2	23.0	19.2	15.2	20.0 Average
pptn (mm)	67.6	41.6	21.3	4.3	2.3	0.0	0.0	0.0	2.3	18.0	47.0	66.5	270.9 Total

(i) Coldest month temperature is 13.3 °C < 18 °C. Therefore it is not A type.

(ii) $\dfrac{r}{t} = \dfrac{27.09}{20} > 1$ but < 2 $\qquad\qquad 1 < \dfrac{r}{t} < 2$ ∴ It is BS style.

(iii) Coldest month temperature 13.3 °C < 18 °C but > − 3 °C.

∴ Summer dry − s

$\overline{T}$ mean annual temperature 20 °C > 18 °C.

∴ h Koppen climate classification is BS hs.

Example

Month	J	F	M	A	M	J	J	A	S	O	N	D	Annual
Temp °C	13.3	15.0	16.2	19.9	23.6	25.0	27.6	28.5	27.0	24.6	21.0	16.2	21.4 Average
plot (mm)	180.0	145.1	23.2	17.5	2.5	0.0	0.0	0.0	0.0	13.3	68.2	170.0	619.8 Total

For the following meteorological data of a station classify Koppen climate.

(i) Coldest month temperature 13.3 °C < 18 °C. ∴ It is not type A.

(ii) $\dfrac{r}{t} = \dfrac{61.98}{21.4} > 2$. ∴ It is not type B.

(iii) Coldest month temperature 13.3 °C < 18 °C but > − 3 °C.

∴ It is type C

(iv) Summer dry. ∴ It is − s

Climate classification is Cs.

Koppens Modified Climatic Classification

Table 27.3 Using mean monthly, mean annual values of pptn (in cm) and temperature (in $^{\circ}$C).

Letter Symbols			Explanation
1st	**2nd**	**3rd**	
A			Average temperature of the coldest month $\geq$ 18°C
	f		pptn in the driest month $\geq$ 6 cm.
	m		pptn in the driest month $\angle$ 6 cm, but
			pptn $\geq \alpha$ where α = 10 – r/25, r = annual pptn in cm.
	w	(g/i)	pptn in the driest month $\angle$ 6 cm, but pptn $\angle \alpha$
B			Evaporation > pptn
	(f)		No seasonal change in pptn or even pptn throughout the year.
			r = 2 (t + 7), where t = annual mean temperature in $^{\circ}$C.
	s		Summer dry and r = 2t*
	W		Winter dry and r = 2 (t + 14)**
		h	annual mean temperature >18 $^{\circ}$C
		k	annual mean temperature < 18 $^{\circ}$C and warmest month >18 $^{\circ}$C
		k'	annual mean temperature and warmest month < 18 $^{\circ}$C
		s	Summer dry season, pptn in driest month of summer < 1/3 (pptn of wettest winter month)
		w	Winter dry season. pptn of the driest month of winter < 1/10 (pptn of the wettest summer month)
C			Average temperature of the warmest month > 10 $^{\circ}$C and the coldest month is < 18 $^{\circ}$C but > – 3 $^{\circ}$C
	f		pptn in the driest month of summer > 3 cm and pptn not fulfilling the conditions of s or w. Cf = no distinct dry season.
	w		pptn in the driest month of winter $< \dfrac{1}{10}$ (pptn in the wettest month of summer) Cw = Winter dry.
	s		pptn in the driest month of summer < 3 cm and < 1/3 (pptn in the wettest winter month). Cs = Summer dry
		a	Average temperature of the warmest month $\geq$ 22 $^{\circ}$C (hot summer)

Table 27.3 *Contd....*

Letter Symbols			Explanation
1st	**2nd**	**3rd**	
		b	Average temperature of the warmest month < 22 °C (Cool summer)
		c	Average temperature of one to three months $\geq$ 10 °C and warmest month < 22 °C (Cc = cool short summer)
D			Average temperature of the coldest month < –3 °C and the warmest month > 10 °C Dw = Cold climate with winter dry Df = no dry season Average temperature of the warmest month > 10 °C and the coldest month is < 18 °C but < –3 °C
	f		pptn in the driest month of summer > 3 cm and pptn not fulfilling the condition of s or w.
	w		pptn in the driest month of winter < $\frac{1}{10}$ (pptn in the wettest month of summer)
	s		pptn in the driest month of summer < 3 cm and < $\frac{1}{3}$ (pptn in the weltest winter month. Ds = Summer dry
		a	Average temperature of the warmest month $\geq$ 22 °C (hot summer)
		b	Average temperature of the warmest month < 22 °C (Cool summer)
		c	Average temperature of one to three months $\geq$ 10 °C and warmest month < 22 °C (Dc = cool short summer) Average temperature of the coldest month < – 3 °C and the warmest month >10 °C
		d	Average temperature of the coldest month < – 38 °C
E	T		Average temperature of the warmest month < 10 °C but > 0 °C, ET = Tundra climate
	F		Average temperature of the warmest month $\leq$ 0 °C EF = Perpetual frost.
H			Temperature of the warmest month < 10 °C and the altitude of the station > 1.5 km.

*** Note** : 70% or more annual ppen occurs in cooler six months (for NH, Oct to March)

****Note** : 70% or more annual pptn occurs in warmer six months (for NH, April to Sept).

Thornthwait's Climate Classification

Like Koppen's climate classification Thornthwait's climate classification has four factors denoted by a letter, so that the complete classification of a station contains four letters, two capital and two lower case with subscripts and superscripts. It is based on potential evapotranspiration (PE) and moisture index. The four factors are :

(i) Moisture index,

(ii) Thermal efficiency,

(iii) Seasonal distribution of effective moisture and

(iv) Summer concentration of thermal efficiency.

(i) Moisture index (Im) is defined as

$$I_m = \frac{100(S - D)}{PE} \qquad .....(27.1)$$

$$I_m = I_h - I_a \qquad ..... (27.1(a))$$

where $\quad$ S = monthly/annual water surplus = P – PE

$\qquad\qquad$ D = monthly/annual water deficiency = PE – P

$\qquad\qquad$ P = pptn amount

$\qquad\qquad$ PE = Potential evapotranspiration.

PE is defined as the amount of water that would evaporate from soil and transpired by vegetation if it were always readily available.

The maximum quantity OR of water which may be evaporated by a uniform cover of dense short gross when the water supply to the soil is not limited. (Penman's definition).

Wet Region : pptn > PE, Dry Region : pptn < PE

$$I_A = \text{Aridity Index} = \frac{100D}{PE} \, ,$$

$$I_H = \text{Humidity index} = \frac{100\,S}{PE}$$

I_A, I_H are expressed as percentage.

If soil moisture assumed to be constant (S = pptn = P, D = 0)

$$\text{Eq. (27.1) reduces to } I_m = 100 \left[\frac{P}{PE} \right] \qquad .....(27.2)$$

$$\text{Annual moisture index } I_A = \sum_{m=1}^{12} I_m \qquad .....(27.3)$$

Using I_A, climatic moisture zones are defined in the Table 27.4.

Table 27.4

Type	Moisture index I_A	
A : Perhumid	≥ 100	Rain forest
B_4 : Humid	80 – 100	
B_3 : Humid	60 – 80	
B_2 : Humid	40 – 60	Forest
B_1 : Humid	20 – 40	
C_2 : moist subhumid	0 – 20	
C_1 : Dry subhumid	– 33.3 to 0	Grass land and Steppe
D : Semi arid	– 66.7 to – 33.3	
E : Arid	– 100 to – 66.7	Desert

Note : Negative values of I_A found in dry climates and positive values in moist climates. Zero I_A separates the dry and moist climates.

Index of Thermal Efficiency

Thornthwait used monthly PE as an index of thermal efficiency.

$$\text{Annual index} = \sum_1^{12} PE$$

The climate types based on thermal efficiency defined as in Table 27.5.

Table 27.5

Thermal efficiency			Summer concentration	
Climate type		ΣPE in cm	Type	Percentage of concentration
A′	Megathermal	≥ 114	a′	≤ 48.0
B_4'	Mesothermal	99.7 - 114	b′	48.0 - 51.9
B_3'	Mesothermal	85.5 - 99.7	b_3'	51.9 - 56.3
B_2'	Mesothermal	71.2 - 85.5	b_2'	56.3 - 61.6
B_1'	Mesothermal	57.0 - 71.2	b_1'	61.6 - 68.0
C_2'	Microthermal	42.7 - 57.0	c_2'	68.0 - 76.3
C_1	Microthermal	28.5 - 42.7	c_1'	76.3 - 88.0
D′	Tundra	14.2 - 28.5	d′	> 88.0
E′	Frost	< 14.2		

Summer Concentration of Thermal Efficiency

Amount of thermal energy received during the three summer months and expressed as percentage of PE.

Seasonal distribution of moisture adequacy derived from aridity (I_a) and humidity index (I_h).

$$I_a = \frac{\Sigma(PE - P)}{\Sigma PE} \times 100$$

$$I_h = \frac{\Sigma(P - PE)}{\Sigma PE} \times 100$$

where

P = monthly pptn (mm)

PE = monthly PE (mm)

Seasons surplus or deficit calculated from monthly data.

Seasonal moisture adequacy defined in Table 27.6.

Table 27.6

Moist Climates	(A, B, C$_2$)	Aridity index (I_a)
r	Little or no water deficit	0 - 10
s	moderate summer deficit	10 - 20
w	moderate winter deficit	10 - 20
s$_2$	large summer deficit	> 20
w$_2$	large winter deficit	> 20

Dry Climates	(C$_1$, D, E)	Humidity index (I_h)
d	Little or no water surplus	0 - 16.7
s	moderate winter water surplus	16.7 - 33.3
w	moderate summer water surplus	16.7 - 33.3
s$_2$	large winter water surplus	> 33.3
w$_2$	large summer water surplus	>33.3

Example

With the following data of a station find the Thornthwait classification of climate.

Given

Annual pptn (cm) 33 cm

Thermal efficiency PE (in cm) 120 cm

Summer concentration (% of PE) 45

We have

Annual surplus S = (P – PE) cm   0

Annual deficit D = (PE – P) cm   87

Humidity index % $I_H = \dfrac{100\,S}{PE} = \dfrac{100 \times 0}{120} = 0$

Aridity index% $I_A = \dfrac{100\,D}{PE} = \dfrac{100 \times 87}{120} = 72.5$

Moisture index % $I_m = 100 \left(\dfrac{P}{PE} - 1 \right)$

$$= 100 \left(\dfrac{33 - 120}{120} \right) = -72.5$$

From thermal efficiency I_A 120 $\geq$ 114

∴ A′ Megathermal

Moisture Index – 72.5, which lies between –66.7 to –100.

∴ it is E Arid (dry climate)

Summer concentration 45< 48.0, ∴ It is a′.

Humidity index I_H = 0, no water surplus ∴ d

Given Station climate is E A′ da′

Example

Find climate type by Thornthwait classification for the following data of a station.

Data

Annual pptn	56 cm
Annual PE (thermal efficiency)	75 cm
Summer concentration (%of PE)	32
$\sum (P - PE)$ surplus	4 cm
$\sum (PE - P)$ deficit	20 cm

(i) From the above data we have thermal efficiency 75 which lies between 71.2 – 85.5 ∴ B_2'

(ii) Summer concentration 32 which < 48.0 ∴ a′

(iii) Humidity index $\dfrac{\sum(P - PE)}{\sum PE} 100 = \dfrac{4}{75} \times 100 = 5.3$

Humidity index 5.3 lies between 0 – 16.7
∴ it is d

(iv) Moisture index % $= 100\left(\dfrac{P}{PE}-1\right) = 100\left(\dfrac{56}{75}-1\right) = -\dfrac{19}{75}\times 100$

$$= -25.3 \text{ which lies between } 0 \text{ to } -33.3$$

$\therefore C_1 = $ dry subhumid.

$\therefore$ the given station climate type is $C_1\, B'_2\, da'$

Example

Find the climate type by Thronthwait classification for the following station data.

Given

pptn	68.0 cm
PE : thermal efficiency	56.8 cm
summer concentration (%of PE)	63.5
$\sum(P-PE)$ surplus	10.4 cm
$\sum(PE-P)$ deficit	5.5 cm

We have

(i)   Thermal efficiency 56.8 which lies between 42.7 to 57.0

   $\therefore$ it is C'_2

(ii) Summer concentration 63.5 which lies between 61.6 to 68.0 .

   $\therefore$ it is $'b'_1'$

(iii) Moisture index I_m $100\left(\dfrac{P}{PE}-1\right) = 100\left(\dfrac{68}{56.8}-1\right) = 19.7$ which lies

between $0-20$ $\therefore$ it is C_2

(iv) Aridity index $I_A = 100\ \dfrac{\sum(PE-P)}{\sum PE} = \dfrac{5.5}{56.8}\times 100 = 9.7$ which lies

between $0-10$. $\therefore$ it is r,

$\therefore$ given station climate type is $C_2\ C'_2\ r\ b'_1$

Radiational Index of Dryness (BI)

Russian meteorologist M.I. Budyko defined Radiational Index of dryness (BI)
using net radiation instead of temperature is given below.

$$BI = \dfrac{R_n}{Lr}$$

where

Rn = net radiation available for evaporation from a wet surface.

Lr = heat required to evaporate the mean annual precipitation

If BI < 1, it is humid area

BI > 1, it is dry area

Using these indices Budyko defined climatic types as in Table A.

Table A

BI	Climate
> 3.0	Desert
2.0 – 3.0	Semi-desert
1.0 – 2.0	Steppe
0.33 – 1.0	Forest
<0.33	Tundra

BI index may be compared with modified Thronthwait index I_m.

World Climatic Regions

World distribution of climatic regions based on moisture and sunshine are broadly four types.

(i)   Controlled by equatorial and tropical air masses. These are 1. Tropical rainy, 2. Tropical monsoon, 3. Tropical wet-and-dry, 4. Tropical arid and 5. Tropical semi-arid climates.

(ii)   Governed by tropical and polar air masses. These are : 6. sub-tropical dry summer, 7. sub-tropical humid, 8. Marine climate 9. Mid-latitude arid climate, 10. Mid-latitute semi-arid climate, 11. Humid continental warm summer climate and 12. Humid continental cool summer climate.

(iii) Governed by polar and arctic type air masses. These are : 13. Taiga, 14. Tundra and 15. polar climates.

(iv)   Governed by altitudes. This type is high land climates

The Geological Time Scale and Climatology Through Ages

According to ancient Hindu legends that the God Brahma ruled the world. He created when awake and destroyed every thing that He made while resting. This was the origin of world fires, floods and other catastrophes. Epochs of destruction alternated periodically with epochs of creation. This was the ancient scientific thinking of the laws of the earth.

Very little is known about climate before the Cambrian epoch. According to some scientists there were at least five precambrian glaciations occurred at an interval of 250 million years. Earth scientists developed the geological time scale based on rock records (rock strata), sediment deposits in lakes and on sea-shores, by studying fossils and recently by radio-active element deposits. Radiocarbon dating used to fix the age of wood, bones, shells, peats etc. It was found by radioactive dating some granite rocks in Africa and Canada are about 3300 million years old. Some meteorites found to be about 4500 million years old. If we assume that the planets were formed at about the same time as earth, then the age of earth is about 4500 million years. The salient features of the geological time or scale is given in Table 27.7. The names of eras were derived from Greek words. Palaes means old, Mesos means middle, Kainos means recent, zoe mean life. Thus Cenozoic means modern life, Mesozoic middle life and Palaeozoic means old or primary life .

Table 27.7 Grological time scale.

Era	Period	Epoch	Time scale in Years
Cenozoic	Quarternary	Recent	10^4
		pleistocene	10^6
	Tertiary	Pliocene	10^7
		Miocene	2.5×10^7
		Oligocene	4.0×10^7
		Eocene	6.0×10^7
		Paleocene	7.0×10^7
Mesozoic	Cretaceous		1.35×10^8
	Jurrassic		1.80×10^8
	Triassic		2.25×10^8
Palaeozoic	Permian		2.70×10^8
	Carboniferous		3.50×10^8
	Devonian		4.00×10^8
	Silurain		4.40×10^8
	Ordovian		5.00×10^8
	Cambrian		6.00×10^8
Precambrian			4.50×10^9

The Precambrian Era

This is the oldest and the longest geological time scale, extended from 600 to 4500 million years. It covers a period of about 3.0×10^9 years, from the time of earth formation to Palaeozoic era. In this period interstellar gas cools down, earth passes through liquid state to form crusts. The atmosphere surrounding the earth was steamy (no oxygen). Water vapour condenses. Rain produces rivers and seas. The first forms of life began in sea about 600 to 1000 million years after the earth formed. Surface of the earth was barren. Life began on planet earth in the form of bacteria, "our ultimate grand parents". It further evolved into cyano-bacteria forms, blue-green in colour. They began to use sun light and produced oxygen and organic materials. Sea weeds begin to appear.

The Earth's atmosphere and oceans were formed from the volatiles of the inner earth, having the same primordial substance. Steam and atmospheric gases generated in the interior of the earth. They rose to the surface by internal heating and volcanic activity. Water and CO_2 constitutes the primordial substance of the solar system. The mass of the hydrosphere is about 0.023% of the mass of the earth, while the mass of the atmosphere is about 0.00009%. The present atmosphere resulted from the evolution of life on earth. All living organisms are primarily composed of C, O_2, H and N. These are also the basic chemical elements of water and air shells of the earth. According to one estimate the earths biosphere (combination of atmosphere and hydrosphere) constitutes 1440×10^{15} t (tons) of water, 233×10^{10} t of CO_2 and 11.8×10^{14} t of oxygen.

Cambrian Period (600-700 million years ago)

Shallow seas covered much of the earths surface. Life exists in sea as sea-weeds, sponges, marine invertebrates, but no life exists on land.

Ordovian Period (500-600 million years ago)

Volcanic eruptions. Seas continue to expand. All life confined to sea (no life on land). Fishes developed. Invertebrates.

Silurian Period (440 to 500 million years ago)

Periodic rise and fall of sea levels. Continuous change on land. Plants began to adopt on land. New species of vertebrate animals formed, first air-breathing animals.

Devonian Period (400 to 440 million years ago)

Increased volcanic activity. Land areas expand. Mountains begin to form. Vertebrate animals develop. A variety of fish appears in sea water. Primitive amphibians. Invertebrates slowly move over to land.

Carboniferous Period (350-400 million years ago)

Seas spread. Most of present Europe and Russia lie under water which slowly, emerge as swampy areas. Coal begin to form in swamp vegetation. Large trees develop over tropical swamps. Amphibian creatures develop. Various marine life and plants develop and spread. Reptiles breed on land.

Permian Period (270 to 350 million years ago)

Warping of earth crust. Northern hemisphere covered with ice. Deciduous plants appear. End of marine creature domination. Creatures increase on land. Insects emerge, spiders and primitive reptiles.

Triassic Period (225 to 270 million years ago)

Shrubs cover mountains and deserts. More development of mountain ranges. arid conditions prevail over northern hemisphere. Fish shaped reptiles, flying fishes and lobster like creatures formed. Rise of dinosaurs. Primitive mammals appear.

Jurassic Period (180-225 million years ago)

Rockies, Endies and Panama develop. High mountains of previous arid period suffer from erosion. Limestone forms. Coniferous forms develop, flowers began to bloom. Aquatic reptiles dominate in sea. On land birds evolve. Giant dinosaurs lived in swamps. Reptiles increase both in size, number and variety. Turtles, egg laying mammals appear.

Cretaceous Period (135 to 180 million years ago)

Swamps, deltas appear. Rivers flow. Chalk deposits appear. Major mountains build up. Deciduous trees and flowering plants spread. Flying reptiles appear. Reptiles continue to dominate in sea. On land big size mammals and great reptiles appear. Dinosaurs become extinct.

Epochs of Paleocene (70-135 million years ago) and Eocene (60-70 million years ago)

Severe volcanic activity. Mountains continue to grow. Warming of climate. Flowering plants dominate. Present species of fishes developed in sea. Big whales and sea-cows appear. Marine reptiles decreased. On land modern animals and giant reptiles appear along with primitive monkeys.

Epoch of Oligocene (40 to 60 million years ago)

Land mass grows. Alps formed. More grass lands spread. In seas crabs, snails evolve. On land animals and plants found in abundant. Primitive anthropoids (apes resembling man) and Mesohippus.

Epoch of Miocene (25 to 40 million years ago)

Earths crust completely formed. Alps and Himalayas formed. Boney fish and Sharks, protohippus, whale develop in sea. On land mammals, water birds, penguins in Antarctica appeared. Equatorial regions cooled by 3 to 4 °C while Europe, America warmed up by 7 to 10 °C.

Epoch of Pliocene (10-25 million years ago)

Continents and oceans develop into present form. European and Asian land masses join. Vegetation limited. Giant sea creatures become extinct. On land number of mammal species decline. Rise of man, Pliohippus.

Epoch of Pleistocene (one to 10 million years ago)

A great Ice age. Ice sheets and glaciers cover most of Europe and America. No change of life in sea. On land rise of modern horse. Man uses fire and makes implements. Vegetation confined to warmer regions. Emergence of Homosapince.

Holocene or recent epoch (10^4 to 10^6 years ago)

Ice sheets retreat. Sea level rises. No major change in sea life. Man's dominance begins. Domestication of animals.

Evolutionary stages of man

According to Hindu cosmological concept the ages of the world is described by four Yugas, called Krita (or Satya) Yuga (1,728,000 yrs), Treta Yuga (1,296,000 yrs), Dwapara Yuga (864,000 yrs) and Kali Yuga (or Iron age 432000 yrs). Kaliy Yuga has began from 8 Feb 3102 BC, and will last for another 427000 yrs. These four Yuga to gether (4,320,000 yrs) is called Mahayuga. These yugas have been described in relation to the existance of man on earth and is believed that the physical stature of man, life span and moral standards have been deteriorating in each yuga compared to the previous. However the knowledge of the man about his existance and environment (Pancha bhootas) being improved proportionately as compared to the previous yuga. This description indicates that the evolution of the primitive man on earth is some (1728000 + 1296000 + 864000 + 3102 + 2005) = 3.9 million years ago.

According to the recent estimates (by carbon dating of fossils) is given below

Primates evolved in the paleocene (about 7×10^7 years ago) epoch. Men and apes had a common ancestor in Miocene epoch (2.5×10^7 years ago) named Ramapithecus. After Ramapithecus, the line of evolution is not clear. An ape like man evolved in Pliocene epoch (more than two million year ago) and called Australopithecus. Australopithecines included paranthropus robustus which may or may not be the ancestors of man. On the evolutionary scale the upright pithecanthropus represents an advance on the Australopithecines. The next in line were Sinanthropus and Homosapiens. Neanderthal man who lived in caves belongs to Homosapines lived in the pleistocene ice age. However Neanderthal man became extinct and Cro-Magnan man took his place.

Scientifically Cro-Magnan man is called Homosapiens, who is the modern man. During Holocene epoch (about 8000 BC) the population of the world was about 8 millions. Farminig, towns, cities came into existence at about 6000 BC. By 100 AD population grew to 300 millions and by 1800 AD it become 1000 millions. At the end of 2000 AD it crossed 6000 millions.

Radioactive Disintegration Rates and Disintegration Constants

The activity of rdioactive substances is due to disintegration of their atoms. The disintegration occurs at random and follows the law of probability or chance. In all radioactive disintegrations, the number of atoms disintegrated is proportional to the number of atoms present.

In the first order disintegration : If N is the number of disintegration atoms (nuclei), then the rate of decrease is given by

$$\frac{-dN}{dt} \alpha N \quad \text{or} \quad \frac{-dN}{dt} = \lambda N \quad \text{Or} \quad \frac{dN}{N} = -\lambda dt$$

where

λ is constant and called decay constant.

Integrating 0 to t

If $N = N_o$ when $t = 0$ and $N = N$ when $t = t$,

$\ln N = -\lambda t + \text{constant}$

then $\ln \dfrac{N}{No} = -\lambda t$ or $N = N_o e^{-\lambda t}$ .....(27.1)

Half life period $\left(\dfrac{t}{2}\right)$ is obtained by putting $N = \dfrac{N_o}{2}$ in eq. (27.1)

i.e.,

$$\frac{N_o}{2} = N_o e^{-\lambda \frac{t}{2}}$$

$$\ln \frac{1}{2} = -\lambda \frac{t}{2} \quad \text{or} \quad \ln 2 = \lambda \frac{t}{2}$$

i.e.,

$$\frac{t}{2} = \frac{1}{\lambda} \ln 2 \qquad ..... (27.2)$$

Carbon Dating

Carbon dating is based on the half life period C_6^{14} is produced by cosmic ray bombardment of Nitrogen in the upper atmosphere $^{14}_{7}N + ^{1}_{0}n \rightarrow ^{14}_{6}C + H_1$

C-14 decays according to the equation $^{14}_{6}C \rightarrow ^{14}_{7}N + ^{0}_{-1}e$.

It has been found that number of cosmic rays is relatively constant and the production and disintegration of C-14 are constant. As a result there is equilibrium between rates of formation and disintegration of C-14 and also between C-14 and C-12 content of the atmosphere. This implies that the concentration of C-14 in the atmosphere has been constant for several million years. C-14 is extremely small but its radioactive properties can be detected and estimated easily. C-14 dioxide is assimilated by plants through the photosynthesis process and plants are eaten by animals. Thus C-14 is incorporated in their systems. Any living material has the equilibrium percentage of C-14 as long as it is alive. As soon as it dies it ceases to incorporate any more C-14. C-14 already present in it begins to decay and C-14 content drops gradually. It is established that half life period of C-14 is 5760 years. In 5760 yrs, 11520 yrs, 17280 yrs..... C-14 contents reduces to 50%, 25%, 12½%... of its initial concentration. This process continues. Thus if we know the equilibrium concentration of C-14 in a dead piece of wood at a particular time the material can be dated. The age of the material is determined using half life constant for C-14. Suppose the equilibrium concentration of C-14 is 1%. Now, let the radioactivity of charcoal measured be 0.125% of the C-14. 1% of original (concentration) reduces to 0.5% after the first half life, 0.125% after the second half life and 0.125% after the third half life. Thus three half-lives must have elapsed after the wood was cut for fire. Therefore the age of charcoal (fixed) is 3 × 5760 yrs i.e., 17280 yrs.

Example

A wooden piece has only 25% as much C-14 activity as a fresh piece of wood. Calculate the age of the wood piece. Given $t_{1/2}$ (half life) for C-14 is 5760 yrs.

Solution

Let $N_o = 100\%$.   Then $N = 25\%$ of $N_o = 25\%$

From eq (27.1)   From eq (27.2)

$$\ln \frac{No}{N} = -\lambda t \qquad\qquad t_{1/2} = \frac{1}{\lambda} \ln 2 = 5760$$

$$\ln \frac{100}{25} = -\lambda t \qquad\qquad \text{or} \quad \lambda = \frac{\ln 2}{5760}$$

or $\ln 4 = -\lambda t$

$$\ln 4 = \frac{\ln 2}{5760} \times t \qquad\qquad \therefore t = \frac{\ln 4}{\ln 2} \times 5760$$

$$t = 2 \times 5760 = 11520 \text{ yrs.}$$

$\therefore$ the age of wooden piece is 11520 yrs.

Some Weather Features of Antarctica

The Antarctica circle (66 ° 33' S) defines the area of Antarctica, where there is continuous sunlight in summer and continuous darkness in winter. Antarctica is surrounded by Indian, Atlantic and Pacific oceans. It has 90% of worlds ice quantity. Weather of Antarctica is predominantly very cold and the worlds lowest temperature of –89.2 °C was recorded at Vostok in 1983, on land - 55.6 °C was recorded in western Siberia in Jan 1999. The mean minimum temperature in winter is – 9 °C and the mean maximum temperature in summer is 0.6 °C Annual snowfall over Antarctica varies 15 cm in the interior to 130 cm on the coast, while precipitation is very small and is less than in any desert over the globe. Similarly high mean wind speed over coastal areas (may exceed 320 kmph) decreases inland toward the polar plateau.

At Dakshin Gangotri (in Antarctica) : The mean annual temperature at Dakshin gangotri (in Antarctica) in 1988, was –15.1 °C while the highest maximum temperature was 9.9 °C (10 Jan) and the lowest minimum temperature was – 42.2 °C (12 Aug). The number of blizzards (blinding storm of wind and snow) recorded 123, (average $\simeq$ 10 per month), mean wind speed 30 kmph (maximum 45 kmph in March, minimum 17 kmph in November). Mean maximum pressure 1015.5 hPa recorded in June while mean minimum pressure of 951.7 hPa (July) with annual mean range 63.8 hPa.

Whiteouts

Inspite of objects being visible during whiteout, snow and ice covered terrain may mixup with the sky and horizon and the depth perception is lost. Good visibility with fine flying weather at one location may be marked by whiteout at short distance away from it. Whiteouts occur when light reflects and refracts both from snow cover surface and from thick cloud ceiling. Surface definition is lost because there are no shadows and vehicle and plane tracks may not be visible. An abandoned box at a short distance away may look like a building. During whiteout travelling becomes very hazardous. In polar regions high frequency radio communications are frequently disrupted for many days during ionospheric disturbances. Because of very cold weather the vegetation is confined to a few hardy lichens, mosses and algae. There are no trees, no edible roots, leaves, berries. Along the coast penguin, skua and seal are found.

Frostbite (freezing of tissues)

At very low temperatures when wind removes body heat faster than the body can replace it, frostbite occurs.

Frostbite or hypothermia occurs either by drop in ambient air temperature or increase in wind speed.

Hypothermia

The lowering of central body temperature is called hypothermia. Normal body temperature is maintained by muscular activity (muscular metabolism) and basal metabolism. The process in which food is converted into living matter and useful form of energy is called metabolism–which depends on energy supplied by food and water as intake. Hypothermia beings when ambient environmental temperature is much lower and body loses heat as compared to what it produced. The symptoms of setting of hypothermia are :

1. Weariness and reluctant to continue moving.

2. Trembling and shivering.

3. False feeling of well-being.

4. Clumsiness and loss of judgement.

Hypothermia progress with fall temperature is given in Table 27.8.

Table 27.8

Temperature		Symptoms
F	C	
99-96	37.2-35.6	Uncontrollable shivering
95-91	35.0-32.8	sluggish thinking and violent shivering, difficult in speaking. Beginning of amnesia.
90-86	32.2-30.0	shivering decreases but shows muscular rigidity. Unclear thinking and dull comprehension of surroundings. Total amenesia begins.
85-81	29.4-27.2	Irriationality, muscular rigidity, lost contact with environment. Pulse and respiration slows down.
80-78	26.7-25.6	Victim does not respond to spoken word. Reflexes mostly cease to function, erratic heartbeat, unconsciousness begins.
Below 78	Below 25.6	Cardiac fibrillation and failure idema and hemorrhage in lungs; death.

Note : In thermal balance, the deep body temperature of man will be 98.6 ° F or 37 °C. The normal skin temperature lies between 31 – 34 °C.

When deep body temperature increases by 2 to 3°C (3.6–5.4 °F) ie. 39–40 °C (102.2 to 104°F) heat stroke occurs - circulatory failure.

at 41 °C (105.8 °F) Coma sets in

at 41–44 °C (106 - 111 °F) death imminent

at 45C° (113 °F) certain death.

Snow Blindness

In Antarctica man cannot (naturally) adjust himself to the reflection of bright sun from snow, ice, water and overcast sky. During overcast days and clear days, under whiteout conditions, snow blindness may be contracted due to high level of sun's UV-radiation. The basic symptoms of snow blindness are pain in the eyes, painful and scratchy eyelids and headache. People have to protect their eyes with dark glasses when they are outside in daylight.

Altitude Sickness

Without proper acclimation (about a week) if people are abruptly transported to high altitude (say about 2 to 3 km from a station level) they feel over exertion, fatigue after arrival. This is called altitude sickness. After arriving at high altitudes even a few minutes of work cause them grasping for breath, chest pain. Over exertion may cause headache, nausea and vomiting, dizziness and weakness. The remedy is rest. In all such cases slow down your work till acclimated. Breath through nose (not through your month) to keep yourself free from dehydration.

Effects of Natural Radiation on Man

Absorption of solar radiation and long IR radiation effects man in two ways. (i) It can heat or cool (thermal effects) and (ii). Photochemical effects. These effects are caused through skin and eyes. Skin of white people reflects visible and near IR radiation to the extent of 40%. Radiation is accepted or rejected by the skin according to its surface properties and colour. Black skin absorbs about 44% more solar energy than white skin, but the horny layer of white skin transmits 3.5 times more UV light than that of Negroes. Irrespective of the skin colour white, yellow or black, human body radiates energy in the long wave (IR) as that of black body at 32 °C. The main wave lengths of emission are for IR at 6 and 9 μm (range 2-14 μm). The important parts of the humans body that react to severe meteorological elements are skin, lungs, throat, nose, eyes and nervous system.

Human heat balance = heat gain (from metabolism + skin absorbed- (solar radiation + long IR radiation) – heat loss (IR radiation from body + convection + evaporation).

UV radiation (near 0.3 μm) causes photochemical inflammation of cornea of the eyes. This damage is caused by reflected solar UV-radiation by snow.

Photochemical action of the skin causes production of Vitamin-D, sunburn, early aging of the skin, strengthening of the horny layer, skin cancer through pigmentation.

Natural sunburn is caused by solar UV-radiation around 0.3 μm, which penetrates deeper into living skin layers. It first causes white skin or tender spots of the skin reddening (after 5-20 minutes of exposure), painful reddening

(30-70 minutes of exposure), larger dose oedema and finally blisters. UV-radiation larger dose causes sunburn under nose, lips, nostrils, upper and inside portions of ears, eyelids, chin and cornea of the eyes.

Protection against sunburn achieved by lowering the dose (shortening the exposure time at high sun and clear sky), window glass and skin creams application (which absorb radiation below 0.32 μm) dry cloths and vegetation completely attenuates. Over dose of UV-radiation leads (after exceeding the minimum dose of reddening) to severe erythema, oedema and blistering. Frequent exposure leads to ageing of the skin and loss of elasticity in skin and causes skin cancer in exposed areas of face, neck etc. (there are about (2.3 million) 2.3×10^6 eccrine sweat gland in man). Sunburn blocks human sweating and hence loses body cooling inbuilt capacity. Skin registers thermal conduction, convection and IR radiation. Human body does not sweat at skin temperature of 28 °C.

UV-radiation Effects

It causes increase in vitamin-D and histamines. Increases gastric acid secretion and protein metabolism. In blood, increased hemoglobin, Ca, Mg and phosphate levels. It causes direct lethal effects on bacteria and indirect effect on man.

Questions

1. What is weather and climate? Write briefly climate classification.

2. Write on Koppen's climate classification (i) temperature criteria, (ii) aridity criteria (iii) mian sub-division (iv) principal sub-classifications.

3. Write briefly about the merits and demerits of Koppen's classification.

4. Write briefly modified Koppen's climate classification.

5. Write the formulae of moisture index and aridity index of Thronathwaits climate classification. Explain the terms.

6. Write briefly thermal efficiency index of Thronathwaits climate classification.

7. Write Budyko's climate types.

8. Write briefly the evolutionary stages of man both Hindu cosmological and recent based on carbondating.

9. Derive an expression of Half life period of carbondating.

10. Write important weather features of Dakhingangotri in Antarctica.

11. Write short notes on : (i) whiteouts (ii) Frostbite (iii) Hypothermia (iv) Snow blindness (v) Altitude sickness.

12. Write briefly the effects of natural radiation an man.

CHAPTER 28

Climate of India

Physiography of Indian Sub-continent

In the north of the Indian sub-continent the mighty Himalayan ranges are spread, roughly in an east-west direction. The ranges are emanating from Pamir Knot in the northwest India. From Pamir Knot another range called Sulaiman passes southwest wards to Baluchisthan from Indus plain. In northeast India, from Assam to Myanmar, Khasi, Jaintia and Garo Hill ranges are spread. South of the Peninsular India is occupied by the Indian ocean, while west of it Arabian sea and east of it Bay of Bengal. In south India, western ghats are spread in a north-south direction along the west coast while eastern ghats are spread along east coast. From the weather and climate point of view the above features are important. The other features rivers, soils etc will be discussed in environment.

A brief description of the average pressure, wind pattern for four seasons 1. Winter (Dec-Feb, represented by Jan), 2. Premonsoon or summer (March-May, represented by April), 3. Monsoon or Rainy season (June-Sept, represented by July), 4. Postmonsoon (Oct-Nov, represented by Oct) is given below and shown in Fig. 28.1 to 28.5, which includes other climatic aspects.

Pressure distribution over Indian Sub-continent has large variation in the four seasons. As compared to January, the July pattern is a complete reversal (High, Lows are replaced by Low. Highs). Consequently the annual mean pressure distribution is of no significance.

Winter (January)

During January north-east trades are prominent with Siberian High (on msl) being located at lat 45 ^{0}N, long 105 °E and India lies at its periphery. Continental cold air blows over central parts of Asia. A part of which comes over to north India while a major part of it is obstructed by the mighty Himalayas. Intense Siberian High persists during Dec, Jan, Feb. In this period the Subtropical High in southern hemisphere is found between equator and lat 10 °S. It extends from north Australia to eastern parts of Africa. A marked trough runs from Kerala to Gujarat and another off Tannasserin coast to northern Myanmar and Assam. A weak ridge runs from NW India to central parts of Bihar.

Summer or Premonsoon (April)

Pressure gradient becomes slack and begins to readjust as the Sun moves north. A feeble high lies over Arabian sea and another over Bay of Bengal. Feeble lows develop over western parts of Rajasthan, Bihar and Myanmar. In this season the Siberian High weakens and shifts north, while the Sub-Tropical High in the southern hemisphere shifts to lat 30 °S. By May summer heat low extends from north Africa to Asia.

Monsoon (July)

Low pressure area extends from north Africa to Siberia Via Baluchisthan and becomes intense. Another trough of low extends from Rajasthan to Head Bay of Bengal (called monsoon trough) which is of great importance to India. The Subtropical High of southern hemisphere become marked and centered around lat 30 ^{0}S and long 60 ^{0}E. A weak ridge extends over Arabian sea off west coast of India and another in the Bay of Bengal off Tannasarin coast to Myanmar. By August Afro-Asian low weakens and by September it becomes east-west trough.

Post Monsoon (October)

By September end monsoon trough shifts to lat 13 °N and pressure field becomes slack. Low pressure belt Africa to west Pacific through Arabian sea and Bay of Bengal runs between equator and Lat 20 °N with centres over Africa, Bay of Bengal and western pacific ocean. The Asian High runs roughly along Lat 50 °N and centered about long 90 °E. The Indian ocean high centered near 30 °S shifts to east would by November and centered at 80 °E. The trough in the south Bay of Bengal persists till November end but disappears in December.

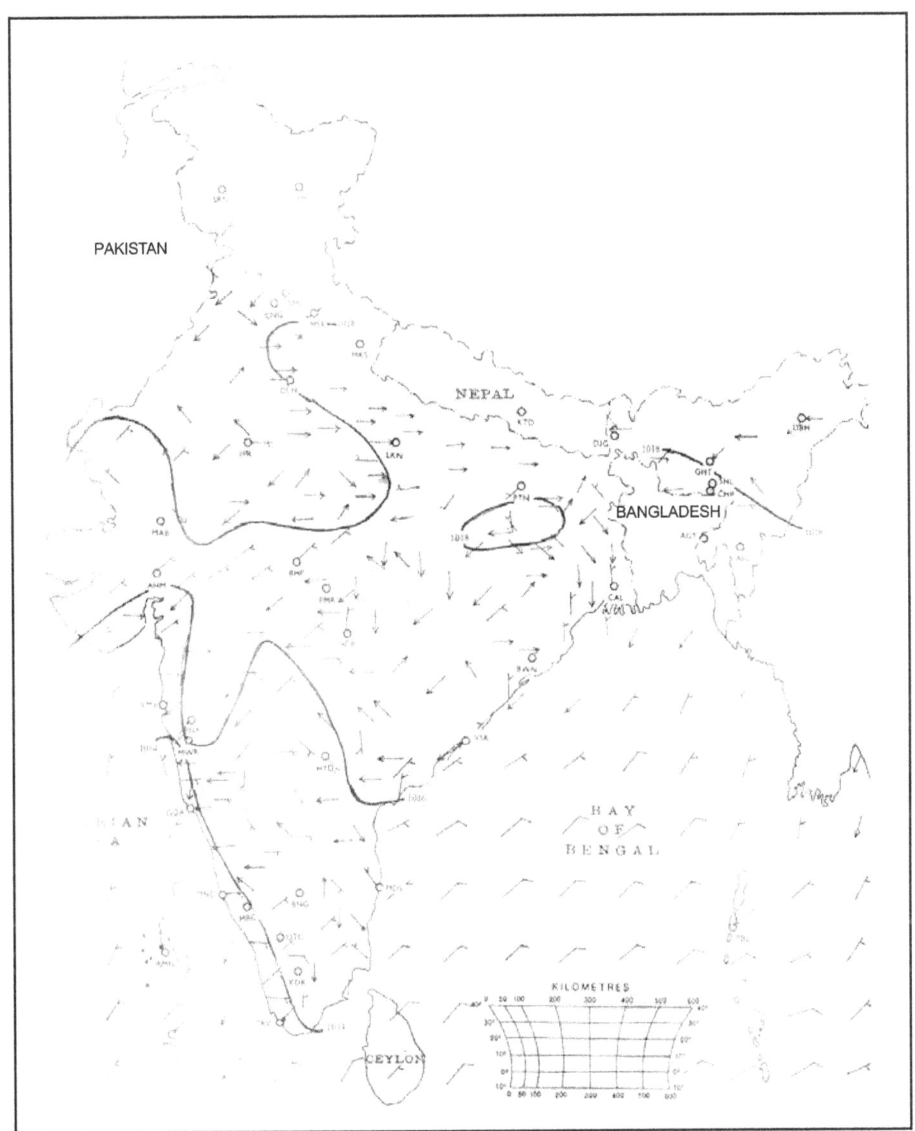

Fig. 28.1 Mean Pressure (hPa) and wind (January).

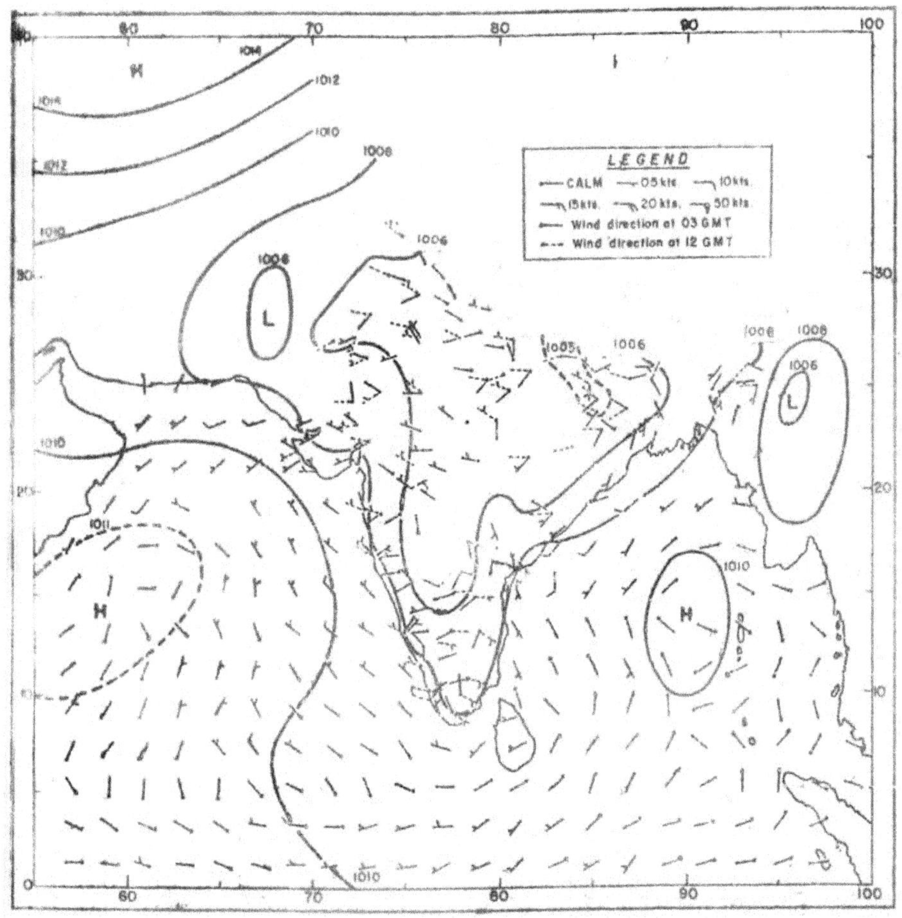

Fig. 28.2 Mean pressure (hPa) and surface winds (April).

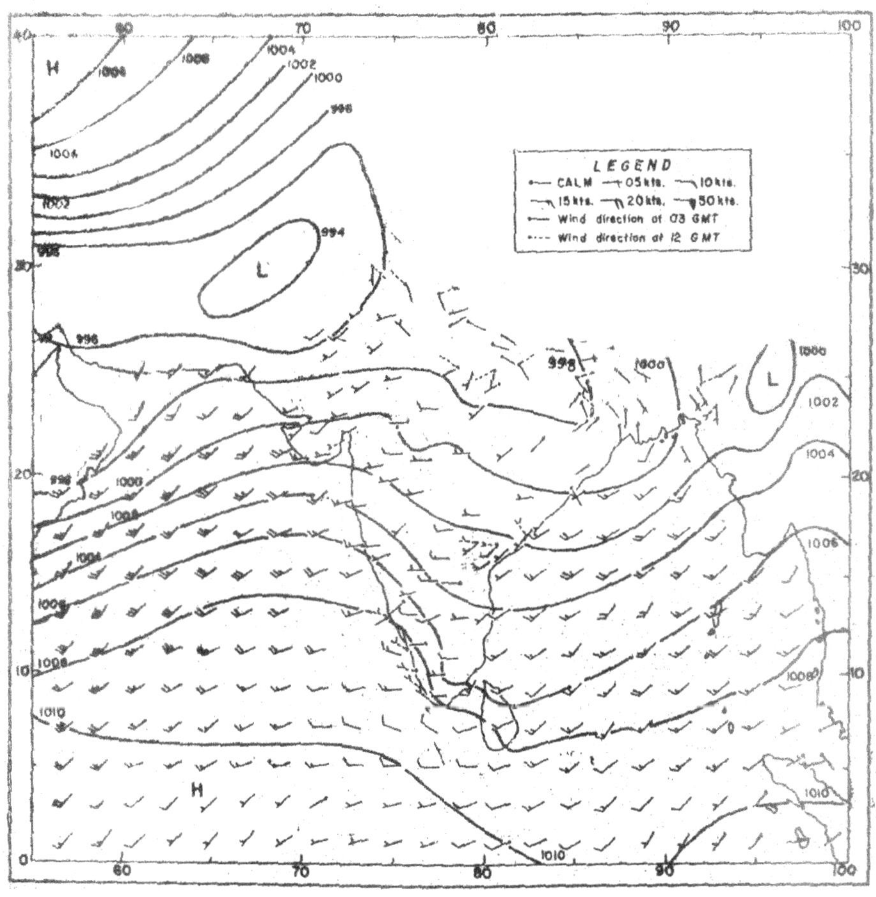

Fig. 28.3 Mean pressure (hPa) and surface winds (July).

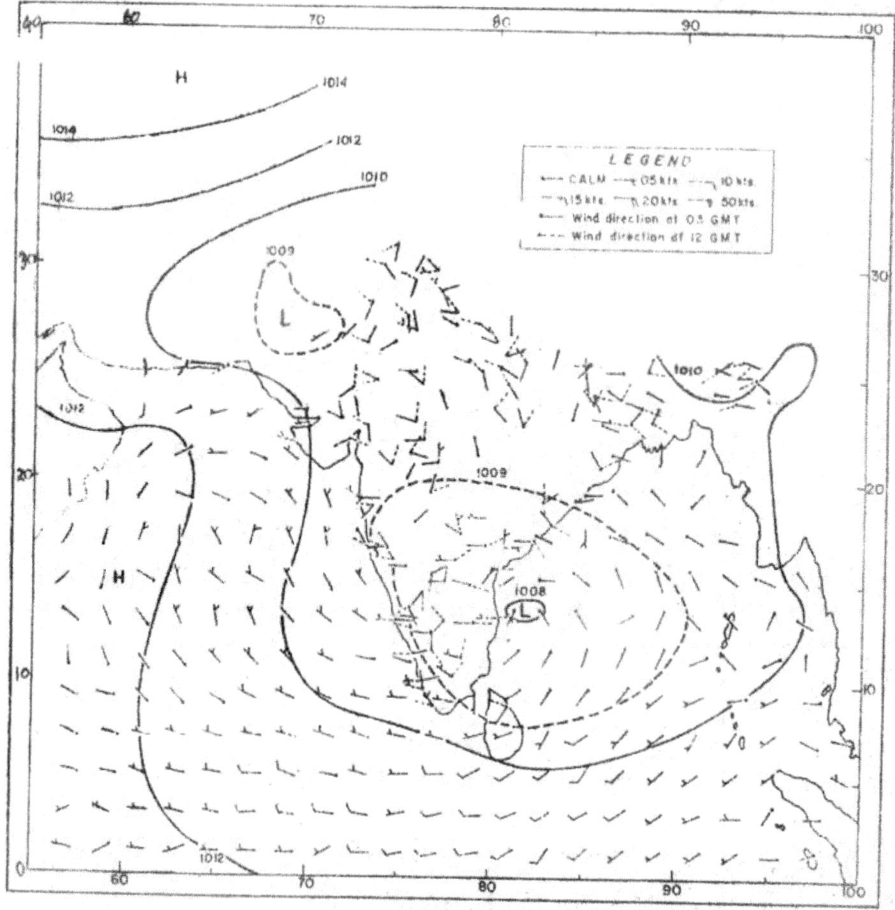

Fig. 28.4 Mean pressure (hPa) and surface winds (October).

MEAN DAILY TEMPERATURE IN °C
(Reduced to sea level at adiabatic lapse rate of 6°C/km.)
JANUARY
APRIL
JULY
OCTOBER

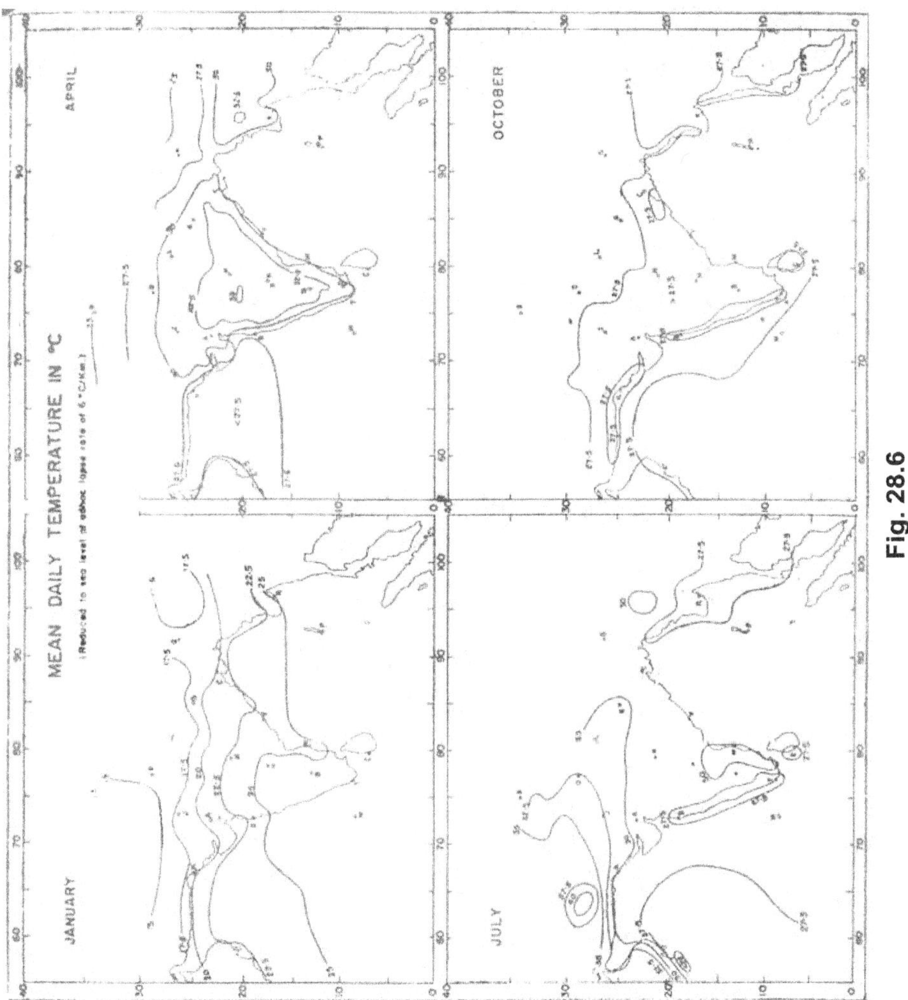

Fig. 28.6

Seasonal Rainfall

Winter (Jan-Feb)

This is the driest season for the country. Except Kashmir, most of India receives an insignificant percentage of annual rainfall. However, this winter rainfall is important for crops in northern India. The highest rainfall of about 20 cm is received in Kashmir and at some places in Nicobar islands. About 5 cm of rain received in Jammu, north Punjab, Haryana, Himachal Pradesh, Uttranchal, Bihar plateau, parts of east MP, Chattisgarh, Orissa and north east India east of long 90 °E. In south India south of lat 10 °N, east coastal strip south of lat 15 °N and island groups receive rainfall. The chief synoptic features are Western Disturbances.

Summer (March-May)

Assam (60-80 cm), Jammu–Kashmirs (20-40 ln), Kerala and neighbourhood (30-40 cm) receives considerable rainfall due to thunderstorm activity. Hail is also associated with thunderstorms, which decreases with the advance of the season. Cherrapunji receives more than 200 cm of rain. Second important area of rainfall activity extends from West Bengal to north coast of Andhra Pradesh. Western parts of India except extreme north and south are driest. Western Rajasthan and adjoining Saurastra and Kutch are practically dry (about 0.5 cm). Norwesters, Kalbaisakhi and Andhi (local names of severe thunderstorms) are confined to this season only. Tropical cyclones are the other synoptic features.

Summer Monsoon (June-Sep)

Summer monsoon or southwest monsoon is the main rainy season for India. Except Kashmir and neighbourhood and Kerala remaining area of the country receives more than 75% of the annual rainfall in this season. Orography and wind direction (wind ward or lee ward) plays an important role in distribution of seasonal rainfall as observed in Western ghats and Khasi-Jaintia hills, where wind blows at right angles to the ghats/hills. In north Indian plains the minimum rainfall belt runs from northwest Rajasthan to the central parts of West Bengal, roughly along the axis of monsoon trough. Rainfall rapidly decreases along west coast, south of lat 9.5 °N to Kanya Kumari. Pamban records only about 4 cm of rain during monsoon. In general July is the rainiest month. Seasons heaviest rainfall zones are Assam (200 - 400 cm), west coast and adjoining ghats north of lat 10 °N (200–300 cm). Ladakh valley in the extreme north, western parts of Rajasthan and southernmost part of Tamilnadu gets very small amount (5 to 10 cm) of rainfall. Chief synoptic features of southwest monsoon: lows, depressions over Head Bay of Bengal and its movement along monsoon trough and mid-tropospheric circulations off Gujarat-Knokan coats and off Andhra, Orissa coasts.

Post Monsoon Period (Oct-Dec)

During this period rainfall activity is mainly confined to south peninsula, the east coast, Assam and parts of Kashmir. It is the wettest season for Tamilnadu (40-80 cm average rain) where northeast-monsoon prevails. Rest of south peninsula gets average rainfall of 5-10 cm. Western India including Jammu and Kashmir gets rainfall less than 5 cm. The chief synoptic features are tropical cyclones both in Bay of Bengal and Arabian Sea.

Rainfall Variation

The average annual rainfall in the plains of India is about 105 cm. During the worst famine years 1899 and 1918 it was 26% and 22% below normal. So far during the wettest year 1917, it was 29% above normal. The coefficient of variation of annual rainfall exceeds 30% over large areas of the country. It is 40 to 50% in parts of Saurastra, Kutch and Rajasthan and in some districts exceeds 100 % (in scanty rainfall zone). The rainfall variability is least 15 to 20% over northeast India and southern coast of Peninsula which may be termed as places of high reliability. The variabilities are high during non-monsoon months. Inspite of the regular behaviour (cyclicity) of monsoon, its activity differs every year. The timing of on set of monsoon season differs in different parts of India and also rainfall distribution including duration of break monsoons. The timing of withdrawal of monsoon and total rainfall of monsoon season varies in different parts of India.

Western Disturbances (Nov- March)]

During winter cloud and precipitation (pptn) belts move from west to east over northern parts of India. These are called western disturbances (W D). The adjective western signifies that the disturbances approach India from west in contrast to the most of the rain giving systems that approach India from east. As a convention in IMD, the term western disturbances signifies a low or trough either at surface or in upper air in the westerly wind regime (north of lat 20 °N) that cause clouding and pptn. In the absence of clouding and pptn it is simply referred to them as troughs. When two or more closed isobars spaced at 2 hPa interval can be drawn on sea level chart, the system is described as" western depression". The terms active and weak are associated depending on pptn amount.

The sequence of pptn associated with WD is : 1. First day over Jammu and Kashmir (north of lat 30 °N) and even Himachal Pradesh, Punjab, Haryana, 2. The next day over east Rajasthan, (north of lat 26 °N and west of long 80 °E) and west Uttar Pradesh (This is the primary system). On many occasions the rainfall may cease at this stage. In some other cases the rainfall now commences over Madhya Pradesh and neighborhood during next two days and then spreads east wards to Assam (this is secondary system). As it spreads east wards its activity of rainfall stops west wards. Some times it happens that the WD may not effect north west India south of Lat 30 °N but it gives rain in north Assam after two to three days (when the WD caused weather over western Himalayas).

After the second rainfall belt moved away east-north-east wards an extension of this system appears southwards or an independent weak low or circulation may develop over Madhya Pradesh and neighborhood which causes third rainfall belt (tertiary system). A weak low or trough may also form over Gulf of Camby (as tertiary system) which while moving in an easterly direction cause weather over Gujarat, north Maharastra. However these tertiary systems are rare. The secondary and tertiary systems that develop over Rajasthan and Madhya Pradesh under the influence of the primary disturbance that moves across north India and Pakistan are now referred to as induced lows. The primary, secondary and tertiary terms have become obsolete.

During (Dec-Feb) winter Mediterranian Sea, Black Sea, Caspian Sea, Lake Balkesh form a zone of cyclogenesis of ETC (extra tropical cyclones). Of these Caspian Sea, Aral Sea cyclogenesis effect India as WD. In this season Extra Tropical Depression/troughs are known to move in a east-north-easterly direction across southern Russia, Turkey, Iran and sometimes even further south. During Jan -Feb ETC originate as wave disturbances and/or cut-off cyclones between Lat 25 to 40 °N. Areas lying north of lat 30 °N may lie at the periphery of these primary western disturbance, which accounts for the first rainfall belt. At this stage a secondary low at Lat 30 °N is formed or occluded fronts (Nascent depression of ETC) enter this part as WD while the primary moves away east-north-east wards across western Himalayas.

WDs in general are lower tropospheric systems. They show variation in strength and vertical extent. These may be superimposed by mid-and upper tropospheric troughs in westerlies. Movement of induced lows are slow, but they move faster when they are superimposed by mid-and upper tropospheric westerly trough. The life period of a simple sequence of WD is 2 to 4 days with induced lows filling up within that time. Before filling up of induced lows, if a fresh WD arrives from west the life period prolongs upto 10 days with the formation of complex trough extending to Lat 20 °N. In such a situation extensive rainfall occurs over the country. WDs that cause rainfall over Punjab and UP arrive in succession of 2 to 5 days interval. At other times these areas are free from weather continuously for one to two weeks. 50% of the occasion the activity of WDs are confined to Jammu and Kashmir, Himachal Pradesh, Punjab, Haryana and hills of west UP or Uttranchal. On an average 6 to 7 disturbances per month move across India in winter. Formation of a depression either in Bay of Bengal or Arabian sea inhibits the occurrence of WD over north India.

Sequene of Weather

Ahead of a WD significant 24 hr pressure fall and rise in the rear is observed. Pressure gradient in the rear is stronger. Ahead of a WD a significant rise in minimum temperature and dew point noticed. In the rear of WD some times cold waves and snow falls. Fog occurs both ahead and rear of WD in Himalayas. Mid-latitude low index cycles seems favour the formation of induced lows over India.

Cold Waves in India

During the period November to March and sometimes in the months of October and April, cold dry air blows from northwesterly/northerly direction and lowers the day and night temperatures significantly. On some occasion temperature falls so much that it leads frost formation and causes damage to crops and loss of life. There is no regular periodicity of occurrence but occurs two to three times in a season. According to IMD convention the 24 hr minimum temperature changes and departure from normal (cold waves) defined as follows.

Cold Waves

A. When the normal minimum temperature of a station is $\geq 10\,°C$.

Magnitude of change °C	Description of 24 hr change	Description of departure from normal
− 1 to 1	Little change	Normal
− 2	Fall	Below normal
− 3 to − 4	Appreciable fall	Appreciably below normal
−5 to −6	Marked fall	Markedly below normal or Moderate cold wave condition
7 or less	Large fall	severe cold wave

B. When the normal minimum temperature of a station is $< 10\,°C$

Magnitude of change °C	Description of 24 hr change	Description of departure from normal
− 1 to 1	Little change	Normal
− 2	Fall	Below normal
− 3 to − 4	Marked fall	Cold wave
−5 or less	Large fall	Severe cold wave

It is obvious from the above definition, only the departure from normal (pentad normal) that defines cold wave condition. For example a station in northwest India on a winter day may record minimum temperature of the order 4 or 5 °C, but it may not be a cold wave at all (because normal itself is low). On the contrary a station in Maratwada or Telangana recorded temperature may be little higher say 6 or 7 °C but it may be classified as cold wave condition (normal being high). In general strong cold winds cause human discomfort. People residing in a place for a long time they get acclimatised to the normal weather conditions of that place. Thus a person from Hyderabad (AP) feel a normal winter day at Delhi to be severe cold wave condition.

The characteristic wave aspects like amplitude, frequency, periodicity etc. are not associated with cold wave situations but it appears cold waves tend to move in a preferred directions. They move east wards and also south wards.

In general cold waves are frequent over northwest India in comparison to other parts of the country. Cold waves do not occur over Bay islands, Lakshadweep, Tamilnadu, coastal Andhra Pradesh, Coastal and South interior Karnataka and Kerala. Jammu and Kashmir is haunted by severe cold waves (on an average four per annum) while neighbouring Punjab, Uttar Pradesh have less frequency (once in two years). This neighbouring phenomena may be attributed to adiabatic compression (warming) of cold air mass during its descent on the mountain slopes. Rajasthan, western parts of Madhya Pradesh, Saurastra and Kutch are affected by severe cold waves once in a year. Ladakh recorded maximum duration of severe cold wave for 30 days.

During October Saurastra and Kutch, Rajasthan, Bihar plateau and neighborhood are affected by a few cold waves (about 1 % of annual). During November from Jammu and Kashmir to west Madhya Pradesh (or west of long 80E) are affected by cold waves of the order of 5% annual incidence. During December, Jammu and Kashmir, Uttar Pradesh, Bihar, Maratwada, Telangana, Orissa, Rayalaseema are affected by cold waves, about 11% of annual.

January and February are the peak periods of cold waves. About 28% of the annual occur in this period and entire country is susceptible to cold waves except Tamilnadu. The main targets in these months are Jammu and Kashmir, Rajasthan, Madhya Pradesh, Gujarat, Maharastra, Telangana, Uttaranchal, Utter Pradesh, Chattisgarh.

During March and April the frequency of cold waves falls, the whole Peninsula is free from cold waves.

The lowest temperature of – 45 °C (departure from normal – 19.7 °C) was recorded at Dras in Jammu and Kashmir on 28 December 1910, world record -89.2 °C recorded at Vostok, Antarctica on 21 July 1983.

The chief synoptic features associated with cold waves are, the rear of active WD and inflow cold air from northern latitudes. Cold waves are associated with cold core troughs in which a pool of cold air mass exists. The cold pool westerly troughs are classified as severe cold core trough, moderate cold core trough and warm core trough depending on the anomaly of temperature (which is the difference of actual dry bulb and mean dry bulb temperature). In severe cold core trough the anomaly is –6 °C or less extending upto a depth of 300 h Pa level. In moderate cold core trough the anomaly is –1 to –5 °C extending upto a depth of 700 hpa level. In warm core trough the anomaly is –1 to + 5 °C extending upto a depth of 600 h Pa level. Cold waves are not generated in warm core troughs.

Heat Waves in India

According to IMD convention the 24 hr maximum temperature changes and departure from normal (heat waves) defined as follows.

Heat Waves

A. When the normal maximum temperature of a station is $\leq$ 40 °C

Magnitude of change °C	Description of 24 hr change	Description of departure from normal
– 1 to 1	Little change	Normal
2	Rise	Above normal
3 to 4	Appreciable rise	Appreciably above normal
5 to 6	Marked rise	Markedly above normal or Moderate heat wave
–7 or more	Large rise	severe heat wave

B. When the normal maximum temperature of a station is > 40 °C

Magnitude of change °C	Description of 24 hr change	Description of departure from normal
– 1 to 1	Little change	Normal
2	Rise	Above normal
3 to 4	Marked rise	Heat wave
5 or more	Large rise	Severe heat wave

During hot whether period (March to July) surface temperatures over many parts of India abnormally shoot up particularly over North India. This heat wave conditions progessively invades the neighbouring region. The incidence of severe heat wave occurs mostly in Uttar Pradesh, but there is no region where reccurs successively every year. Bay islands, Lakshadweep, Tamilnadu, Kerala, Coastal and South Interior Karnataka are not affected by heat waves. Rest of the country is prone to the incidence of heat waves.

During March Saurashtra and Kutch are prone to the incidence of severe heat waves (about 17% of annual). In April the frequency of heat waves fall (7%) and mostly occur in Jammu and Kashmir, Punjab, Rajasthan, Madhya-

Pradesh and to thence Konkan, Bihar plains, West Bengal, south Assam. In May the heat waves activity slightly increases (10%) and mostly occurs over Jammu and Kashmir, Telangana, Rayalaseema, Vidarbha and Orissa. The month of June is the peak period of heat wave activity (about 54%) particularly in places where the onset of monsoon not taken place\delayed. Uttar Pradesh, Madhya Pradesh, Chattisgarh experience severe heat waves which are locally called 'Loo". Rajasthan is the hottest part of India but rarely records heat waves. Entire west coast, Saurastra and Kutch to Kerala is free from heat waves.

The month July also records heat waves but they are confined to Uttar Pradesh and Punjab.

During severe heat wave conditions maximum temperature departure from normal goes up to 8 to 10 °C, particularly over Uttar Pradesh and Bihar. Alwar (in east Rajasthan) recorded the highest day temperature of 50.6 °C on 10 May 1956 (world record of highest day temperature is 58.0 °C recorded at Alazizyah, Libya on 13 September 1922).

Heat wave conditions in any part of the country does not last more than 5 to 6 days and severe heat waves one to two days. The maximum period of 15 days heat wave lasted over Jammu and Kashmir. Heat waves generally develop over Punjab, Saurastra and Kutch and spread to east and south wards but not west wards.

Favourable conditions for the formation of heat waves : Prevalence of dry air over the region with little or no moisture in the lower troposphere. The area should be cloud free and lapse rate may be dry adiabatic in the lower troposphere.

Summer Thunderstorms Over India

During hot weather period (March-May) rapid rise of surface temperature, fall of pressure, intensification of southern Indian ocean anticyclone, Northward movement of equatorial trough, Nor westers, Dust storms, Dust raising winds, Hail storms are common features over India neighborhood. Thunderstorm activity continues in monsoon season but less marked. The main thunderstorm activity observed during the period are : (i) The area from Northeast India to east Madhya Pradesh, east Vidarbha and adjoining Andhra pradesh, (ii) Southwest Peninsula (Kerala and neighborhood), (iii) northwest India excluding Rajasthan. Squalls associated with summer thunderstorms are very strong/violent. They uproot trees and there were instances in the past that trains being thrown out of tracks. On an average 15 to 20 thunderstorm days recorded in the season. Assam and adjacent states, south Kerala have the highest frequency (30 to 40 days). Entire India is susceptible to thunderstorm activity but Gujarat state has the lowest frequency (about five days in whole of the season).

Thunderstorm activity progressively increases from March. During March the areas of maximum activity (frequency more than 6-days) are Assam and adjacent States and Kerala. The other areas are foot hills of the Himalaya (J & K), Punjab, Himachal Pradesh, east Madhya Pradesh, Bihar, Bengal. The lowest activity lies in Saurastra, Kutch, Gujarat and Konkan coast.

In April thunderstorm activity increases. Maximum frequency (12-14 days) lies over Assam and Kerala. The other area of high frequency (about 8 days) are east Madhya Pradesh, Vidarbha, Deccan Plateau. The lowest frequency (about one or two days) lies in south coastal Andhra Pradesh, coastal Tamilnadu, Saurastra, Kutch, Gujarat and KonKan coasts.

May is the month of peak thunderstorms activity. The maximum frequency zone runs from Assam to south Kerala. Assam and Bengal have frequency of about 16 days, while Gujarat, Saurastra and Kutch have the least frequency or no activity.

The chief atmospheric features during this season are : (i) a number of meteorological parameters show large diurnal variation, (ii) formation of heat low over central/interior parts of the country, (iii) weak to moderate westerly wind flow in the lower troposphere and (iv) intense convective activity associated with conditional convective instability (steep lapse rate).

Thunderstorm activity continues in monsoon and post monsoon season and becomes least in winter. The highest frequency of thunderstorms (more than 80 days in a year) extends from Assam to east Madhya Pradesh. Assam and neighbouring Bengal have frequency of more than 100 days, Silchar has 101 days, Mohanbari 106 days. Gauhati 102 days and Sibsagar, Krishnanagar have 108 days. The other high frequency zone is central Kerala and hilly regions of extreme north India. The lowest frequency, less than 10 days in a year, are in coastal Saurastra and Kutch.

Norwesters

The severe Thunderstorms of hot weather period over Bengal and Orissa are known as Norwesters. They are locally called 'Kal Baisakhi' or the fateful thing of the month Baisakh (15 April to 15 May). These severe thunderstorms usually approach a station from northwest direction (and hence the name Norwester) and burst over a station with great violence raising clouds of dust. The squalls associated with these Norwesters reach tornadic violence (120 to 150 kmph). They cause considerable damage to life and property in their path. Tornadoes rarely occur in north east India. Assam and adjacent states are more susceptible to tornadoes locality called Hatishmura (frequency once in a few years). Generally the direction of the squalls is Northwesterly direction.

Favourable Synoptic Situations

(i) Induced low over Punjab, Haryana, Rajasthan, West Uttar Pradesh or northwest Madhya Pradesh (ii) conditional convective instability in the lower-mid troposphere, (iii) adequate moisture supply in the lower troposphere (Note: Horizontal distribution of moisture at the ground can be ascertained by dew point temperature of surface observations while vertical distribution of moisture is known through Radiosonde ascents) (iv) An extension of trough east-south-east wards from induced low to Bihar Plateau and west wards (v) Penetration of easterlies upto 850 h Pa level over Bihar plains, (vi) In mid-and upper troposphere mainly westerly wind flow over north-east India.

Thunderstorms in Gangetic West Bengal, Bihar, Orissa have definite sequence of time. They generally develop over Bihar plateau, south Madhya Pradesh or west Orissa. They travel east or south-east wards towards Gangetic West Bengal or Head Bay of Bengal.

The severe thunderstorms of north-east India are classified into four types.

Type A : They develop over Bihar plateau and adjoining area in the after noon and move with fury in a southeasterly direction.

Type B : They originate in submontane districts of north Bengal and move south wards. They generally form and move during night or early morning.

Type C : They originate in the hills of Nagaland, Manipur and Mizoram and move west wards.

Type D : They develop near the Khasi hills and move south wards.

In fact Perturbation in the westerlies caused by a trough or Jet stream or a cut-off low. Deep westerly trough causes widespread thunderstorm activity over north-east India.

Thunderstorms and Dust-storms over north-west India

During March-June (in all months), thunderstorms occur in north-west India and Uttar Pradesh. The activity is more in western Himalayas as compared to the plains. Thunderstorm activity progressively increases from March and attains peak in May and June. Dust storms occur in plains of UP, Punjab, Haryana and north Rajasthan. They begin in April and attain peak in June. The dust storms locally called Andhi–meaning blinding.

Dust-storms are of two types .(i) Pressure gradient type and (ii) convective type. In pressure gradient type, strong winds raise loose dust from the surface of the earth into the atmosphere. In this case no Cb is involved. In convective type dust is raised into the atmosphere by strong downdrafts from Cb cloud. In convective type dust storm proceeds a thunderstorm.

The maximum frequency of dust-storms occur in Rajasthan. In pressure gradient type strong surface winds are built up due to steep gradient around low. Wind raised dust suspends in the atmosphere and reduces the visibility to below one km. Pressure gradient of 1 to 1.5 h Pa or more per degree latitude prevails over a longer period. Winds of the order 30 to 50 Kt prevail upto 1.5 km (850 hpa level). Dust raising winds may commence in the morning and continue for the whole day with maximum intensity in the afternoon when the lapse rate near the ground reaches superadiabatic. In convective type dust-storms the duration of strong winds is of short duration (few minutes or fraction of an hour). Even after duststorm/dust raising winds subside, the raised dust in the atmosphere may go upto 700 hPa level and remain suspended for days together and reduce the visibility.

Favourable Synoptic Situation

Shallow heat low forms over central and northwest India. Quasi Stationary wind discontinuity in low levels run in north-south direction to Peninsula and another south-east Madhya Pradesh to Assam. A shallow anticyclone over Bay of Bengal feeds moisture from the south on one side of the wind discontinuity and contrasting dry northerly or north-westerlies prevail on the other side of the discontinuity line. Over north india in the mid-and upper troposphere sub-tropical jet stream runs, which provides upper divergence. Near the surface of the earth and in planetary boundary layer superadiabatic temperature lapse rate favours convective activity.

Thunderstorms Over South Peninsula

During March-May high frequency of thunderstorm activity prevails over Kerala (about 15 days each month) and neighborhood. Orography plays an important role. Abundant moisture feed comes from the neighbouring sea. Kodaikanal: seasonal 34 days, annual 75 days; Trivendrum: seasonal 36 days, annual 70 days.

Hailstorms

Hailstorms in India occur in northern parts of the country. They are much less common over Peninsula. Hailstorms are always associated with Cb cloud only. Hailstorms are common in winter (Dec-Feb) over northwest India, Uttar Pradesh, Madhya Pradesh, and in premonsoon period over north-east India, Peninsula, Vidarbha and Madhya Pradesh. No hailstorm reports from any part of the country during monsoon (June-September) period. During post monsoon (October-November) there is little activity in isolated pockets north of Lat 20 °N and west of long 90 °E.

The annual frequency of hailstorms over northeast India is about 10 days, over Uttar Pradesh, east Bihar and Agartala about 6 days, over east Madhya

Pradesh, Chattisgarh, Jharkhand is 2 to 3 days. Maximum activity over south Peninsula is 1 to 2 near Madurai. The entire west coast, south coastal Andhra Pradesh are free from hailstorm activity.

Favourable Synoptic Features

Intense convective conditional instability, large Cb cloud extending to upper troposphere (12 to 14 km). In lower and mid-troposphere high moisture content, strong vertical wind shear surface to mid-troposphere, low freezing level, lowering of trough in the jet core.

Questions

1. Describe the salient features of fours seasons of India.

2. Describe the seasonal distribution of rainfall and its variability. or Describe the annual and monsoon season rainfall and its variability.

3. Write briefly about WD over north India and its effect over Agriculture and in Aviation.

4. Write short notes on : (i) coldwaves over India, (ii) heat waves over India, (iii) Norwesters (iv) Duststorms over northwest India (v) Hailstorms over India.

CHAPTER 29

Air Pollution

Definition

The presence of solid, liquid and gaseous contaminants in the atmosphere in such quantities which damage materials and property, and cause injury to human, animal or plant life or interfere unreasonably with normal human activity and enjoyment of life is called air pollution. In short any undersirable concentration of contaminants in air may be called air pollution.

Natural Contaminants

SO_2 (Sulphur dioxide) H_2S (Hydrogen sulphide), HF (Hydrogen fluoride), HCl (Hydrogen chloride) which emanate from volcanoes, due to thunderstorms, Putrefication. These are considered harmful only when the concentration exceeds certain limit. Toxic gases like Nitrogen dioxide (N_2O), Ozone (O_3) form or evolve due to electrical discharges in the atmosphere. Natural particulate matter, such as dust from the deserts and volcanoes, ashes from forest fires, salt particles from sea, meteoritic dust, pollen from flowers are important.

Artificial Pollutants

Emission from chimneys, exhaust from automobiles, aeroplanes and other man made activities. Agricultural insecticidal, pesticidal dusting and spraying, domestic and municipal incinerators, burning of vegetation introduces pollutants into the atmosphere. Thermo-nuclear explosions introduce radioactive pollutants.

The Important Air Pollutants

CO (Carbon monoxide), Hydrocarbons, SO_2, Oxides of nitrogen, Aldehydes, Ammonia etc., photochemical process lead to further generation of harmful gases and aerosols; Hydrocarbons formed by the incomplete combustion of liquid petroleum products. Pollutants that enter the atmosphere by human activities are called Anthropogenic pollutants.

Some basic definitions are given below.

A gram molecule or mole (kilogram molecule or kilomole)

It is the amount of chemical substance whose mass expressed in grams (kilogram) is numerically equal to its molecular weight.

The volume of one mole of a substance is called its molar volume.

A gram atom (kilogram atom)

It is the amount of a simple chemical substance (an element) whose mass expressed in grams (kilograms) equals to its atomic weight.

Avogadro number

For all substances

number of molecules in a gram molecule

$$= \text{the number of atoms in a gram atom} = N_A.$$

This number N_A is called Avogadro number.

$$N_A = 6.023 \times 10^{23} \text{ mol}^{-1}$$
$$= 6.023 \times 10^{26} \text{ k mol}^{-1}$$

Molar ratio (or volume ratio) : The number of moles of the compound per mole of air.

Mass ratio : The mass of the compound per unit mass of air.

Mass Concentration : The mass of compound per unit volume.

Concentration (or molecular concentration)

The number of moles of the compound per unit volume.

$$\text{Mass ratio} = \text{molar ratio} \times \frac{M_j}{M}$$

Where $M \simeq 29 =$ the average relative molecular mass of air

$M_j =$ Relative molecular mass of the compound considered.

$\rho =$ density of air

$$\text{Mass concentration} = \text{Mass ratio} \times \rho$$

$$\text{Concentration} = \frac{\text{Mass concentration}}{M_j}$$

$\text{ppm} = \text{parts per million } (10^{-6})$

$\text{ppb} = \text{parts per billion } (10^{-9})$

$\text{ppt} = \text{parts per trillion, } (10^{-12})$

At temperature 15 °C, pressure $p = 1013$ h Pa, the air density $\rho = 1.225 \text{ kg/m}^3$

Notation : (m) in bracket denotes mass ratio, (v) in bracket denotes volume ratio.

Examples

Conversion

(i) Given molar ratio of oxygen in dry air is 21% (v)

$$\text{Mass ratio of oxygen} = \text{Molar ratio} \times \frac{M_j}{M}$$

$$= 21 \times \frac{32}{29}$$

$$= 23.17 \text{ (m)}$$

(ii) Given mass ratio of $SO_2 = 1.1$ ppb (m)

Molar ratio or volume ratio of SO_2

$$= \text{Mass ratio of } SO_2 \times \frac{M}{M_j}$$

$$= 1.1 \text{ ppb (m)} \times \frac{29}{64} \qquad \begin{cases} \text{Mol wt of S} & = & 32 \\ \text{Mol wt of O} & = & 16 \\ \text{Mol wt of } SO_2 & = & 64 \end{cases}$$

$$= 0.5 \text{ ppb (v)}$$

Mass concentration and molar concentrations of minor constituents of air are expressed in units $\mu g \ m^{-3}$ (micrograms per cubic meter) or μ mole m^{-3}.

Example

Consider x molecules cm^{-3}.

$$x \text{ molecules } cm^{-3} = x \ \frac{\text{mole}}{N_A} \ cm^{-3}$$

$$= x \ \frac{\text{mole}}{N_A} \ 10^6 \ m^{-3}$$

$$= \frac{x}{N_A} \, 10^6 \, \mu \text{ mole m}^{-3}$$

$$= \frac{x}{NA} \, 10^{12} \, \text{m mole m}^{-3}$$

Example

Consider 145 ppb (υ) of CO as mass concentrations and molar concentration

$$145 \text{ ppb } (\upsilon) = 145 \times \frac{28}{29} \text{ ppb (m) at T} = 15 \, ^0\text{C, P} = 1013 \text{ h Pa}$$

$$= 140 \times 10^{-9} \times 1.225 \text{ kgm}^{-3}$$

$$= 140 \, \mu\text{g} \times 1.225 \text{ m}^{-3}$$

$$= 171.5 \, \mu\text{g m}^{-3}$$

$$= \frac{171.5}{28} \, \mu \text{ mole m}^{-3}$$

$$= 6.12 \, \mu \text{ mole m}^{-3}$$

$$= 6.12 \, \frac{N_A}{10^{12}} \text{ molecules cm}^{-3}$$

$$= \frac{6.12 \times 6.03 \times 10^{23}}{10^{12}} \text{ molecules cm}^{-3}$$

$$= 3.69 \times 10^{12} \text{ molecules cm}^{-3}$$

According to WMO estimates the world industrial activity during last century increased by 20 to 25 folds and this is contributed a great deal to the atmospheric pollution. During this period world population grown more than doubled , 2.5 billions to more than 5.5 billons. More land was brought under cultivation in this period as compared to the earlier human history. This resulted in the destruction of green lungs of the earth (plant kingdom). Water consumption increased by four folds, fossil fuel (oil, coal , natural gas) consumption increased by 30-35 times. The later added CO_2, SO_2 and oxides of nitrogen emissions to the atmosphere together with toxic chemicals such as lead and mercury. Some of these are responsible for (i) the destruction of ozone layer in the stratosphere, (ii) acid rain and (iii) photochemical smogs. Increased agricultural activity added methane, toxic pesticides to the atmosphere. During the same period globally the green house gases (namely CO_2, CH_4, nitrous oxide, chlorofloro carbons), tropospheric ozone have been added into the atmosphere, which are responsible for the global warming of the atmosphere.

If N_2 and O_2 were the only two constituents of the atmosphere, the average air temperature near the earths surface would have been -18 °C that is 33 °C colder than the at present 15 °C. The gases CO_2, N_2O, O_3, CH_4, CFC^S, water vapour warm the earth's surface like that of a glass house by allowing solar radiation coming to the earth and preventing the terrestrial radiation escaping from the atmosphere. These gases are called green house gases. The principal sources of green house gases are man made, such as fossil fuels, aerosol sprays, refrigeration, agriculture, deforestation. Natural sources are volcanoes and forest fires which are insignificant. During last 100 years barring water vapour all other green house gases increased in the atmosphere. In totality during last 150 years the green house gases increased in the atmosphere by an amount radiatively equivalent to a 53% increase in CO_2, although the gas (CO_2) itself increased by only 26%. The main impact of the green house gases are climate change, sea level rise and adverse effect on health.

Except the $CFC's$ the green house gases cycle (through atmosphere, biosphere-earth system) with sources and sinks are adding to or substracting from the atmospheric concentrations. On the whole man has achieved in reducing the sinks and adding to the sources. The life time of the green house gases in the atmosphere vary from few hours or weeks to more than 100 years. Tropospheric ozone has a life period of few hours or days, $CFC's$ have about 75 to 110 years, nitrous oxide (N_2O) 150 years, methane 7-10 years, CO_2 50-200 years. The concentration of CO_2 increased from 275 ppmv before industrialisation to 354 ppmv by the end of 20th century. The major sources of emission of this gas is burning of fossil fuels, deforestation and change in land use. The concentration of tropospheric ozone increased 15 ppbv to 35 ppbv by the end of 20th Century. Ozone formed from vehicle exhausts and industrial pollutants in sunshine (photo chemical action). Ozone contribution to the green house effect is about 8%. The increased effect of CFC's to the green house effect during last 100 years is about 25%. It was non-existent in atmosphere before industrialization and it is found to the order of 0.25 to 0.45 ppbv by the end of 2000 AD. The sources of $CFC's$ are man made chemicals used as solvents, spray can propellants making of foam and refrigeration. Nitrous oxide contributes 7-8 percent in green house effect. It is increased 228 ppbv before industrialisations to 320 ppbv by the end of 2000 AD. It (N_2O) is found in fossil fuels and burning of biomass, fertilizers and land use change. Methane contributes about 13% to the green house effect. Its concentration roughly trebled (0.7 ppmv to 2.0 ppmv) during last 100 years. The major methane producing sources are : Swamps, rice paddies, ruminants (animal cud-chewing) and fossil fuel extraction. Overall man has achieved to add green house gases and reduced the sinks.

According to WMO, No. 735 (-1990) the annual carbon fluxes were : photosynthesis on land removes 100 Gt [(1) Gt (gigaton) = one million metric tons)] of carbon from the atmosphere annually in the form of CO_2, plant and soil respiration together returns 100 Gt (each 50 Gt). Fossil fuel burning and deforestation releases carbon into atmosphere 5 and 2 Gt respectively. Physicochemical processes at the sea surface releases about 100 Gt into the atmosphere and absorb about 104 Gt. The net gain by the atmosphere is about 3 Gt annually.

Of the green house gases increased by human activities 70% accounts to energy sector. Vegetation is called the green lungs of the earth is also a carbon sink. Deforestation reduced one third of forested land on the globe (six billion hectares to four billion hectares).

In addition to release of green house gases, many metals and chlorinated hydrocarbons are also released into the atmosphere by industrial activities. Some of these substances are toxic in high concentration to human beings, animals and plants. According to WMO estimates the anthropogenic toxic substances that are introduced into the air in kilotons per year are: lead 332, Cadmium 7.6, Copper 35, Nickel 56, Zinc 132, Arsenic 18 while the natural process introduces these substances 19, 1.0, 19, 26, 46 and 8 Kilotons/year respectively.

Group of Experts on the Scientific Aspects of Marine Pollution (GESAMP) established that all oceanic areas over the globe are affected by man made pollutions. Though concentrations are not high, but these toxics are deposited in the seas and on land over long periods and bio-accumulate in the food chain. From the atmosphere to sea, fluxes of metals such as Pb, Hg, Cd, Zn are estimated to be greater than or comparable in magnitude to direct discharges and transport by rivers.

By burning Coal, oil and metal smelting industry introduces mainly SO_2 and oxides of nitrogen into the atmosphere, which are largely responsible for acidic rain. Acid rain is mainly confined to local, regional scale than global. Rain and snow are acidic and have a global pH of about 5.6.

pH Value

Concept of Acids and Bases

A substance which can donate or give a pair of electrons to form a coordinate bond is called a base. A substance which can accept a pair of electrons to form a coordinate bond is called an acid. Thus base is an electron donor while an acid is an electron acceptor. pH of a solution is the negative logarithm of the concentration (in moles per liter) of hydrogen ions which it contains. pH = – log [H].

A neutral solution has pH = 7

If pH is less than 7 the solution is acidic

If pH is greater than 7 the solution is alkaline.

The pH of lemon juice is about 2.2, Vinegar 3.0, pure rain 5.6, Distilled water 7, Ammonia 12.

Buffer Solution

A solution which resist change in its pH value on addition of an acid or a base is called Buffer solution.

LeChatelier's Principle

The state of equilibrium is affected by concentration, temperature and pressure. According to LeChateliers principle if a system at equilibrium is subjected to a change of concentration, temperature or pressure the equilibrium shifts in the direction that tends to undo the effect of the change. Thus in a chemical equilibrium, when concentration of reactants is increased the equilibrium shifts in favor of the products. Conversely when the concentration of the products is increased the equilibrium shifts infavour of the reactants.

Thus a liquid changes into vapour by absorption of heat. The vapour pressure of a liquid increases with increase of temperature. The boiling point of a liquid increases with increase of pressure. The solubility of substance increases with increase of temperature.

Methods of measurements of important pollutants are given below.

SO_2 : Measurements are made by UV-pulsed fluorescence, flame photometry, dilution or permeation tube calibrations.

CO : Measured by Non-dispersive IR tank gas and dilution calibration.

O_3 : Measured by ozone UV-generators, UV-spectrometers, gas Phase titration (GPT) calibrators.

NO_2 : Measured by Chemiluminescence method, permeation, or GPT caliberation.

Pb (lead) : Measured High Volume Sampler and atomic absorption analysis.

Sulphates and Nitrates : Measured by High volume Sampler and chemical analysis.

Hydrocarbons : Measured by flame ionization method or Gas Chromatograph, caliberation with methane tank gas.

TSP (Total Suspended Particulate)

Measured by High Volume Sampler and weight determination method

Based on pollutants effects, pollutions standard index (PSI) is determined. Corresponding to PSI air quality, pollutants concentration are given below in Table 29.1.

Table 29.1

PSI	Air Quality 24-hr	Pollutants concentration				
		TPS 24-hr $\mu g/m^3$	SO_2 24-hr $\mu g/m^3$	CO 8-hr $\mu g/m^3$	O_3 1-hr $\mu g/m^3$	NO_2 1-hr $\mu g/m^3$
0		0	0	0	0	0
50	50% NAAQS	75	80	5	80	–
100	NAAQS	260	365	10	160	–
200	Alert	375	800	17	400	1130
300	Warning	825	1600	34	800	2260
400	Emergency	875	2100	46	1000	3000
500	Significant harm	1000	2620	57.5	1200	3750

Note : NAAQS : National Ambient Air Quality Standards. μg = Micrograms NAAQS – 1970 given below in Table 29.2.

Table 29.2

Pullutant	Period of measurement	Primary standard	
		$\mu g/m^3$	ppm
CO	8–hr	10,000	9
	1–hr	4000	3
Hydrocarbons (non-methane)	3–hr	160	0.24
NO_2 (Nitrogen oxide)	1–yr	100	0.05
O_3	1-hr	240	0.12
SO_x	1–yr	80	0.03
	24-hr	365	0.14
	3–hr	none	none
TSP	1–yr	75	–
	24–hr	260	–
Pb	3–months	1.5	–

The Table 29.3 gives PSI, corresponding health effects and precautions.

Table 29.3

PSI	Air quality	Health effects and Precautions
200	Alert	Generally unhealthy. Irritation symptoms in healthy persons. Milds aggravations of symptoms in susceptible persons. Persons suffering form heart and respiratory ailments reduce physical exertion and out door activity.
300	Warning	Persons suffering from heart and lung diseases symptoms of significant aggravation and decreased exercise tolerance. Healthy persons lose excercise tolerance. Elderly persons with heart and lung ailments should stay indoors and lessen their physical activity.
400	Emergency	In ailing persons aggravation of symptoms. In eldery persons primature anset of certain diseases. In healthy persons decreased excercise tolercence. Ill and elderly persons stay indoors, avoid physical exertions. Others avoid outdoor activity.
500	Significant harm	In ill and elderly persons onset of premature death. Healthy persons experience adverse effect in their normal activity. All persons should remain indoors. Doors and windows should be closed. Minimise physical exertion. Avoid traffic.

A The effects of Air Pollutants

The effects of air pollutants depends on the toxicity of the substance its concentration, period of exposure and individual personal/animal/plants internal resistivity.

B Effects of Gases Compounds

Effects of some important air pollutants are given below.

SO_2 : Affects health, damages to materials and ecosystem, aids acid precipitation and affects climate indirectly.

CO : Directly affects health and indirectly affects climate.

CO_2 **and CFC**[s] : Directly affects climate and indirectly affects stratospheric ozone.

O_3 : Directly affects health and climate and damages materials and ecosystem.

Hydrocarbons : Directly affects health and indirectly affects climate, causes damages to materials and ecosystem.

NO, NO_2 : Directly and indirectly affects health, damages to materials and ecosystem, aids acid pptn, directly affects stratospheric ozone layer and indirectly affects climate.

NO_2 : Indirectly affects stratospheric ozone layer and directly affects climate.

B. Particulate matter

Aerosol particles directly affects health and climate. Sulphate and nitrate containing particles directly affects health and climate and damages to materials, aids acidic rain.

So it affects climate. Particulate matter of heavy metals and radionuclides directly affects health and causes damage to ecosystem.

The above discussed pollutants are directly introduced into the atmosphere by industrial activities. However a part of N_2O and CO_2 are let into the atmosphere by agriculture and forestry activities. The primary species, like SO_2, NO and NO_2 aid for the secondary formation of tropospheric ozone, sulphate particles and some particulate matter. It may be noted here that all of these compounds barring CFCs and some radionuclides are natural constituents but they should be considered pollutants only when their anthropogenic emissions effect their concentration in air significantly.

It has been estimated that on average in a year the atmosphere is polluted by more than 200 million tons of CO, more than 50 million tons of various hydrocarbons, about 146 million tons of SO_2, 53 million tons of oxides of nitrogen, 200-250 tons of dust and 120 million tons of ash and soot. Daily 10000 tons of cosmic dust falls on the earth. Its origin is still a mystery. It may be the result of the activity of the sun or originate in zodiacal nebulae.

Suspended Particulate Matter

It was mentioned earlier that particulate mater in the atmosphere effects human beings and also weather system. In high concentration many kinds of air borne particles were found to be toxic. The hazards of Chernobyl radioactive fallout and Bhopal isocyanide gas leakage were well documented. The accident of Chernobyl (USSR), nuclear power plant radioactive fallout moved with the winds to Scandinavia. The fallout was scavenged by rains to render lichen and the reindeer browsing on lichen radioactive. Animals were found to carry too high a burden of radioactivity to be consumed by the people of the region. The radioactive debris were carried over the whole of northern hemisphere by westerly wind circulation.

Dust and ash particles that spewed out of volcanoes or industrial smoke stachs has cooling effect on the atmosphere. The eruptions of Tambora (1815 which took a human toll of 12000 lives) spewed out about 80 cubic kilometers of ash and rock pieces into the atmosphere. Some particles entered into stratosphere and remained there. This caused cooling. As a result there was no summer in 1816 in northern parts of US and it caused extensive crop failure. The volcanic dust, ash, water vapour reached high altitudes and were carried by winds around the globe. The precipitation of this dust back to the earth took years. The ejected particles into the atmosphere, dispersed sunlight and temporarily increased the earth's albedo. The Krakatova (in East Indies) eruption on August 27, 1883, was the most powerful volcanic eruption in the history of mankind till that date. The eruption of clouds reached the heights about 30 km. It turned day into night in Batavia, about 150 km away. A Tsunami wave, caused by the fall of broken away part of Krakatova island travelled around the world. Rare optical phenomena were observed in Europe in the end of November 1883. The sky remained purple for several hours during sunsets. This was caused by the dispersion of sunlight by a layer of dust particles (size about 2 μm) injected by the volcano into the stratosphere. The dust/ash particles did not settle down on earth for several years. Successive years of weather over the entire earth was cooler than the normal. This was due to the increase of earths albedo and partly due to the mixing of ocean waters by Tsunami waves. The violent erruption of volcano Gunung Agung in Bali in 1963 injected particulate matter into the upper troposphere and stratosphere which gave rise to the spectacular sunsets throughout the world. The eruption caused the rise in stratospheric temperature by about 5 °C in the height range of 18-20 km (60-80 hPa). The volcanic particles consisted of fragments of lava and crystalline material (Calcium and Ammonium sulphate). The ash and pumice discharges of a few reknowned volcanic eruptions are given below in Table 29.4.

Table 29.4

Name of volcano	Year	Estimated Amount of discharge (ash and pumice) (Cubic kilometers = km³)
Mt. Mazama	4600 BC	42
Mt. St. Helens	1900 BC	4
Vesuvius	79 AD	3
Tambora	1815	80
Krakatova	1883	18
Mt. Katmai	1912	12

Airborne particulate matter acts as a cloud condensation nuclei and freezing nuclei. It scatters and absorbs solar radiation and this influences the weather. Seadust (spray) caused by the surf rises into PBL and on evaporating in dry air leaves sea salt and material particles in it. Since major part of the earth's surface is covered with oceans, the concentration of salt particles in the atmosphere are found to be about 10^8 particles /m³ over oceanic area and 10^6 particles/ m³ over land. The fallout of salt in coastal areas and islands range 10-150 kg per acre per annum with extremes as high as 1400 -1800 kg per acre per annum. Atmospheric concentrations of aerosol particles vary from less than 0.1 µg/kg in polar regions to 100 µg/kg or more in dusty continental air . On an average 1 µg/kg have been noted in the middle and upper troposphere.

Airborne particles inhaled deposit in lungs passage and lungs. This may effect the tissues and cause or aggravate lung diseases. Carcinogenic substance present in the air cause lung cancer. Airborne particles may stick to any surface and cause soiling or corrosion. In addition to the scattering of sun light, air-borne particles reduce insolation reaching earth when present in bulk and under favourable atmospheric conditions (inversion of temperature close to the surface of the earth). It causes haze and smog. This deteriorates the visibility and becomes hazardous in airports and other surface transports. Airborne particles are subjected to all sorts of chemical reactions such as photochemical action, oxidation, reduction, polymerization, condensation and catalysis. It is also possible that some of these reactions lead to the formation of substances which are more toxic.

Note : *Precipitaion scavenging* means removal of any particulate or gaseous matter from atmosphere and depositing on the earth's surface by cloud droplets, rain, snow flakes, fog droplets (atmospheric hydrometeors) e.g., In Hyderabad (India) normal day TSPM at important junctions found to be 220-290 $\mu g/m^3$. After continuous 3 days raining (13-15 July 04) it was found to be 150-200 $\mu g/m^3$ due to rain washout

Cycles : denotes the circulation of the element between the reservoirs, that is individual atoms after sometime return to the same reservoir again.

Resident time is the length of time that an atom or molecule spends in each reservoir. Resident times of aerosol particles depend on the size of the particles and on meteorological factors.

Reservoir is defined by its physical characteristics such as the atmosphere, the oceans, the biosphere etc. However it also refers to the chemical form in which the element occurs. For example sulphur occurs in the atmosphere in the form of SO_2. [Most of the elements or compounds in different environments are transferred by physical or chemical transport processes between different reservoirs].

Airborne particles (aerosols) may have different size, shape, chemical properties and density. Consequently their atmospheric residence time and removal processes differ widely. Normally sea spray particles, fine dust particles size any from 2 to 20 µm, larger particles (called coarse type particles) raised by gusty strong winds have smaller residence time–minutes to hours because they are heavy and have rapid sedimentation. Coarse type particles are removed from the atmosphere mainly by sedimentation, dry deposition at the surface or by precipitation scavenging. Fine airborne particles have very low rate sedimentation and dry deposition. Their residence time in the atmosphere is governed by precipitation scavenging. Residence time of aerosol particles is a function of the size of the particles and precipitation frequency. On average the residence time of these particles vary from a few days to a few weeks. In polar regions the average residence time extend to many weeks because of low rate of precipitation and stable stratification of atmosphere. Similarly is the case of with the residence time of aerosol particles in the upper troposphere.

Non-radioatcive trace gases pollutants contain mainly Carbon, Sulphur, Hydrocarbons, Nitrogen and Ozone. On global scale the natural sources of many of these trace gas pollutants and particles immensely exceed the anthropogenic emissions. However in case of industrial and urban community air pollution, anthropogenic emission are much more than natural ones. Thus hazardous air pollutants are confined to these population areas. Trace pollutant substances may be of organic or inorganic in nature. As regards radionuclid, they have three sources; (i) Natural substances from the soil and oceans, (ii) Cosmic dust, (iii) man-made nuclear bomb-debries.

Compared to anthropogenic non-radioactive trace gas pollutants, radioactive pollutants are very few but they are potentially health hazards. The consequence of nuclear radiations can be realised from the world war nuclear explosions over Hiroshima and Nagasaki. However radioactive bomb debris are used in the study of understanding stratospheric processes (they affect the electrical conductivity).

The important natural sources (earth's crust) of radio nuclides found in the atmosphere are : Uranium-U^{238}, Radon-Rn^{222}, Thorium Th^{232}, Radium -Ra^{226}, Thoron-Tn^{220}. These have half life period 4.5×10^9 yr, 3.83 days, 1.65×10^{10} Yr. 1.58×10^3 Yr and 54.5 Sec respectively. Tropospheric residence time of U^{238}, Ra^{226}, Th^{232} is 5-30 days, Rn^{222} has 3.8 days and Tn^{220} has 50 sec. The important cosmic dust radionuclides are Beryllium - Be^{10} & B_e^7 Carbon-

c^{14}, Sulphur-S^{35} and phosphorus-P^{32}. Their half-life period is 2.7×10^6 yr and

Be^7 53 days, 5.7×10^3 Yr, 87 days and 14.3 days respectively. Tropospheric residence time of Be^{10} and Be^7 is 5-30 days and that of C^{14}, S^{35} and P^{32} is not known.

Important man-made sources of radionuclides in the atmosphere are nuclear weapons testing, waste from nuclear reactors and nuclear power fuel cycle. The principal emissions are Strontium-Sr^{90}, Cesium-Cs^{137}, Carbon-C^{14}, Zirconium -Zr^{95}, Barium - Ba, their half-life period is 27.7 yrs, 28.8 yrs, 5.7×10^3 Yr, 65 days, 12.8 days respectively and tropospheric residence time 5-30 days for all except Ba which has 13 days. In addition to the above Tritium - H^3(t),

Krepton $-Kr^{35}$ also have anthropogenic sources and transuranium species-Plutonium, Americium and Curium.

Note

1. Radioactive form of carbon is called Carbon - 14 ($^{14}_{C}$). $^{14}_{C}$ is readily detectable with Geiger counter and hence it is used as a tracer. $^{14}_{C}$ combined with oxygen form a tell-tale carbondioxide which is used in biological research.

2. Radioactivity Effects of radiation on living tissue are very complex. It mainly destroys the blood cells when exposed to it. White blood cells form the marrow of bone. Consequently radioactive emission destroys bone and marrow. Radioactive emission destroys/influences gases and cause mutation. Human mutations are fetal. It causes death of a baby before it born. People who were exposed to nuclear radiation due to atomic explosion over Nagasaki and Hiroshima became the victims of Leukemia, Cancer of Thyroid, breast, lung and stomach. It caused damage to bone marrow, the spleen, the gastrointestinal track lining and the central nervous system. Mental and physical damage was found in unborn and born infants. It was found radiation exposer causes destruction of reproduction cells, sudden increase in white Blood cells count and then drop, slight drop in Red Blood Cells count, fatigue, fever, sore throat, vomiting, diarrhea, and cataract of eye.

3. Radiation energy received by an organism is called the absorbed does. It is expressed in units of rads.

   Radiation shows effects in man when it exceeds 100 rads. Radiation commonly measured in millirads. When a group of people exposed to a single dose of 350 to 450 rads, about half of them lose their lives. This is expressed as LD 50 (Lethal Dose-50) for humans. Another unit rem (roentgen equivalent man) is defined as the product of rad and a qualifying factor, where qualifying factor is a function of tissue exposure, ionising radiation and energy associated with radiation. In general radiation absorption expressed in milli rems. An average man receives about 1 m rem per year from energy production.

   Absorbed radiation dose =

   $$\frac{\text{Radiation energy absorbed}}{\text{The mass of the substance exposed to radiation}} \quad (J/kg)$$

   $1 \text{ rad} = 10^{-2} \text{ J/Kg}$

It was estimated that in America, an average man receives total whole body radiation doses from all sources (natural/environmental, diagnostic and other sources) is about 180 m rem per year per person. Human body normally recovers from a maximum of an acute absorption of 250 rems.

4. Geiger-Muller counter/detector is used for radiation measurement.

$$1r = 2.58\ 0 \times 10^{-4}\ \text{Coulomb Per Kilogram} \quad (\text{where } r = \text{roentgen})$$

The roentgen is the exposure does which produces a total of 1 esu (esu = electro static unit.) ions of each sign in one cubic centimeter of air under standard conditions.

5. *Radioactivity* is define as the spontaneous disintegration (or transmutation) atoms of unstable isotopes of one element into isotopes of another element. The disintegration is accompanied by the emission of certain particles (helium, α, β or neutran particles, or gamma radiation).

Carbon and its Compounds

The second most pollutant of atmosphere after dust particles is carbon and its compounds. The majority of earth's living matter is contained in the green plants (or the green lungs of the planet earth) which entrap solar energy and prepare food for itself (carbohydrates and some proteins) by photosynthesis process. In this process plants absorb CO_2 and water and give out oxygen. This forms the basis for life on earth.

$$x\ CO_2 + y\ H_2o \underset{\text{respiration}}{\overset{\text{Photosynthesis}}{\rightleftharpoons}} C_x(H_2O)_y + xO_2.$$

In this process one gram molecule of CO_2 absorbs 112 k cal of energy. The first photosynthesising organism on earth probably blue-green algae. It is estimated that about 248 billion tons (matric tons) of O_2 is released into the atmosphere every year by the green lungs of the earth. Earth's biosphere contains 1.44×10^{18} tons of water, 1.18×10^{15} tons of O_2 and 2.33×10^{12} tons of CO_2.

The first geochemical history of carbon begins with volcanic eruptions of CO_2, CO and CH_4 from the mantle of the earth. Most of the atmospheric carbon occurs as CO_2 and in smaller quantity as CO, CH_4. Any living organism of the earth is mainly composed of carbon, oxygen, hydrogen and nitrogen which are also the basic chemical elements of water and atmosphere.

CO_2 (Carbondioxide)

It enters the atmosphere through fuel combustion, biological decay, deforestation, volcanoes and release from ocean waters. The mass of CO_2 in global oceans is estimated to be about 1.4×10^{17} kg while its mass in atmosphere is 2.33×10^{15} kg (the farmer is about 60 times the later). The annual fluxes of Carbon globally are as follows : Plant and soil respiration adds (100 GT) 10^{14} kg to the atmosphere, fossil -fuel burning, deforestation adds 5×10^{12} kg, 2×10^{12} kg respectively. Physio-chemical processes at sea surface releases (carbon source) about 10^{14} kg to the atmosphere while it absorbs (carbon sink) about 1.04×10^{14} kg from the atmosphere. Photosynthesis removes about 10^{14} kg of carbon in the form of CO_2 from the atmosphere. Thus in totality 3×10^{12} kg of carbon per year is added to the atmosphere which contributes to the global warming. It is estimated that during last 250 years (from middle of eighteenth century) 26% of CO_2 increased in the atmosphere which is radioactively equivalent to 53% increase in CO_2 in the atmosphere. The life time of CO_2 in atmosphere varies 2-4 years to 50-200 years. The pre-industrial concentration of CO_2 is about 275 ppmv while it was 354 ppmv during the last decade of 20th century. The back ground concentration of CO_2 is 320 ppm and water solubility 1.64.

CO (Carbon monoxide)

It enters the atmosphere mainly through auto exhaust and incomplete combustion, forest fires. In small amount from oceans and volcanoes. Anthropogenic pollution mass is about $(4.5 \text{ to } 6.4) \times 10^{11}$ kg per year while total emission into atmosphere is about $(2.2 \text{ to } 6.5) \times 10^{12}$ kg per year. The global sink is $(3.3 \text{ to } 5.6) \times 10^{12}$ kg per year. Atmospheric back ground concentration is 0.1 ppm and residence time about 3 years. As compared to natural emission anthropogenic emission of CO is much less.

CO is colorless and odourless gas does not support combustion, sparingly soluble in water (Water solubility is 0.00284). CO has more affinity with hemoglobin of the blood than oxygen (about 250 times) and forms. Carboxyhemoglobin (CoHb) and leads to carbon monoxide poisoning. COHb prevents oxygen meeting with body tissues and thus starves tissues from oxygen and causes tissue suffocation. CO poisoning is dangerous as symptoms are not readily seen. When COHb level exceeds 5% level it starts effecting the body. It starts with burning of eyes followed by headache, dizziness, throbbing of temples, weariness, nausea, ringing in the ears, fast heart beating, buckling of the knees. Some victims show irritability, obstinacy, soon paralysis sets in and death fallows due to tissue suffocation. When first symptoms are noticed victim should be shifted and oxygen be given immediately. Heart attacks likely with people suffering from angina pectoris when CO level exceeds 5%.

Hydrocarbons (CH_4)

Gasoline evaporation is the main source of hydrocarbon pollution. Tne other sources of hydrocarbon emission into atmosphere are combustion, biological decay, agriculture and release from ocean waters. The estimated emission of anthropogenetic origin is about 8.8×10^{10} kg per year while natural emission is about 4.8×10^{11} kg per year. Atmospheric background concentration of CH_4 is about 1.5 ppm and non-methane hydrocarbons less than 1 ppb. Atmospheric residence time of methane is about 16 years. It is removed slowly from atmosphere by photochemical reactions with oxides of nitrogen and ozone.

Sulphur (S)

According to one estimate of global sulphur cycle (mainly gaseous H_2S, dimethyl sulphur and SO_2) about 6.5×10^{10} kg of S per year emitted into the atmosphere by athropogenic sources while the natural emission is about 4.0×10^{10} kg, from volcanoes (3×10^9 kg), volatile sulphur from water logged soil (3×10^9 kg) and biological decay (3.4×10^{10} kg). Sea spray emission is about 4.4×10^{10} kg. This shows that other than sea spray emission anthropogenic emissions of sulphur is more than the natural sources. The atmospheric back ground concentration of SO_2 is about 0.2 ppbv, residence time 1-4 days, water solubility is 11.3. It is removed from the atmosphere by photochemical oxidation by ozone. Sulphur is removed from the atmosphere by wet and dry deposition by adsorption (7.0×10^{10} kg) and gaseous adsorption (3.5×10^{10} kg).

SO_2 is colourless and non-combustible gas and has bleaching properties. Most of sulphur in the atmosphere is found as SO_2. Its toxic effects are : irritates the mucous membrane of nose, throat and eyes. Sometimes causes emphysema and swelling of throat. In acute cases it causes paralysis of respiratory track. It destroys green plantation.

Atmospheric Nitrogen and its Compounds

Nitrogen is present in the atmosphere about 78% by volume and by mass 3.865×10^{18} kg. Its atmospheric life time is about 2×10^7 yrs, consequently its global variation is negligible. In combined forms it occurs in the atmosphere as NO (Nitric oxide), N_2O (Nitrous oxide) NO_2 (Nitrogen peroxide) and NH_4 (Ammonia). Nitrogen is an essential constituent of animal and vegetable matter. It is non-poisonous, does not support combustion and respiration. It does not combine with metals and non-metals. However it combines with oxygen in (lightning) the atmospheric electrical discharges and forms oxides of nitrogen.

Fixation of Nitrogen

Combining atmospheric nitrogen to form commercial compounds nitricoxide (then nitric acid and nitrates), ammonia (ammonium formate, bicarbonate, and sulphate), cyanides and cynamids, nitrides (finally obtain ammonia) is called fixation of nitrogen.

Natural emission of oxides of nitrogen consists of decomposition of nitrite in soils, fixation by lightning and conversion from ammonia, while anthropogenic emissions involve mainly high temperature combustion processes which are associated with transportation and energy production. A large part of ammonia emission is through urea from domestic animals.

Nitrogen Cycle

Atmospheric nitrogen is a chief source for numerous nitrogen compounds. Free ammonia occurs in small quantities in air which is washed down along with rain water. Nitrous and nitric oxides are formed in the atmosphere by photochemical reactions in sunlight between nitrogen and oxygen and also during electrical discharges in air. These oxides from nitrous and nitric acids with rain. When they come down to earth they form calcium nitric and calcium nitrate in the soil which is absorbed by the plant as food. There are some bacteria in the roots of leguminous plants which directly take atmospheric nitrogen. Azotobacter (one kind of bacteria found in soil) algae, fungi and mosses also use atmospheric nitrogen directly. The dead plants give out ammoniacal compounds to the earth by nitrosifying and nitrifying bacteria. Animal wastes and urea on decomposition give out ammonia and ammonium compounds. These are formed in the animal body by the process of peptic and tryptic digestion of the nitrogenous plant proteins taken by the animals as food. There are some denitrifying bacteria which breaks up soil nitrates and nitrites and ammoniacal compounds to nitrogen and thus completes the nitrogen cycle, By photochemical and bacterial processes ammonia to nitrate, nitrite to nitrate and again nitrate to nitrogen completes the nitrogen cycle.

N_2O

It enters the atmosphere through natural process of biological action in soil. Its estimated emission of mass is about 5.9×10^{11} kg per year, back ground concentration is 0.25 ppm, residence time in atmosphere is about 4 years. Its water solubility is about 0.121. It is removed from the atmosphere by photodissociation and biological action in soil. When inhaled with oxygen causes nervous excitement. Pure N_2O causes unconsciousness and may be fetal.

NO/NO_2

These gases enter the atmosphere by combustion and bacterial action in soil. The estimated mass of pollution NO is 5.3×10^{10} kg and back ground concentration 0.2 to 2 ppb, while the estimated pollution mass of NO_2 is 5.3×10^{10} kg and back ground concentration is 0.5 – 4 ppb. However their natural emission is 4.36×10^{11} kg (for NO), 6.58×10^{11} kg (for NO_2) per year, atmospheric residence time about 5 days. The water solubility of NO is 0.00618. It is removed from the atmosphere by photochemical reactions and oxidation.

$\overset{+}{N}$ **(Active Nitrogen)**

It exists in metastable state and not an allotropic modification of nitrogen. It shows band spectrum. Active nitrogen is produced under electrical discharges and shows bright luminescence after electrical discharge stopped. It has short life and emits phosphorescence in the presence of sulphur and sodium.

Note : Both natural and anthropogenic trace pollutant gases are removed from the atmosphere by hydrometeors viz cloud droplets, raindrops, snow flakes, fog droplets. Removal process are: Wet removal process, dry removal process and chemical transformation.

Ozone (O_3)

In the composition of the atmosphere it was briefly discussed about stratospheric ozone. Here we shall consider about its formation in stratosphere and near the earths surface. Incoming solar radiation consists (about 99%) in the wave length (λ) range of $0.15 - 4.0$ μm with peak energy at (λ) 0.5 μm while terrestrial radiation consists (about 99%) in λ - range of $4 - 80$ μm with peak at 10 μm. Visible radiation consists in the range $0.4 - 0.7$ μm. Ozone absorbs UV - radiation (UVB) between λ-range $0.2 - 0.3$ μm and IR - radiation at $\lambda = 9.6$ μm.

Ozone is found in the atmosphere between altitudes 10-45 km with a maximum concentration (one ozone molecule to 10^6 molecules of normal oxygen) between altitudes 18-20 km in a globe encircling ozone layer. The temperature of top ozone layer varies $- 20$ to $+10$ °C (at altitude of about 45 km) due to powerful absorbing of UVB by this layer. Ozone is formed in the atmosphere in two stages. In the first stage NO_2 is dissociated by short wave UVB [$NO_2 + h\,\upsilon \rightarrow NO + O$ where hυ is the energy of a solar photon] into NO and atomic oxygen. In the second stage a recombination of atomic oxygen and oxygen molecule of air ($O + O_2 \rightarrow O_3$). Ozone is also generated in small amounts in the atmosphere by the electrical discharges (thunderstorms). In free atmosphere the life period of ozone varies from a few weeks to months. According to WMO assessment (1989) there is a decrease in stratospheric ozone content with the evidence of substantial decrease in stratospheric ozone over Antarctica in spring time (the later one is described as Antarctica ozone hole). Pre-industrial period background ozone concentration was 15 ppbv which shot up to 35 ppbv in 1990. Further it was estimated that the ozone contribution to the greenhouse effect was about 8%.

Ozone shows diurnal variation in concentration in air, over land with maxima in the day time and minima at night. Near the surface of the earth ozone is found in urban polluted atmosphere as photochemical smog which is a complex mixture of ozone and other pollutant gases. In smog ozone is produced by the oxides of nitrogen and hydrocarbons in the presence of sunlight. These gases are emitted from vehicle exhaust and other industrial plants. Thus anthropogenic pollutants are the cause of smog and its adverse effect on nature.

Ozone has a special characteristic that it absorbs the sun's harmful UVB and prevents it from reaching the surface of the earth. UVB induces skin Cancers, cataract of the eyes, suppression of immune system in humans and effects the productivity of aquatic and terrestrial ecosystems. It has been estimated that a decline of 1% ozone content in stratosphere causes 3% rise in the incidence of skin cancers in the humans. If ozone concentration exceeds 800 $\mu g/m^3$ in one hour it is harmful, particularly to elderly and ailing-persons. Ozone and photochemical smog attacks and damages rubber and various synthetic materials. Ozone concentration of 20 ppm or more is poisonous. Ozone causes impairment of lung function and acceleration of aging. The vertical distribution of ozone is changing due to the effects of man-made emissions of oxides of nitrogen and hydrocarbons. As a result there is a net decline in total ozone column which is a danger signal of decreasing protection to the life matter on earth from the lethal effects of UVB. The chlorofluorocarbons (CFC[s]) are powerful green house gases and they are the main cause for the decline of the stratospheric ozone.

In pursuant to the global convention for the protection of the ozone layer (Vienna, 1985), the Montreal Protocol held in 1987 to control substances (and related compounds) CFC_s that deplete the ozone layer in the first global agreement for the protection of earth's atmosphere. In amending the protocol, in 1990, signatory countries have committed themselves to a completes phase out the use of chemicals that deplete the stratospheric ozone layer by the year 2000 AD.

Elementary Ways of Pollution Control

Pollution control may be accomplished: (i) by the use of substitute materials which are less pollutants (ii) by diminishing the use of pollutant sources. and (iii) by treatment for the removal of pollutants. For example we can use natural gas instead of coal. We can use natural gas/electricity driven autos instead of mineral oil autos. We can convert the pollutant to a harmless substance such as exhause catalytic unit to convert CO to CO_2.

Control of Suspended Particulate Matter

Incomplete burning is the cause of carbon particulates (smoke and soot), and flyash. Dust and other particulate matters are produced in industrial manufacturing processes such as grinding, crushing, cement/asphalt plants, foundaries, construction and demolition works.

Particles from combustion and dust can be filtered by air cleaning equipments such as cyclones, scrubbers, baghouses and electrostatic precipitators which are briefly given below.

In *cyclone* dust collector dust-laden air is introduced at the top outer edge to whirl around and around inside the cylinder. Centrifugal force flung the dust particles against the outside wall where it slides down to the bottom and removed periodically. The dust cleared air escapes from a duct located in the centre of the cylinder. By this process 50 to 80% of the dust particles of size 10 μm or more are removed. Small size particles still remain in the cleared air.

Scrubbers are simple screens of water/liquid spray from which air moves up. It removes large particles of air. Scrubbing towers are used to remove particles and gaseous pollutants. Water is used if the gas, is soluble in it otherwise a liquid is used in which the gas is soluble.

Bag houses

Bag houses consists of bags or long sleeves of fabric which withstand against high temperatures.

As the air passes through the fabric to the otherside particles in the airstream are filtered out. The sleeves are periodically cleared. To prevent condensation the temperature of the bag house is maintained to be higher than dew point temperature of water. Bag houses are very effective for removal of fine particles of air.

Eletrostatic Precipitator

It consists of positively charged wires centred between plates which are negatively charged and grounded. When high electric charge is passed on wires, they create charge on air stream particles which are attracted and settled on grounded plates. Settled particles are removed periodically. This system is efficient when particles in the air are electrically charged.

Control of Oxides of Sulphur

Oxides of sulphur mostly emitted into the atmosphere by smelters, oil refineries, paper industry and burning of fuels containing sulphur. Combustion of fuels containing sulphur evolves sulphur dioxide (SO_2) and sulphur trioxide (SO_3), the farmer being in bulk. Parts of SO_2 converts into SO_3 by photochemical processes. Moisture converts SO_3 into sulphuric acid (H_2SO_4). Sulphur can be undressed from oil but with the declining of oil reserves it is wise to use coal fuel or hydropower for power generation plants. Coal contains sulphur in organic form as iron pyrite. By crushing and washing about 30% of iron pyrite can be removed. Rest of sulphur is chemically bound in organic form to the coal. The later can be removed by gasification. Large scale coal use requires flue gas desulfurisation.

Control of CO

Most of CO enters the atmosphere through internal combustion engines. A diesel engine provides more complete combustion and emits less CO as compared to gasoline engine. Autos operating on combustion engines are to be fitted with catalytic converters to reduce CO and hydrocarbons exhaust.

Control of Oxides of Nitrogen, Hydrocarbons and Ozone

Reduction of oxides of nitrogen in the atmosphere is very difficult. Most of the hydrocarbons enter the atmosphere through evaporation of gasoline which can be reduced by proper usage of floating roofs on storages. In autos it can be reduced by the use of (PCV) Positive Crank case Ventilation systems.

Acid Rain

Man's industrial activities changed the composition of atmosphere and precipitation both locally and regionally. In industrialised region the chemical composition of precipitation is found to be acidic with pH value around 4.5.

The gas phase chemical equations of atmospheric acidic rain are given below.

$$O_3 + h\upsilon \rightarrow O_2 + O\ (D') \qquad\qquad .....(29.1)$$

Where $h\upsilon$ = energy of solar photon.

$$O\ (D') + H_2O \rightarrow OH + OH \qquad\qquad .....(29.2)$$

$O\ (D')$ atomic oxygen (by photo dissociation)

$$NO + O_3 \rightarrow NO_2 + O_2 \qquad\qquad .....(29.3)$$

$$NO_2 + OH\ (+\ M) \rightarrow HNO_3\ (+\ M) \qquad\qquad .....(29.4)$$

$(+\ M)$ indicate molecule

$$SO_2 + OH\ (+M) \rightarrow HSO_3\ (+M) \qquad\qquad .....(29.5)$$

$$HSO_3 \text{ is oxidised to } H_2SO_4 \qquad\qquad .....(29.6)$$

In liquid phase oxidation processes of conversion of nitrogen species into nitrate is not known clearly. Gas phase converation to nitric acid (HNO_3) is significant. However HNO_3 (gas) $\leftarrow \rightarrow HNO_3$ (liquid).

Oxidation of SO_2 to sulphate (SO_4) in the liquid phase proceeds via dissolved compounds, particularly ozone and hydrogen peroxide (H_2O_2).

Most of the acidity in precipitation may be attributed to oxides of sulphur and oxides of nitrogen which enter into the atmosphere mainly from the combustion of fossil fuels and industrial processes. SO_2 oxidises to sulphuric acid in rain drops.

Sulphuric acid and nitric acid accounts most of the acidic precipitation, the former accounts two-thirds while the later one-third.

Atmospheric acid is causing erosion of priceless statues, marvellous buildings through dry deposition of sulphate particles (which become acid with rain). Similarly aquatic and terrestrial ecosystem (relationship of organisms to its environment) is being damaged through long term deposition of acidic rain and snow. A number of fresh water lakes without natural buffer are suffering from acid rain or deposition. It was found during 1960 the pH of rain water was 6 to 7 (neutral value is 7). By 2000 AD it changed to 4.5 to 4.0 in some lakes in Europe and America. This acidification resulted in heavy losses of commercial fish population, and gradual disappearance of fauna. Fat head minnows, some zoo plankton disappear at pH 5.9, algal forms, lake trout disappear at a pH value of 5.6 or less. Ground water in shallow wells and aquifers also affected due to contamination by toxic metals such as mercury, cadmium. On land forests were reported dying in Europe and north America due to acid rain, high ozone concentration and smogs. Acidity of the soil increased by ten times in Europe to a depth of about one meter.

BAPMoN

In the mid-1960s, WMO established Background Air Pollution Monitoring Network (BAPMoN) to measure the changing background chemical composition of the atmosphere near the ground, away from cities and strong polluting sources (industries/plants). A complementary Urban network is coordinated by the World Health Organisation (WHO). This network of observatories collect samples for measuring the chemical composition of rain and snow. Under Global Atmospheric Watch (GAW) it provides information on (i) global increase of greenhouse gases (namely CO_2, CH_4, Nitrous oxides, CFC's, tropospheric ozone), (ii) regional distribution of sulpher, nitrogen compounds - which result in acid rain, fog, smog and other changes in the lower level (PBL) chemical composition and (iii) radiation transmitted through the atmosphere and aerosols that were ejected from volcanoes, factories to assess their contribution in cooling the atmosphere. At a few locations toxic metals (mercury, lead etc) and toxic organic compounds (pesticides, herbicides) are also determined. Complementary data on air pollution within cities and by the side of heavy polluting sources are coordinated through WHO and United Nations Environmental Programme (UNEP). In mid-1990's there were about 350 GAW stations operating in 70 countries. 164 stations collect sampling for measurements of precipitation chemistry (rain and snow), 95 BAPMoN stations record turbidity (transparency of air), 78 stations suspended particulate matter, 52 stations CO_2, 22 stations surface ozone, 9 stations methane (CH_4), and 5 stations measure CFC's. The atmosphere is sampled according to well defined criteria, instruments and procedures by trained personnel. Through the global ozone observing system continuous measurements of atmospheric ozone in total column, its vertical distribution and changes are made, This network has about 140 stations over the globe-complemented during 1980's by satellite remote sensing.

UNO, through its agencies WMO and UNEP, urged to monitor the quality of air and environment. WMO coordinating with national services for setting the network of air pollution monitoring stations to monitor both in time and space. Horizontal scale ranges 100 to 1000 km, vertical scale ranges 1 to 10 km and time scale few hours to a year or more. Each individual station is expected to represent space radius 5000 km, time one year. A regional station may provide hourly values representative over a circle of radius 50 km. Station should be located away from Urban areas (cities/industries), high ways, power generation station etc ; however it should experience frequent natural phenomena like heat/cold waves, cyclones, forest fires, sand/dust storms, volcanoes etc. Local influence should be as least as possible. When such conditions are satisfied the monitoring station represent the back ground values. An example is given below.

The following table gives emissions of Singrauli thermal power plant of 2000 MW capacity, situated at Shakthinagar Uttar Pradesh (India), during September 1992, measured over 11 days. The thermal power plant burns about 1000 tones of coal per hour and emits about 9680 kg SO_2 per hour from four smoke stacks (height about 225 m). The plant is equipped with eletrostatic precipitators for removal of particulates from the plume emission. Measurement of trace gases –SO_2, NO_2, NH_3, O_3 and TSP were made upwind of the power plant at a distance of one km north of the power plant. Daily three samples of trace gases for 3 hrs duration were collected (morning, afternoon and evening), one sample of TSP 24 hrs duration collected on Whatman 41 filter paper using high volume air sampler with a flow rate of 1.2 m^3/min.

Table 29.5 Average Concentrations (μgm m^{-3}) of trace gases and TSP at Shaktinagar.

	SO_2	NO_2	NH_3	O_3	TSP
Sept 1992	33.2	8.5	22.5	7.5	134

Table 29.6 The average chemical (ionic) composition of TSP ($\mu gm\ m^{-3}$)

Cl	SO$_4$	NO$_3$	NH$_4$	Na	K	Ca	Mg
1.16	2.47	1.30	0.41	0.64	0.61	2.55	0.86

The conclusions drawn based on the above data was that the average concentrations of NO_2 and O_3 are of the order of back ground values. The maximum ground level concentration of SO_2 was due to the burning of coal in the thermal power plant. The concentrations of TSP and its water soluble components (SO_4 and Ca) are high at Shakthinagar.

A number of WMO member countries provide data to the centralized data collection centres for GAW. Canada operates the WMO World Ozone Data Centre and publishes ozone data every other month. Amercia provides a base for data on precipitation chemistry analyses, acid rain and atmospheric turbidity measurements. Russia (earlier soviet union) is responsible for the collection of solar radiation and atmospheric electricity data and Japanese Government to operate the World Data Centre for Greenhouse gases.

Smog

In the Meuse Valley (Belgium) a mysterious yellow–brown fog was formed in December 1930. It caused irritation and many become ill the next day. In 1944, an air pollution episode in Los Angeles area damaged vegetation and when it became severe it caused eye irritations and tearing. It was later found that auto exhaust gases oxidised by the sun's rays (photochemical oxidation) formed complex compounds which were responsible for irritation. This phenomena later termed of smog. Smog is a combined word of smoke and fog, represents the mixture of smoke and fog. In 5-9 Dec 1952 another air pollution episode over London took a death toll of 4000 people. This attracted the world scientist to find the cause. It was found that there was temperature inversion in the atmosphere close to the ground (PBL) and SO_2 concentration was six times than usual. Subsequently it was confirmed that all air pollution death cases were associated with temperature inversion close the ground (that is warm air above and cold air near the ground) and smog. Further it was found that these incidence also affected the vegetation.

Smog is of two types. (i) Photochemical smog and (ii) Coal-burning smog. Photochemical smog results from the action of sun rays on contaminated organic vapours particularly unburned gasoline, and oxides of nitrogen. Coal-burning smog is a combination of coal smoke and fog. Present day smog in most cities is a combination of these two types in varying proportions. Dense coal-burning smog contains carcinogenic compounds, polynuclear hydrocarbons is one which is effective carcinogen; and 3,4-benzypyrone is another carcinogen. This exists in particles.

In photochemical smog the following reactions take place,

$$NO_2 + h\upsilon \rightarrow NO + O \text{ (atomic oxygen or nascent oxygen)}$$
$$\text{(photochemical dissociation)}$$
$$O + O_2 (+ M) \rightarrow O_3 (+ M)$$
$$O_3 + NO \rightarrow NO_2 + O_2$$

Atomic oxygen reacts with various hydrocarbons (such as olefines of auto exhaust) forms free radicals

$$O + \text{olefines} \rightarrow R + R'O \text{ or}$$

Where R, R' are free radicals

$$O_3 + \text{olefines} \rightarrow \text{products}$$
$$R + O_2 \rightarrow RO_2$$
$$RO_2 + O_2 \rightarrow RO + O_3$$

Thus organic compounds such as aldehydes, acrolein or peroxyacetyl nitrate are formed. Probably these compounds are responsible for eye irritation in smogs. Smog is notorious in corrosive action. It corrodes plant cells, rubber fabrics, nylon stocking, buildings and monuments. In India Tajmahal, Redfort and other historical buildings in Agra are being targeted by smog. In 1966 Tokyo was alarmed for smog 154 days. After air pollution death episodes smog alarms are introduced in some countries.

Toxic Air Pollutants

Air pollutants that are health hazards are: asbestos, arsenic, antimony, cadmium, copper, lead, nickle, zinc, mercury, vinylchloride, benzene, radionuclides, polycyclic organic matter, ethylene chloride, methylchloroform, toluene, trichloroethylene, benzopyrene, pesticides/chlorinated hydrocarbons. Some are carcinogens. They are found in urban run-off, industrial discharge of waste water. EPA restricted the use of DDT, DDD, heptachlor, endrin, lindane, chlorodane, 3, 4- benzopyrene is a carcinogen, asbestos when inhaled may cause cancer. Toxic metals and chlorinated hydro carbons/pesticides are released into atmosphere as by products from industries and agriculture. During 1970's and 1980's it was found minute quantities of pesticides in melted snow, in the tissues of polar bears and in breast milk at the Arctic; and in birds at the Antarctic.

Toxic some trace metals of natural processes and anthropogenic sources with global values of emission are given in Table 29.7.

Table 29.7

Metals	Sources		Remarks
	Natural in 10^6 kg/year	Anthropogenic 10^6 kg/year	
Lead	19	332	White lead is very poisonus
Cadmium	1.0	7.6	
Copper	19	35	
Nickle	26	56	
Zinc	46	132	
Arsenic	8	18	Arsenious oxides are drastic poisons, a dose 0.125 – 0.25 gm is fetal.

In the Arctic haze it was found high concentration of aerosols of sulphate, man-made pesticides like lindane, dieldrin, DDT and toxic heavy metals lead and mercury. WMO Marine pollution studies shows the presence of a number of chlorinated hydrocarbons over marine areas around the world but their concentrations are not high. According to its estimates 80% of the global chlorinated hydrocarbons in air are spread over oceanic areas. The contribution to the sea water pollution through atmosphere is of comparable magnitude (if not more) with that of contamination of sea water by highly polluted river waters particularly of fluxes of metals such as lead, mercury, cadmium and zinc whose sources are located far away from the sea areas. Examples are the pollution of Mediterranean sea, Great lakes of North America. It was found that fish in some lakes contained PCBs and were unfit for human consumption. Water pollution transmitted through atmosphere is difficult to address as pollution sources involved several countries and States that are far away.

Dispersal of Air Pollutants

Dispersal of pollutants in air depends on wind and environmental air stability and turbulence. Precipitation acts as scavenger. Thus weather effects the dispersal of air pollutants and inturn air pollutants effect the weather (particularly precipitation, visibility, acid rain, fog, smog, insolation).

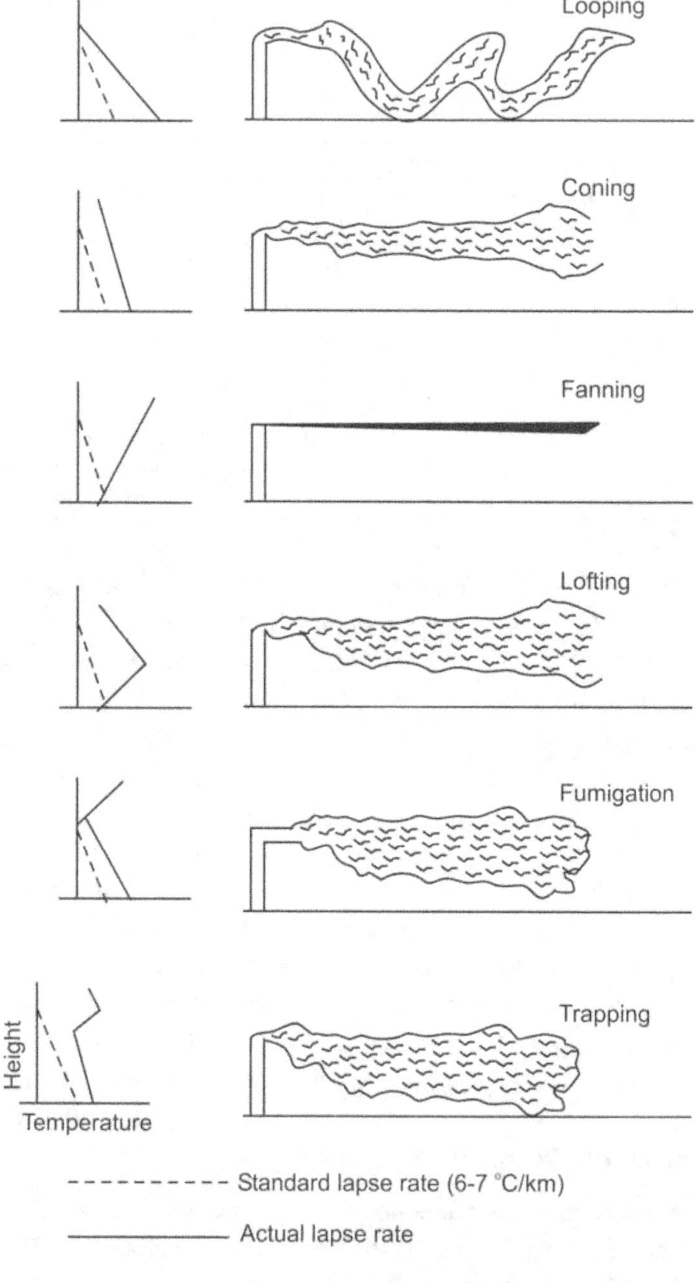

Fig. 29.1 Plume classes under different stability conditions.

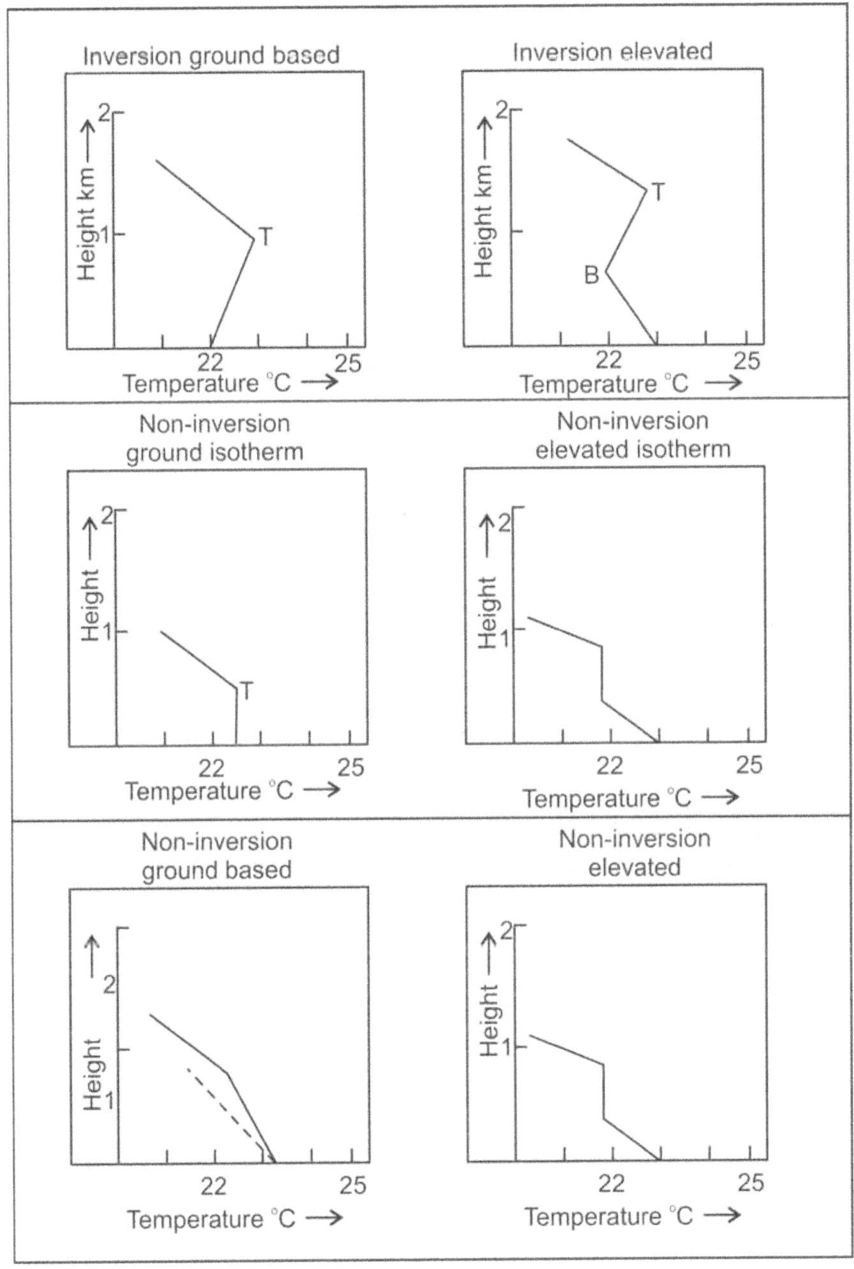

Fig. 29.2 Explanations for stable layers.

Wind direction and speed at surface and also at the chimney stack level has a great bearing on the dispersal of air pollutants from stacks. Stronger the wind speed quicker the dispersal and the pollutants are carried away farther before they are settled. When the wind is light or calm pollutants concentration near the chimney stack is dense and not carried away to great distance. Gustiness dispersal covers a wide area of the chimney plume. Stable conditions of environmental air causes concentration of pollutants locally and unstable environmental air conditions causes pollutants dispersal over a large area leading to less concentration of pollutants. Based on environmental stability/instability (lapse rates) we observe six types of plumes. Looping, coning, fanning, lofting, fumigation and trapping. These cases illustrated in Figs. 29.1 and 29.2.

Precipitation and Humidity Effects

Natural scavenging of air pollutants occur by gravity fallout, rainout and washout. In the absence of turbulence fine particles settledown by impaction with objects. Heavy particles settledown by gravity. In rain out small particles form condensation nuclei. On condensation they fall out to ground with rain drops. In wash out, rain drops while coming, sweep the pollutants in its way.

Table 29.8 Measurement of Air pollutants

Types Pollutants	Sampling equipment	Anaytical Method
Dust fall	Dust fall Jar	Gravimetric
Suspended particulates	High volume sampler	Gravimetric
Total sulphur compounds	Lead-candle	Gravimetric
Sulphur dioxide	Air sampling kit	West and Gaeka method
Oxides of nitrogen	Air sampling kit	Jacob and Hochneissr method
Hydrogen shulphide	Air sampling kit	Methylene blue method
Any other gaseous pollutants	Air sampling kit	–
Wind direction and speed	Wind vane recorder D.P.T recorder	Recording chart
Temperature and humidity	Stevensan screen with thermometer and evaporimeters	Instrument reading Thermographs Hydrographs

The main objectives of air monitoring : (i) collection of basic data of air pollutants (quantitatively and qualitatively), (ii) Meteorological factors aiding pollution, (iii) Topographic influence, (iv) climate study (v) population.

Methods of Estimation of Particulate Matter in Air

Dust-fall Jar

It is an open mouthed polyethylene jar or glass. It contains water at the bottom which is exposed to the atmosphere for a period of one month. The contents of the jar analysed for total deposits. From which deposits per unit area calculated. Dust-fall jar method used to measure the amount of settleable particles such as soot, flyash, smoke particles in the air. Dust fall is expressed as tonnes per square kilometer per month. (tonnes/km^2/month).(see Fig. 29.3)

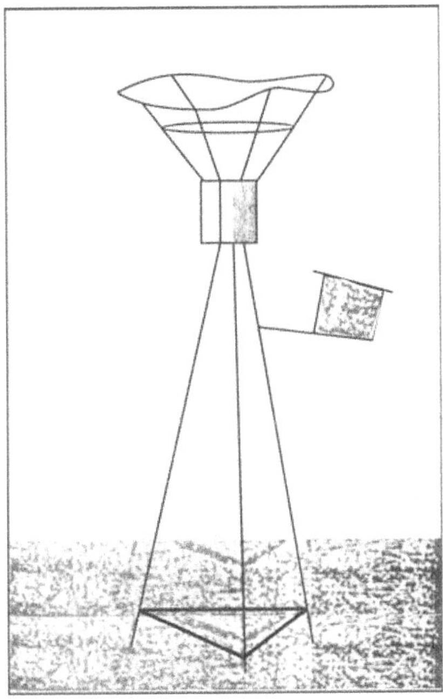

Fig. 29.3 Dust-fall jar and lead candle fixed to a tripod stand.

High Volume Sampler

It contains a 20 cm × 25 cm glass fibre filter, which collects sample over a period of 24 hrs. The glass fibre filtrates 100% of all particulate matter of 0.3 μm diameter or more. Air is drawn from vertically upward flow at the average speed of 19 m/min. It sucks all particulate malter upto diameter of 100 μm. Filtered particles over a period of 24 hrs is measured by weight and expressed as μgm/m^3 of air. Generally air is drawn at the rate of 1 – 1.5 m^3/min.

Methods of Estimation of Sulphur Compounds in Air

Lead Peroxide Candle Method

Gaseous Sulphur compounds + Solid lead peroxide
= lead sulphate

$$PbO_2 + SO_2 \rightarrow PbSO_4$$

A piece of tapestry cloth size 10 cm × 10 cm is wound round on a glass cylinder or polyethylene tube of 100 cm^2 curved surface area. Tragacanth (gum) and lead peroxide paste is made and is evenly applied on tapestry cloth surface. This is dried in a desiccator to make it a candle. This lead peroxide candle is erected in a wooden stevenson screen like box, which protects it from rain but allows the air to circulate over it. In this way the candle is exposed to air for a period of one month. Then the lead sulphate formed over the candle is estimated gravimetrically after converting into barium sulphate. The results expressed as milligrams SO_3/100 cm^2/per day.

The above lead peroxide candle arrangement generally attached to one of the legs of dust fall jar tripod.

Multi-gas Sampling Kit

It is used for sampling measurements of different gases simultaneously. Selective absorbents are kept in different impingers (generally four) to trap respective gaseous pollutants. A vacuum pump operates to suck atmospheric air which circulates over the impingers containing different absorbents and the purified air leaves the kit. Gaseous pollutants react with the absorbents. Generally air is sucked into the kit at the range of 0.1 to 3 liters per minute. Using this kit O_3, SO_2, H_2S, oxides of nitrogen, ammonia can be sampled. The results are expressed as ppm (ml/m^3) or as $\mu g/m^3$. (μg = micro grams).

Questions

1. Define Air pollution, state natural and artificial pollutants.

2. Write briefly about green house gases and their principal sources.

3. Write the annual carbon fluxes budget.

4. Define pH value, using pH values define a solution is acidic and alkaline.

5. Using PSI and acidquality write 24 hr TPS concentrations.

6. Write briefly the particulate matter and suspended particulate matter.

7. What were the effects of volcanic eruptions of Tambora and Krakatova on weather.

8. Write briefly on : (i) Resident time of atom/molecule (ii) Reservoir (iii) Airborne particles.

9. What are the effects of radiation. Explain lethal dose-50.

10. Write briefly about the carbon and its compounds in the nature/atmosphere.

11. Write briefly about atmospheric ozone and explain ozone hole.

12. Write briefly on acid rains.

13. What is BAPMON?

14. Write briefly an smog pollution.

15. Write briefly an Toxic air pollutants.

16. Write short notes an dispersal of air pollutants with atmospheric stability.

CHAPTER 30

Water and Its Pollution

All life on earth is sustained by water. It is the second most vital matter for life after air and then follows food, clothing, shelter etc. Man can survive without food for a few weeks but without water he cannot survive even a few days (a week) and without air even a few minutes. It is estimated nearly 70% of man by body weight is water. Water is inorganic and does not provide energy to tissues. Much of water is in body-photoplasm and in spaces between cells. Plasma (fluid part of the blood) contains 91-92 percent water. Water is essential for digestion of food, blood circulation and removal of waste products from body. Man's average daily loss of water is 2500 CC. Of which 1500 CC as urine, 500 CC in perspiration, 400 CC in exhaled air and 100 CC in faeces. Any loss of water more than 250 CC (10%) is fetal. On an average man drinks about 1500 CC of water and the remaining he gets it from food. Fruits and vegetables contain about 80% water. Kidneys remove excess water along with cellular waste.

On an average man requires 2.5 to 3 liters of water per day for drinking and cooking or one cubic meter per year. However to improve his living conditions he is using 100-200 liters per day for his household needs and a thousand times for industrial consumption. For example, production of one ton sugar requires about 100 cubic meters (m^3) of water, one ton of paper requires 250 m^3 of water, processing of 1000 liters of milk requires 5000 liters of water. Production of one ton steel requires 150 m^3, Nickel 800 m^3, Aluminium 1500 m^3 of water. In recent times man has been mining ground water for agriculture, domestic and industrial use. This is causing depletion of ground water resource which was reliably dependent through the ages. Unknowingly man has made many blunders in water use and conservation. The worst part is pollution of lakes and rivers. Pollution is the main factor that is threatening the exhaustion of fresh water resources. One cubic meter of sewage dump contaminates more than 12 cubic meters of clean water in rivers and lakes. Use of polluted water is a severe health hazard. Pollution of water is the foremost concern of mankind. Experts are of the opinion that water famine may arise only through inefficient use of natural water resources rather than lack of water in the world. Further they concluded that the water resources on the earth fully sufficient to meet all the growing needs of man for an indefinite time period provided man must practice to avoid water pollution, extend recycling of suitable water resources.

In 1950's the world population was about 2.5 billion which shot upto 6 billions by the end of 2000. In fifties about one-third of the population lived in cities which rose to one-half by the end of 20th century and it is projected to rise to two-thirds by 2025. Mega cities (population size exceeding 10^6) are emerging in advanced countries where 80% of worlds urban population would live. During 20th century the global demand for water has increased enormously because of the growth of industries, irrigation and domestic use. During this period water withdrawl from existing source was more than six fold, which was more than double the rate of population growth in the same period. It is estimated per capita consumption of water by urban dwellers is 150 liters per day or 55 m^3 per year while the village dwellers consumption 50 liters per day or 18 m^3 per year. Consumption of water is a sign of country's economy and civilization, more use means more developed. A country's economic development largely depends on the water resources and its geographic distribution.

Water is essential to all lives of animals and plants. It has crucial role in industries and steam generation plays important role in engineering. Water is used in large quantities in the production of steel, sugar, paper, rayon, chemicals, textiles, ice, drinking, bathing, washing, irrigation, fire fighting, power generation and atomic energy.

Water in the World

The volume of the global water is estimated to be 1455×10^6 cubic kilometers (km^3). About 1370×10^6 km^3 or 94% of the global water is estimated to be in the seas, which occupies 361 million km^2 (square kilometers) of the global surface area or 70.8%. 85×10^6 km^3 or 6% of global water is fresh water, which accounts water of rivers (1200 km^3 or 0.0001% of global water), lakes (750×10^3 km^3 or 0.05% of global water), glaciers (2.4×10^6 km^3 or 4.11% of global water). Water in the atmosphere is 14×10^3 km^3 or 0.001% of global water.

It is estimated that about 4.5 million km^2 or 3% of land area is occupied by inland water bodies and 16.5 million km^2 or 11% (land) area occupied by glaciers. The volume of fresh surface water is about 751200 km^3 which is very small amount compared to the volume of the oceans and seas water, yet it plays a vital role in the life of man.

Hydrological Cycle

Definition

Hydrology may be defined as the science that deals with the processes governing the depletion and replenishment of surface water and ground water resources of the earth. Consequently it may be viewed as part of physical geography and hydrometeorology.

Hydrometry is the technology of water measurement spanning all aspects of water movement within hydrological cycle.

The hydrological cycle is a continuous process of movement of water from surface of earth to atmosphere (as evaporation), from atmosphere to ground (as precipitation), then to rivers, lakes, underground reservoirs and to the sea. (see Fig. 30.1). Like lithosphere, the bulk composition of hydrosphere is oxygen. The composition of water has 88.9% oxygen and 11.1% hydrogen by weight. Waters of oceans, rivers, lakes have very small amounts of almost all elements of the earth's crust. Sea water has 3.5% dissolved minerals, of which NaCl (about 2.6%) is the most abundant. Consequently sea water has salty taste.

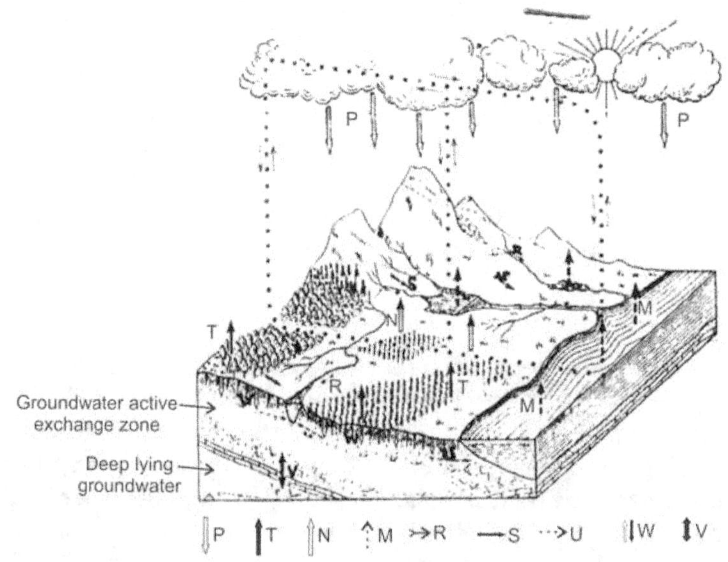

Fig. 30.1

About 80% of the total water vapour in the atmosphere comes from the evaporation of sea water.

Note : Steam occupies a volume of 1600 times greater than that of the liquid water. In gaseous state molecules are about 12 times apart than they are in liquid state.

Thus hydrological cycle is a God's gift to all living beings because fresh water resources on land are formed from rain water. In a sense it is a gigaintic desalination plant which converts saline sea water into water vapour and then into fresh water that falls on earth as precipitation. It may be noted that all minerals/matter transforms to other forms but water remains as water. In sansakrit water is called Amritha, meanining deathless. At any one time the lakes and rivers contain 751200 km^3 of water which is little more than 0.05% of global water. An estimate of annual water balance of the world by M.I. LVOVICH is given in Table 30.1.

Table 30.1

		Volume in km^3	Average depth in mm
A.	Over periphery of land area (116.8 ×10^6 km^2)		
	Precipitation	106000	910
	River discharge	41000	350
	evaporation	65000	560
B.	In land area (32.1 × 10^6 km^2)		
	Precipitation	7500	238
	Evaporation	7500	238
C.	World ocean (361.1 × 106 km^2)		
	Precipitation	411600	1140
	Inflow river water	41000	111
	Evaporation	452600	1251
D.	World total (510 × 10^6 km^2)		
	Precipitation	525100	1030
	Evaporation	525100	1030

Global volume of rain water $= 4\,\pi\,R^2\,h,$

$$= 4 \times \frac{22}{7} \times (6370)^2\, h$$

$$= 510 \times 10^6\ km^2 \times h$$

$$= 510 \times 10^6 \times \frac{1030}{1000} \times \frac{1}{1000} \cong 525100\ K m^3$$

Where $\qquad$ R $\;=$ Mean radius of the earth 6370 km.

$$h \;= \text{height of pptn} = \frac{1030}{1000} \times \frac{1}{1000} \; \text{km.}$$

Water Resources

Water resources can be divided into two categories. (i) Surface water and (ii) ground water. Surface water resources are : Rivers, Lakes, Swamps, Glaciers and sea water. Underground water resources are wells and spring water.

Rain water is the purest form of natural waters. However it contains traces of dissolved material, gaseous compounds. Rivers, lakes, swamps waters are all accumulated rain water. These have some dissolved minerals and suspended matter. Sea water contains dissolved salts about 35 gm/kg or 3.5% by weight. Six ions of sea salts together will be more than 99%. They are ions of Chloride (55%), Sodium (30%), Sulphate (8%), Magnesium (4%), Calcium (1%) and Potassium (1%).

Water, in the form of vapour, is exchanged between sea and air. More than 80% of water vapour in the atmosphere comes from the evaporation sea surface water. Along with water vapour energy is also transferred from sea to atmosphere in the form of latent heat of water vapour. On condensation it releases latent heat. Condensed water returns to the sea in hydrological cycle but energy remains in the atmosphere. Part of this heat energy is converted to mechanical energy (wind). Two-thirds of the precipitation on land returns to the atmosphere by soil evaporation and plants. The remaining one-third precipitation either percolates into soil or runs off on the land surface. The bulk of the fresh water on land is stored in the form of ice in glaciers (volume 24×10^6 km^3) which is about 20000 times the worlds river waters, Ice melted water flows into rivers and lakes. Water stored in lakes (750×10^3 km^3) is about 625 times the water in rivers. This water is of great importance in many parts of the world. They are the fresh water source for cities and agriculture. Because of importance artificial lakes are created by damming rivers.

Percolation of water through large pores into the soil is called gravity water. This gravity water accumulates at some depth over rocky layer and fills all cracks and pore spaces. The level below which soil and rocks are saturated with water is called water table. Water below water table is called ground water. The depth of water table varies from place to place and with season. In dry season the depth will be more than in wet season. During good rains water table slowly increases and this results in ground water flow into near by streams. When there are persistant heavy rains stream channels will be overflowing into adjacent areas. This causes floods. Surface run-off of water depends on precipitation and varies with the intensity of precipitation. Ground water flow on the other hand is more steady. The ground water (volume $=60 \times 10^6$ km^3) is about 50,000 times the volume of river water.

Artesian Wells

When rain occurs on mountain slopes water passes through the porous rock and then flows under ground. If this water flows between two layers of low permeability artesian system (like water flow through a pipe) develops. The under ground aquifer (water saturated zone) flow carry large quantities of water. This can be drawn through digging wells on plane area, where water comes to the surface by pressure. This type of well is called Artesian well.

Water impurities can be categorized as : (i) Suspended impurities (ii) Dissolved inorganic impurities and (iii) organic impurities.

1. Suspended impurities can be removed by filtration or settling. Suspended impurities include clay silt, bacteria, algae, protozoa.

2. Dissolved inorganic impurities include (i) calcium, magnesium and sodium carbonates, bicarbonates, sulphates, chlorides, fluorides (ii) Metal and oxides e.g., magnesium, iron oxide, lead, Arsenic (iii) gases eg. Oxygen, CO_2, H_2S.

3. Organic impurities include (i) Suspended (e.g., vegetables, dead animals). (ii) Dissolved (e.g., vegetables and animals).

The essentials of driking water standard in India are given below in Table 30.2.

Table 30.2

Item	Desirable	Maximum permissible limit (in the absence of alteranate source)
pH	6.5 – 8.5	- - -
Alkalinity (mg/l)	200	600
Total hardness (as $CaCo_3$)	300	600
Dissolved solid material (mg/l)	500	2000
Calcium (mg/l)	75	200
Chlorides (mg/l)	250	1000
Chlorine (mg/l)	1.2	- - -
Fluroid (mg/l)	1.0	1.5
Nitrate (mg/l)	45	100
Iron (mg/l)	0.3	1.0
Copper (mg/l)	1.5	1.5
Manganese (mg/l)	0.1	0.3
Zinc (mg/l)	5	15
Arsenic (mg/l)	0.5	- - -
Cyanide (mg/l)	0.5	- - -
Lead (mg/l)	0.5	- - -
Total Coliforms	nil/100ml	10 counts/100ml
Faecal coliforms	nil/100ml	- - -

mg = milligrams   l = liter

Hardness of Water

Water that does not produce lather easily with soap but produces white curd like form is called hard water, while water that produces lather easily with soap is called soft water. Hardness of water is of two types : 1. Temporary hardness and 2. Permanent hardness. Temporary hardness is caused by the presence of dissolved bicarbonates of calcium, magnesium and heavy metals and the carbonates of iron in the water. Temporary hardness can be removed by boiling of water which decomposes bicarbonates into insoluble carbonates and carbondioxide. Carbonates can be removed by filtration while carbondioxide escapes out. Permanent hardness of water is caused by the presence of dissolved chlorides, sulphates of calcium and magnesium, iron and other heavy metals in the water and it cannot be removed by boiling.

Hardness of water is generally measured as parts per million (ppm). Parts per million is the parts of $CaCO_3$ equivalent hardness per 10^6 parts of water.

1 ppm = 1 part of $CaCO_3$ equivalent hardness in 10^6 parts of water.

Equivalent of $CaCO_3$ is defined as below.

$$\text{Equivalent of } CaCO_3 = \frac{\text{Mass of hardness causing substance} \times 50}{\text{Chemical equivalent of hardness causing substance}}$$

Where 50 = Chemical equivalent of $CaCO_3$.

Hardness of water classified as below in Table 30.3.

Table 30.3

Hardness in ppm	0-50	50-100	100-150	150-200	200-250	more than 250
Nature of hardness	Soft	Moderately soft	Slightly hard	Moderately hard	Hard	Very hard

Example

What is the temporary hardness in ppm and total hardness in ppm of a sample of water which have the following composition. Ca $(HCO_3)_2$ = 16.2 mg/l, Mg $(HCO_3)_2$ = 7.3 mg/l, $MgCl_2$ = 9.5 mg/l, $CaSO_4$ = 13.6 mg/l.

Solution

Where molar mass of $Ca(HCO_3)2$ = 40 + 2 (1 + 12 + 48) = 162 etc

Substance	Mass mg/l	Molar Mass	Multiplying Factor	Equivalent of $CaCo_3$
$Ca(HCO_3)_2$	16.2	162	100/162	$16.2 \times \dfrac{100}{162} = 10$
$Mg(HCO_3)_2$	7.3	146	100/146	$7.3 \times \dfrac{100}{146} = 5$
$MgCl_2$	9.5	95	100/95	$9.5 \times \dfrac{100}{95} = 10$
$CaSO_4$	13.6	136	100/136	$13.6 \times \dfrac{100}{136} = 10$

Temporary hardness caused by dissolved bicarbonates of Ca and Mg.

$\therefore$ temporary hardness $= 5 + 10 = 15$ ppm.

Total hardness $= 10 + 5 + 10 + 10 = 35$ ppm.

Example

What is the total hardness and permanent hardness in ppm of a sample of water containing $Ca(HCO_3)_2 = 16.2$ mg/l, $Mg(HCO_3)_2 = 7.3$ mg/l, $MgCl_2 = 9.5$ mg/l and $CaSO_4 = 13.6$ mg/l.

Solution

Substance	Mass mg/l	Molar Mass	Multiplying Factor	Equivalent of $CaCO_3$
$Ca(HCO_3)_2$	16.2	162	100/162	$16.2 \times \dfrac{100}{162} = 10$
$Mg(HCO_3)_2$	7.3	146	100/146	$7.3 \times \dfrac{100}{146} = 5$
$MgCl_2$	9.5	95	100/95	$9.5 \times \dfrac{100}{95} = 10$
$CaSO_4$	13.6	136	100/136	$13.6 \times \dfrac{100}{136} = 10$

Total hardness $= 10 + 5 + 10 + 10 = 35$ ppm.

Permanent hardness $=$ hardness of $MgCl_2$ and $CaSO_4 = 10+10$

$$= 20 \text{ ppm}$$

There are several disadvantages in use of hard water for domestic purpose, industries and in steam generation.

In Domestic Use

In washing and cleaning, bathing, cooking, drinking hard water causes lot of wastage of soap, skin becomes dry and dark, consumption of more fuel, gives unpleasant taste, pulses and beans etc are not properly cooked and digestive system may be affected.

In Industrial Use

It casts bad effects in industries such as textile, sugar, dyeing, paper, laundry, concrete etc. Use of hard water in steam generation causes boiler scale, sludge formation, corrosion, priming and foaming, Caustic embrittlement, *boiler corrosion* etc. These cause adverse effects on industrial production. Some of these terms are explained below.

Boiler Scale

When hard water is boiled, dissolved material in it forms a hard deposit on the inner surface of the boiler. This is called boiler scale.

Sludge

It is loose and slimy soft precipitate formed in the boiler, which can be removed by scrapping with brush.

Boiler Corrosion

It is the decay of boiler material caused by chemical or electro-chemical attacks by environment.

Priming

When hard water steam produced rapidly, some liquid water particles jump out along with steam. This is called priming.

Foaming

When boiling hard water contains some oily substances, it produces foam or bubbles which do not break.

Caustic Embrittlement

It is a form of boiler corrosion formed by alkaline water.

Softening of Water

To avoid ill effects of corrosion, boiler scale etc., water used in industries must be sufficiently soft. Methods used for removal of dissolved salts from water is called softening.

Lime Soda Process

In this method lime $Ca(OH)_2$, Magnesium hydroxide $Mg(OH)_2$, Soda Na_2CO_3, Calcuim carbonate $CaCO_3$ are added to the hard water. Soluble salts of Calcium, Magnesium are converted into insoluble compounds which are removed. By this method water with 10-15 ppm hardness is obtained.

Zeolite

Zeolites are complex hydrated silicates of Al, Ca, Na, K or Fe.

Permutite or Zeolite Process

In this method hard water is slowly percolated through Zeolite bed of sodium-alumino silicates. This bed converts calcium, magnesium salts into calcium, magnesium-zeolites and sodium salts. Filtered water becomes soft water with 10 ppm hardness and contains soluble sodium salts, while Ca, Mg salts are removed.

Drinking Water

Water for drinking purpose must be soft and clean. It should be colourless, odourless and have pleasant taste. Turbidity should be less than 10 ppm and dissolved solids should be less than 500 ppm. It should be free from micro-organisms. Natural water from rivers, lakes and underground do not satisfy these qualities and requires purification

Suspended impurities are removed by screening (passing water through screens) and by sedimentation (keeping water in big tanks without disturbing to settle down suspended particles at the bottom by gravity). Sedimentation with coagulation is used to remove suspended finer clay particles, colloidal matter. In this process chemicals, called coagulants (Alum, Ferrous sulphate etc) are added to the water before sedimentation. Coagulants are also useful for removing colour and odour and to provide pleasant taste. Alum is most widely used as coagulant. Sodium aluminate, ferrous sulphate are also used as coagulants depending on water alkalinity.

With the application of screening, sedimentation and coagulation all impurities are not removed, some will be left over which will be removed by special filtration beds. Special filteration bed removes colloidal matter, microorganisms and most of the bacteria. These filters remove 98% of bacteria and almost all suspended impurities. A typical filter bad is shown in Fig. 30.2.

Disinfection and Sterilization

Nearly killing or destroying of all micro-organisms pathogenic bacteria to make water potable is called disinfection, while sterilization means total destruction of all living organism in water.

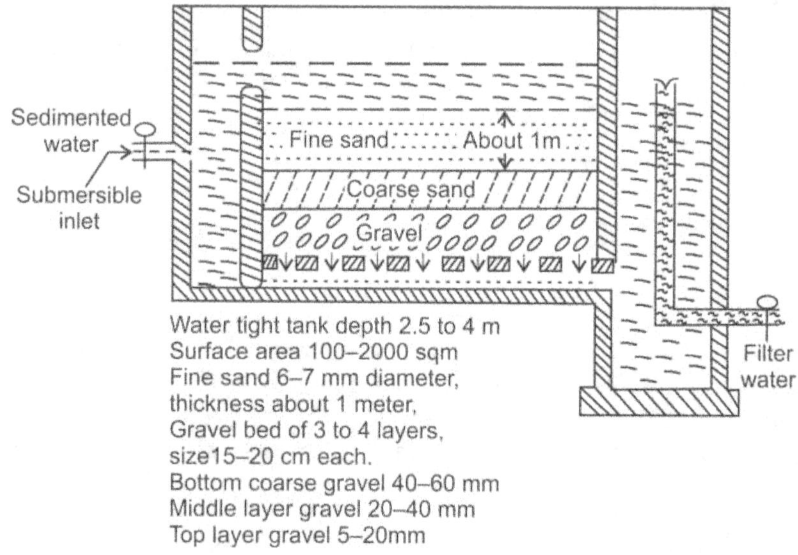

Fig. 30.2 A typical filter bed.

Removal of Micro-organisms from Water

After filteration the residual pathogenic bacteria is killed by disinfection. Water is disinfected by adding bleaching powder, by chlorination, by using chloramine, ozone, potassium permanganate and by ultraviolet light.

Chlorination

Generally level of 0.2 mg/liter of free chlorine will kill bacteria. However to kill viruses it requires 3 to 4 mg/liter and to kill protozoa it may require a level of 500 mg/lit. Disinfection by chlorination is widely used and considered safe.

Sewage

All waste waters or liquids–domestic waste, industrial waste, ground waste, storm waste, human waste–is called sewage. Sewage emanates gases, such as hydrogen sulphide, ammonium sulphide, phosphine etc. Some of these gases have dirty odour. Sewage contains aerobic and anaerobic bacteria which cause oxidation of organic compounds in it.

Aerobic

Living only in the presence of free molecular oxygen gas or dissolved oxygen in water.

Anaerobic

Living in the absence of free oxygen as gas or dissolved in water.

Prototrophic bacteria takes food from minerals (like nitrites, carbonates etc) present in the sewage.

Metatrophic bacteria takes food from organic compounds nitrogenous and carbonaceous.

Important sewage characteristics are : Physical, Chemical and Biological.

Physical characteristics include colour, odour, temperature and turbidity of sewage.

Sewage contains colloidal matter, dissolved gases and suspended matter. Fresh sewage is odourless and has gray colour. In about 4 hours time it becomes stale (oxygen being exhausted) and starts emitting offensive gases like hydrogen sulphide, ammonia, methane etc. Its colour becomes dark. Temperature of sewage in general slightly higher than the water supply. Sewage contains about 99% water and the remaining solid matter such as faecial solids, matches, bits of paper, twigs, grease, vegetable matter.

Chemical Characteristics

Fresh sewage is little alkaline while stable sewage is acidic in nature. Solid sewage consists of organic (about 40%) in, inorganic (about 55%) matter and it contains dissolved gases like CO_2 H_2S, CH_4 etc. It is estimated that sewage consists of 0.045% solids of which 0.0225% in dissolved form, 0.0112% in suspended form and 0.0113% in settleable.

Biological Characteristics

Sewage contains large quantity of bacteria such as algae, fungi, pathogen, protozoa and other microorganisms.

Biological Oxygen Demand (BOD)

BOD of a sewage is the amount of free oxygen required for biological oxidation of the organic matter in aerobic condition at 20 °C for a period of 5 days. BOD is expressed as mg/l or ppm. An average sewage has BOD of 100-150 mg\l.

Sewage Treatment

To dislodge its harmful effects sewage must be treated to public health or aquatic life before it is letoff into natural water course (rivers/ponds or on land). Sewage is passed through bar screens or mesh screen which separate the large suspended floating matter and coarse solid silt, gravel etc. It is then let off into sedimentation tanks where it will be was treated with chemicals before settlement. Generally alum, ferrous sulphate are used to coagulate, which also removes colloidal matter. It is then aerated or subjected to aerobic

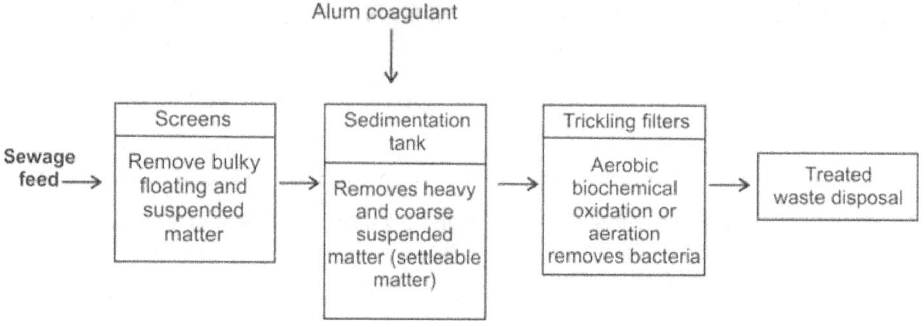

Fig. 30.3 Flow chart of sewage treatment.

biochemical oxidation. In this process sewage is passed through special sprinklers to maintain aerobic conditions. In aerobic condition organic carbon material is converted into CO_2, the nitrogen into NH_3/nitrites/nitrates (– this forms salts). Filtering media removes micro-organisms. Treated waste sewage is removed from the bottom which is used for agriculture fertiliser etc. See flow chart in Fig. 30.3

Desalination of Brine

Sea water contains about 3.5% of dissolved salts and it is called brine or brackish water. Brackish water is completely unfit for drinking purpose. The removal of salts from brackish water is called desalination. Over sea areas and in coastal areas desalination is essential for getting potable water. See Fig. 30.4.

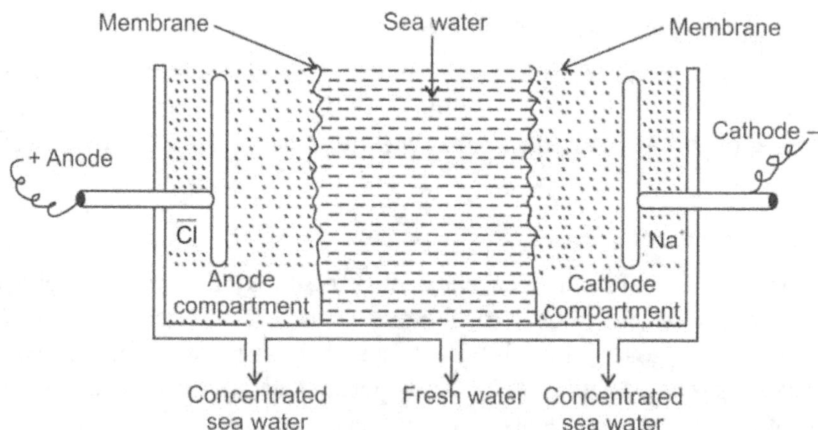

Fig. 30.4 Desalination of brine.

An age old method of desalination is boiling of sea water and then condensing the water vapor to obtain fresh water. Where electricity is easily available, the desalination is achieved by electro-dialysis. In this method direct electric current is passed through brine. By this sodium and chlorine ions are attracted at cathode and anode respectively. Fitting special compartments of permiable memebrane around cathode and anode, the central compartment gets accumulated with desalinated water; which is removed.

Reverse Osmosis

If two solutions of different concentrations of a solute are separated by a semi-permeable memebrane, by Osmosis process, dilute solution flows into concentrated solution through the membrane. This flow continues till the two sides of the membrane attains equal concentrations. If a hydrostatic pressure in excess of osmotic pressure is applied on the concentrated solution, the solvent flow reverses, that is, flow starts from concentrated to less concentrated solution across the membrane. This is called reverse Osmosis. Reverse Osmosis method is used for desalination of brine. This method is shown in Fig. 30.5

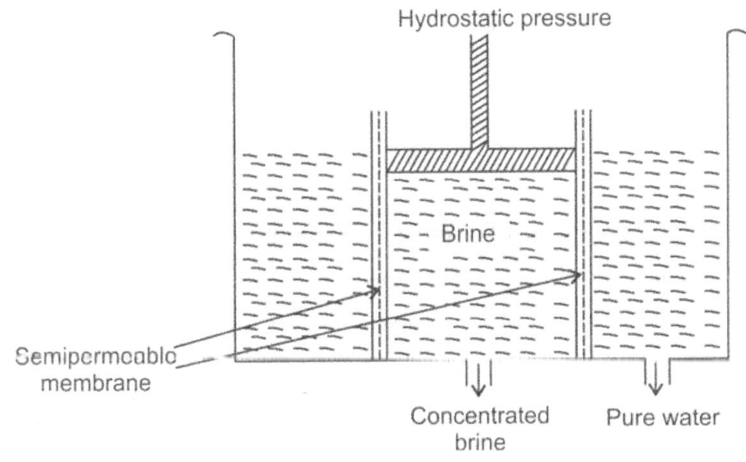

Fig. 30.5 Reverse osomosis.

Solid Wastes

Solid wastes include any garbage, refuse, sludge, air pollutants, any discarded material from domestic houses, industrial operations, mining, agriculture, construction and demolition of buildings. Solid wastes may be biodegradable or combustible. Industrial wastes include organic and inorganic chemicals, pesticides, explosives, paints and allied products, petroleum, refining materials, rubber, plastic, waste oil. Some of these are toxic in nature. Municipal wastes include garbage, fats, paper, leaves, grass, wood, plastics, rubbers, rags, ash, glass, ceramic etc.

Generation of solid wastes cannot be stopped but they can be reused or recycled for use. Community refuse may contain reusable materials such as steel, aluminium, glass, paper. These materials may be collected separately and reused, which helps is reducing the cost of manufacture.

Solid waste disposal mainly carried by (i) land fill (ii) incineration and (iii) compost.

Toxic solid waste disposal by land fill may create health problems. Incinerators reduce the volume of solid waste by burning combustible materials. In this case also one must take care of air pollution. Land fill consists of depositing refuse in a low place or excavated trench. The refuse after dump is covered by earth about one feet depth.

Incinerators generally burn 85 to 90% of combustible material. Heat generated in incinerators are used for many purposes and now considered as a source of energy.

Dumping of food and agricultural waste, sewage sludge in pits to decompose by biological action is called composting. The residue of composting material is used as manure in agriculture. Micro-organisms break down the degradable material into powdery material called compost.

Questions

1. Write briefly the water requirements of a man. It is foretold that there will be water famine over the globe, comment?

2. Write short essay on global water, including hydrological cycle.

3. Write briefly the water resources over the globe.

4. Write briefly on water impurities and hardness of water including its classification and disadvantages in the use of hardwater.

5. A sample of water contains $Ca(HCO_3) = 16.2$ mg/L, $Mg(HCO_3) = 7.3$ mg/l, $MgCl_2 = 9.5$ mg/l and $CaSO_4 = 13.6$ mg/l. Find total hardness and permanent hardness?

6. Write briefly on drinking water including chlorination?

7. Write briefly on : (i) Sewage (ii) BOD and treatment (iii) Desalination of brine (iv) Reverse osmosis (v) Solid wastes.

CHAPTER 31

Environment

It is now known that all forms of life is made up of protoplasm and carry identical processes but they live in totally different surroundings/environments. There exists a critical relationship between living things and its physical environment. Biosphere is the area near the earth's surface which encompasses living organisms. It includes parts of atmosphere, hydrosphere and Lithosphere. Environment of an organism is its surrounding media. Ecology is the relationship of an organism to its environment. Ecosystem is the functioning of living and non-living components of environment.

The physical factors of an environment includes soil condition, temperature, inclination of the sun or light, water, atmospheric conditions and topography.

Soil conditions have great impact on plant and animal life. Soil is classified as loam, clay loam or sandy loam. The chemical nature of soil may be acidic or alkaline. Temperature conditions affect both the plant and animal life. The variations of temperature are mainly caused by the inclination of the sun and soil conditions. Water is essential for all living things. Plants and animals that live in fresh water are called aquatic, while that live in saline water are called marine. Transfer of life from aquatic to marine environments or vice versa may be fetal. Atmospheric conditions and topography play vital role on variation of life. It may be noted that certain life exists in all environmental conditions. As environment changes, change in life is obvious and this process is going on ever since the formation of life on earth. However it is felt that during last century man had unreasonably interfered with the nature by his activities and damaging it. In this regard the study of environmental pollution and its effects are of great significance.

The main factors of environmental pollution constitutes (i) Air pollution (ii) Water pollution (ii) Solid waste, (iv) Sound pollution (v) Radiation hazards and (vi) Toxic or hazardous substances.

We have already studied all these except sound pollution which will be discussed presently.

Noise Pollution

Sound is a form of energy which is emitted by a vibrating body. Human ear is sensible to sound waves in the frequency range (16Hz to 20 KH_z. In this range sound waves cause sensation on the tympanic membrane (ear drum). Sound waves with frequencies less than 16Hz are called infrasonic and more than 20 KHz are called ultrasonic. Ear's response is nearly proportional to the logarithm of sound intensity or sound (accoustic) pressure which are given below.

$$\text{Sound intensity level (SIL)} = B = 10 \log \frac{I}{I_o}$$

where

I = Intensity

I_o = reference intensity = 10^{-12} watts/m^2

log = Common logarithm (base 10)

$$\text{Sound pressure level (SPL)} = B = 20 \log \frac{P_{eff}}{P_o}$$

where

P_{eff} – Pressure or effective acoustic pressure produced by sound waves

P_o = Standard/reference air pressure = 2×10^{-5} N/m^2

B is expressed in units of decibels (dB) which is one-tenth of a Bell (unit). The unit Bell was named in honour of Alexandar Graham Bell.

If $\qquad\qquad I = I_o = (10^{-12} \text{ w/m}^2)$

$$0 \text{ dB} = 10 \log \frac{I_o}{I_o}, \text{ this is threshold of hearing}$$

less than zero dB is not audible.

If $\qquad\qquad P_{eff} = Po \; (= 2 \times 10^{-5} \text{ N/m}^2)$

$$0 \text{ dB} = 20 \log \left(\frac{P_o}{P_o} \right)$$

If $I = 100\,I_o$, then SIL $= 10 \log \dfrac{100\,I_o}{I_o} = 20$ dB.

If $P_{eff} = 10\,P_o$, then SPL $= 20 \log \left(\dfrac{10 P_o}{P_o}\right) = 20$ dB

when $\qquad\qquad I = 10^{12}\,I_o\ (= 1\ w/m^2)$

$$SIL = 10 \log \left(\dfrac{10^{12} I_o}{I_o}\right) = 120\ dB.$$

when $\qquad P_{eff} = 10^6\,P_o\ (=10^6 \times 2 \times 10^{-5}\ N/m^2)$

or $\qquad\quad P_{eff} = 20\ N/m^2$

$$SPL = 20 \log \left(\dfrac{10^6 P_o}{P_o}\right) = 120\ dB$$

120 dB is the threshold of pain. If it exceeds 120 dB the normal auditory perception becomes impossible and causes pain in the ear drum. Some normal noise levels are given below.

Table 31.1

Source of noise	dB	Intensity w/m^2	effective accoustic or rms pressure N/m^2
Rocket engine	180	10^6	2×10^4
Threshold of hearing	0	10^{-12}	
Rustling of leaves in breeze	10	10^{-11}	2×10^{-5}
Wishper or radio broad cast	20	10^{-10}	2×10^{-4}
Radio at home/living room	40	10^{-8}	2×10^{-3}
Conversation	60	10^{-6}	2×10^{-2}
Traffic/office with machine noise	80	10^{-4}	2×10^{-1}
Jet aircraft take off (600 m)	100	10^{-2}	$2 \times 10^0 = 2$
Threshold of pain	120	$10^0 = 1$	$2 \times 10 = 20$

Example

If B_1, B_2 are intensity levels in dB of sound intensities I_1, I_2 respectively, then

$$B_1 = 10 \log \dfrac{I_1}{I_o}, \quad B_2 = 10 \log \dfrac{I_2}{I_o}$$

Then
$$B_2 - B_1 = 10 \left(\log \frac{I_2}{I_o} - \log \frac{I_1}{I_o} \right)$$

$$= 10 \log \left(\frac{I_2}{I_o} \times \frac{I_o}{I_1} \right) = 10 \log \frac{I_2}{I_1}$$

If P_1, P_2 are accoustic pressures of two sounding waves.

then $B_1 = 20 \log \dfrac{P_1}{P_o}$, $B_2 = 20 \log \dfrac{P_2}{P_o}$

$$B_2 - B_1 = 20 \left[\log \frac{P_2}{P_o} - \log \frac{P_1}{P_o} \right]$$

$$= 20 \left[\log \frac{P_2}{P_o} \times \frac{P_o}{P_1} \right]$$

$$B_2 - B_1 = 20 \log \frac{P_2}{P_1}$$

Example

If $I_o = 10^{-12}$ w/m^2, then the intensity 10^{-6} w/m^2 has the intensity level in dB

is given by $B = 10 \log \dfrac{I}{I_0} = 10 \log \dfrac{10^{-6}}{10^{-12}} = 10 \left[\log 10^{-6} - \log 10^{-12} \right]$

$$= 10 \left[-6 + 12 \right] = 60 \text{ dB}$$

Threshold Hearing

Intensity of sound just sufficient to cause sensation of hearing is called Threshold of hearing. It depends on frequency of sound with minimum value 2×10^{-5} N/m^2 at frequencies 700-6000 Hz.

Hearing Loss

When there is a shift in threshold hearing–that is, sounds of louder and louder required to hear is called hearing loss.

Acoustic Trauma (more than 120 dB)

Sudden noise or blast or explosion causes severe damage to interior of the ear is called acoustic trauma. It depends on the severity of damage. Damage may be or may not be recovered. Permanent effect of an acoustic trauma may be determined only after six to eight months period.

Shift in Hearing Threshold

Continuous exposure to noise (more than 80 dB) causes shift in the hearing threshold, which is much milder than acoustic trauma. Threshold shift may be temporary and may recover as soon as the noise stops. However repeated or continuous exposure to noises may lead to permanent shift in hearing threshold.

Note

(i)   If noise intensity level is 100-120 dB, it is categorised Discomfort and more than 130 dB it is categorised Pain.

(ii) Hearing loss with age is called *presbycusis*.

(iii) According to US National Environmental Health Association; standard intensity level during day 50 dB and during night time level 40 dB

Loss of hearing risk criteria given in Table 31.2.

Table 31.2

Maximum Permissible noise during day (hours)	Noise level dBA, Slow response
8	90
6	92
4	95
3	97
2	100
1.5	102
1.0	105
0.5	110
0.25 or less	115

Effects of Noise

Noise is a kind of pollution. It effects man and animals in a variety of ways. Supersonic aircrafts sonic booms cause cracks in buildings, may beak windows, and even rock falls. Human physiological effects of noise are (i) Disturbed or loss of sleep, irritability, anxiety, nervousness, neuro muscular tension. Small blood vessels of the body constrict which causes decrease of blood flow. Dialation of the pupils, tensing of the voluntary and involuntary muscles, increase in diastolic blood pressure and heartbeat rate. *Impulsive* noises cause more damage to hearing than continuous noise. Noise interferes in speech and work.

Movement of Energy in Ecosystem

For all living matter the sun is the main source of energy. Plants derive their food through photosynthesis process, whose equation is

$$6CO_2 \text{ (carbondioxide)} + 6H_2O \text{ (water)} + \text{Sunlight energy} \rightarrow$$

$$C_6H_{12}O_6 \text{ (Sugar or glucose)} + 6O_2 \text{ (Oxygen)}$$

In photosynthesis the plant organelles called chloroplasts absorb solar energy, break CO_2 into C and O_2. Subsequently C joins with hydrogen and oxygen of water to form sugar and free oxygen is liberated.

In respiration the sugar produced by plants is broken into energy by plant organells which is used for its growth, repair and reproduction. in a sense it is reverse process of photosynthesis. The respiration equation may be written as

$$C_6 H_{12} O_6 \text{ (sugar)} + 6O_2 \rightarrow 6CO_2 + 6H_2O + \text{energy}$$

Note : Respiration referred here is not inhaling oxygen and exhaling carbon dioxide, which is body respiration.

Green plants and some species of bacteria which use sun's energy for production of their food (through photosynthesis) are called autotrophs while other forms of life which depend on autotrophs for their food (lifes energy) are called hetrotrophs. Thus green plants are primary producers. Animals which depend on plants for their food are called herbivores, while those depend on animals for their food are called Carnivores or flesh eaters. Ominivorses feed on both plants and animals.

Food Chain

Energy flow from autotrophs or green plants through consumer organisms in each trophic level is called food chain. Interlinked or complex food chains (in which one population feeds on a number of other population) are called food webs. Through food, energy uses move from one organism to another or food is the mechanism by which energy is moved. Thus all green plants (energy producers) are categorised first trophic level. All herbivores categorised as second trophic level. Animals that feed on herbivores are third trophic level, animals that feed consumers of the trophic level just 3 and 4 are categorised as trophic level four and five respectively.

Food webs are of two types. (i) grazing, (ii) detritus, (wormdown matter). In grazing food web energy and minerals move from green plants to herbivores and to carnivores. Phytoplankton forms the primary grazing food web in a aquatic life. Phytoplankton are categorized by zooplankton (small floating animals). Zooplankton in turn are food for small fish and filter feeders.

Productivity

Formation of biomass (weight of living organism) by the use of sun's energy is called productivity. Conversion of solar energy by green plants (by photosynthesis process) is called primary productivity. The highest productivity occurs in some esturies, springs coral reefs and terrestrial communities where the order is $10 - 25 \times 10^3$ k cal/m^2/year. In summary three things can happen to energy assimilated at each trophic level.

1. It can be used for respiration of organisms on that trophic level and lost as beak.
2. It can become part of the detritus food chain either when the organisms of that level die or decay or as it passes through the bodies of other animals without being assimilated.
3. It can be passed on to the next trophic level when animals are consumed and assimilated.

During transfer of energy from one trophic level to the next, 80 to 90% of the energy is lost through respiration or decay leaving only 10 to 20%. Thus energy available at each trophic level is shown as decreasing pyramids. Energy however is basic in examination of trophic levels, for it provides the power necessary to sustain life.

Sun's energy provides the fuel for the operation of food chain. During photosynthesis this heat is lost. At each trophic level (T_1, T_2 and T_3) fuel is lost because of respiration (R). Some energy available to the next trophic level (E_A) is removed by decomposers (D) and some energy taken in by the next trophic level is not used (U).

Inter Relation of Grazing and Detritus Food Chains

Producers and consumers are part of the grazing food chain, while scavengers and decomposers are part of the detritus food chain. Producers are green plants. Primary consumers are herbivores; secondary and tertiary consumers consists of carnivores like lions, human beings, sharks and some mosquitoes. Scavengers such as vultures and crabs eat the remains of other organisms. Macro decomposers are primarily earthworms and insect larvae, and micro-decomposers are bacteria and fungi.

Primary Productivity (amount/source)

(i) $\angle\, 0.5 \times 10^3$ k cal/m^2/year, in deserts
(ii) $0.5 - 3.0 \times 10^3$ k cal /m^2/year, in grass lands, deep lakes, mountain forests, some agriculture.
(iii) $3{-}10 \times 10^3$ k cal/m^2/year, in moist forests and secondary communities, shallow lakes; moist grass lands; moist agriculture,
(iv) $10{-}25 \times 10^3$ k cal/m^2/year in some esturies; springs, coral reefs; terrestrial communities.

Biomes

A large recongnisable communities in different parts of the world are called biomes. Biomes are the biological expressions of the interactions of organisms with the physical factors in different regions of the world. Similar environmental conditions create similar biomes in different parts of the world.

Man can live almost everywhere except in a few areas with extreme environmental conditions. People are an important part of the ecosystem besides other living things such as plants and animals life. Non-living substances such as air, water, soil and minerals form an another important segment of the ecosystem. Both animates and inanimates (living and non-living objects) closely related to each other via energy flow and mineral cycles. Nothing exists in vacuum. All organisms depend on the physical surroundings and on the other life.

Man is the highest form of life in an environment but he is not free from the other life and environment. Man has the ability to destroy an environment or burn a whole forest and can cause severe change in the ecosystem, yet he is very closely dependent on the natural system such as atmospheric oxygen for breathing, animal and plant kingdom for his food.

Role of Atmospheric Pressure

Atmospheric pressure measurement and influence on life is based on the air pressure at sea level. At any altitude, all life are exposed to the same pressure. Pressure decreases exponentially with increase in altitude. Animal feel difficulty in breathing with increase in altitude. Similarly when we move into ocean depths pressure increases. Rapid change in pressure causes severe effects. When pressure falls as in the case of Tornadoes birds loose their plumes, human beings suffer several impacts on body. When pressure increases rapidly eardrums can burst. Thus rapid atmospheric pressure change indicate impending danger of severe weather such as cyclones, hurricanes, tornadoes. and alerts animals.

Evolution

Some members of a population develop adaptation or characteristics that makes them better siuted to their environments. Human ancestors, small rodentlike creatures known as prosmians adopted to the extensive forests. Adopted individuals to the environment produce more offsprings and contribute more genetic materials to the future generations. Random genetic forms appear regularly but the survival of the new individual depends on the adoptation to the environment. The dominance of a new genetic form as a consequence of environmental change is called naturalisation. A change in the genetic make up of a population over a time through natural selection results in evolution. Evolution takes place over hundreds or millions of years, but micro-evolution can be a very rapid process. For example housefly population is sensitive to DDT and die. The few flies unaffected by DDT reproduce rapidly and in a period of few years the new house fly population will be immune to DDT.

Questions

1. Define an environment state the factors which controls environment.

2. Write briefly how energy moves in ecosystem.

3. Write short note on : (i) food chain (ii) productivity (iii) food webs (iv) biomes (v) evolution.

CHAPTER 32

Wind Energy

Introduction

Wind Energy has been in use in some form or the other as a source of power from the dawn of civilization. This is an ideal example of renewable natural energy source without causing any pollution and interfering with nature. As early as 5000 BC, Egyptians used wind energy to drive boats and ships across the river Nile. Several centuries BC, Chinese developed wind mills for pumping water. In Persia, in 200-100 BC, vertical axis wind mill developed with woven reed sails and used for grinding grain. With the passage of time wind energy use was spread to Mediterranean region and from there it spread over Europe in 11th century, mainly by merchants. By the beginning of 20th century, the Dutch perfected the wind energy machines and then spread all over the world. However at that time fossil-fueled steam engined ships came into existence and slowly took over the wind driven wooden sailing ships because of improved technology. Similar replacement observed in America. Thousands of American farmers who had steel-towered wind mills for pumping water were replaced by rural electrification by fossil fuels. Present day research and development programmes aim at producing electricity (of the order 3 MW or more) by using wind mills. For this purpose wind speeds of the order 6 mps or more required, which is a restriction in selecting wind mill site. According to WMO estimates the world wide wind energy potential of generation of electricity is about 20 million MW.

Wind Generated Electricity

Towards the end of 19th century wind mills were the principal sources of power in agriculture in Denmark. The power generated was of the order 200 MW. In about 1890 -1908, Prof. P. La Cour errected a wind mill at Ashov (Denmark) and poineeringly generated electricity. The wind will had four blades of diameter 23 meters, errected on a steel tower of height 24.4 m. This wind mill was used to drive two 9 KW D.C. generators. This was the first recorded wind generated electricity. By 1944 wind generated electricity reached the peak with about 88 wind mills. This electricity was converted to AC in 1955, when it generated about 2000 kwh/per year. In 1941, the world's largest wind mill (Smith-Putnam) was errected at Grandpa's Knob in central Vermont (America). This wind mill had two blades of diameter about 53.5 m, errected on a tower of 33.54 m high. It generated electricity about 1.25 MW. However it failed after about 18 months due to breakage of main bearing, which could not be replaced. Technical problems of converting wind energy into electricity gradually overcome and at present any country with suitable wind climate can establish wind power electricity.

In 1931, the Russians errected their first wind mill to generate electricity at Yalta near the Black Sea. It had three blades of diameter 30.5m, mounted on a tower of 30.5 m high but was provided with an inclined strut to bear the high wind thrust. It was connected to an A.C generator and used as a supplementary power source. The site had annual mean wind speed of 15 kmph and its annual output was 279000 kwh.

Wind Energy Estimates

Air in motion is called wind. It has mass and its source of energy is the sun. The mass of the atmosphere is about 5.6×10^{18} kg. Horizontal wind speed (V_h) in general is much greater than the vertical wind speed (V_v). Vh>>Vv. The total energy of the atmosphere is present as potential energy (PE) and kinetic energy (KE). According to one estimate the total atmospheric PE = 2242.3×10^6 J/m^2 and KE = 1153.4×10^3 J/m^2. This shows that the average amount

of atmospheric KE = $\dfrac{1153.4 \times 10^3}{2242.3 \times 10^6} \simeq \dfrac{1}{2000}$ of the PE. The winds are the

result of conversion of atmospheric PE into KE. This conversion is mainly carried out by the work of atmospheric pressure force.

Atmospheric engine is mainly driven by insolation. We have studied that there is an excess of insolation as against the out going terrestrial radiation in equatorial region (equator to 38 °N or S). Where as there is a net loss of radiation from the ground between lat 38 and poles. As a result there is a transfer of energy from equator to poles at the surface level of the earth. This transfer of energy is effected by the atmospheric winds and ocean currents.

It is estimated that ocean currents transfer about 30% of excess heat from equator to pole wards and the remaining is carried out by latent heat of vaporisation (cyclones, thunderstorms or convective cells), that is through different scales of weather and winds.

$$\text{The KE of a moving air stream of unit mass} = \frac{1}{2}V^2 \qquad .....(32.1)$$

where V is the speed of air.

The mass of air that passes through cross section area A with a speed V, ρ density in unit time $\rho\,A\,V$ $\qquad .....(32.2)$

($\because$ mass = volume $\times$ density = AV $\times \rho$)

$\therefore$ The wind power (P) of a air stream that passes through a cross section area A, speed V, density ρ in unit time is equal to the product of (32.1) and (32.2).

$$\text{i.e.,} \qquad P = \frac{1}{2}V^2 \times \rho\,A\,V = \frac{1}{2}\rho\,A\,V^3 \qquad .....(23.3)$$

Let A be the circular area swept by a rotor of diameter D

$$\text{then} \qquad A = \pi\,\frac{D^2}{4} \qquad .....(23.4)$$

substituting 32.4 in 32.3 we have

$$P = \frac{1}{2}\rho.\,\pi\,\frac{D^2}{4}.\,V^3 = \frac{\pi}{8}\rho\,D^2\,V^3 \qquad .....(23.5)$$

Power available (P_A) conventionally expressed as

$$P_A = K_r\,D^2\,V^3 \qquad .....(23.6)$$

Where K_r = Constant, depends on wind dynamics and efficiency of rotor.

Table 32.1 The empirical value of contant K_r is given below.

Unit Power	Unit of area	Unit of wind velocity or speed	Value of Kr
K W (Kilowatts)	m^2	mps	0.00064
K W	m^2	Kmph	0.000014

In 1927, a German engineer A.Betz of Gottingen showed that maximum energy that can be extracted is 0.59259 P. This efficiency can be achieved by suitable blade design. In practice this factor may not exceed by 0.4

$$\text{i.e.,} \qquad P \simeq 0.4\,KAV^3$$

Taking air density $\rho = 1.201 \, Kg/m^3$ at normal atmospheric pressure 1000 hPa, and temperature $T = 290 \, °K$. and assuming a rotor conversion efficiency of 75%, $K_r \simeq = \dfrac{\pi}{8} \times \dfrac{1.201 \times 0.593 \times 0.75}{1000}$

$$K_r \simeq 0.0002$$

The effect of the height of the wind mill tower on the performance can be significant. Empirical power law of indices have established which will be discussed subsequently.

The actual mechanical or electrical output would be less than $0.0002 \, D^2 \, V^3$. The energy produced per annum by a wind mill is given by

$$Ey = K_r \, D^2 \, V^3 \times K_s \, t \, (KWh) \qquad ......(32.7)$$

where
$\qquad t =$ average number of hours in a year

$\qquad K_s =$ Semi empirical factor associated with the statistical nature of wind energy recovery.

The value of $K_s = 2.06$, was suggested by Pontin (1975) after computer analysis.

$$\text{Thus } K_r \, K_s = 0.0002 \times 2.06 \simeq 0.0004.$$

$$Ey = \sum Em,$$

where
$\qquad$ Em monthly power available

$$Em = K_r \, K_s \, D^2 \, V^3 \times t \quad Kwh \qquad .....(32.8)$$

$\qquad t =$ no. of hours in a month.

G. Hellman derived an empirical formula relating wind speed (V_z) at an altitude Z and wind speed (V_{10}) at 10 m agl (above ground level) is given below.

$$V_z = V_{10} \, [\, 0.2337 + 0.656 \log_{10} (Z + 4.75)] \qquad .....(32.9)$$

The above formula was further modified by N . Carruthers as

$$V_z = K \, Z^{0.17},$$

where K is proportionality contant

Much work done by using formula

$$V_z = V_o \, Z^{\alpha}$$

where
$\qquad V_o =$ wind speed at standard height

$\qquad \alpha =$ exponent changes with height, temperature, time, nature of ground

From the last formula we have

$$\frac{V_2}{V_1} = \left(\frac{Z_2}{Z_1}\right)^\propto$$

.....(32.10)

In this $\propto$ is called power law index.

where V_2 = mean wind speed at height Z_2

V_1 = mean wind speed at height Z_1

Taking logarithms of (32.10) we have

$$\ln V_2 - \ln V_1 = \propto [\ln Z_2 - \ln Z_1]$$

or $$\propto = \frac{\ln V_2 - \ln V_1}{\ln Z_2 - \ln Z_1}$$

.....(32.11)

Power Law or Wind Field in Urban Areas

Devenport gave a similar empirical formula called power law relating surface wind ($\bar{u}$) and gradient wind (V_g) as follows.

$$\bar{u} = Vg \left(\frac{Z}{Z_G}\right)^\propto$$

.....(32.12)

where $\bar{u}$ = surface mean wind speed

V_g = gradient wind speed

Z = height of surface wind measurement

Z_G = height of gradient wind measurement

Devenports values of $\propto$, Z_G in different environments are given below in Table 32.2.

Table 32.2

Environment	$\propto$	Z_G meters
Flat open country	0.16	270
Suburban area	0.28	390
Urban centres	0.40	420

$\propto$ - changes with pressure, temperature, height, time and nature of ground. $\propto$ is determined empirically from the observational data.

The power law $\dfrac{V_2}{V_1} = \left(\dfrac{Z_2}{Z_1}\right)^\propto$ is valid upto 100-150 m agl. This law is used to estimate mean wind speed at any height (Z_2), if the mean wind speed at another height (Z_1) is known.

Density

Mass per unit volume is called density. Mathematically density of a body is defined as the limit of the ratio of the mass Δm of an element of the body to its volume Δv as $\Delta v \to o$.

i.e.,
$$\rho = \underset{\Delta v \to 0}{Lt} \frac{\Delta m}{\Delta v} = \frac{dm}{dv}$$

or
$$dm = \rho \, dv.$$

Integrating we have $m = \int_0^v \rho \, dv$

The units of density : $1 kg/m^3$ in SI units

and $1 \, gm/cm^3$ CGS system.

Typical values of densities at room temperature are given below

Mercury $13.6 \, gm/cm^3$, water $1.00 \, gm/cm^3$, sea water $1.03 \, gm/cm^3$, Dry air at $0 \, °C$ and 1000 hPa pressure has density $1.277 \times 10^{-3} \, gm/cm^3$ or $1.277 \, kg/m^3$

The density of air changes with height and atmospheric conditions (pressure, temperature, RH etc).

$$\text{For dry air } (\rho \, gm/m^3) = \frac{3 \, 48.8 \times \text{Atmospheric pressure in hPa}}{\text{Temperature (in } °K)}$$

Put $T = 273 \, °K$, $P = 1000$ h Pa we have

$$\rho \, (gm/m^3) = \frac{348.8 \times 1000}{273} \simeq 1277 \, gm/m^3$$

or
$$\rho = 1.277 \, kg/m^3$$

The density of air changes in space and time and at a given location with temperature, pressure, humidity etc. We can find the changes in ambient air density due to atmospheric changes using ideal gas equation.

$$\rho = \frac{P}{R \, T_v}$$

where
$$\rho = \text{density in } gm/cm^3$$

P = Pressure 10^3 hPa

p = atmospheric pressure dynes/cm^2

T_v = Virtual temperature of air in $^\circ$K

R = gas constant,

for dry air   = 2.8703×10^6 ergs/gm $^\circ$K

$$T_\upsilon = \frac{T}{\left(1 - \dfrac{3e}{8p}\right)} \simeq T\left(1 + \frac{3e}{8p}\right)$$

Where e = partial pressure of water vapour

        T = temperature of moist air in $^\circ$K

For practical use, air density may be assumed to decrease by 1% of the surface value for every 100 m of height from ground level. According to Putnam, the density of air at 300 m asl 600 m asl and 3 km asl are 1300 gm/m^3, 1200 gm/m^3 and 900 gm/m^3 respectively.

DPTA

The principle of measurement of wind direction and velocity by Dines pressure tube anemometer (already discussed). The instrument devised by W.H. Dines mostly used as standard. DPTA provides a continuous record of wind direction and velocity. The records serve as a basis in the search for windy districts and for locating winds power sites.

Wind Energy Estimates :

Examples

Wind energy that can be produced by a wind mill in year is given by the formula

$$E_y = K_r \, K_s \, D^2 \, V^3 \, t \text{ (kwh/year)}$$

Where $K_r \simeq 0.00020$, constant; $K_s \simeq 2.06$ semi empirical factor, suggested by Pontin in 1975. The product $K_r K_s \simeq 0.0004$, t = number of hours in a year.

The following examples Table 32.3, 32.4, 32.5 give the wind energy estimates of a wind mill based on average wind speed for various months. Data extracted from IMD climatological tables (based on surface observation data at 0300 UTC and 1200 UTC for the period 1931 - 60).

Note : Em (Kwh/month)

Table 32.3 Station: Begumpet (Hyderabad-India) Lat 17° 27 N, Long 78° 28 E Height amsl 545 m.

Months	Jan	Feb	Mar	Apr	May	June	July	Aug	Sep	Oct	Nov	Dec	Annual kwh/year Ey
Total no. of hrs	744	672	744	720	744	720	744	744	720	744	720	744	
mean wind speed $\overline{V}$ = Kmph	8.1	8.9	9.6	10.9	12.4	23.8	22.1	18.3	12.6	8.9	8.0	7.4	
$\overline{V}$ = mps	2.25	2.47	2.67	30.3	3.44	6.61	6.14	5.08	3.50	2.47	2.22	2.05	
V^{-3}	11.39	15.07	19.03	27.82	40.71	288.8	231.47	131.10	42.87	15.07	10.94	8.61	
E_m for D = 3.65 m	46.5	55.6	77.7	109.9	166.3	1141.4	945.3	535.4	169.4	61.5	39.5	35.2	3343.7
D = 5 m	87.3	104.3	145.8	206.3	312.0	2141.7	1773.8	1004.6	317.9	115.5	81.1	66.0	6306.3
D = 7 m	171.1	204.4	285.8	404.4	611.5	4197.8	3476.6	1969.1	623.1	226.3	159.0	129.3	12458.4

Table 32.4 Station:Puri, India, Lat 19° 48'N, Long 85° 49'E, Height amsl 6 m.

Months	Jan	Feb	Mar	Apr	May	June	July	Aug	Sep	Oct	Nov	Dec	Annual kwh/year Ey
Total no. of hrs	744	672	744	720	744	720	744	744	720	744	720	744	
mean wind speed Kmph $\bar{V}$ = Kmph	11.9	15.9	20.5	24.0	26.2	23.2	23.3	19.7	15.9	12.3	10.2	0.5	
V = mps	3.31	4.42	5.69	6.67	7.28	6.44	6.47	5.47	4.42	3.42	2.83	0.14	
V^3	36.26	86.35	184.22	296.74	385.83	267.09	270.84	163.67	86.35	40.0	22.67	0.003	
E_m for D = 3.65 m	143.7	309.2	730.3	1138.3	1529.4	1024.6	1073.6	648.8	331.3	158.6	87.0	0.01	6895.8
D = 5 m	269.7	580.3	1370.6	2136.5	2870.6	1923.0	2015.0	1217.7	621.7	297.6	163.2	0.02	13465.9
D = 7 m	528.8	1137.3	2686.4	4187.6	5626.3	3769.2	3949.5	2386.7	1218.6	583.3	319.9	0.0	26393.6

Table 32.5. Dwaraka-India , Lat 22° 22'N, Long 69°.05 E, Height amsl 11m.

Months	Jan	Feb	Mar	Apr	May	June	July	Aug	Sep	Oct	Nov	Dec	Annual kwh/year Ey
Total no. of hrs	744	672	744	720	744	720	744	744	720	744	720	744	
meanwind speed $\overline{V}$ =Kmph	13.5	14.8	16.1	16.7	20.0	23.5	27.1	21.9	15.0	11.8	12.4	12.3	
$\overline{V}$ = mps	3.75	4.11	4.47	4.64	5.56	6.53	7.53	6.08	4.17	3.28	3.44	3.42	
V^3	52.73	69.43	89.31	99.90	171.88	278.44	426.96	224.76	72.51	35.29	40.71	40.0	
E_m for D = 3.65 m	215.3	256.1	364.7	394.7	701.8	1100.2	1743.3	918.5	286.5	144.1	160.9	163.3	6449.3
D = 5 m	404.1	480.6	684.4	7409	1317.1	20649	3271.9	723.9	537.7	270.4	301.9	306.5	12104.3
D = 7 m	492	941.9	1341.4	1452.1	2851.6	4047.2-	6412.9	3375.9	1054.0	530.1	591.7	600.8	23721.5

Questions

1. Write short note on the historical development of wind energy and electricity from it.

2. Write wind energy formula and explain each term in it.

3. Write wind power law developed by Davenport and explain each term in it. Where this formula is a applicable?

CHAPTER 33

Brief Description of the Physical Environement of India

India is enclosed between the latitudes 8° 4′ 28″ N and 37° 17′ 53″ N, longitudes 68° 7′ 33″ E and 97° 24′ 47″ E. The north-south distance is about 3200 km, east-west distance 3000 km, covers area about 327.6×10^6 hectares, coast line length about 5700 km, Land frontier about 12500 km with Pakistan, Afghanistan, China, Nepal, Bangladesh and Mynmar. Andaman, Nicobar group islands located in Bay of Bengal and Lakshadweep group islands in Arabian sea.

The main land may be divided into three regions – The Himalayan region, Indo-gangetic plain and the Deccan Peninsula.

The Himalayan Region

The Himalayan mountain ranges are spread in east-west direction with scattered plateaus and valleys which include Kashmir and Kulu basins with fertile land. The important passes across the Himalayas are Jelepla, Nathula (on the Indo-Tibet trade route) through Chumbi valley northeast of Darjeeling. The mountain ranges are spread 2400 Km long, 240-320 Km wide. The Garo, Khasi, Jaintia, Naga hills aligned east-west and the join of Lushai, Arakan hills running north-south.

The Indo-gangetic plain extends 2400 Km long and 240-320 Km wide formed by the combin basins of the Sindh, the Ganga and the Brahmaputra. It is very fertile and formed the cradle of Indian civilization.

Geologically India is divided into three regions -

(i)   Peninsular region which occupies 70% of the land area includes north-east area of Shillong plateau (in Assam), Kutch-Kathiawar in Gujarat region,

(ii)   Extra peninsular region, which includes the Himalayas, Kashmir, Nagaland, Manipur, Mizoram, Tripura, Bay Islands (area = about 15% of land area),

(iii) Indo-Gangetic plain in between the above two (occupies the remaining 15% of land area). The peninsular India is the oldest area of the world (Gondawana) and has granite base which date backs to pre-cambrian rocks. This region is considered to be stable land mass of great rigidity. The Deccan Trap is made of lava flows of the Cretaceous-Teritiary period.

The extra peninsular region is made of weak and flexible portion of the earth's crust which was folded, faulted and over-thrust. The Himalayan region rose from the Tethyan sea. The steepness and the deep Himalayan gorges formed by the upheaval and folding of the range. The Himalayan ranges contain lot of marine strata of Cambrian Tertiary period.

Gangetic plains contain alluvium of recent age received from the Himalays and rivers flowing from the central plateau.

Rivers of India

There are four systems of rivers in India. They are : (a) The Himalayan, (b) The Deccan Plateau, (c) The Central and (d) Inland drainage basin.

The Himalayan rivers (Ganga, Yamuna, Brahmaputra) are perennial are fed by rainfall and snow melt and flood prone. The Ganga is the largest basin, has five tributaries - the Yamuna, the Kali, the Ghagra, the Gandak and the Kosi. They originate in the Himalayas and fall into Bay of Bengal. However the three tributaries- the Chambal, the Betwa, the Sone rise in the Vindhyas flow northward and join the Ganga or Jamuna. Brahmaputra has Teesta and Manas tributaries, while Indus (Sindu) has five tributaries Jhelum, Chenab, Ravi, Beas and Satluj, which falls into Arabian sea.

Narmada and Tapi run east-west direction in Vindhyas and fall into Arabian sea. Mahanadi flows in Chathisgarh, Orissa states and fall into Bay of Bengal. These rivers have fertile alluvial soil.

Godavari and Krishna rivers originate in western ghats, flowing eastward fall into Bay of Bengal. These two rivers cover most part of the Deccan plateau are fed by rain water. Together with their tributarias virtually dry up in summer. Streams in western Rajasthan are short lived, which fall into salt lakes like the Samber or lost in the sands. Only Luni flows into Rann of Kutch.

Soils

Soils constitutes the topmost layer of the earth's crust. Soils are formed by the mechanical and chemical weathering of rocks and partly by the decomposition of organic matter, under the action of weather and climate. More than hundred years are required to form a few centimeter of top soil. A vertical section of soil presents a succession of natural layers, which are formed as a consequence of leaching and transfer of material (called profile). Soils differ in colour, texture, reaction, consistency and porosity. A large portion of clay and organic matter increases the water holding capacity of a soil. The important soil varities are described below.

(i) **Alluvial soil :**   These are generally found along rivers and lowlying tracts. They show stratification of sandy, clay and loam.

(ii) **Sandy soil :** These are found along sea coast and along major river banks. They vary from sandy loam to pure sand. They are highly porous, deficient in nitrogen, phosphorus, potassium and in some cases lime.

(iii) **Desert soil :** These are found in Rajasthan and adjoining parts of Gujarat. They are poor in organic matter nutrients but they have sufficient minerals.

(iv) **Peaty soils :** Peat means decomposed vegetable matter. Peaty soils are found in humid regions and contain large organic matter. They are black in colour, contain high clay, nitrogen and sufficient potash but low phosphorus.

(v) **Saline and alkaline soils :** These are found in drier parts of north-west India, over Deccan plateau. In summer they bring dissolved salts to the surface by capillary action which form a white crust. These soils are unproductive for tree growth.

(vi) **Laterite :** Laterite soil is a mixture of hydrated oxides of Alluminium, Iron with small quantities of Manganese oxide, Titanium etc., These are found in regions of heavy rainfall zones and moist climate. They are poor in nutirents, phosphorus, potassium and calcium and dificient in nitrogen. The pH ranges 4.5 to 5.5 and base exchange capacity is low.

(vii) **Red soil :** They are deficient in organic matter and poor in plant nutrients. The presence of ferric oxide gives the soil red colouration. Base status low with poor exchange capacity.

(viii) **Black soil :** These soils are called regdi or black cotton soils, developed from basaltic rocks. The black colouration is due to the presence of iron salts. They are loamy to clayey, expand on wetting and shrink on drying. They are poor in organic matter, phosphorus nitrogen but have sufficient calcium. They are found over most part of the Deccan plateau.

(ix) Forest/mountain meadow soils : They are rich in nitrogen and contain organic matter in surface layer. They have variable base status depending on the degree of washing by percolation (leaching).

(x) Skeletal soils : They are found in hills of drier regions of the peninsula and in the Himalyas. These are isolated and have local importance.

Soils under different forest vegetation differ considerably in their physical and chemical characteristics.

Indian Land Use (1987-88) Classification of Geographical Area

(i)   Total area 328.72×10^6 hectares.

(ii)   Forest area 64.06×10^6 hectares   (19.4% of total area)

(iii) Agricultural/cultivated area 152.27×10^6 hectares (46.3% of total), uncultivate land area 42.18×10^6 hectares (12.9% of total).

(iv) Non-agricultural use area 26.76×10^6 hectares (8.1% of total)

(v)   Barron and non-cultivable land 43.45×10^6 hectares (13.3% of total).

(vi) According to statement of forest report 1991, the total area 3287263 km^2, forest area 770078 km^2 (23.4% of total).

Distribution of Forest Types in India

Based on vegetation, geographical location and climatic features forests in India classified into 16 types, which are briefly described with a typical meteorological station.

$$L \text{ stands for Lakh} = 10^5$$
$$M \text{ stands for Million} = 10^6$$

1. Tropical wet evergreen forest :   Found along the wetern side of the western-ghats, south-west strip from upper Assam to Cachar and in the Andamans. Ex : Mangalore, Silchar

     Extent of area = 5.12 M ha.(8% of total)

2. Tropical Semi-evergreen forest : Found on western coast of Assam, lower slopes of the eastern Himalayas, Orissa and Andamans. Ex : Karwar, Jalparguri

     Extent of area = 2.62 M ha.(4.1% of total)

3. Tropical moist deciduous forest : They contain predominant deciduous trees. Found in Andamans, moist parts of UP, MP, Gujarat, Maharashtra, Karnataka and Kerala. Ex : Portblair, Chanda, Dehra Dun

     Extent of area = 23.68 M ha.(37% of total)

4. **Littoral and swampy forest :**   Evergreen species of varying density. They are found all along the coast and swamp forests in the deltaic area of the bigger rivers. Ex : Kakinada, Calcutta

    Extent of area = 3.84 L ha.(0.6% of total)

5. **Tropical dry deciduous forests :** Upper canopy closed. They are found in an irregular wide strip running north-south from the foot of the Himalayas to Kanyakum except in Rajasthan, Western ghats and Bengal. Ex : Nagpur, Kadapa, Roorkee

    Extent of area = 18.3 M ha.(28.6% of total)

6. **Tropical thorn forest :** Thorny, Leguminous species are dominant. They are fond in South Punjab, Rajasthan, Upper gangetic plains, the Deccan plateau and the lower peninsular India. Ex : Solapur, Jaipur

    Extent of area = 16.64 L ha.(0.2% of total)

7. **Tropicaldry evergreen forest :** Low forest (height about 12 m) with full canopy of ever green trees. They are found in coastal Karnataka.

    Ex : Machilipatnam.

    Extent of area = 1.28 L ha.(0.2% of total)

8. **Sub-tropical broad leaved Hill forest :**Thick and green forests with evergreen species. They are found in lower Himalayan slopes, in Bengal and Asam (and other hill ranges), Khasi, Nilgiri and Mahabaleshwar. Ex : Mercara, Pachmarhi, Kalimpong

    Extent of area = 2.56 L ha.(0.4% of total)

9. **Sub-tropical Pine Forest :** Found in north-west Himalayas. Ex : Ranikhet

    Extent of area = 42.24 L ha.(6.6% of total)

10. **Sub-tropical dry evergreen forest :** Low scrub forest with evergreen trees. Found in Bhabar, the Siwaliks and the Western Himalayas. Ex : Jammu

    Extent of area = 16 L ha.(2.5% of total)

11. **Montane wet temperature forest :** Closed evergreen forest. Found in Kerala, Tamilnadu, eastern Himalayas in Bengal, Assam, Sikkim and Nagaland. Ex : Ooty, Darjeeling

    Exten of area = 23.04 L ha.(3.6% of total)

12. **Himalayan moist temperate forest :**Coniferous forest of evergreen with mosses and ferns (grow abandantly) on trees. They are found in Kashmir, Himachal Pradesh, Punjab, UP, Darjeeling and Sikkim. Ex : Chakrata

    Extent of area = 21.76 L ha.(3.4% of total)

13. **Himalayan dry temperate forest :** Mostly coniferous forest found in Ladakh, Lahaul, Chamba, Kinnaur Garhwal and Sikkim. Ex : Kolba.

   Extent of area = 1.92 L ha.(0.3% of total)

14. **Sub-alpine forest :** Dense growth of small trees and large shrubs. They are found in the upper limit of forest in the Himalayas. Ex : Dras

15. **Moist alpine scrub :** Low evergreen dense growth of Alpine shrubs. Mosses and ferns on the ground with these shrubs and flowering herbs. These are found along the entire length of the Himalayas above 3000 m height to the snow line. Ex : Leh.

   Extent of area = 13.44 L ha.(2.1% of total)

16. **Dry alpine scrub :** The uppermost limit of scrub xerophytic, dwarf shrubes. They are found in dry zones of Himalaya above 3500 m.

World's Tallest Trees

According to National Geographic Magazine (1961) the tallest trees are found in California and in Southern Oregon USA. 97 m Douglas Fir at Ryderwood in Washington, a 97 m Eucalyptus regnans in the Styx River valley of Tasmania

World's Oldest Living Trees

Three Bristlecone pines which are about 4000 years old, and Methuselh is about 4600 years old which are found in California, Nevada, Colarado, New Mexico. The largest volume of timber got from a tree in California was about 1412 m^3 and maximum weight of a Redwood tree estimated to be over 6000 tonnes (from Sequoia National park). A single Banyan tree (found in India and Africa) with its adventitious roots covers largest area (may shelter about thousand people). Banyan tree considered immortal and called Aksaya Vat. [From Dendro climatology i.e., by counting the annual rings of a cross section of a tree, one can estimate age of the tree and ring width provides a clue for the estimation climate, particularly dry, wet years the life period of the tree in its region].

Some protected Indian trees recorded maximum height and girth are given in Table 33.1 .

Table 33.1

Species	Location	Height (meters)	Girth (cm)
Pinus roxburghii	UP	65.5	411
Pterocarpus dalbergiodes	Andamans	51.5	602
Dipcerocarpas gracilis	Andamans	43.9	724
Shorea robusta	UP	41.4	734
Eucalyptus globulus	Kodaikanal	36.6	960

(Ref. Indian forester 1977)

The maximum recorded height and girth of trees in India are given in Table 33.2.

Table 33.2

Species	Location	Height (meters)	Girt (cm)
Chinar	Bijbehara, Kashmir	–	2150
Chir Pine	Garhwal, U.P	65.5	411
Deodar	Tons Divison, U.P	64.9	808
Spruce	Garhwal, U.P	64.0	732
Blue gum	Niligiri, Tamilnadu	64.0	541
Paduak	North Andamans	51.5	602
Sal	Haldwani, U.P	51.2	264
Teak	S. Chanda, Maarashtra	43.0	396
Laurel	Bilaspur, M.P	43.3	561
Mango	Malda, West Bengal	36.6	686

The worlds tallest trees found in California height 110.34 to 105.7 m.

Forest Products

These are classified as Major and Minor products. Major forest products consists of timber, small wood, fuel wood or charcoal. The estimated demand of raw materials of wood in 2000 AD were :

Industrial wood	$50 - 65 \times 10^6$ m³,	
Fuel wood	$320 - 350 \times 10^6$ m³,	
Printing paper	1.78×10^6 tonnes	
Writing paper	7.0×10^5 tonnes	(7 lakh tonnes)
Industrial paper	7.8×10^5 tonnes	(7.8 lakh tonnes)
News print	7.6×10^5 tonnes	(7.6 lakh tonnes)

Minor or non-wood forest products consists of Fibres and Flosses, Grasses, Bamboos/canes, Essential oils (includes those from grasses); Tans and Dyes, Gums, Resins and Oleoresins, Drugs, Spices, Poisons and Insecticides; Edible products, leaves and Animal, Mineral and Miscellaneous products.

Uses

Fibres are obtained from tissues of certain woody species, used for rope making.

1. Flosses obtained from certain fruits, used for stuffing pillows, mattresses etc.
2. *Grasses*, bamboos and canes used as foodder, for making paper.

Tall grass used for making chicks, stools, chairs etc, leaves twisted for making strings. Bamboos used for rafters, roofing basketry etc. Bamboo seed is eaten as grain, its fibre used for making paper. Bamboos are used for making bridges. Canes used as plaiting material, ropes, for furniture, sports goods, walking sticks etc.

Essential oils are used in soap making, cosmetics, parmaceutical preparations, confectionary, incense etc. Eucalyptus globulus oils used in medicine, Rusa grass, Khus and Sandal wood are most valuable in commercial field.

3. *Lemon grass* oil contains 80% citral which is used for manufacture of vitamin A. Also used in aromatic chemicals, perfumery, soap, cosmetics, making pain balms and disinfectants.

   Sandal wood used in valuable art carvings, distilling oil which is used in Indian medicine, perfumery, toilet soaps etc.

   Rusa (palmrosa) oil extracted from varmotia and certain other grass, which contains 90% geraniol - used in perfermery, cosmetics, as rosecent in soaps, flavouring tobacco.

   Khus oil distilled from the roots of vetiveria ziziniodes grass - used in perfumery (Itra of Khus).

4. **Tans and Dyes :** Tannins are secretion products of plant tissues. Tanning materials are got from Harra tree, Babul bark, Avaram bark and Wattle bark.

   Dyes are derived from Red sanders which bright red. Khair wood gives chocolate colour (is a preservative) used for dyeing canvas, fishing nets, mail bags, sail cloth etc. The flowers of palas (boiled in water) gives orange colour. Fruits of Mallotus Phillipensis and roots of Morinda tinctoria used as dyes.

5. **Gums, Resins and Oleoresins :** Gums and resins are oozed by plants (due to injury to the bark or wood). Resions when mixed with high percentage of essential oil is called oleoresins. The gum is used in textile trade, cosmetics, dentrifices, cigar and food industry.

6. **Drugs, Spices, Poisons and Insecticides :** They are derived from trees, shrubs, climbers, herbs, primitive plants, fruits, flowers, leaves, stems and roots.

Important Drugs are

Root and underground parts : Ipecac from Cephaelis ipecacuanha cultivated in Darjeeling, shillong and the Niligirs. It gives emetine which is effective for treating amoebic dysentery. Liquorice derived from glycyrrhiza glabra, is a tonic, expectorant, demulcent and mild laxative.

Sarpagandha are the roots Rauwolfia serpentina contains reseprine which is very effective in high blood pressure. Kuth or costus are the roots of saussaria lappa, used as tonic, stomachic, stimulant and spasmodic.

Bark Drugs

Quinine is the most important Indian vegetable drug. Quinine is extracted from the bark of Cinchona ledgerina cincona hybrida, effective in treatment of malaria.

Kurchi is the bark of Holarrhena antedysenterica, a shrub. It is a remedy against dysentery.

Wood Drugs

Alkaloidephedrine used in the treatment of bronchial diseases.

Leaf Drugs

Hemp is derived from cannabis sativa, grows wild in the foot hills of the Himalayas. It is a source for well-known narcotics - Ganja, Charas and Bhang. The flowering tops are Ganja, the resinous exudation from flowers and leaves is Charas, (the basis for Hashish). Bang is the drink made out of pounded Ganja.

Drugs from Fruits and Seeds

Kala-zira is a seed cultivated at attitudes 3 to 3.5 km asl. It is stomachic Carminative and Lactagogue, has strong flavour.

Chaulmogra is a common tree in Western Ghatas. Its seed oil is specific for Leprosy treatment.

Spices

Spices are aromatic vegetable products, used in cooking for flavour. The important spices of forest produce are :

(i)   *Galangal/Alpine galangal* is found in the Himalayas and western ghats. It has aromatic pernnial rhizomes which are used for treatment of rhematism and catarrhal infection.

(ii)   *Dalchini or Cinnamon* is a bark of a tree found in western ghats south of Konkan. It is stimulant, carminative. Used in candy, gums, incense and perfumes. Recent American research indicates it is very effective in the treatment of blood sugar (diabaties) and high blood pressure.

(iii)   *Choti Ilaychi and Badi Ilaychi* grown in ever green forests of South India and in East Himilayan forests respectively. The seeds are carminative and used as flavouring for sweet dishes and in cooking.

Poisons

Many forest plants such as strychnine, aconite Datura etc., are poisonus but in small regulated doses it acts as good medicine. Barks of various trees are used as fish poisons. Gunja seed paste inserted into the flesh of cattle kills them.

Insecticides

Several of Derris and Tephrosia contain rotenone, which is an insect poison.

Edible Forest Products

Several forest frutis, flowers and leaves, tuberous roots are eaten. A few important species of fruits are given below.

Aam (mango), Bel, Ber, Jamun, Kathal, Khirni, Phalsa, Sitaphal, Tendu etc.

Species of Kernals Eaten

Achar, Akhrot or Walnut, Cashewnut, Chilgoza, Kamal, Surghara etc.

Parts of Species used as Pickles and Vegetables

Amla, Amra, Anar or wild pomegranate, Imli or tamarind, Karaunda, Kokam, Munga (drumstick), Kachnar, Kaith (or wood apple)

There are several other tree products such mushrooms, palmyrah, mahuva, Bhilva etc., used for many purposes making liquors etc.

Management of State Forests

In India planning commisision, under five year plans, emphasized the need to rehablitate the depleted forests. It stressed for the creation of plantations of valuable species and conservation of wild life by establishing wild life sancturies. Further stressed to preserve natural environment and ecology for scenic beauty, sports, recreation, scientific study and conservation, preservation of biologically diverse floara and fauna. Wild life conservation include the protected areas to be linked with its corridors to maintain genetic continuity between artificially sub-sections of migrant wild life. Tribal people be put in protection of and development of forests, in raising trees and raw materials.

Wildlife

Wildlife connotes the plant and animal communities coexist in dynamic equilibrium or simply means the animals and birds of the wild or the entire animal and plant kindgdom of a locality. They are jointly called the eco-complex, which varies in mass and composition both in time and space. The variations depend on climatic and edaphic factors.

India has a rich heritage of wildlife beacause of traditional compassion for all life. Emperor Ashoka's edicts prohibited killing of animals and preservation of forests. Now the national animal of India is 'Tiger' and the national bird 'Peacock'. We have briefly learnt about the Indian flora and now we shall learn about the Indian fauna.

Based on climate and physiography India provides a wide range of habitat wich accounts for large variety of wildlife. The fossils of extinct animals in Sivalika range reveals that wildlife belongs to Tertiary period of geological times. There were elephants, hippopotamuses, rhinocerses, fourhorned ruminants (the Siva thereums), giraffies, large and pigmy horses, camels, wild oxen, buffaloes, bisons, deer, antelope, wild pigs. Fossil beds inicate the existence of chimpanzees, orang-utans baboons, langurs and macaques. The carnivores include a type of cheetah, sabre-toothed tigers, wolves, jackals, foxes, also civets, martens, ratels and otters porcupines, hares, bears etc.

The present day Indian fauna has about 500 species of mammals. The important is the elephant, one-horned rhinoceros, the Arna (the wild buffalo), the Gaur (the Indian bison). Important mammals in India are distributed in three zones – the Himalayas, the Sub-Himalayan Terai, and the peninsula.

The animals found in the first two zones are :

The Himalayan yak (the snow Ox) found in Ladakh. The Himalayan sheep are shapu (or Urial), bhoral (the blue sheep), nayan (huge sheep with curved horns); Goat-antelopes are Serow and Goral found in the eastern Himalayas. The Himalayan wild goats found in Kashmir are markhor, the ibex, the Himalayan thar, the beautiful kashmir stag, (counter part of European red stag) and Kastura (or the musk deer). Thamin is a pretty deer found in Manipur. The tigers, panthers, snow leopards, the brown and black bears, the red panda, the Indian otters, the beech marten, the yellow-throated marten, the Himalayan Weasel, Marmots, a few species of squirrels, bats, the elephant and the tiger. The one-horned rhinoceros, the hog deer, the clouded leopard are all the typical animals that are found in the Himalayan foot hills. The other main carnivores found are sloth bear, the fox, the wild dog, feline cats. The chinese pangolin is found in Assam besides elephants and Arna, deers-barasurgha, samber, the chital (spotted deer), the muntiac (barking deer), giant and flying squirrels, Indian porcupine and wild pigs.

Animals Found in Peninsular India

The elephants, gaur, chinkara (or the Indian gazelle), the black buck (Indian antelope), the Neelgai (blue bull), the mouse deer, the chowsinga (four horned antelope), the nilgiri thar, the Indian lion, cheetah (hunting leopard). All deer species of the Himalayan foothills except hog deer are also found.

Carnivores

Tiger, Panther, Sloth bear, Cats, Wild dogs, the fox, the jackal, the hyena, the mongoose, civet cat, Ratel. A few arna found in Bastar district of Chattisgarh, giant and flying squarrels, hares, the pangolin are also found.

Avifauna

The Indian bird life is rich. There are hundreds of species of aquatic, gallinaceous and arboreal birds. The chief aquatic birds are : storks, herons, egrets, cormorants and flamingoes, ducks. The waders and shore birds are : the snipes, ibises, cranes and lapwings. The main ground birds are : the great Indian Bustard, peafowl, jungle fowl, partridge and quoil.

The species of perchers and song birds are : Babblers, barbits, bulbuls, cuckoos, mynas, doves, parakeets, pigeons, rollers, beaters, flycatchers, hoopoes, drongos finches, finchalkars, wagtails, wagtails, warblers, orioler are important. pied Malabar hornbill is a rare arboreal bird. A variety of owls, eagles, kites, falcons, kestrel, the shikra are main birds of prey, the scavenging species of vultures and crows. The famous Siberian crane visits Bharatpur sanctury every year.

Reptiles

India has a large variety of snakes – poisonus ones are Cobra, Dhaman large snake. phythons of large size are found amongst rocks along streams. Crocodile (Mugger) and gharial are famous for their skins. Lizards include chemeleon, monitor lizard or varanus.

Fishes

A variety of fishes are found in ponds, marshes, streams, inland rivers, esturies and the sea. The important river species are : Mahseer, Chilwa, Olive Carp, Lesser Barils, Seetul, Batchwa, Murral, Eel, Carnatic carp.

Tanks fish : Rohu

Esturies fish : Bahmin, Surmai, Horse mackeral, Queen fish, Morwasa, Gobra, Rock perch, Banasia and Woli Herring.

Sea fish : Shark

The rainbow tront common in Himalayan rivers. During twentieth century and earlier there has been human and cattle population explosion. To meet their needs extreme pressures have been put on natural resources particularly on forests, pasture lands, forest lands to agriculture, grazing cattle. This resulted iin encroachment of forests and wildlife habitat and their destruction. Many species have become locally extinct, and our wild life slowly decimated. In order to improve the wildlife Government of India constituted Indian Board for Wildlife in 1952. Under the advisory of Board conservation and protection of wildlife, constitution of national parks, sancturies, zoological gardens and promoting public interest and awareness in conservation of wildlife in harmony with natural environment was taken up by the Central and State Governments on priority. The wildlife (protection) act 1972, was enacted by parliament in September 1972. In April 1973, Project Tiger was launched. Under this project Tiger reserve (area 400- 4000 sq. km) created under different types of habitat which are given below.

1. Manas (Assam), 2. Palamaru (Bihar), 3. Simlipal (Orissa), 4.Corbett Park (UP), 5. Ranthambor (Rajasthan), 6. Kanha (MP) 7. Melghat (Maharashtra), 8. Bandipur (Karnataka), 9. Sundarbans (West Bengal), 10. Sariska (Rajasthan), 11. Periyar (Kerala), 12. Dudhwa (UP)

In each tiger reserve sufficient area has been set apart as the core, where no forestry operations will be done, rigid protection against poaching, stopping grazing cattle, creating water bodies are important. This helped not only in wildlife protection but also wildlife tourism and recreation value. The important National Parks and Sancturies are given below.

Manas (Assam/north Kamrup) – for wild buffalo and rhinoceros.

Kazirnaga (West Bengal) – for rhinoceros, hog, deer, barasinga

Jaldapara (Jalpaiguri district) – for rhinoceros

Hazaribagh National Park (Bihar)

Palamau National Park (Bihar) – for gaur

Gir National Park (Gujrat) – for Indian Lion

Bandipur National Park (Karnataka) – for elephant, gaur

Ranganthattu – for bird sanctury

Dachigam National Park (Kashmir) – for Kashmir stag

Periyar National Park (Kerala) – for elephants and gaur

Kane National Park (M.P) – for Tiger, Barasinga, gaur, black buck etc.

Shivpuri National Park (M.P) – for Chinkara

Bandhavgarh National Park (M.P) – for tiger, sambar, neelgai

Sanjay National Park (M.P) – for tiger, deer etc.

Taroba National Park (Maharashtra) – for tiger, bear

Bharatpur (Keolaghana) (Rajasthan) – for bird sanctury

Sariska National Park (Rajasthan) – for tiger, panther

Ranthambor National Park (Rajasthan) – for sambar

Corbett National Park (UP) – for tiger, elephant, hog, deer, gharial

Rajaji National Park (UP) –

Dudhwa National Park (UP) – for tiger, elephant

Conservation of Forest Soil

Soil formation is a very slow process caused by the combined effect of integration of sub-soil and the action of micro-organisms. It takes about 100 years to form one centimeter layer soil, while it may take few hours or days to erode soil through the action of rain or wind.

Removal of soil, rock material by water, wind and gravity action is called soil erosion. Soil erosion principally caused by land misuse, faulty agricultural practices, inadvertent cutting of trees, uncontrolled cattle grazing, forest fires and adverse biotic factors. Erosion, raincut deep gullies (called ravines) cause soil loss. The reclamation of such lands may be achieved by treating them with growing grass along with diary farming, afforestation, banning of cattle grazing. It may be noted here that forest fires not only destroys standing timber but destroys the future forest by burning the seeds, humus of the forest floor and takes a toll of animal life. The common causes of forest fire are : incendiaries, debries of burners, such as matchsticks, remanents of burning cigars, automobiles, cloud lightning, burning of coal from camp fires, railroads, lumbering and frictional fire during summer.

A central soil conservation Board was set up in 1953 for development of soil and water conservation. As the result of five year plans by 1985-87, a total area of 2.5 million hectares had been treated with various soil conservation measures such as closure of cattle grazing, water disposing devices, contour trenching and bunding, gully plugging, errection of check dams and silt detention dams. During fifth five year plan land management and soil conservation integrated to the schemes : (i) Drought prone area programme, (ii) Flood prone area programme and (iii) Rural development programme.

Conservation of Natural Resources

This includes:

    (i)   the conservation of soil,

    (ii)   the conservation of water,

    (iii) the conservation of forest wealth and

    (iv) the conservation of wild life (which includes animals and plant species)

Morethan 60% of the world population is living in cities. The prosperity of a city depends on industries, farms, bussiness and the availability of water, while the prosperity of country side depenpds mainly on agriculture, soil, water resources and forest wealth. Natural disasters that mostly effect the cities are flash floods, earthquakes, heat waves/cold waves, cyclones, acute scarcity of water, while the country sides are affected by the above and in addition affected by severe floods. Agriculture is badly affected by floods and droughts and soil erosion. We have already learnt that conservation of soil and wild life are related to forest conservation.

Domestic Animals in India

In environmental study it is very opt to know something about domestic animals, poultry besides wild life. Keeping this in vew a brief sketch about domestic animals and poultry is given.

India has large domestic animal population (about 3700 lakhs) which includes buffaloes, sheep, goats, pigs, horses, mules, donkeys, camels and elephants. There are about 1600 lakh poultry birds but their contribution to the national economy is very little inspite of being an agricultural country. This is attriburted to poor climatic and environmental conditions, poor genetic breeds and non-nutritious cattle poultry feed. At the dawn of civilization cattle were domesticated and considered them as beasts of burden. Buffaloes are domesticated mainly in India, China, Thailand, Indonesia, Myanmar, Nepal, Bangladesh, Pakistan and Phillipines. The American buffalo in reality is a bison, which in European countries generally it is not seen. In India cow is worshipped as Gow Mata and particularly fed during Pitra Masa (Ancestrol month). The buffalo gives more milk (about three times) than the cow. In rural India even now bullocks are a source of power in agricultural operations. Cattle are the peasants valuable possession after land. They are used for plougling, harrowing, thrashing,harvesting, transporting agricultural produce and lifting water from agricultural wells. The estimated milk production in India in 2000 AD was about 820 lakh tonnes. According to 1982 livestock census there were about 2622.3 lakh head of cattle, of which buffaloes were about 697.8 lakhs. Cattle deserve gentle treatment with kindness. Regular feeding, grazing, milking pays better dividents in case of milch animals. Cattle must be protected from inhospitable weather conditions by parking them in pucca cattle sheds. Cattle urine and dung used as manure. Gobar gas prepared from cattle dung, which is used for cooking and domestic lightning.

According to FAO of UN, there were 11900 lakh sheep population over the world in 1991 and there were about 6630 lakh sheep existed in India. There were about 200 breeds of sheep over the world. India has three main types of sheep which belong to the temperate Himalayan region, the dry western region and the southern region. They provide us is milk, meat and wool. Sheep grow well on natural grases, herbs, and farm wastes.

Goat is considered as a poor man's cow. It adopts itself to all weather conditions, eats a variety of food which other animals do not touch. According to 1991 estimates, India produced about 20 lakh tonnes of goat milk a compared to 100 lakh tonnes of global production (about 20%). According to FAO estimates (1991) India produced 5860 lakh metric tonnes of goat and sheep flesh. Goat skins are used for making shoes, gloves and leather articles. Goat hair is used for making rope. The goat, nibbles the buds of trees and damages all varities of plants and hence considered a great enemy of vegetation and soil erosion. In 1991 the world population of goats was about 5740 lakh. of which 20% (about 1120 lakh) goats were in India. The averge milk ingradients of domestic animals are given in Table 33.3.

Table 33.3

	Name of animal	Protein%	Fat%	Sugar%	Minerals%	Water%
1.	Cow	3.39	3.68	4.94	0.74	87.27
2.	Buffalo	4.37	7.65	4.82	0.94	82.22
3.	Goat	3.76	4.07	4.64	0.85	86.88
4.	Sheep	5.15	6.18	4.17	0.93	83.57
5.	Mares	2.05	1.14	5.87	0.36	90.58
6.	Ass	1.37	1.85	6.19	0.47	90.12
7.	Camel	4.0	3.07	5.59	0.77	86.57
8.	Elephant	2.51	9.10	8.59	0.50	79.30
9.	Average	3.325	4.593	5.60	0.695	85.81

Pigs

Pigs are most productive breeders. Pork (pigs meat) has great demand in the manufacture of several products. Pork is a high protein food and compared to other meats it has higher energy content. In India pig rearing and pork industry is very backward. In America pig production is a great business while China leads in the world in hog production. According to 1991 FAO estimates the global pig production was 8590 lakh of which India had only 104.5 lakh pigs.

Horses, Donkey and Mules

Horse is one of the earliest domesticated animal. Horses, mules and donkeys are the first cousins. They are all single-toed mammals and their toes are enclosed in hoofs. They are highly useful to man in work, warfare and sports. In mountainus/hilly regions they are the only means of land transport. They are engaged in riding, racing and in polo games. According 1982 census India had 9 lakh horses and ponies.

Donkey was domesticated around 4000 BC in Mesopatamia and Egypt. Donkey is a hard working animlas and it works continuously even with poor food. It is used as a pack animal (with sure footedness) but it is known for its stupidity and obstinanacy. India estimated to had 10.2 lakh donkeys in 1982. The milk of an ass in considered as a substitute for human milk in India, Italy, Spain and in France. The poitou donkey is the largest in the world, while the Indian ass is the smallest. Mule is a cross breed of a male donkey and mare.

Mule has great capacity of endurance in adverse climate and food. It is the most favourite pack-animal in rugged hilly tracks. It is assumed that mule has interited its intelligence from the horse and sure–footedness from donkey. According 1982 census India had 1.31 lakh mules.

Camels

Camel is described as the ship of the desert. It is a very important beast of burden and transport in arid, sandy desert regions. It is called a mystery animal because of its ability to withstand to thirst and hunger. It can live on thorny shrubs in a desert. It chews the cud like cattle, sheep and goats. Its thick foot-pads help it to walk on hot sand and it can close its nostrils during snadstorms.

In India camels are used for riding, carrying loads, ploughing land, thrasing grains, driving carts, water-wheels and crushing (sugarcane, oil seeds) mills. Camel hair is used for making ropes, blankets and other warm garments. The flesh of camel is eaten in many countries. Its milk is a rich drink for humans. Its hide is used for making saddle. According to 1982 census there were 10.8 lakh camels in India. They are widely used in Rajasthan.

Elephants

Elephant is the greatest and the most mejestic animal on the Earth. It is mostly found in the forest districts of India, Srilanka and Myanmar. It is found scattered in Malay Peninsula, Thailand, Cochin Chaina, Sumatra and Borneo. They are also found in large numbers in Africa.

In India elephants are mostly found along the foot hills of the Himalayas, locally found in the forests between Ganges and Krishna rivers, in Western ghats, in the forests of Karnataka and Assam. According to rough estimates there are about 7000 wild elephants in India of which about 3000 are in South India. They are essential in timber industry, used for riding in thick forest, in tiger hunting and in festival processions and in circus. The tuskers are valued for their ivory. The vision of elephpant is very sharp, has thick (about 2 cm) skin, acute sense of smell and hearing. The optimum average weight that it can carry is 450-550kg. The weight of an elephant varies from about 1050-4200kg.

Poultry

Poultry includes many kins of birds such as chicken, ducks, geese, turkeys, guinea fowls etc., In India chicken or domestic fowl is important which was domesticated about ne thousand years B.C. Before the middle of last century it had considerable importance in rural economy but now poultry farming is important in sub-urban cultivators economy. Poultry farming provides eggs, meat, feathers and good manure. Eggs and poultrymeat are rich sources of protein and vitamins. Eggs contain phosphorus, Iron, Lipids, Phosphpolipids Vitamin A,B & D besides rich in proteins (about 15% of whole egg). Egg yolk contains about 16% protein and white 10% of their own weight. It may be noted that there is no loss of protein in boiled eggs, while there is 3% loss in omelettes, 9% in fried and 14% in scambled eggs.

According to FAO estimates for 1991, the world chicken population was 1103.4 crores, of which India had 38 crores, China had the largest chicken population of 207.7 crores. In 1991, the annual egg production in India was more than 2000 crores and pooultry meat 328.20 crore tonnes. On an average an Indian hen lays 60 eggs/year, while the American brees Rhode island Red, Leghorn lay 193, 212 eggs per annum respectively.

The normal rectal temperature of a chicken is about 42 °c or 107.6° (varies 40.5 to 43 °c), respiration rate is 15-30 per minute and heart beat is 300/minute. Hens also lay eggs in the absence of a cock, such eggs are called infertile or vegetarian eggs. They do not hatch but have better preserving quality.

Hatcheries

In recent times artificial hatching by incubators has become popular in India. It is also economical for large (100-1000) number of eggs hatching. Incubators temperature is maintained between 38.3 to 39.4 °C and humidity 60 to 65%.

Chicken appear be eternal hungry and always look around for something to eat. A complete (wholesome) poultry feed should contain (by weight) 30% yellow maize or mixture of cereals, 20% polish bran, 10% barley/oats, 10% wheat bran, 15% decorticated groundnut cake, 4.5% corn gluten meal, 4% steamed fish meals 3% steamed meat meal, 1% steamed bonemeal, 2% calcium powder, 0.5% common salt. For every 100 kg of this mash mixture 4.4gm of vitamin A, 0.5gm vitamin B_2, 0.6gm vitamin D_3 and 22gm manganese sulphate should be added. Birds may be given sufficient greens.

Ducks

Ducks thrive welll in semi-aquatic conditions. They bred for eggs, meat and ornamental purposes. In a year a duck lays eggs 80 to 150, which are heavier than hen eggs (averge weight about 60gm each). In India there were about 14.84 lakh ducks (estimated) in 1982. A balance feed for duck should contain 35% maize or cereal, 20% wheat or rice bran, 30% groundnut cake, 5% fish meal, 4% ground limestone or oyster shell, 5% dried greens, 0.5% each of salt and shark liver oil. The normal temperature of a duck is about 42 °C (varies 41.5 to 42.5 °C), respiration 15 to 48 per minute, heart beat 120 to 160 per minute.

Geese

Geese are bred for meat (weight about 3 to 3.5kg each). They lay eggs to10 in the morning on alternate days. The average weight of geese egg is about 125-130gm each. The normal temperature of geese is about 42°C, respiration 20 to 25 per minute, heart beat 110 per minute.

Turkeys

Turkeys are mainly reared for meat, like the broiler. In America turkey meat is on par with mutton, pork and chicken. Turkey's lay eggs about 70 in the first year. The weight of an egg is about 85gm. Most of the turkeys in India are Broad Breasted Bronze Breed, which is popular in America. They are hardy and weight about 7-11kg each, their normal temperature is about 42 °C.

Guinea Fowls

Guinea fowls are reared mainly for meat and considered like chickens. The normal temperature is about 42 °C.

Questions

1. Briefly describe the Himalayan Region and rivers of India.
2. Write briefly soil types found in India and type and distribution of forests.
3. Write briefly word's oldest, tallest trees and forest products.
4. Write briefly wildlife in India with national parks and sanctuaries.
5. Write briefly on : (i) conservation of forest soil (ii) conservation of natural resources.
6. Write briefly about the Indian domestic animals.
7. Write briefly on (i) poultry (ii) hatcheries.
8. Write briefly on : ducks, geese, turkeys.

CHAPTER 34

Aeronautical Meteorology

Introduction

Aeronautical meteorology began with Leonardo da Vinci, who was very fond of air masses, flights of birds and visualised the possibility of human flights like birds. In 1500 AD he designed the sketches for a helicopter and other flying machines. In the later part of the 18th century balloonists in France understood about the favourable and unfavourable winds for their direction of flights. In the later part of the 1800's, "Otto Lilienthal" of Germany developed gliders and successfully made about 2000 glides with a flight range of about 100 meters. In 1896 he died in a flight crash of gusty winds. In December 1903, the Wright Brothers, at Kill Devil Hills in North Carolina USA, made four successful powered engine aircraft flights in the head winds of about 20-25 knots. Of this one flight covered a distance of 582 feet (on land) in a strong head wind. This flight established the relationship between aircraft speed, wind speed and ground speed (that is, vector displacement). After this successful powered aircraft flight aeronautics spread all over the world. A number of atmospheric studies were made, particularly of turbulence to understand its effect on the safety of flights. Meteorologists were specially flown to make studies of atmospheric turbulence over Point Loma near San Diego in California. As a result of these flight investigations it was found that were updrafts of air surge at 4000 ft on hill tops. Even now the atmospheric turbulence has great importance in aeronautics.

In Aviation flight operations the meteorological support is required in all the three phases. The take-off (Climbout), Cruising, Landing and Taxing phases.

Surface wind variations are considered significant in take-off, landing if it has head wind of 10 kt, tail wind of 2 kt and cross wind of 5 kt.

Horizontal and vertical visibility over airfield is a significant meteorological element. Vertical wind shears within the final approach, take-off and initial climbout areas, particularly in the lowest 60 m (600 ft) is also crucial/hazardous Jet stream, CAT, Cb enroute of flight is significant.

In view of the special needs of meteorological service in aviation this chapter has been included.

Meteorology in Aircraft Performance

The flight of an aircraft depends on four factors.

1. Total weight (W) of the aircraft (effect of gravity).
2. Lift (L) (or buoyancy) of the aircraft developed by the airflow over the wings.
3. Drag (D), which is the resistance or opposition to the forward motion caused by air.
4. Thurst (T), which is supplied by the engines of the aircraft.

The performance characteristics of an aircraft depends on the three phases of the flight, namely (i) Take-off (ii) Cruise and (iii) Landing. Depending on the phases of flight different meteorological observations at the aerodrome take-off in air at flight level and surface observations at aerodrome landing are required. Inspite of all technological improvements, most of the aircraft severe accident still are confined to the take-off and landing phases of the flight. These accidents are attributed to wind shear, severe convection, microburst, lightning strike, runway flooding and fog (visibility).

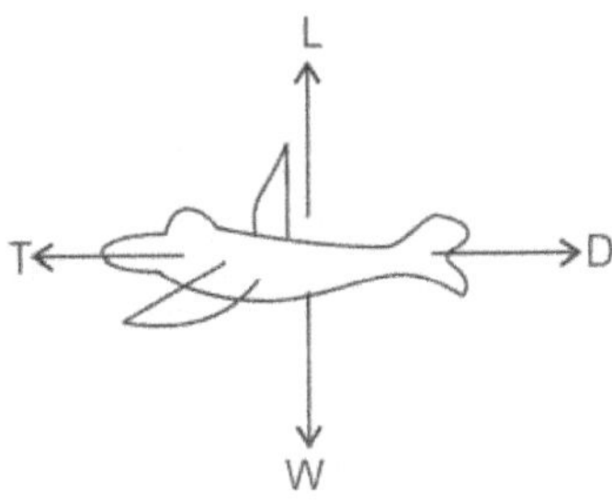

Piston-engined aircraft or small aircrafts (like Dakota, Cissna) fly below an altitude of 3 Km (flight levels -050, 070, 100) at speed about 200 kt. The aircraft speed of take-off or landing comparatively low. These flights encounter all weather conditions that are operated for local and short distance flights. Turbo-propeller aircraft fly between 4 to 8 Km altitudes (flights levels 140, 185, 205, 235) at moderate speed 300 kt (about 500 kmph). Take-off and landing speeds are moderate/relatively low.

These aircrafts are operated in short distance routes encounter all weather conditions.

Jet-engine aircrafts are the most frequent, they fly above 10 Km altitude (flight levels, 340, 390 430 etc) for long distance routes (3000 to 6000 Km) with cruise speed of about 500 Kt. At these levels they do not face bad weather but take advantage of cold temperatures to improve engine efficiency.

Aircrafts, however encounter all weather conditions in the critical phases of take-off or landing.

Surface Wind

From the aerodynamics

we have	$L = K_1 \rho V^2$
and	$L = W$
where	K_1 = aerodynamic constant
	ρ = density of air
	V = speed of aircraft
	L = lift
	W = gross weight of aircraft.

For inflight conditions for a given weight there is a critical speed V_s below which aircraft will not be supported in the air. This speed V_s is called stalling speed. Aircraft requires this stalling speed for take-off or landing. If the surface head wind speed is X kt then the ground speed of take-off or landing is $V_s - X$. This slower speed ($V_s - X$) is safer in operation.

Surface wind speed is required in planning the take-off weight. If there is a strong head wind the take-off run is reduced, for a given run way length which permits a high value of pay load. On the contrary if there is light head wind or calm the take-off run is increased, hence for a given runway the aircraft has to reduce its pay load.

It may be noted that large cross-wind is sometimes hazardous for landing and take-off. Gustiness of the wind seriously affects to the pilots touch in countering the effects of gustiness.

Upper Winds

For air navigation between two places, upper winds at cruising level required. In a steady flight aircraft moves forward in a straight line (with respect to the earth) in relation to the air in which it is flying. If there is wind (cross wind) the path of the aircraft will be displaced in relation to the earth. Suppose the aircraft is moving north PQ, and let QR is the relative drift, because of cross wind. Though the aircraft is moving north P to Q, it will drift to R (P to R) due to cross wind. (See Fig. 34.1).

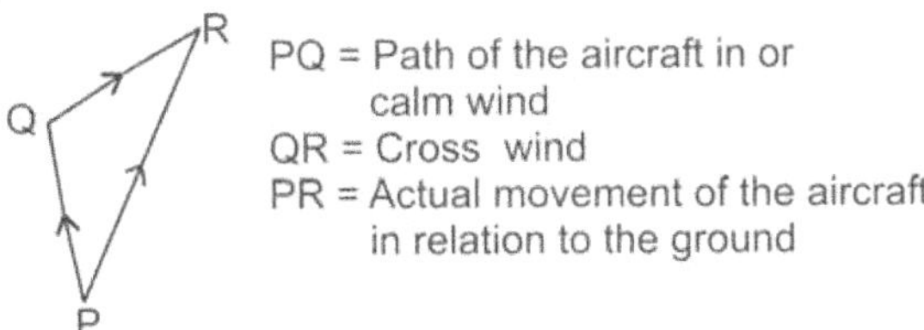

Fig. 34.1 Wind draft.

Upper wind information is required for planning the fuel load. In case of strong head wind, aircraft requires more flight time as compared to light wind, hence requires more fuel. This means reducing pay load. On the contrary if there is strong tail wind, the aircraft flight time decreases, which means reduced fuel load. This enables to have increased pay load.

Vertical Wind Shear

Wind shear is very important in climbout or descent area. Vertical wind shear is defined as the change in wind velocity with height. In simple terms changes in wind speed and direction or both with increased height.

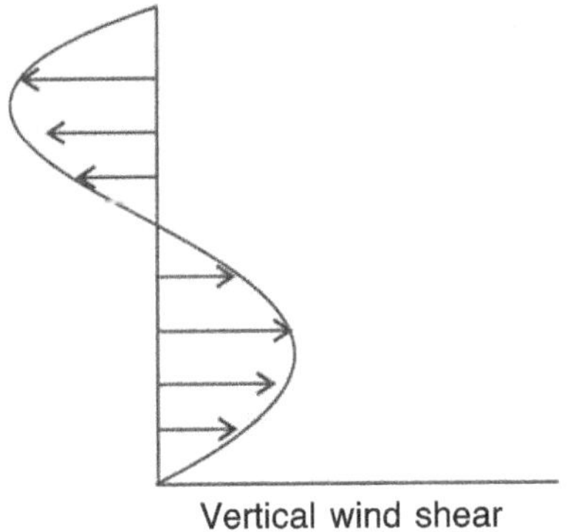

Vertical wind shear

When an aircraft descends on an airfield where wind speeds are decreasing downwards aircraft may under shoot. (See Fig. 34.2).

When an aircraft descends on an airfield where wind speeds are increasing downwards aircraft may overshoot. (See Fig. 34.3).

In take-off (or climbing) of an aircraft into decreasing wind speed the angle of climb decreases with height (θ decreases). (See Fig. 34.4).

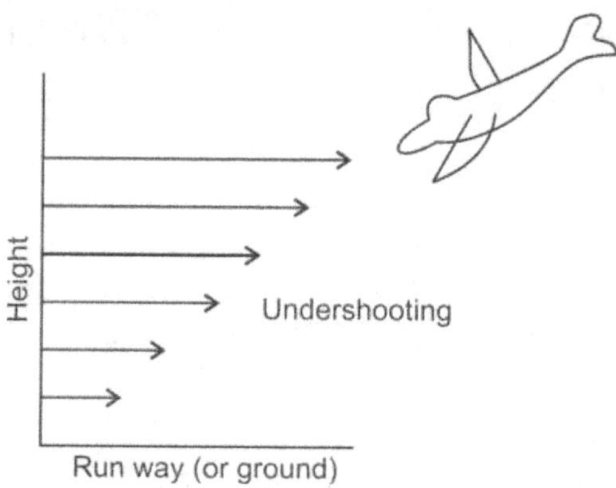

Fig. 34.2

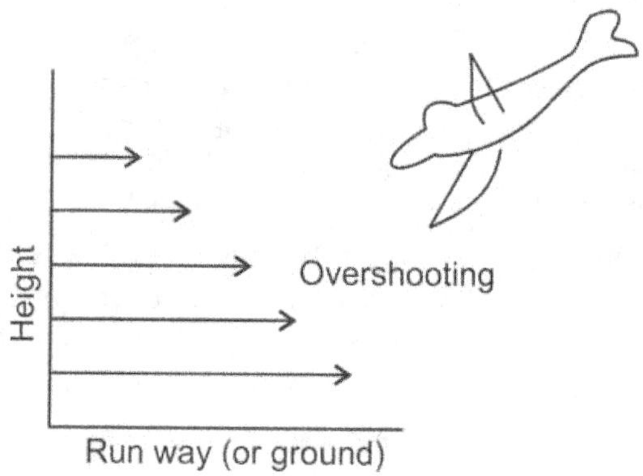

Fig. 34.3

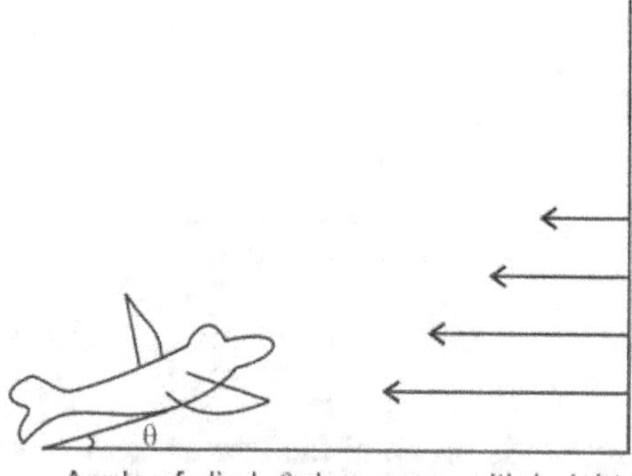

Fig. 34.4

In take-off or climbing of an aircraft into increasing wind speed, the angle of climb increases with height. (See Fig. 34.5).

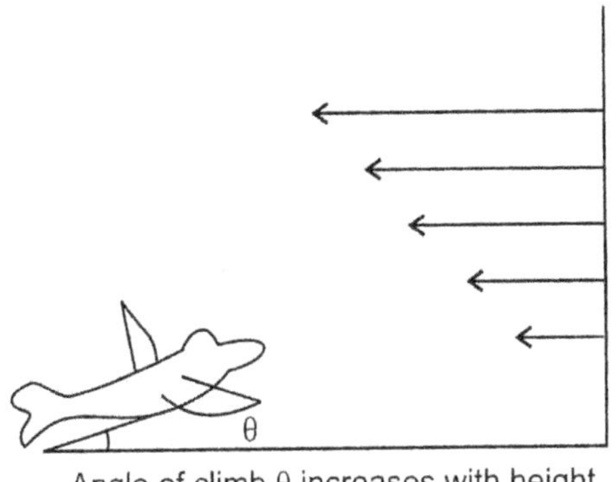

Angle of climb θ increases with height

Fig. 34.5

Wind Shear

Wind shear implies a change in wind direction and/or wind speed, including both updrafts and downdrafts. Thus variation in the wind vector along the flight path of an aircraft with respect to pattern, intensity and duration is called wind shear. This will cause the displacement of aircraft from its predetermined flight path. So to correct this displacement substantial control has to be applied by the pilot in command of the aircraft.

Vertical wind shear = change of wind vector with height

$$= \frac{\vec{V}_2 - \vec{V}_1}{Z_2 - Z_1}$$

where $\vec{V}_1$, $\vec{V}_2$ are wind vectors at height Z_1, Z_2 respectively, and $Z_2 > Z_1$

Horizontal wind shear = change of wind vector with horizontal distance

$$= \frac{\vec{V}_2 - \vec{V}_1}{d_2 - d_1}$$

where $\vec{V}_1$, $\vec{V}_2$ are wind vectors at d_1, d_2 which are in the same horizontal plane and $d_2 > d_1$

Updraft or downdraft shear $= \dfrac{\vec{W}_2 - \vec{W}_1}{Z_2 - Z_1}$

where $\overrightarrow{W}_1$, $\overrightarrow{W}_2$ are vertical components of wind vector at Z_1, Z_2 respectively.

Wind shears are associated with (i) thunderstorms (ii) frontal passage (frontal zones with marked temperature difference of the two masses ($\Delta t = 5$ °C or more), change in wind direction with wind speed (equal to or exceeding 30 kt in the frontal zone) (iii) inversion (of temperature) zone.

Visibility

Meteorological visibility is defined as the lowest visibility in any direction of the observing aeronautical station.

Surface visibility is one of the critical meteorological element in landing and take-off phases in all types of aircraft which are not equipped with ILS (instrument landing system).

Fog, mist conditions effect visibility and they are associated with inversions of temperature. When visibility is less than 5000 m, RVR (Runway Visual Range) will be reported.

Visibility reduces in precipitation, (rain, shower etc) dust and sand storm.

We shall consider here some aspects of visibility in relation to aircraft operations. Visibility means transparency of atmosphere both in horizontal and vertical directions. Visibility is afftected by the meteorological paramters haze, mist, fog, precipitation, that is by the presence of hydrometeors and lithometeors suspended in the atmosphere. In aviation horozontal and vertical visibility over an aerodrome and its vicinity affects the landing and take off operations.

In poor horizontal visibility over aerodrome : Refer to the Fig. 34.6 when aircraft was located at position A in its landing operation the aerodrome is clearly visible in mist or fog of shallow thickness over the aerodrome but when it is in the position of B the aerodrome is not clearly visible due to poor slant visibility, thickness being much more than vertical visibility in position A.

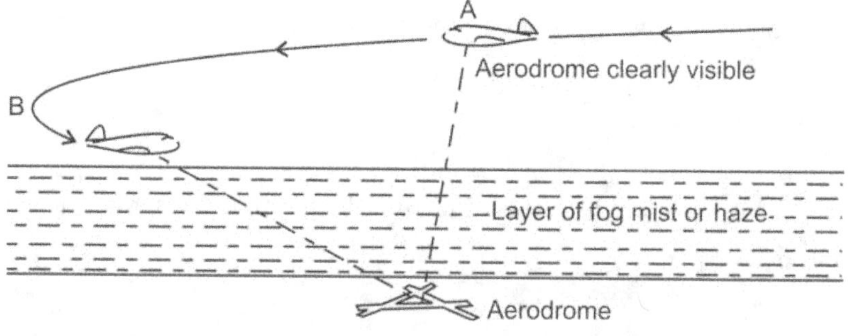

Fig. 34.6 Variation in visibility during descent.

In good horizontal visibility at aerodrome : Refer to Fig. 34.7 when mist/fog lifted and suspended in the air, aircraft in landing operation will not be able to see aerodrome lifting due to lifting fog/mist while the horizontal visibility at aerodrome is good.

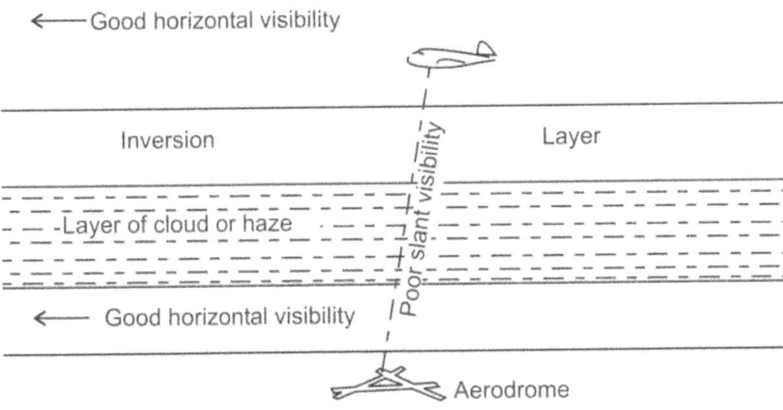

Fig. 34.7 Effect on visibility of cloud layer beneath an upper inversion.

Runway Visual Range

In view of the above two visibility situations of hindrance to aircraft operations observations of RVR are required for safe landing. RVR observations are made when visibility is less than 1500 m, with the help of object markers or lights provided along the runway. A transmissometer is used for the estimation of RVR.

Definition of RVR

The maximum (of distance) in the direction of take-off or landing at which the specified lights or surface markings delineating the runway can be identified from a position above a specified point on runway's centre line at a height (about 5 meters) corresponding to the average eye-level of the pilots at touch down.

Note : (i) The height of average eye-level of the pilots at touch-down is taken as five meters. (ii) RVR observations are made only when the meteorological visibility is 1500 m or less (iii) for a day RVR observations marked objects are placed along side the runway (iv) night RVR observations lights along side the runway are used.

The Transmissometer

It is really measures the turbidity of atmosphere between one end of the runway to the other end. It consists of an intense source of focussed light at one end of the runway and a photocell receiver at the other end. Using modulated light source with a tuned telephotometer day and night time observations are taken. The light transmission can provide continuous record of the turbidity of the concentration of hydrometeors/lithometeors on the specified runway, that is transparency of air on that runway.

Cloud

When the low cloud coverage over an airfield is 5 oktas are more, the height of low cloud base 450 m or less is a critical meteorological element in landing and take-off. The presence of convective clouds over an airfield indicates the presence of turbulence. The presence of Towering Cu and Cb portends the danger due to thunderstorms, updrafts, downdrafts, electrical discharges, in some cases hail, squalls, downbursts and wind shear. The effects of these clouds already discussion in convective clouds.

Surface Air Temperature

Engine efficiency lowers with higher temperature. Density of air is a function of temperature and pressure. Lift (L) is a function of density and speed of aircraft. Take-off is made at a given aerodrome pressure. If temperature is more than normal the aircraft requires a higher un-stick speed, (V_s), which means a longer run. The runway length is fixed and may not be sufficient for an aircraft to take-off at normal pay load. This requires reducing pay load.

Upper Air Temperature

As said earlier upper air temperatures are required for flights planning of fuel load. If the temperatures are higher than normal at cruising altitude more fuel is required to maintain cruising power. Similarly if the temperature are lower than normal at cruising altitude less fuel is required. Upper air temperatures are also required to assess aircraft icing.

Atmospheric Pressure and Air Density

We have already studied that the average atmospheric pressure at sea level is 1013.25 hPa, and density 1.225 Kg/m^3. The global average temperature 15 °C, while in the equatorial belt (between Tropic of Cancer to Tropic of Capricorn) 27 °C. Pressure and air density varies with height which is discussed in chapter-4. The ICAO ISA, variation of pressure, density and temperature with height are given in Table 34.1.

Table 34.1

Height amsl Km	Ft	Pressure hPa	Density gm/m^3	Temperature °C
32.00	1,04,987	8.9	(0.011ρ) 13.5	− 44.7
30.48	1,00,000	11.1	(0.014 ρ) 17.1	− 46.2
27.43	90,000	17.3	(0.022 ρ) 26.9	− 49.2
24.38	80,000	28.0	(0.036 ρ) 44.1	− 52.2
21.43	70,000	44.9	(0.058 ρ) 71.1	− 55.2
20.00	65,620	56.7	(0.072ρ) 88.2	− 56.5
15.24	50,000	116.6	(0.153 ρ) 187.4	− 56.5
13.71	45,000	148.2	(0.195 ρ) 238.9	− 56.5
11.78	38,662	200	(0.263 ρ) 322.2	− 56.5
11.00	36,090	228.2	(0.297 ρ) 363.8	− 56.5
9.16	30,065	300	(0.368 ρ) 450.8	− 44.4
5.51	18, 289	500	(0.564 ρ) 690.9	− 21.2
3.05	10,000	696.8	(0.738 ρ) 904.1	− 4.8
3.01	9,882	700	(0.741 ρ) 907.7	− 4.6
1.46	4,781	850	(0.873 ρ) 1069.4	− 5.5
0	0	1013.25	ρ = 1225	15

The ICAO ISA the pressure altitude (Zp) and corresponding pressure (p) are given in Table 34.2.

Table 34.2

Z_p meter	Pressure hPa
− 1000	1139.29
− 500	1074.77
0	1013.25
500	954.608
1000	898.745
5000	540.199
20,000	54.7487
30,000	11.7186

The variation of density is given below.

Altitude Km	Density (ρ) Kg/m³
0 = msl	Kg/m³
5.5	0.66
30	0.013
6 Km (20,000 ft)	0.5 ρ
12 Km (40,000 ft)	0.25 ρ
18 Km (60,000 ft)	0.1 ρ

Density will change by about 1% for every 3 degrees change in temperature or 10 hPa change in pressure.

At surface level density increases with increase in latitude, i.e., at equator it is minimum and it is maximum at poles. At about 7.8 Km asl the density decreases with latitude and maximum deviation from standard occurs at about 15 Km.

The performance of an aircraft will be better above tropopause at equator as compared to poles.

We know from the ideal gas law $\rho = \dfrac{p}{RT}$, i.e., density is a function of pressure and temperature and $L = K_1 \rho V^2$

$$\text{or} \qquad L = C_L^{1/2} \rho V^2 S$$

where L = Lift,

              V = Taxing Aircraft speed (TAS)

              C_L = Coefficient of lift,

              ρ = Density of air

              S = Wind of area

It follows from these relations, that keeping other factors equal, an aircraft should fly at higher speed to maintain height when the density of the air is decreased. Higher speed induces greater drag, which has to be equalised by the engine thurst, in turn this requires more fuel (load). Because of this high speed Jet aircrafts require more fuel. Decrease in pressure or density requires increased runway length in order to attain take-off speed (as in the case of rise in temperature). When troughs, low pressure systems develop and move, pressure variation has to be taken into account, in addition to the diurnal pressure change in flight planning of take-off. Associated with decrease in pressure air density decreases which implies increase in aerodrome altitude. At high altitude aerodromes a longer runway is required for take-off run. This aspect has to be taken care while designing aerodrome. Decrease in air density results in decrease in engine power, hence affects the climbing power of the aircraft. If the air density falls below a certain value it is necessary to reduce the gross weight of the aircraft.

Aircraft Icing

When an aircraft flies at an altitude of freezing level or just above it and through supercooled water droplets, aircraft icing occurs. (Liquid water whose temperature is below 0 °C is called supercooled water. Water droplets can remain in liquid state upto –40 °C). Between 0 and –20 °C both large and small supercooled water droplets can exist but between –20 to –40 °C only supercooled small water droplets can exist. Below –40 °C, water exists only in ice form. Flight through supercooled droplets forms an icy layers on the propeller, edges of wings and tail.

Aircraft icing causes : (i) increase in weight, (ii) decrease in lifting power and (iii) increase in drag. The first two are countered by more power from the engine and the last by extra thrust (stalling speed). All these require increased fuel. If engine lacks to provide this extra power, aircraft cannot remain airborne. A thin film of icing about 20 μm (sand paper thickness) can reduce the lift by 30% and increase the drag by 40%. Severe icing about 1 cm thickness in minute can distabilise the aircraft by displacement of centre of gravity. Uneven weight of ice on propeller, front edges of the wing and tail can cause vibrations (may be felt by the occupants). This will cause error readings in ASIs, VSI, altimeters and Mach meters.

ASI = Air Speed Indicator.

The presence of hydrometeors (fog, mist), sand storm, dust close to the surface of the earth in large quantities cause visibility problems. These were dealt in visibility and fog.

Thunderstorms and Squalls

We have earlier studied about thunderstorms, hail and lightening, squalls downburst and micro-burst. In aviation these meteorological events play hazardous role, In brief, the penetration of an aircraft into thunderstorm cell, downburst, microburst may be confronted with turbulence, hail, aircraft icing and lightening. In severe turbulence/downburst/microburst aircraft may breakup. Squalls/microburst/downburst are associated with severe turbulence and are most critical on an approach or take-off. It may cause uncontrollable attitude and may lead to aircrash. To avoid such hazards it is advised not to fly into it or in the vicinity of thunderstorms.

Temperature

Temperature is one of the fundamental units–meter, kilogram, second and Kelvin degree. A Kelvin degree °K is the unit of thermodynamic temperature and is approximately equal to 1/273 of the thermodynamic temperature of the triple point of water (where water coexists in three phases–solid, liquid and gas). Temperature of a body simply means the degree of hotness or coldness. Temperature is measured indirectly by making use of certain physical properties of bodies that depend on temperature. In setting a temperature scale phase change values are selected as reference point. Generally melting point of ice (or ice point) and boiling point of water (or stream point) at standard pressure 1013.25 hPa or 760 mm of Hg are taken as reference points. Let these reference points be denoted by T_o, T_b. The most commonly used temperature scales are :

(i)   Celsius (or centigrade) temperature scale :

$$T_o = 0\ ^\circ C$$

$$T_b = 100\ ^\circ C$$

(ii)   Kelvin temperature scale T is measured from absolute zero $(T = -273.15\ ^\circ C)$ is called the absolute temperature scale.

$$T^\circ k = T\ ^\circ C + 273.15\ ^\circ C$$

or $\qquad\qquad T_c = T_k - 273.15\ ^\circ K$

(iii)   Fahrenheit temperature scale

$$T_o = 32\ ^\circ F$$

$$Tb = 212\ ^\circ F$$

$$\frac{T\ ^\circ C}{100} = \frac{T\ ^\circ F - 32}{180}$$

The reference points of some substances along with heats of fusion and vaporisation are given in Table 34.3.

Table 34.3

Substance	Melting point		Heat of fusion	Boiling point		Heat of vaporisation
	K	°C	J.gm	K	°C	J.gm
Hydrogen	13.84	−259.30	58.6	20.26	−252.89	452
Nitrogen	63.18	−209.97	25.5	77.34	−195.81	201
Oxygen	54.36	−218.79	13.8	90.18	−182.97	213
Ethylacohal	159	−114	104.2	351	78	853
Mercury	234	−39	11.8	630	357	272
Water	273.15	0.00	335	373.15	100.00	2256

A thermometer is a temperature measuring instrument and based on the physical properties of matter such as : (i) expansion of solids, liquids and gases, (ii) electrical resistance with temperature. High temperatures are measured with pyrometers e.g., radiation.

Clear Air Turbulence (CAT)

Turbulence in the atmosphere is caused in four ways (i) Mechanical or frictional turbulence, (ii) Thermal turbulence, (iii) Orographic ascent or mountain waves, (iv) Clear Air Turbulence (CAT).

We have studied the first three processes in methods of formation of clouds. The airflow near the ground surface is effected by friction, which is associated with the roughness caused by tall buildings, trees, big rocks, hillocks, steep coast etc. Thermal turbulence is caused by temperature (thermal currents). When strong wind blows roughly perpendicular to orographic barrier (like the western ghats), the wind is forced to climb up and creates turbulence on wind ward side of the mountain ridge and standing waves on leeward side. CAT is associated with stable atmosphere with horizontal and vertical shears in wind and temperature. We shall study CAT in detail.

The mechanisms that are responsible for CAT, are (i) standing waves in the lee of a mountain barrier, and (ii) strong vertical shear in a statically stable layer. These mechanisms are pronounced during winter months when the wind speed and horizontal temperature gradient are maximum. CAT is a micro-scale sub-synoptic scale phenomena.

Mountain waves are developed in stable atmosphere when strong wind (speed more than 10 mps) blows nearly at right angles to the mountain barrier. These lee waves are found between the ridge and 200 km downstream. CAT is a patch phenomena with intensity less than the turbulence encountered in thunderstorms. The typical dimensions of CAT are : 30 km by 5 to 15 Km across, between heights 1 to 4.5 Km (FL150). Thickness is of the order 300 to 500 meters, life period 30 minutes to 3 hours. Individual patches often found in close proximity to one another both in horizontal and vertical direction. High level turbulence is associated with Kelvin Helmholtz instability waves . These waves are caused by vertical wind shear in a statically stable layer which amplify, break and tumble over into chaotic motion like that of a breaking ocean waves. The values of Ri (Richardson number) <1 (about 0.6). The trump frequency is greatest at higher levels than at lower levels. CAT is prominent between 1300 meters above and below the tropopause in the proximity of high level jet stream core just downstream of the mountain ridge line. Vertical wind shear induced CAT mostly found in sloping baroclinic zones or internal fronts and in the tropopause region. The dimensions of shear-induced CAT in horizontal direction is 10 to 500 Km. It has a patchy structure, occurs in thin layers of 200 m to 1500 m thickness, with life period 30 minutes to 24 hours, where Kelvin-Helmholtz instability is found.

Important features of CAT

1. CAT is mostly a patchy phenomena. Intensity of CAT is less than the turbulence associated with thunderstorms.
2. There is strong association between Jet-streams and CAT
3. High level turbulence is ansotropic with stronger horizontal gusts.
4. Bump frequency is higher at higher levels as compared with lower levels.
5. Over land area CAT is frequent on the cyclonic side, while over water bodies CAT is frequent on the anticyclonic side.
6. CAT is frequent in the vicinity of tropopause.
7. High terrain has positive influence on CAT
8. CAT is associated with cold advection in upper trough in the vicinity of jet core.
9. CAT is associated with an upper level ridge.
10. CAT is associated with surface cyclogenesis.
11. CAT is associated with confluence zone of jet-stream cores.
12. CAT is associated with the formation zone of a upper level Low.
13. CAT is associated with the shearline in the throat of an upper level cut-off Low.
14. CAT is associated with the formation of difluent upper flow pattern.
15. CAT is associated with close isotherm packing at 300 hPa level in the tropics.

Hazards of CAT in Aviation

CAT may cause temporarily aircraft stalling, loss of control and airframe damage. Depending on the changes in altitude and/or attitude of a aircraft CAT is defined as follows.

Light Turbulence

Instant turbulence that cause slight changes in aircraft altitude and/or attitude without fluctuation in Indicator of Air speed (IAS). IAS may fluctuate 5-15 Kt. Changes in accelerometer readings less than 0.5 g at the aircrafts centre of gravity. Passengers (or occupants) may feel slight strain against seat-belts but no difficulty in walking. Loose objects may be slightly displaced.

Moderate Turbulence

Instant turbulence that cause moderate changes in aircraft altitude and/or attitude but the aircraft remains in positive control at all times. IAS may fluctuate slightly 15-25 Kt. Changes in accelerometer readings of 0.5 g to 1.0 g at the aircrafts centre of gravity. It may cause jolts and rapid bumps in aircraft. Passengers experience strain against seat belts, difficulty in walking loose, objects will move about or displaced considerably.

Severe Turbulence

Large and sudden changes in aircraft altitude and/or attitude and momentarily aircraft may be out of control. IAS may fluctuate in excess of 25 Kt. Changes in accelerometer readings more than 1.0 g at the aircrafts centre of gravity. Passengers are violently forced against seat belts, great strain in walking. Loose objects are tossed about.

Detection

An air borne radar can detect CAT at a distance of 30 Km with a 20 dB improvement in sensitivity. Vertical wind shear (which is the best indicator for the presence of CAT) can be detected with Doppler beat frequencies between spaced laser impulses. For this purpose 100 fold increase in sensitivity and coherence of laser beam are required. Infrared and microwave radiometers measure temperature fluctuations, temperature anomalies, associated with CAT along the flight path.

Note : There is as yet no clearcut way of predicting CAT. As such it is essential that all pilots who encounter CAT should report to ATS (Airtraffic Services) unit with which they have radio contact, giving details of : (i) time of occurrence, location, flight level, type of turbulence, type of aircraft.

Ex :   (i)   over western ghats, at 0900 UTC, intermittent moderate CAT FL330 B 747.

(ii)   from 80 Km east of Jaipur, 1005 UTC, intermittant moderate/severe CAT FL 330 B 737

Empirical Rules Favourable to the Occurrence of CAT

In the vicinity of :

(i)   Vertical wind shear greater than 4 kt/300 meters

(ii)   Horizontal temeperature shear 5 °C/150 Km

(iii)   Moderate turbulence if horizontal wind shear is greater than 25 kt/150 Km and severe turbulence if the shear is 50 kt/150 Km.

(iv)   Polar side of the jet stream and next below the tropical tropopause.

(v)   In the convergence zone of two jet-streams

(vi)   In inversion layers next to leeward side of mountain ranges.

(vii) Over upper air sharp troughs and ridges.

Incidence of CAT Over India

1. Over central and north India the CAT frequency is highest during October to May, while over southern India the CAT frequency is highest during July-August.

2. In association with the peak activity of sub-tropical westerly jet-stream, the maximum frequency of CAT is from December to February.

3. CAT zones are generally patchy in nature.

   Average dimensions of CAT zones are 150 Km in north-south direction and 300 Km in east-west direction. Vertical thickness may be 300 to 600 m.

4. Severe CAT occurs in during Dec-Feb while light to moderate CAT during rest period.

5. The high chances of CAT are in a zone of 300 Km south of the axis of sub-tropical jet-stream.

6. In western Himalayas CAT is most frequent in October when the sub-tropical jet-stream appears over the area.

7. In eastern Himalayas CAT is most frequent in mid-winter months.

8. In peninsular India, CAT is conspicuous with easterly jet-stream during southwest monsoon.

Avoidance of CAT

If CAT is encountered in flight, the following evasive action recommended.

1. Turn towards north or south to get out of CAT zone.

2. Climb or descend to a level where the vertical wind shear (rate of change of wind speed with height) is known to be less.

3. Climb or descend to a level where the lapse rate is known to be less.

   A layer next above cirrus or cirrostratus is likely to be free from CAT.

4. Avoid flying below the tropopause.

5. Flight above one km of tropopause may be free from CAT.

6. Fly on the windward side of the high ground.

Altimetry

Pressure decreases with height. Pressure-height relation is the basis in the construction of aircraft altimeters. Pressure altimeters are aneroid barometers, scale graduated in terms of height instead of pressure. ICAO altimeters are constructed and calibrated to read in the ISA (international standard atmosphere). The scale is linear. Altimeters indicate correct height under specified conditions used in caliberation. In case of deviations in pressure and temperatures corrections and settings are required to obtain true height. We have already studied about ICAO standard atmosphere wherein altitude corresponds to any given pressure.

The pressure altitude (Z_p) to a given pressure in ICAO ISA is given below.

Z_p meters	Pressure hPa (mb)
− 1000	1139.29
− 500	1074.77
0	1013.25
+ 500	594.61
+ 1000	898.75
+ 5000	540.20

The day to day atmosphere is not the same as ICAO ISA which is an agreed model. The difference of Z_p determined by ICAO ISA from true altitude (Z) is called D-factor.

$$D = Z - Z_p$$

The cause of difference is : (i) due to the difference in actual msl pressure and ICAO ISA 1013.25 hPa, (ii) the temperature of the atmosphere between the surface and the altitude in question differs from that of ICAO ISA.

Altimeter Setting

Aircraft altimeter consists of a main circular scale calibrated in hundreds and thousands of feet (or meters) and a sub-scale calibrated in hPa (mb). The sub-scale is operated by a Knob to set any pressure value as the base. Corresponding to this base value (pressure) the main scale shows altitudes. The sub-scale setting is made by QFE or QNH.

QFE : Barometric pressure at aerodrome elevation or at runway threshold.

QNH : Barometric pressure at the aerodrome elevation (QFE) reduced to mean sea level under ICAO ISA.

QFE Altimeter Setting

When the sub-scale of altimeter set to QFE the altimeter main scale reads zero when the aircraft is on the aerodrome (runway). This is called zero setting. The main scale will read the height of the aircraft above aerodrome if it is in the air.

QNH Altimeter Setting

When the sub-scale of altimeter set to QNH the altimeter main scale will indicate the elevation of the aerodrome when the aircraft is on the runway and it will indicate the elevation of the aircraft when it is in the air.

In the lower levels the pressure-altitude relation is 27 ft per hPa (the difference of one hPa causes 27 ft height difference).

The indicated altitude (altimeter reading) on altimeter will be correct only as long as the airfield pressure set on the sub-scale remains unchanged. On landing back to the same airfield after a while the aircraft altimeters will read high (over-read) if the pressure has fallen in the mean while and will read low (under-read) if the pressure has risen (compared to setting before take-off). In a similar fashion the aircraft altimeters will read high (more) at a destination airfield where pressure is lower and read low if the pressure is high as compared to the pressure set at the take-off airfield.

Example 34.1

Given sub-scale pressure (altimeter setting), corresponding indicated altitude, find true altitude of a place where QNH is given.

	QNH (hPa)	Altimeter setting (hPa)	True altitude (ft)	Indicated altitude (ft)
1.	1012	1010	–	4036
2.	999	1013	–	8500

Solution

1. Pressure difference 1012 – 1010 = 2 hPa
   Corresponding height difference would be 2 × 27 = 54 ft
   ∴ the true altitude is 4036 + 54 = 4090 ft
2. Pressure difference 1013 – 999 = 14 hPa
   Corresponding height difference 14 × 27 = 378
   ∴ the true altitude 8500 – 378 = 8122 ft.

Example 34.2

An aircraft over an airfield elevation 540 ft, has an altimeter reading 783 ft with a setting of 1002 hPa. What is the actual QNH.

Solution

The difference of altimeter reading and airfield elevation = 783 – 540

= 243 ft (more), corresponding pressure difference would be $\dfrac{243}{27} = 9$ hPa.

∴ the actual QNH = 1002 – 9 = 993 hPa

Example 34.3

What is the altimeter reading if the setting is 1005 hPa, QNH 1010 hPa when the airfield elevation is 220 ft.

Solution

Pressure difference 1010 – 1005 = 5 hPa

Corresponding height difference 5 × 27 = 135 ft.

Altimeter reading 220 – 135 = 85 ft.

Example 34.4

Fill in the blanks dash place (answer given in bracket).

QNH	Altimeter setting	True altitude ft	Indicated altitude (altimeters reading) ft
1012	1010	—(4090)	4036
999	1013	8122	—(8500)
1015	1010	5135	—(5000)
1017	1027	—(3300)	3570
1012	1000	— (324)	0
1024	— (1009)	405	0
1025	— (1015)	4760	4490
— (1020)	1013	10689	10500
— (1015)	1010	5135	5000

Example 34.5

The regional pressure setting is 1012 hPa, the altimeter setting is 1020 hPa and the indicated altitude is 4046 ft. Ahead there was some high mond of altitude 3500 ft. Will the aircraft clear the mond, and if so by how much height?

Solution

Given QNH = 1012 hPa, setting 1020 hPa, indicated altitude 4046 ft.

Pressure difference 1020 – 1012 = 8 hPa, corresponding to this height difference would be 8 × 27 = 216 ft.

∴ the elevation of the air field is 4046 – 216 = 3830 ft.

The aircraft will clear of the mond by 3830 – 3500 = 330 ft.

Thumb Rules

High – Low – High   (QNH,   setting   true altitude compare to indicated)

Low – High – Low (QNH,   setting   true altitude compare to indicated)

For high pressure (QNH)   :   True altitude > indicated altitude

For low pressure (QNH)   :   True altitude < indicated altitude

For colder than ISA   :   True altitude < indicated altitude

For warmer than ISA   :   True altitude > indicated altitude.

Altimeter Temperature Error Correction

Pressure altimeters are calibrated to indicate true altitude under ICAO IAS conditions. Any deviations from ICAO IAS temperature will show wrong (erroneous) readings. The error is proportional to the difference between the actual and ICAO IAS temperature and the vertical distance of the aircraft above altimeter setting datum i.e., height above touchdown. The error is approximately 4 ft/1000 ft for every 1 °C difference.

Example 34.6

ISA temperature deviation °C		Height above touch down or Height above aerodrome in ft		
		200	**500**	**1000**
1.	− 5	4	10	20
2.	− 10	8	20	40
3.	− 15	12	30	60
4.	− 25	20	50	100
5.	− 35	28	70	140

Example 34.7

ISA deviation by Radiosonde.

Height m	Height (ft)	Actual temperature °C	ISA temperature °C	ISA deviation °C
450	(1500)	28	12	+ 16
1500	(5000)	15	5	+ 10
2850	(9,500)	− 5	− 4	− 1
5250	(17,500)	− 18	− 19.65	+ 1.65
7200	(24,000)	− 35	− 32.5	− 2.5
11100	(37,000)	−45	− 56.5	+ 11.5

Transition Altitude

This is the highest altitude below which an aircraft will always fly on local QNH i.e., below which the vertical position of an aircraft is controlled with reference to the height above aerodrome.

Flight Levels

Surfaces of constant pressure at or above transition level separated by pressure interval corresponding to 500 ft with msl pressure as 1013.25 hPa.

Transition Level

This is the lowest flight level above which the aircraft will always fly on standard QNH of 1013.25 hPa or the lowest flight level (1013.25 hPa) available for use above the transition altitude.

Transition Layer

The air space between transition altitude and transition level.

Computation of these Parameters

TA = Transition altitude = E + 1500 ft or E + H + 1000 ft

   where   E = Elevation of the airfield

               H = Height of the highest obstruction within 25 nautical miles around the aerodrome.

Transition altitude thus calculated is rounded off to the next 100 ft and the higher of the two values taken.

Transition level = TA + Pressure correction + Temperature correction

where

    Pressure correction = 27 (1013.25 − lowest QNH) ft.

$$\text{Temperature correction} = (\text{TA} + \text{pressure correction})\,\frac{\left(T_{ISA} - T_{min}\right)}{T_{min}}$$

   where T_{ISA} = ICAO ISA temperature at aerodrome level

              T_{min} = Lowest surface minimum temperature

Transition level is rounded off to the next 500 ft. If $T_{min} > T_{ISA}$, no temperature correction need apply.

Standard Procedure of Altimeter Setting

ICAO ISA standard QNH is 1013.25 hPa

Before take-off the altimeter is set to the current local QNH (altimeter setting for the aerodrome). During climb below transition altitude, the aircraft remains on this local setting. After crossing the transition altitude, the aircraft altimeter has to be set to standard QNH 1013.25 hPa. This change over should be completed within the transition layer. After this the aircraft remains at the flight level allotted (cruising level) on the same standard setting. To assess terrain clearance latest and nearest QNH may be used.

Before initial approach for landing at the destination aerodrome, the transition level has to be obtained from ATC (Air Traffic Control). In crossing the transition level the altimeter setting must be changed to local QNH. A pilot can use local QFE setting instead of local QNH setting.

Some Features of Observational Organisation of IMD

Meteorological observational stations are called observatories. Surface observations are made by an observer stationed on the ground, sea level or on board ships. These observations mainly consists of atmospheric wind (direction and speed), visibility, weather (thunderstorms, rain, fog, mist, hail duststorm etc). pressure, temperature, humidity, rainfall, cloud (type amount and base height) etc. Some of these observations are made by using our own (human) senses–called sensory observations, others based on instrumental measurements–called instrumental observations. Meteorological observatories (stations) are spread all over the country. Similar net work maintained by other countries over the world.

The basic observational stations of IMD consists of :

(i)   Surface observatories

(ii)   Upper air observatories,

(iii) Aeronautical meteorological observatories

(iv)   Climatological stations for hydrological purposes.

(v)   Special stations (like Storm Detection Radar Stations, Cyclone detection Radar stations, Radiation and Ozone observatories, Environmental meteorological stations, Satellite meteorological stations

(vi)   Agricultural meteorological stations,

(vii) Marine meteorological stations.

In addition to the above IMD has established Flood Meteorological offices, for Flood Forecasting, Seismological observatories net work for routine determination of earthquakes epicenter, preparation of bulletins, study and research.

Surface and Upper air observatories together called synoptic observations, which are taken at standard times. Surface observations are taken at 00, 03, 06, 09, 12, 15, 18 and 21 hrs UTC while upper air observations are taken at 00, 06, 12 and 18 hrs UTC. Depending on the class of observatory (which is related to automatic recording instruments) daily one, two, four, six or eight observations are taken. The WMO (World Meteorological Organisation) principal synoptic observations are taken at 00 and 12 hrs UTC, however in India the principal surface observations are taken at 03 and 12 hrs UTC while upper air observations are taken at 00 and 12 hrs UTC. The observations are plotted on geographical (Mercator projection in tropical areas) maps, analysed and used for forecasting purposes. The charts are called synoptic surface, upper air charts. Meteorological stations which take observations for climatological purposes (for long term aspects of weather) are called climatological stations.

Meteorological stations which serve the special needs of aviation are called Aeronautical meteorological observatories. Meteorological stations which serve the needs of the Agriculture, Horticulture, Animal Husbandry and Forestry are called Agricultural Meteorological stations.

Similarly meteorological stations which serve the particular meteorological events termed special meteorological stations.

WMO Recommended the Density Net Work of Stations

Basic surface observatories for synoptic purposes should not be spaced more than 150 Km apart while upper air observatories should not be more than 300 Km apart.

Synoptic Observations include the following

1. Wind (direction and speed)
2. Surface visibility
3. Present and part weather
4. Air temperature (screen temperature at a height of 1.2 m agl)
5. Dew point temperature
6. Atmospheric pressure reduced to msl
7. Amount, type, height of base of cloud.
8. Rain fall

    In addition to the above the following observations are also made :

9. Daily maximum/minimum temperature
10. 24 hr pressure change (P_{24} P_{24})
11. Direction of cloud movement (Neph analysis)
12. Special phenomena (like squalls, size of hail etc), soil temperature.

    Ocean weather stations (ship based or anchored (bouys)) include the following observations

13. Sea surface water temperature
14. Duration of movement of waves and its height
15. Ice
16. Special phenomena (like Tsunami waves)

    Climatological stations record 1-12 observations + snow cover + sunshine.

Aeronautical Meteorological Observations

In addition to the synoptic and climatological observations, special meteorological observations are taken to serve the needs of aviation over specified areas, such as take-off, landing and approach of a particular runway for operational requirements. There are four types of meteorological offices for providing meteorological services to international air navigation.

(i)   Class I meteorological office,

(ii) Class II meteorological office,

(iii) Class III meteorological office and

(iv) Meteorological watch office (MWO).

Aviation meteorological main observations include :

(i)   Aviation routine weather report, called METAR.

(ii) Aviation selected special weather report called SPECI.

(iii) Upper air report from an aircraft.

(iv) Forecast upper wind and temperature for aviation.

(v)   Aerodrome forecast – TAFOR

(vi) Area forecast for aviation – ARFOR

(vii) Route forecast for aviation – ROFOR

These will be discussed briefly along the codes.

There are about 90 to 100 aviation meteorological stations maintained by IMD. These maintain current weather watch at aerodromes or important locations along and near air routes which provide meteorological service for the use of aircraft and air traffic control units. About 25 of these function at class I meteorological offices, 45 at class III meteorological offices at aerodromes and the remaining at class III meteorological offices not located at aerodromes. These stations are equipped with usual meteorological instruments and maintained by departmental staff. They take all surface observations at synoptic hour's issue hourly or half-hourly weather reports as required together with special and selected special weather reports of deterioration/improvement of weather.

Aviation meteorological observatories at international airports are equipped with remote indicating and recording equipment such as current weather instrument pannel system, Skopograph, Ceilograph for continuous recording of meteorological conditions prevailing near touch-down points of the runways. The meteorological parameters include temperature, dew point, wind speed and direction, cloud ceiling, runway visual range (RVR), visibility (horizontal transparency). At present these facilities are available at Mumbai, Chennai, New Delhi and Calcutta. Most of the national airports have distant indicating wind equipment while a few of them equipped with electrical anemographs.

The pilots of all aircrafts are required to record and report at specified locations, observations of meteorological elements and weather encountered during their flights. Any significant weather encountered by the pilots, special observations made in between locations have to be transmitted by them. These reports are made by the pilots on AIREP forms. Routine and special aircraft reports are used by the forecasting offices in their daily routine forecasting work and by the MWO's for issuing SIGMET information in respect of their FIR (flight information region) for which the MWO's are responsible.

Meteorological (MET) Services of IMD for Air Navigation

IMD provides its services to air navigation for safety, regularity and efficiency of aeronautical operations. Meteorological services are provided to both national and international air navigation through its offices located in civil aerodromes in India.

The meteorological branches of Indian Airforce and Indian Navy also provide Met-services for national air navigation operating through airfields controlled by them.

It is the responsibility of the flight operators to notify sufficiently early to obtain flight documentation and met-briefings as given below.

For National Flights

At class I Met-offices notification must be given at least-3 hours in advance of the time of flight departure.

At class III Met-offices, at least 18 to 24 hours in advance of the time of departure of the flights.

For International Flights

(i) At class I Met-offices (at Mumbai, Kolkatta, Delhi, Chennai, Nagpur, Banglore, Gauhati, Hyderabad, Trivandrum) at least 3-hours in advance of the time of departure of the flight.

(ii) At other class I Met-offices, at least 12 hours in advance of the time of departure of the flight.

(iii) At class III Met-offices, at least 18-24 hours in advance of the time of departure of the flight.

The above notice is required in respect of non-scheduled flights. For scheduled flights sufficiently in advance.

Area Forecast Centre : Under the ICAO Area Forecast System an Area Forecast centre is functioning at the Northern Hemisphere Analysis Centre (NHAC) New Delhi.

Area Forecast Centre : Under ICAO Area Forecast System an Area Forecast Centre is functioning at Northern Hemisphere Analysis Centre (NHAC) New Delhi.

NHAC caters the needs of aviation forecasts in the form of Prognostic charts (PROG) for the area bounded by long 30 °E to 125 °E and latitude 40 °N to equator. The prognostic charts made are Facsimile broadcasts as per schedule given below.

(i) Surface chart (ii) Significant weather chart for below flight level 460, (iii) Tropopause/maximum wind, (iv) Winds and temperature for isobaric levels 500 hPa, 300 hPa, 250 hPa and 200 hPa.

Area Coverage of the Charts

Map Area A : Lat 25 °S to 450 °N, long 30 °E to 125 °E

Map Area B : Lat 0 – 40 °N, long 30 °E to 125 °E

Projection : Mercator

Upper air PROGS are valid for the fixed time 24 hrs after the time of observation.

Map B : Significant weather PROG valid 03-15 time of observation 1800.

Significant weather PROG, 09-21, time of observation 0000

Significant weather PROG 15-03 time of observation 0600

Significant weather PROG 21-09 time of observation 1200

Aeronautical Meteorological Offices

IMD has set up to cater needs of air navigation 20 class I, 57 class III Met. Offices 4 MWO's in addition Area Forecast Centre at New Delhi. MWO's are located at Mumbai, Kolkatta, Delhi and Chennai.

A The responsibilities of class I Met-offices are :
 (a) Preparation and/or obtaining forecasts and other relevant information for flights operating from their aerodromes.
 (b) Preparation of forecasts of local meteorological conditions.
 (c) Keeping continuous watch over the Met-conditions over their local aerodromes and over aerodromes served by their associated class III Met-offices.
 (d) Arrangement of briefing, consultation and flight documentation to flight crew members and/or other flight operators.
 (e) Supply of any other Met-information to aeronautical users.
 (f) Display of relevant latest Met-information.
 (g) Exchange of Met-information with other Met-offices
 (h) Issue of landling and take-off forecasts
 (i) Supply of winds and temperatures for flight planning

(j)   Supply of aerodrome forecasts of relevant aerodromes

(k)   Supply of forecasts to their associated class III Met-offices in respect of flights operating from their aerodromes.

The other Met-information supplied in addition to the above are :

(i)   Hourly/half-hourly current weather observations and special reports.

(ii)   RVR observations

(iii)   Landing and take-off reports on request for the required elements (RVR, QFE Alt, wind shear etc)

(iv)   Met-information for Volmet broadcasts, ATIS (Air Traffic Information Service) broadcasts, VOR broadcasts etc.

(v)   Pressure data (QFE/QNH).

(vi)   Radar observations reports (Rareps)

(vii)   SIGMETs of other FIRs.

(viii)   AIREPS available.

(ix)   METAR'S/SPECIE's of other stations as appropriate.

(x)   Low level wind shear and temperature inversions.

B The responsibility of class II Meteorological offices are :

(i)   Obtain forecasts from the associated class I meteorological office together with other relevant flight information

(ii)   Provide Met-briefing and documentation to the concerned flight operator (staff)

(iii)   Display of relevant meteorological information

(iv)   Exchange of meteorological information with other meteorological offices.

C The responsibilities of class III Meteorological offices are :

(i)   For aeronautical users, supply of current weather observations of their own stations and those of other stations as required, by procuring them from the concerned stations

(ii)   Providing documentation for flights originating from their stations after procuring the forecast from their associated class I Met-office(s).

D The responsibilities of Meteorological Watch offices are :

(i)   Maintain continuous watch of the meteorological conditions over their respective FIRs.

(ii)   Prepare SIGMET information messages for their FIRs.

(iii)   Exchange of SIGMET information with other MWOs in the neighbouring countries situated within 1100 NM from boundaries of its own FIR.

(iv) Supply to their associated FIRs, SIGMETs received from MWOs within a radius of 1100 NM from the boundaries of their own FIR.

(v) Disseminate their SIGMET information to the other forecasting offices in India.

Explanation of Some Related Terms

1. **Aeronautical Meteorological Office :** An office designated to provide meteorological service for international air navigation.

2. **Meteorological Information :** This includes meteorological reports, forecasts, surface and upper air analysis, existing or expected meteorological condition at the information centre or its meteorological offices.

3. **Meteorological Report :** A statement of observed and/or expected meteorological conditions related to a specified period/time and location/ specified area/air space.

4. **SIGMET Information :** These are originated at MWOs in respect of their FIRs. SIGMET information are prepared for the occurrence or expected occurrence of one or more of the following weather phenomena.

   (a) For subsonic cruising level flights
   - Active thunderstorm area
   - Tropical revolving storm
   - Severe line squall
   - Heavy hail
   - Severe icing
   - Marked mountain waves
   - Widespread sand storm/dust-storm

   (b) For transonic and supersonic cruising levels
   - Moderate or severe turbulence (CAT)
   - Cumulonimbus cloud
   - Hail

5. **Meteorological Briefing :** Oral brief explanation, discussion with the pilots or flight operators of existing and/or expected meteorological conditions enroute the flight and at departure/landing airport.

6. **Flights Levels :** Surfaces of constant atmospheric pressure which are related to a specific datum 1013.25 hPa and are separated by specific pressure intervals.

E.g.,

FL – 50	Corresponds to	1500 m (5000 ft)
FL – 100	Corresponds to	3000 m (10,000 ft)
FL – 140	Corresponds to	4200 m (15,000 ft)
FL – 185	Corresponds to	5600 m
FL – 300	Corresponds to	9200 m
FL – 340	Corresponds to	10400 m

In lower levels FL $\times$ 100 = Pressure altitude in feet

FL $\times$ 30 = Pressure altitude in meters

7. **Operator** : A person or organisation engaged/offering to engage in aircraft operation.

8. **FIR** : Flight Information Region

   ATS : Air Traffic Services

   IFR flight : A flight conducted based an instrument flight rules.

9. **Aerodrome Meteorological Minima** : Limiting meteorological conditions are prescribed for the purposes of determining the usability of an aerodrome, either for take-off or for landing. These limiting conditions are known as aerodrome meteorological minima.

10. **Controlled Airspace** : An air space with defined dimensions within which air traffic control service is provided to controlled flights.

11. **Synoptic charts** : Analysed msl synoptic charts indicate surface weather over a specific space and time. They are prepared for standard time of synoptic observations. Pronostic msl synoptic charts indicate expected weather during next 12 to 24 hours. Upper air soundings are plotted on aerological diagrams (T-ϕ grams). These charts are generally prepared at area forecast centres and transmitted to aeronautical Met-offices by facsimile.

12. **Warnings** : Warnings for aviation hazardous phenomena like thunderstorms, snow, dust/sandstorm, gales, squalls, hail, fog etc, issued to listed recipients.

Surface Synoptic Weather Report Code

After recording the meteorological observations the weather reports are transmitted in a suitable coded form which was developed for easy, rapid and efficient transmission. There has been international agreement concerning the codes for exchange of weather information between national meteorological services within a region. Each code form bears a number preceded by the letters FM (Forecasting Manual of WMO). Here we shall present a SYNOP code form which is used by IMD.

SYNOP

Section 0	Mi Mi Mj Mj	YY GG i_w

Section 1 $I_R i_x h$ VV   I I i i i   N dd ff $1 s_n$ TTT $2 s_n T_d T_d T_d$

$3 P_o P_o P_o P_o$ 4 P PPP   7 WW $W_1 W_2$ $8 N_h C_L C_M C_H$

Section 3 333 $1 s_n T_x T_x T_x$ $2 s_n T_n T_n T_n$ $56 D_L D_M D_H$

$57 C D_a e_c$ $58 P_{24} P_{24} P_{24}$ $59 P_{24} P_{24} P_{24}$ $6 R R R t_R$

$8 N_s C h_s h_s$ $9 S_p S_p s_p s_p$

Section 5 555 $O R_{24} R_{24} R_{24} R_{24}$ $1 R_T R_T R_T R_T$ $2 D_q F_q q q_t$

Explanation

Section	Number	
0	–	Data for identifying type, date and time of observation.
1	–	Data for international exchange.
2	222	Maritime data pertaining to a coastal station.
3	333	Data for Regional exchange
5	555	Data for National exchange

Note : All stations in India with elevation between 500 to 800 m will report 4 RRRR only.

Stations with elevations 800 m or more will report 4 PPPP and report 4 a_3 hhh giving the geopotential height of standard isobaric level.

The group 7 WW $W_1 W_2$ will be omitted if both present weather and past weather are insignificant. The codes 00, 01, 02, 03 for WW and 00, 01, 02, for $W_1 W_2$ are considered insignificant.

The meaning of the symbols in the above code :

a_3 : Indicator giving the standard constant pressure level, of which the geopotential is reported.

C   : Type of cloud

C_H : Type of High cloud

C_M : Type of Medium cloud

C_L : Type of Low cloud

D_a : Direction in which orographic clouds or C_b, C_u (with vertical development) are observed.

D_L : Direction of Low cloud movement

D_M : Direction of Medium cloud movement

D_H : Direction of High cloud movement

D_q : Direction of squall

dd	:	Direction of surface wind
e_c	:	Angle of elevation of the top of the cloud reported by C
F_q	:	Maximum of squall wind force in Beaufort scale
ff	:	Wind speed in knots
GG	:	Actual time of observation to the nearest whole hour in UTC or barometer to be read at exact time.
h	:	Height agl of the base of the lowest cloud
hhh	:	Geopotential of the standard pressure level given by a_3 in standard geopotential meters.
$h_s h_s$	:	Height agl of the cloud base reported by C.
i_R	:	Indicator for inclusion or omission of precipitation data.
i_w	:	Wind indicator.
i_x	:	Indicator for type of station operation (manned or automatic).
$M_i M_i$	:	Identification letters of the report e.g., a SYNOP report.
$M_j M_j$	:	Identification letters of the part of the report. SYNOP reports from land stations is identified by AAXX.
N	:	Amount of cloud coverage (oktas).
N_h	:	Amount of low cloud coverage (Oktas).
N_s	:	Amount of individual cloud layers of type C.
$P_{24} P_{24} P_{24}$	:	Change (fall or rise) in station level pressure in last 24 hours in tenths of hPa.
PPPP	:	Pressure reduced to msl in tenths of a hPa omitting thousandth digit.
$P_o P_o P_o P_o$	:	Station level pressure in tenths of a hPa.
q	:	Nature of squall.
q_o	:	Time of occurrence of squall.
RRR	:	Amount of precipitation in mm since 0300 UTC.
$R_{24} R_{24} R_{24} R_{24}$	:	Amount of precipitation since 0300 UTC in the tenths of mm.
$R_T R_T R_T R_T$	:	Progressive seasonal total precipitation in full mm.
S_n	:	Sign of temperature. '0' stands for positive temperature and '1' stands for negative temperature.
$S_p S_p S_p S_p$	:	special phenomena.

$T\,T\,T$	:	Screen air temperature in tenths of a degree Celsius.
$T_d T_d T_d$	:	Dew point temperature in tenths of a degree Celsius.
$T_n T_n T_n$	:	Minimum temperature in tenths of a degree Celsius.
$T_x T_x T_x$	:	Maximum temperature in tenths of a degree Celsius.
t_R	:	Duration of precipitation.
VV	:	Horizontal visibility
$W_1 W_2$	:	Past weather
WW	:	Present weather
YY	:	Day of the month (Date)

Example of SYNOP

AA xx	:	SYNOP of Inland station
180 34	:	Date 18, time 03 UTC, wind indicator anemometer instrument
43 128	:	Station index 43 block international station number 128 (Hyderabad - India)
$214\,96\;i_R = 2$	:	Rainfall inclusion, $i_x = 1$ manned, h = 4 (300 – 600 m) $VV = 96$, visibility coded 96
727 10	:	Total cloud coverage 7 oktas, wind direction, speed 270 deg. 10 kt.
10 248	:	$s_n = 0$ positive temperature, D.B. temperature 24.8 °C
2 0 2 1 0	:	Dew point temperature 21.0 °C
3 9448	:	Station level pressure in first decimal 944.8 hPa
400 47	:	MSL pressure in tenths of hPa 1004.7 hPa
7 0 5 2 2	:	Present weather 05 Haze, past weather 22 no significant weather.
8 5 7 3 0	:	Amount of low cloud 5 oktas, low cloud type 7 Stratus, medium cloud type 3 Altocumulus.
3 3 3	:	Section three
2 0 2 2 7	:	$s_n = 0$ positive temperature, minimum temperature 22.7 °C
5 6 6 9 0	:	Direction of movement of low cloud 6 = west, medium cloud 9 = not known
5 8 0 3 2	:	24 hr pressure rise 03.2 hPa
6 00 3	:	Rainfall since 0300 UTC 003 mm, duration not reported.
8 4 7 1 0	:	Cloud amount 4 oktas, type of cloud Stratus (7) or genus, height of base of cloud 10 = 300 m

8 3 6 1 5	:	Cloud amount 3 oktas, type of cloud stratocumulus (6), height of base of cloud 15 = 450 m.
8 5 3 60	:	Cloud amount 5 oktas, type of cloud Altocumulus (3) or genus, height of base of cloud 60 = 3000 m
555	:	Section five
00026	:	24 hr rainfall (intenths of mm) 2.6 mm
1 0 4 6 2	:	Seasonal total rainfall (in full mm) 462 mm

Example 2

AA XX	:	Synop of inland station.
16 124	:	Date 16, time 1200 UTC, wind indicator instrument.
4 3 1 2 8	:	Station index Hyderabad. (Block 4 3, station number 128 for Hyderabad.
3 1 4 9 6	:	i_R = 3 rainfall included. i_x = 1 manned. h = 4 height of base of lowest cloud 300 to 600 m. VV = 96, visibility coded 96.
7 2710	:	Total cloud coverage 7 oktas, wind : 270°/10kt.
10 294	:	'0' positive temperature' DB 29.4 °C.
20 222	:	'0' positive temperature' dew point 22.2 °C.
3 9381	:	station level pressure 938.1 hPa
499 66	:	MSL pressure 996.6 hPa
705 22	:	Present weather 05 = haze, past weather 22 = not significant.
8 59 30	:	Amount of low cloud 5 oktas Type of low cloud (9) cumulonimbus. Type of medium cloud (3) Altocumulus
3 3 3	:	Section 3
10 304	:	'0' positive temperature, maximum temperature 30.4 °C
5 66 90	:	Direction of low cloud (6) westerly, Direction of medium cloud (9) not known
5 79 73	:	9-Cb, direction of CB(7) northwesterly, angle of elevation (3) 3°
590 15	:	24 hr pressure (0) rise 1.5 hPa

8 3615	:	3 oktas low cloud (type 6), Stratus, base height (15) = 450 m
8 3825	:	3 oktas low cloud (type 8) Cumulus, base height (25) = 750 m
8 1 9 3 0	:	okta low cloud (type 9) Cumulonimbus, base height (30) 900 m
8 5 3 6 0	:	5 oktas medium cloud (type 3) Altocumulus, base height (60) = 3000 m

		C_H	
	TT	C_M	PPP
VV	ww	Ⓝ	$P_{24} P_{24}$
	$T_d T_d$	C_L N_n/h	$w_1 w_1$	$R_{24} R_{24} R_{24}$

SYNOP Station Plotting Model

Above two examples plotted below.

Example 1

Date 18, time 0300 U T C

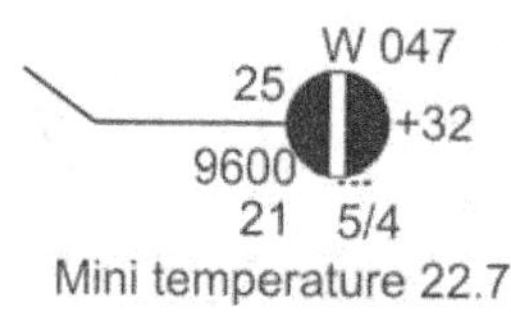

Mini temperature 22.7

Station : Block 43 St.no. 128 Hyderabad

Example 2

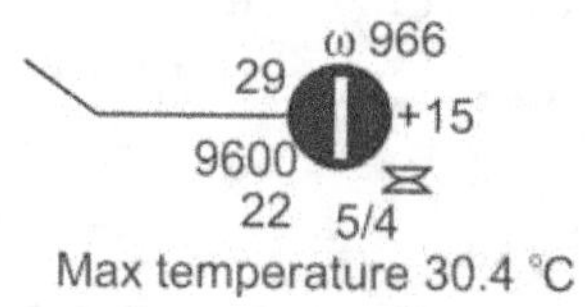

Max temperature 30.4 °C

Date 16, time 1200 UTC

Station Hyderabad

N	0	1	2	3	4	5	6	7	8	9	/
Plotting symbol	○	◐	◐	◐	◑	◓	◕	◕	●	⊗	⊗

Important SYNOP Plotting Symbols are given below.

1. N – Total cloud amount

   N = 0 represents clearsky

   N = 9 Sky obscure (with fog etc)

   N = / Sky cloud coverage not known.

2. ddff – wind direction and wind speed in kt.

   Direction is plotted by shaft of an arrow showing the direction of wind blowing.

   Wind speed is plotted by barbs – half barb 5 kt, full barb 10 kt, pentad 50 kt

Example

In northern hemisphere easterly 15 kt

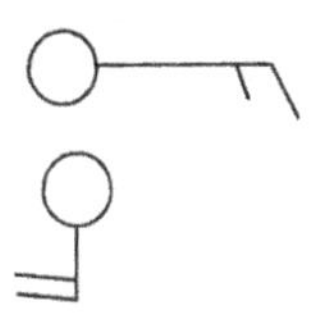

Southerly 20 kt

Southwesterly 45 kt

Westerly 60 kt

Northwesterly 6 kt

(code 99 =) variable 04 kt

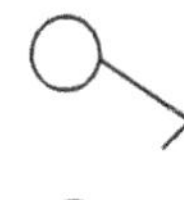

Southeasterly 10 kt

Southerly 20 kt

3. Plotting cloud symbols for C_L, C_M, C_H (8 N_h, C_L C_M C_H)

	C_L	C_M	C_H
0			
1			
2			
3			
4			
5			
6			
7			
8			
9			

4. Plotting symbols of past weather W

Code	0	1	2	3	4	5	6	7	8	9
Symbol	not	not	not		$\equiv$	,	●	*	∇	R
	used	used	used	Sand/dust storm	fog	Drizzle	Rain	Snow	Showers	Thunder storm

5. Plotting symbols for present weather ww

WW	0	1	2	3	4	5	6	7	8	9
00										
10										
20										
30										
40										
50										
60										
70										
80										
90										

Code ww (00 - 99)

00 – 49 :   When there is no pptn at the station at the time of observation.

50 – 99 :   There is pptn at the station at the time of observation.

Details

01, 02, 03, No change in the state of sky.

04   –   Visibility reduced due to smoke

05   –   Haze

06   –   Widespread dust suspended in the air

07   –   For land station : Dust/Sand raised by wind at the time of observation.

For Ship : Blowing spray at the observation site.

08   –   Well developed dust/sand whirl(s) observed at the station or nearby.

09   –   Dust/sandstorm at the station or nearby at the time of observation or during the preceding hour.

10   –   Mist

11   –   Patches of shallow fog

12   –   Roughly continuous shallow fog

13   –   Fog only lightning visible

Code 14 – 16   pptn within sight, but not at the station

17   –   Thunderstorm without pptn at the time of observation

18   –   Squall at/within sight at the time of observation or during preceding hour.

19   –   Funnel cloud at/within sight of the station at the time of observation or during the preceding hour.

Code 20 – 29   Refers to pptn, fog, ice fog or thunderstorm at the station during preceding hour.

These codes are not used where there is pptn at the time of observation.

28   –   Fog or ice fog.

29   –   Thunderstorm without pptn.

Code 30 – 39   Refers to duststorm, sand storm, drifting or blowing snow at the station at the time of observation.

30 – 35   Dust and sand storm.

36 – 39   Drifting or blowing snow.

Code 40 – 49   Refers to fog or ice fog at the time of observation.

40 – Fog or ice fog (not at the station).

41 – 47 Fog or ice fog.

48 – 49 Fog depositing rime.

Code 50 – 59 Refers to drizzle.

Code 60 – 69 Rain (at the station).

Code 70 – 79 Refers to solid precipitation not in showers.

Code 80 – 89 Showery pptn.

Code 90 – 99 Showery pptn with or without thunderstorm .

91 – 94 Thunderstorm during preceding hour.

95 – 99 Thunderstorm at the time of observation.

6. Past weather

  0 – 2 No significant change.

  3 Sand/dust storm or blowing snow.

  4 Fog/ice fog or thick haze.

  5 Drizzle.

  6 Rain.

  7 Snow or rain and snow mixed.

  8 Showers.

  9 Thunderstorm with or without pptn.

7. N_h All low clouds (C_L) sky coverage in oktas.

  $N_h = 9$ when the sky is not discernible

  If there are no low clouds, N_h should be reported by medium cloud coverage irrespective of their height.

  $N_h \leq N,$ C_L – Clouds are Sc, St, Cu, Cb

      C_M – Clouds are Ac, As, Ns

      C_H – Clouds are Ci, Cc, Cs

8. Symbols for pressure change 'a'.

a	0	1	2	3	4	5	6	7	8
	∧	⌐	/	✓	—	∨	∖	\	∧

9. N 0 Clear sky

   1 1 Okta or less

   2 2 oktas

3	3 oktas
4	4 oktas
5	5 oktas
6	6 oktas
7	7 oktas or more but not overcast
8	8 oktas overcast
9	Sky obscured/sky not discernible.

10. $h_s h_s$ Height of base of cloud layer indicated by C

The code is direct multiple of 30 for 0 – 50

hs hs		
00		< 30 m
01		30 m
..		..
..		..
..		..
20		600 m
..		..
..		..
..		..
50		1500 m

51 – 55   not used

56 – 80   Increments of 300 m after substracling 50

$h_s h_s$ = 60   (60 – 50) × 300 = 3000 m

= 70   (70 – 50) × 300 = 6000 m

= 80   (80 – 50) × 300 = 9000 m

81 – 88   Increments of 1500 m from code 80

$h_s h_s$ = 81   (80 – 50) × 300 + (81 – 80) 1500 = 9000 + 1500

= 10500 m

= 82   9000 + (82 – 80) 1500 = 12000 m

= 86   9000 + 8 × 1500 = 21000 m

= 89   is used if the height is greater than 21000 m

Code figures 90 – 99 used under regional instructions.

Tables of meteors and their symbols (same as on Forecast form)

11. C Genus of cloud/type of cloud.

Code fig.	Genus	Symbols
0	Cirrus	
1	Cirricumulus	
2	Cirrostratus	
3	Altocumulus	
4	Altostratus	
5	Nimbostratus	
6	Stratocumulus	
7	Stratus	
8	Cumulus	
9	Cumulonimbus	
/	Cloud not discernible due to darkness, fog, dust/sand storm etc.	

12. Horizontal Visibility VV

   Code 00 – 50   Direct reading in units of 100 meters

   e.g.,   05 stands for 5 × 100 = 500 m

   12 stands for 12 × 100 = 1200 m

   50 stands for 50 × 100 = 5000 m

   51 – 55 not used.

   56 – 80   Substract 50, the remaining figure direct reading in Km

   e.g.,   65 stands for 65 – 50 = 15 Km

   70 stands for 70 – 50 = 20 Km

81 – 89	Upto 80 is 30 Km, thereafter increments in 5 Km

$$e.g., \quad 81 \text{ stands for } 30 + (81 - 80) \times 5$$
$$= 30 + 5 = 35 \text{ Km}$$
$$85 \text{ stands for } 30 + (85 - 80) \times 5 = 55 \text{ Km}$$
$$89 \text{ stands for } 30 + 9 \times 5 = 75 \text{ Km}$$

13. Height of base of lowest clouds 'h' (agl = above ground level)

Code fig.	Height in meters
0	0 – 50
1	5 – 100
2	100 – 200
3	200 – 300
4	300 – 600
5	600 – 1000
6	1000 – 1500
7	1500 – 2000
8	2000 – 2500
9	2500 or more or no clouds
/	Height of cloud base not known.

14. i_3

Code fig.	Specification
0	Wind speed estimated
1	Wind speed taken from anemometer, recorded in meters per second.
3	Wind speed estimated, recorded in knots.
4	Wind speed taken from anemometer and recorded in knots.

Aviation Weather Codes

Aviation weather codes are used for the exchange of observed and forecast weather information for aviation services. The codes have been envolved by the WMO (World Meteorological Organisation) in consultation with the ICAO (International Civil Aviation organisation). Each code form bears number preceded by the letters FM (Forecasting manual of WMO). Here the codes are given as used by IMD, which closely follows the manual on codes WMO. No. 306

We shall learn some of the important codes.

METAR is Aviation routine weather report (with or without trend forecast).

Code Form

$$\left.\begin{array}{l} \text{METAR} \\ \text{or} \\ \text{SPECI} \end{array}\right] \quad \text{CCCC YY GG gg z} \quad (\text{Auto}) \text{ ddd ff G } f_m f_m \left[\begin{array}{l} \text{KT or} \\ \text{MPS or} \\ \text{KMH} \end{array}\right] \quad d_n d_n d_n V d_x d_x d_x$$

$$\left[\begin{array}{l} \text{VVVV } D_V \\ \text{or} \\ \text{CAVOK} \end{array}\right. \quad V_x V_x V_x V_x D_v \left[\begin{array}{l} R\, D_R D_R / V_R V_R V_R V_R I \\ \text{or} \\ R\, D_R D_R / V_R V_R V_R V_R V . V_R V_R V_R V_R I \end{array}\right] \quad w' \; w' \left[\begin{array}{l} N_s N_s N_s h_s h_s h_s \\ \text{or VV///} \\ \text{or} \\ \text{SKC} \\ \text{or} \\ \text{NSC} \end{array}\right.$$

$$T'T'/T_d' T_d' \qquad Q P_H P_H P_H P_H \qquad RE\ W'W' \left[\begin{array}{l} \text{WS RWYD}_R D_R \\ \text{or} \\ \text{WS ALL RWY} \end{array}\right.$$

$$W\, T_s T_s / SS \qquad (R_R R_R E_R e_R e_R B_R B_R)$$

$$\left[\begin{array}{l} (\text{TTTTT} \\ \text{or} \\ \text{NOSIG)} \end{array}\right. \quad \text{TTGGgg ddd ff G} f_m f_m \left[\begin{array}{l} \text{KMH or} \\ \text{KT or} \\ \text{MPS} \end{array}\right. \left[\begin{array}{l} \text{VVVV} \\ \text{or} \\ \text{CAVOK} \end{array}\right. \left[\begin{array}{l} W' \; W' \\ \text{or} \\ \text{NSW} \end{array}\right. \left[\begin{array}{l} N_s N_s N_s h_s h_s h_s \\ \text{or} \\ \text{VV///} \\ \text{or} \\ \text{SKC} \\ \text{or} \\ \text{NSC} \end{array}\right.$$

$$(\text{RML} \ldots\ldots)$$

Explanation

1. METAR is the name of the code for an aviation routine weather report. SPECI is the name of the code for aviation selected special weather report. A METAR report and a SPECI report may have a trend forecast appended.

2. The groups contain a non-uniform number of characters. When an element or phenomena does not occur, the corresponding group or the extension of a group, is omitted from a particular report. The groups enclosed in brackets are used in accordance with regional or national decision. Groups may have to be repeated in accordance with the detailed instructions for each group.

3. The code form includes a section containing the trend forecast identified either by a change indicator (TTTTT = BECMG or TEMPO as required) or by the code word NOSIG.

General Regulations

METAR or SPECI has to be included at the beginning of an individual report, followed by the location indicator of the station and the time of observation in hours and minutes UTC followed by without a space, the letter indicator Z.

When there is deterioration in one weather element with simultaneous improvement in another element (e.g., lowering of clouds and improvement in visibility) a single SPECI report is sufficient.

Group CCCC

The identification of reporting station in every report has to be with ICAO location indicator.

e.g., VABB (Bombay airport), VOHY (Hyderabad), VECC (Calcutta) RJIT (Tokyo Homchul. Japan), ASSY (Sydney, N.S.W. Australia), EGIL (Heathrow London), LFPO (Paris/Orly), RPMM (Manila), EDDW (Breman), EHAM (Amsterdam Schiphol).

Group YY GG gg Z

Date, time of observation in hours and minutes UTC followed by without a space the letter Z indicator in each METAR report.

In SPECI reports, this group will indicate the time of occurrence of the changes, which required to issue of the report.

e.g., 200925 Z   Date 20th, time of observation   0925 UTC

Code Word (AUTO)

The code word AUTO is optional. It is used before wind group, which indicates that the report is fully automated observations without human intervention. Any element not observed is reported by appropriate solidi e.g., in case of visibility four solidi, present weather two etc.

Groups

$$dddff\ Gf_m f_m \begin{bmatrix} KT\ or \\ MPS\ or \\ KMH \end{bmatrix} d_n d_n d_n V d_x d_x d_x$$

The mean wind direction (ddd) in terms of degrees (average wind over 10 minutes preceding observation) from which it is blowing, speed (ff) followed by (without a space) one of the abbreviations KT, MPS or KMH depending on reporting wind speed in Knots, meters per second or kilometers per hour. Wind is to be reported from true north (360 °).

Ex : 31008 KT, in case of gust it is reported as 30015G28 Kt.

In case of variable wind direction ddd is to be encoded as VRB when the wins speed is 3 Knots or less

e.g.,   VRBO3 KT (variable with mean wind speed 3 Knots)

VRB12 KT implies speed is 12 Knots but the wind direction change is $180°$ or more.

In case direction change is more than $60°$ but less that $180°$ and direction change is clockwise, this is to be reported as e.g., 300 V080 implies. Clockwise variation of wind from $300°$ to $080°$. Calm wind is reported as 00000 KT (or MPS or KMH).

Note : ddff – is the mean direction and wind speed taken over a period of 10 minutes immediately preceding of the report prepared. The direction of the wind is reported in terms of degrees. Thus the third figures is always zero except in case of variable wind reported as 999 followed by wind speed.

Gusting

If the maximum gusting wind exceeds the mean speed by 10 KT (5 ms^{-1} or 20 KMH) or more then G $f_m f_m$ reported immediately after ddd ff G $f_m f_m$, otherwise G $f_m f_m$ is not included.

e.g., 12015 G 28 KT

Groups VVVV $D_V V_x V_x V_x V_x D_V$

VVVV   :   Surface horizontal visibility in meters, kilometers as per ICAO specifications.

When horizontal visibility not markedly different in different directions, only VVVV is to be reported and D_V is to be dropped e.g., 4500

Directional variation in visibility not considered unless the variation is 50% or more of the minimum visibility. This is not required when the minimum visibility value is 5000 meters or more.

DV   :   Direction of observation/consists of the letters of eight points of the compass N, NE, E, SE, S, SW, W, NW.

When visibility is not the same in all directions and if the lowest visibility is observed in more than one direction then D_v represent the most operationally significant direction.

e.g., 3000 NE (visibility 3000 meters of the northeast).

$V_x V_x V_x V_x D_V$: Directional variation of visibility.

When the minimum visibility is less than 1500 meters while the visibility in another direction is more than 5000 meters, the group $V_x V_x V_x V_x \, D_V$ is to be reported with the value and direction of maximum visibility.

e.g., 1400 SE 6000 N (Visibility one thousand four hundred meters to the south east and six thousand meters to the north.

Reporting Visibility Steps are as Follows

(i)   Upto 800 meters reported rounding down to the nearest 50 meters.

e.g., 670 and 690 to be reported as 650.

(ii) Between 800 to 5000 meters reported rounding down to the nearest 100 meters

e.g., 3770 and 3730 is to be reported as 3700 m.

(iii) Between 5000 meters and 9999 meters reported round down to the nearest 1000 meters.

e.g., 6900 meters, 6400 meters to be reported as 6000 meters

(iv) 10000 meters or 10 Km or more report 9999.

e.g., 9999 indicates visibility 10 Km or more.

CAVOK is reported when the following simultaneous conditions exist.

(a) Visibility 10 Km or more

(b) No cloud with base 1500 meters or less or the highest minimum sector altitude whichever is greater.

(c) No CB

(d) No significant weather phenomena (like fog, smog, drizzle, rain, thunderstorm, squall, hail, dust/sandstorm, tornado, waterspout etc).

Groups $\left[\begin{array}{l} RD_R D_R / V_R V_R V_R V_R \, i \\ \text{or} \\ RD_R D_R / V_R V_R V_R V_R V \, V_R V_R V_R V_R i \end{array} \right.$

where $D_R D_R$ runway designator

$V_R V_R V_R V_R$ runway visual range

R indicator of runway

When the visibility is less than 1500 meters on runway this group has to be included in the report. The letter indicator R followed by runway designator $D_R D_R$ and to precede the RVR reports.

e.g., R27/1200 (RVR on runway 27 one thousand two hundred meters).

i – tendency of runway visual range

i = U stands for increasing tendency

i = D stands for decreasing tendency

i = N stands for no distinct change.

$V_R V_R V_R V_R$ i = Mean values and tendency of runway visual range over ten minutes period immediately preceding the observation.

In case of sudden sharp variation over short interval only the latest period mean value to be reported (the period to consider after sudden sharp change).

Note : A marked discontinuity occurs with sharp/abrupt change $RD_R D_R / V_R V_R V_R V_R$ V $V_R V_R V_R V_R$ i – significant variation in Runway.

In case of significant variation visual range in RVR, the one-minute mean extreme values if values by more than 50 meters or more than 20% of the mean value (which ever is greater) the one minute mean minimum and the one minute mean maximum values is to be reported in the form $RD_R D_R /$ $V_R V_R V_R V_R$ V $V_R V_R V_R V_R$ instead of the 10 minute mean. Extreme RVR values shall be reported.

Extreme Values of Runway Visual Range

$PV_R V_R V_R V_R$:   When RVR is greater than the maximum value $PV_R V_R V_R V_R$ reported. When RVR is to be more than 1500 meters it shall be reported as P1500.

e.g., R27/P1500 (RVR on Runway 27, greater than one thousand five hundred meters)

$MV_R V_R V_R V_R$:   When RVR is below the minimum value $MV_R V_R V_R V_R$ reported. When the RVR is assessed to be less than 50 meters it shall be reported as M 0050.

e.g., R09/M 0150 (RVR on runway 09, less then one hundred fifty meters.

Group W'W' :   Significant present and forecast weather one or more groups of W'W' but not exceeding three to be reported in respect of present weather phenomena.

The	groups to be ordered as follows		
	– Ra	is light rain	– SH light showers
	RA	moderate rain	SH moderate showers
	+ RA	heavy rain	+ SH heavy showers
–	light	(– is qualifier)	+ SHRA (heavy shower rain)
	moderate	(no qualifer)	
+	heavy	(+ is qualifer)	

After the qualifer for intensity the abbreviation for the observed weather phenomena or combinations are used as shown above.

Intensity shall be indicated only with precipitation, associated with showers and/or thunderstorms, blowing dust/sand or snow, dust storm or sandstorm.

The intensity of present weather phenomena ($W'W'$) shall be determined by the intensity at the time of observation.

If more than one form of precipitation is observed the appropriate letter abbreviations to be combined in a single group with dominant precipitation to be reported first.

e.g., light drizzle and fog be encoded as DZ FG. Moderate rain and snow dominant) should be encoded SNRA.

The qualifier SH shall be used for shower type precipitation.

The qualifer TS shall be used for thunderstorm. The qualifier FZ shall be used only yo indicate supercooled water droplets or supercooled precipitation.

An fog consisting mostly of water droplets at temperature below (zero °C) 0 °C shall be reported as freezing fog (FZ FG).

Criteria for Issuing SPECI (selected special report)

SPECI is issued for significant change in (i) surface wind, (ii) visibility, (iii) present weather (iv) cloud base height of 5 oktas or more and (v) surface temperature. The details of the criteria are given below.

Surface Wind

When mean surface wind speed with respect to the latest (preceding) Met. observation when mean surface wind speed changed by 10 KT or more routine Met observation changed by 10 kt or more, (ii) mean wind direction changed by 60° or more along with mean wind speed before and/or after change being 10 kt or more. (iii) variation in mean surface wind speed of 15 kt in gusts exceeds by 10 kt or more (to 25 kt or more).

Surface Visibility

With respect to the latest (preceding routine Met report visibility changes to or passes to 5000 m, 3000m, 800m.

RVR (Runway Visual Range)

RVR issued when visibility which is measured by transmission is reduced to 1500 m or less. With respect to the latest routine Met report RVR changes to or passes to 800 m 600 m and 350 m.

Present Weather

With respect to the latest routine met report onset cessation or change in intensity of any of the following weather phenomena or their combination occur.

(i) freezing pptn (ii) freezing fog (iii) moderate or heavy pptn, (iv) low drifting dust/sand/snow and (iv) blowing dust/sand/snow (v) sandstorm (vi) Duststorm (vii) thunderstorm (with or without pptn) (viii) Squall (ix) funnel cloud (tornado or water spout).

Significant Amount Cloud Base

With respect to the latest Met report height of the base of significant low cloud of 5 oktas or more changes to or passes 450 m, 300 m, 150 m, 60 m, 30 m (1500 ft, 1000 ft, 500 ft, 200 ft, 100 ft).

Surface Air Temperature

With respect to the latest met report if air temperature increased by 2 °C or more.

SPECI to be issued both for deterioration/improvement. If deterioration of one weather element accompanied by improvement in other element, a single selected special report is issued and treated as deterioration report.

Landing Forecast/Trend Forecasts

Trend type of landing forecasts are issued in India.

A brief expected significant changes in wind, visibility, present weather and clouds given METAR/SPECI is called Trend forecast or landing forecast. Trend forecast is valid for two hours from the time of issue of METAR/SPECI to which it is appended. The change indicaters are BECMG; TEMPO used with abbreviations FM (for from), TL (for till) or AT (for at) which are followed by time group; however AT is not used with TEMPO. In case no change expected during next two hours trend is given by NOSIG.

Probability occurrences of forecast elements is also indicated by "PROB" followed by probability percent (in tens) and time period during which it is expected to occur.

e.g., 1500 PROB 30 1214 0800 FG this implies visibility. 1500 m reduces to 080 m in fog during 12 to 14 UTC with probability of 30%. In aviation forecast probability of 30 or 40% are used less than this is not relevant.

For 50% or more is indicated by BCMG or TEMPO.

Examples of METAR and SPECI reports appended with trend.

Surface Wind

METAR   VOHY   0310Z   25010KT   8000

SCT020 SCT 100   28/16   Q 1004   BECMG

TL0330 250 25 KT

SPECI   VOHY   0335Z   25025G40 KT 8000

SCT020 SCT 100 28/16   Q1004   BECMG FM0440

22010KT

Visibility

METAR VABB   0240 Z 00000 KT   6000

SCT015 25/20   Q1008   TEMPO TLO340

2000 BR

SPECI   VABB   0300Z   0000KT 4000HZ   SCT 015   25/20

Q1008   TEMPO TL0340 2000 BR

SPECI   VABB   0325Z   00000 KT 1400 BR   R27/P1500

SCT015 25/20 Q1008   BECMG 0400 4000

Present Weather

SPECI   VOMM 0905 Z 22015 KT   4000

TSRA SCT020 SCT025CB 26/25   Q1008   BECMG NSW

METAR VECC   0950 Z 18010KT   6000 SCT025

FW 030CB 26/22   Q1009   TEMPO TL 1100 15025 + TSRA

Clouds

METAR VOTP 0340Z 09005 KT   6000 BKN012

OVC 100   24/22   Q1008   TEMPO TL 0430 RA

BKN 008   OVC 080

SPECI   VOVZ0400 Z 27005KT 6000 RA BKN 008

OVC100 24/22   Q1006   TEMPO 4000 +RA

Aerodrome Forecast

Code Form

TAF   CCCC   YYGGgg Z $Y_1Y_1 G_1G_1 G_2G_2$ ddd ff G $f_m f_m$ $\begin{bmatrix} KT \\ Mps \\ KMP \end{bmatrix}$

$\begin{bmatrix} VVVV \\ or \\ CAVOK \end{bmatrix}$ $\begin{bmatrix} W'W' \\ or \\ NSW \end{bmatrix}$ $\begin{bmatrix} N_sN_sN_s\ h_sh_sh_s \\ or \\ VV////\ \ or \\ SKC\ or\ NSC \end{bmatrix}$ $PROBC_2C_2$ $GG\ G_eG_e$

$\begin{bmatrix} TTTTT & GG\ G_eG_e \\ or \\ TT\ GG\ gg \end{bmatrix}$ $(T \times T_FT_F/G_FG_FZ\ \ TN\ T_FT_F/\ G_F\ G_FZ)$

Explanation

TAF : is the name of the code for an aerodrome forecast.

CCCC : identification of the station.

YY GG gg Z : date (YY), time of origin GG (UTC hrs) gg (minutes)

$Y_1Y_1 G_1G_1 G_2G_2$: date (Y_1Y_1) time of validity G_1G_1 to G_2G_2 hours UTC.

ddd ff G $f_m f_m$: wind direction (ddd) speed (ff), gusting (G) wind speed $(f_m\ f_m)$

VVVV : lowest surface visibility in meters

W'W' : present weather. Significant weather same as in METAR

NSW : no significant weather

$N_sN_sN_s\ h_sh_sh_s$: cloud amount $(N_sN_sN_s)$ [given FEW (few 1 to 2 oktas), SCT (scattered 3 to 4 oktas), BKN (Broken 5 to 7 oktas), OVC (overcast)]. $h_sh_sh_s$ height of base cloud in flight levels

CB TCU : to be reported as SCT020 CB or BKN025 TCU etc.

VV/// : when sky is expected to be obscured, cannot forecast in lieu of $N_sN_sN_s\ h_sh_sh_s$

SKC : clear sky expected

CAVOK : is used when expected (i) visibility in 10 Km or more (ii) No clouds below 1500 m, (iii) no CB and (iv) No significant weather.

TTTTT : change indicator of trend forecast (BECMG, TEMPO)

$G G G_e G_z$: change period GG to $G_e G_e$ (hrs)

PROB C_2C_2 : probability C_2C_2 generally 30% or 40%

For 50% or more BECMG or TEMPO to be used.

TX $T_F T_F$ / $G_F G_F$ Z : maximum temperature (TX)

TN $T_F T_F$ / $G_F G_F$ Z : minimum temperature (TN)

$T_F T_F$ expected at the time of $G_F G_F$ Z

Temperatures between –9 °C to + 9 °C should be preceded by 0; temperature below 0 °C be preceded by the letter M that is minus.

Amended aerodrome forecast be coded TAFAMD.

Example

TAF VOHY   170600Z   170918 29010 KT   6000
SCT020   SCT100 TEMPO 1015 FEW025TCU/CB

TAF VECC   170600Z 170918 12005 KT   500   HZ
FEW 020   SCT 100 BECMG 1112 16008 KT   4000 HZ
TEMPO 0918 1500 TSRA SCT010 FEW 025 CB BKN 090

TAF VIDP 170600 Z   170918 08010G 20KT 4000
HZ   SCT 030   SCT100 TEMPO 0912 1500 TSRA
FEW 030CB   BECMG 1315 06008KT   3000 HZ   TEMPO
1215 30030 KT   0800 DS/TSRA   FEW 060 CB

Route Forecast for Aviation

Code form :

Section 1   ROFOR   (YYGGggZ) $Y_1Y_1G_1G_1G_2G_2$ $\left[\begin{array}{l}\text{KMH or}\\ \text{KT or}\\ \text{MPS}\end{array}\right]$

CCCC $(QL_aL_aL_oL_o)$ CCCC OI_2zzz

(VVVV) $(W_1W_1W_1)$ $N_sCCh_sh_sh_s$ $7h_th_th_th_fh_fh_f$ $61_ch_ih_ih_it_L$

$5Bh_Bh_Bh_Bt_L$ $(4h_xh_xh_xT_hT_h$ $d_hd_hf_hf_hf_h)$ $(2h_ph_pT_pT_p)$

Section 2 $(11111$ $QL_aL_aL_oL_o$ $h'_j h'_j f_jf_jf_j)$

Section 3 $(22222$ $h'_m h'_m f_mf_mf_m (d_md_mVV))$

Section 4 $9i_snnn$

Notes

1. ROFOR is the name of the code for an aviation forecast in figure code prepared for a route between two specified aerodromes.

2. See notes (2) under FM 51 TAF.

3. The code form is divided into four sections as follows :

Section number	Symbolic figure group	Contents
1	–	Code identification and time groups; route forecast
2	11111	Jet-stream data (optional)
3	22222	Data of maximum wind ans vertical wind shear (optional)
4	–	Supplementary phenomena

Section 2, 3 and 4 are not transmitted separately.

Where

YY	Date
i_2	Zone indicator or route divided into sections.
VVVV	Visibility
W1W1W1	Forecast weather
CC	Genus of cloud
Ns	Amount of cloud
t_L	Thickness of layer
La La	Latitude in whole degrees
Lo Lo	Longitude in whole degrees
$h_f h_f h_f$	Altitude if freezing level (0 °C isotherm)
$h_i h_i h_i$	Height if lowest level of icing
$h_t h_t h_t$	Altitude of cloud layer or mass
$h_B h_B h_B$	Height if lowest level of turbulence
$h_x h_x h_x$	Altitude to which temperature and wind refers
$h_j' h_j'$	Height of the level of jet stream core
$h_p' h_p'$	Height of tropopause level
$h_m' h_m'$	Height of the maximum wind
$d_h d_h$	Direction of wind at height $h_x h_x$
$T_h T_h$	Temperature in whole degrees corresponding to the height $h_x h_x$
$f_h f_h f_h$	Wind speed at height $h_x h_x$

Speciman route forecast progcharts and significant weather charts given below.

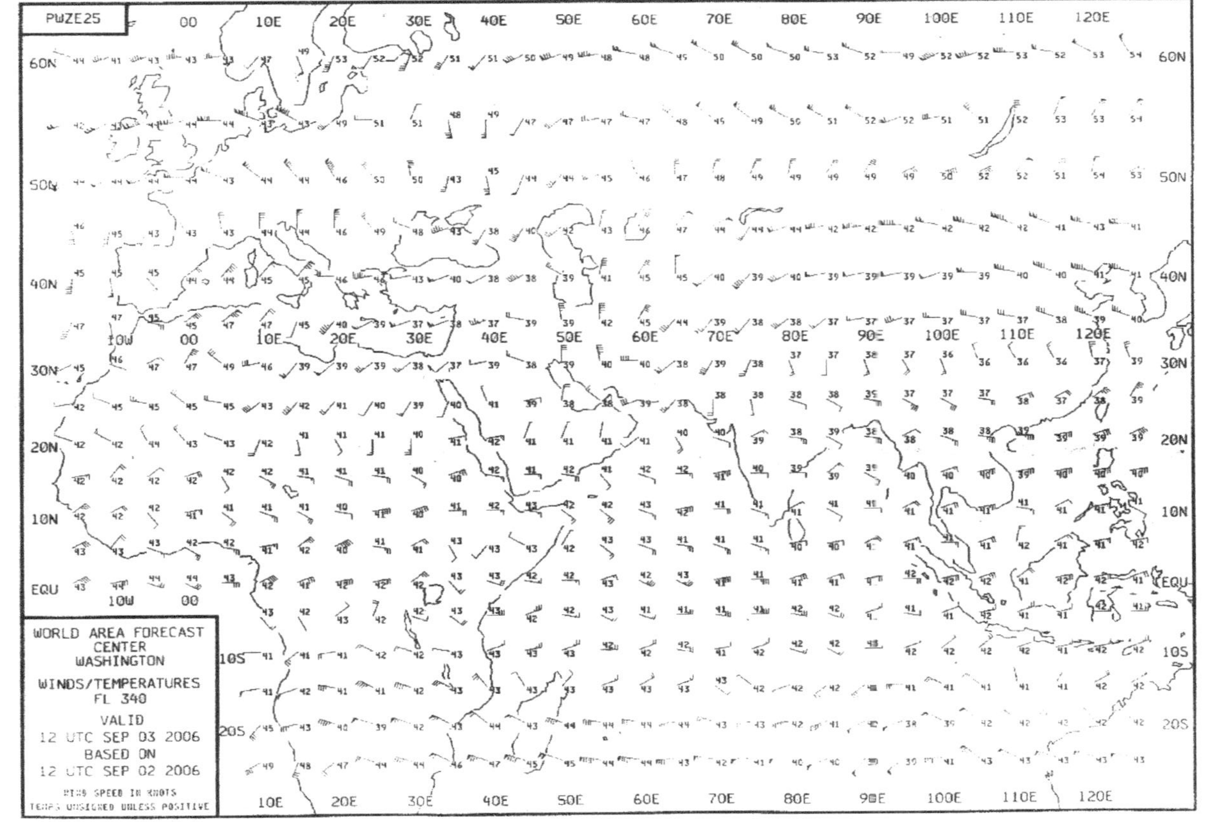
PWZE25
WORLD AREA FORECAST
CENTER
WASHINGTON
WINDS/TEMPERATURES
FL 340
VALID
12 UTC SEP 03 2006
BASED ON
12 UTC SEP 02 2006
WIND SPEED IN KNOTS
TEMPS UNSIGNED UNLESS POSITIVE

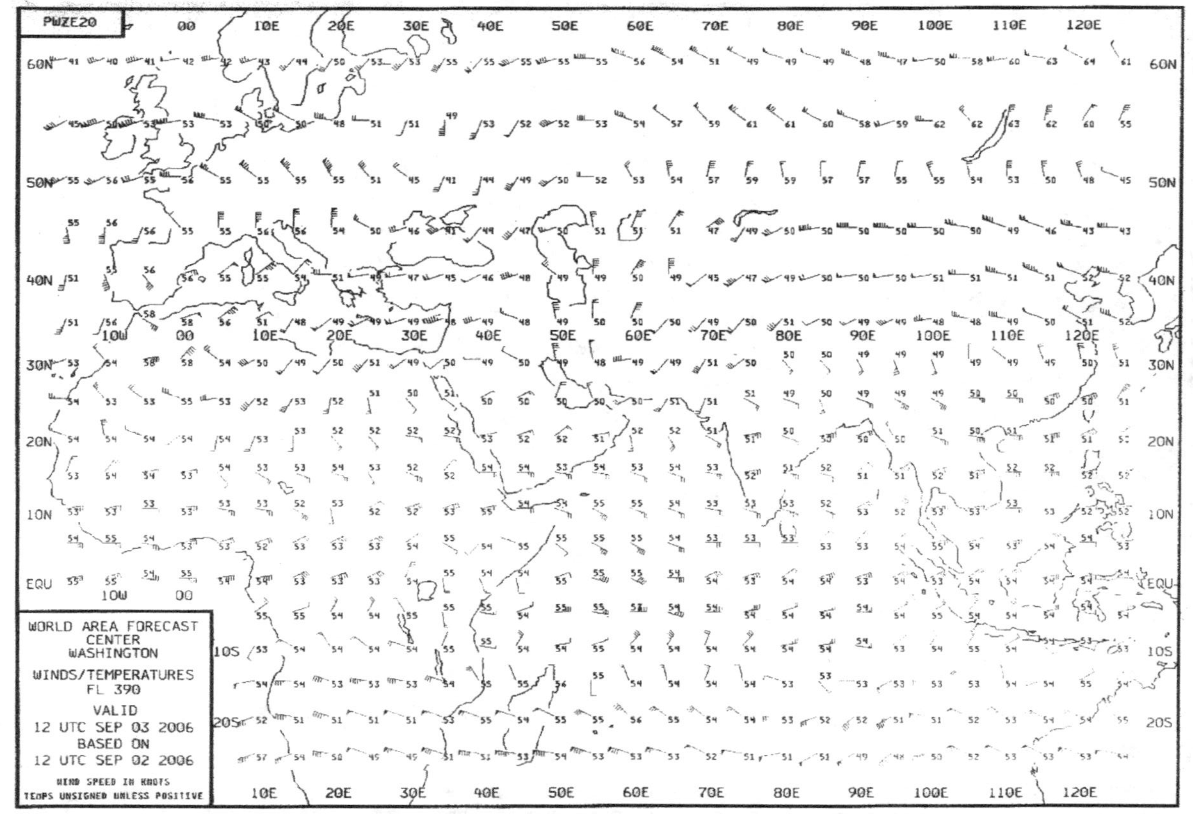
PWZE20
WORLD AREA FORECAST
CENTER
WASHINGTON
WINDS/TEMPERATURES
FL 390
VALID
12 UTC SEP 03 2006
BASED ON
12 UTC SEP 02 2006
WIND SPEED IN KNOTS
TEMPS UNSIGNED UNLESS POSITIVE

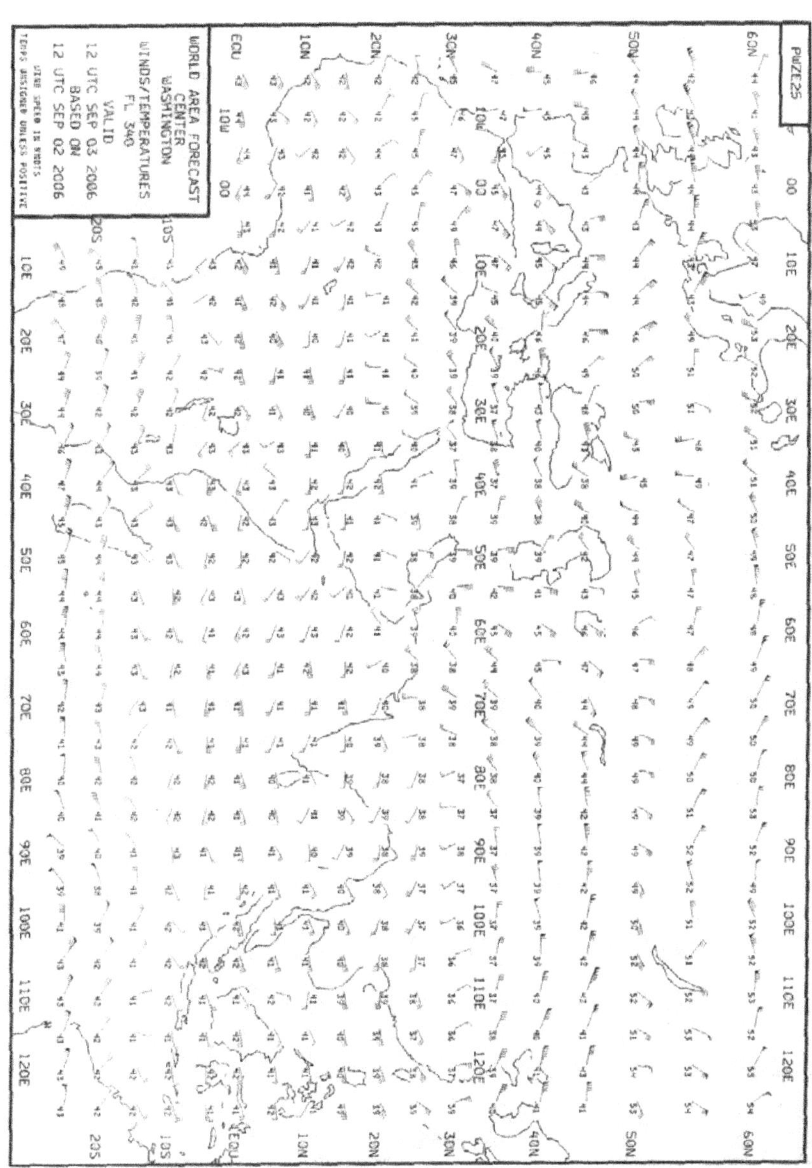
PWZE25
WORLD AREA FORECAST
CENTER
WASHINGTON
WINDS/TEMPERATURES
FL 340
VALID
12 UTC SEP 03 2006
BASED ON
12 UTC SEP 02 2006
WIND SPEED IN KNOTS
TEMPS ASSUMED NEGATIVE UNLESS POSITIVE

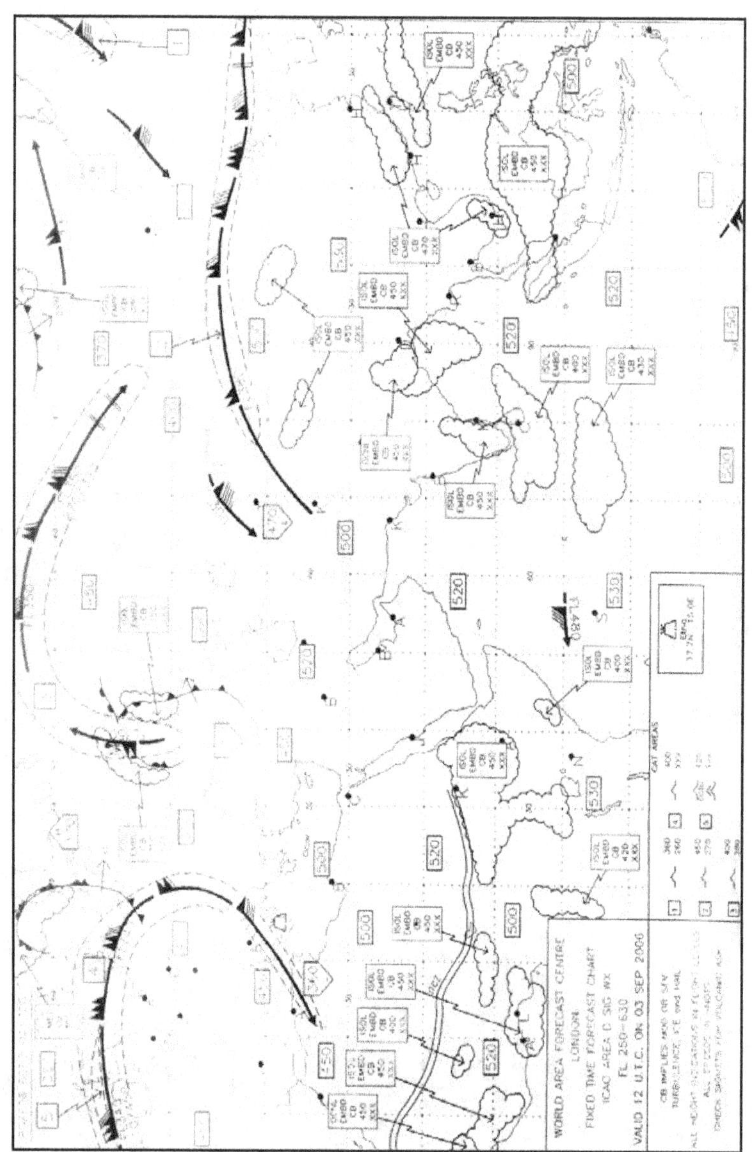

Significant Present and Forecast and Forecast Weather

Qualifier		Weather Phenomena		
Intensity or Proximity	Descriptor	Precipitation	Obscuration	Other
– Light **Moderate** (no qualifier) + Heavy {well developed in the case of dust/ sand whirls (dust devils) and funnel clouds} VC In the Vicinity	**MI** Shallow **BC** Patches **PR** Partial (covering part of the aerodrome **DR** Low drifing **BL** BLowing **SH** Showers (s) **TS** Thunder- storm **FZ** Freezing (super cooled)	**DZ** Drizzle **RA** Rain **SN** Snow **SG** Snow- grains **IC** Ice Crystals (diamond dust) **PL** Ice Pellets **GR** Hail **GS** Small Hail and/or snow pellets	**BR** Mist **FG** Fog **FU** Smoke **VA** Volcanic ash **DU** Wide spread dust **SA** Sand **HZ** Haze	**PO** Dust/sand whirls (dust devils) **SQ** Squalls **FC** Funnel clouds(s) (tornado or water spout) **SS** Sandstorm **DS** Duststorm

Clouds

Type

CI	–	Cirrus	**SC**	–	Stratocumulus
CC	–	Cirrucumulus	**ST**	–	Stratus
CS	–	Cirrostratus	**CU**	–	Cumulus
AC	–	Altocumulus	**TCU**	–	Towering culmulus
AS	–	Altostratus	**CB**	–	Cumulonimbus
NS	–	Nimbostratus	**LYR**	–	Layer or Layered

Amount

Clouds except CB

SKC	–	Sky clear (0/8)
FEW	–	Few (1/8 to 2/8)
SCT	–	Scattered (3/8 to 4/8)
BKN	–	Broken (5/8 to 7/8)
OVC	–	Overcast (8/8)

CB only

ISOL	–	Individual CBs (isolated)
OCNL	–	Well separated CBs (occasional)
FRQ	–	CBs with little or no separation (frequent)
EMBD	–	CBs embedded in layers of other clouds or concealed by haze (embedded)

Weather Symbols

Thunderstorm

Tropical cyclone

Severe squall line

Moderate turbulence

Severe turbulence

Mountain waves

Slight aircraft icing

Moderate aircraft icing

Severe aircraft icing

Widespread fog

Hail

Volcanic eruption

Visible ash cloud

Mountain obstruction

Drizzle

Rain

Snow

Shower

Widespread blowing snow

Severe sand or dust haze

Widespread sandstorm or duststorm

Widespread haze

Widespread mist

Widespread smoke

Freezing precipitation

Clear air turbulence

Radioactive materials in the atmosphere

Fronts, Convergence Zones and Other Symbols

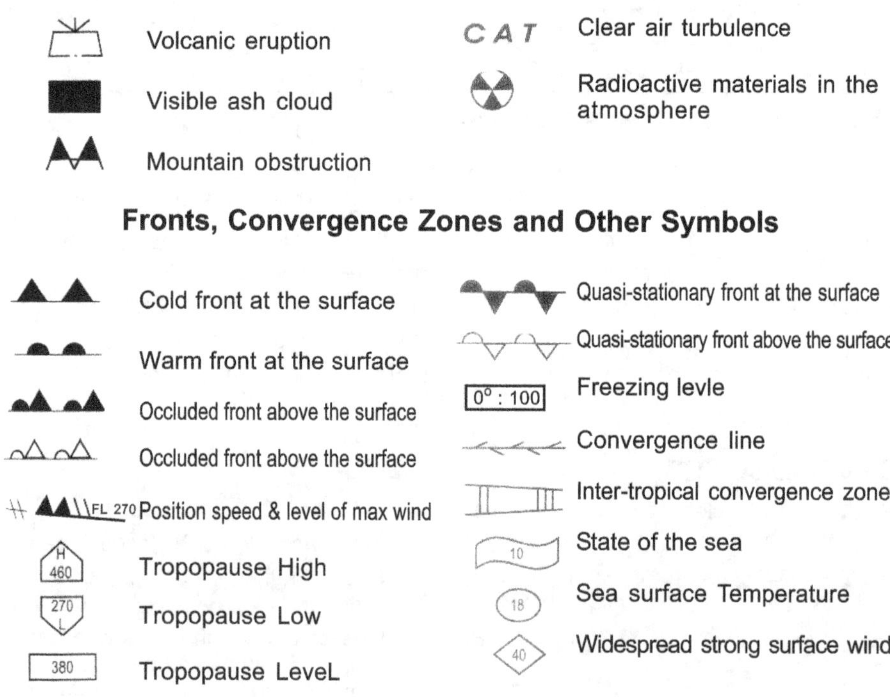

Cold front at the surface

Warm front at the surface

Occluded front above the surface

Occluded front above the surface

Position speed & level of max wind

Tropopause High

Tropopause Low

Tropopause LeveL

Quasi-stationary front at the surface

Quasi-stationary front above the surface

Freezing levle

Convergence line

Inter-tropical convergence zone

State of the sea

Sea surface Temperature

Widespread strong surface wind

- Wind arrows indicate the aximum wind in jet and the fight level at which it occurs. Significant chnages (speed of 20 Knots or more, 3000 ft less if practicable in flight level) are marked by the double bar. in the example at the double bar the wind speed is 225 km/h (120 kt).
- The heavu line delineating the jet axis begind/ends at the points where a wind speed of 150 km/h (80 kt) is forecast.
- This symbol refers to widespread surface wind speeds ecvedding 60 km/h (30 kt).

Boundaries

Boundaries of Significant Weather

CAT Boundary

References

SAARC-Training seminar cum workshop, *on monsoon forecasting January 8-19, 1990, IMD.*

Compendium lecture notes for class III meteorological persons.

Compendium of meteorology *Vol-1 WMO No. 364*

Tropospheric chemistry and Air Pollution *WMO TN . 176 WMO No. 583.*

Weather analysis and forecasting *Vol-1 S. Pettersen, MC Graw-Hill book company INC New York, Toronto, London - 1956*

General climatology *Howard J. Critch field, Prentic Hall of India Pvt Ltd New Delhi - 1968*

Atmosphere, weather and climate *R.G. Barry R.J. Chorley, third edition– 1976 ELBS and Methuen and Co. ltd.*

Physical meteorology *S.L. Hess*

Compendium of meteorology *Vol-1 part - 2*

Compendium *Vol II, part -1 General Hydrology*

Drought and agriculture *WMO T.N. No. 138*

IMD – *Tracks of cyclonic storms 1891 – 1970.*

Introduction to Dynamic meteorology *JR. Holton Academy Press.*

Climatology of India and neighbourhood *Y.P. Rao, K.S. Ramamurthy FMU No. 1-2.*

Western Disturbances and associated weather *FMU No. III Y.P Rao, V. Srinivas.*

Northeast Monsoon *part-IV V.S. Srinivasam K. Ramamurthy.*

Summer Norwesters and Andhis and largescale.

Convective activity over peninsular and central parts of the country. Srinivasan K. Ramamurthy and Y.R. Nene.

WMO - No. 345 *one hundred years of International cooperation in meteorology 1873-1973.*

WMO - No. 113 TP 50 *Geneva, Weather and Food.*

WMO No. 143 TP 67 *Geneva 1964, Weather and Man.*

WMO - No. 624 *Geneva 1984 Meteorology Aids Food Production.*

WMO No. 204 TP 107 *Geneva 1966, Weather and Water.*

WMO - No. 220 TP 117 *Geneva 1967, Harvest from Weather.*

WMO No. 183 TP - 92, *Geneva 1966 World Weather Watch.*

WMO – 1956 *International cloud Altas*

WMO No. 735 *– Geneva 1990, the Atmosphere of the living planet earth.*

The world's water. *MI Lvovich, Mir. Publishers, Moscow – 1973.*

Our Planet the Earth. *A.V Byalko, Mir. Publisher Moscow – 1987.*

Origin and chemical evolution of the Earth. *G. Voitkevich, Mir. Publisher, Moscow 1988.*

A Planet of Riddles. *Translated by David sobolev, Mir Publishers Moscow.*

Satellite and Typhoon Eye-to-Eye, *S. Baibakov, A.Martynow Mir Publishers Moscow 1986.*

The Mystery of the Earth's Mantle. *Translated by David Sobolev, Peace Publishers Moscow.*

Tropical Meteorology. *G.C. Asnami 822, Sindh Colony Aundh, 1993, Pune - 411007 (India).*

Domestic Animals. Harbans Singh, *Revised by B.P.S. Puri, NBT India – 1997.*

Forest and Forestry. *K.P. Sagreiya, Revised by Dr. SS Negi NBT India - 1997.*

Modern Biology - *Trumam J. Moon, Paul B. Mann, James H.otto, Henry holt and company New York 1956.*

Environmental Science, *Purdom and Anderson.*

Thirteenth Indian expedition to Antarctica. *G. Sudhakar Rao IMD Technical Publication No. 11.*

"Storm surges in the Bay of Bengal" *J.R. Met. Society., 100, 437 - 449.Das, P.K., M.C. Sinha and V. Balasubramanyam 1974.*

Avalanche and Lanslide forecasting and mitigation Intromet-2004, *May S.S. Sharma, KC.VSM and Snehmani.*

Index

D

E

F

Features of CAT 530
Ferrel cell 230
FIR 543
Floods 310
Fog 142, 150, 153, 155
Fog dispersal 155
Fohn 126
Food chain 482, 483
Food webs 482
Forest products 503, 506
Forest soil 510
Freezing nuclei 169
Freezing rain 142
Frictional force 343
Front 183, 185, 251
Frontal waves 188
Frontogenesis 188
Frontolysis 188
Frost 103, 107, 143

G

Gas laws 41
Geese 515
General circulation of the atmosphere 184
Geological time scale 400
Geostationary satellite 357, 375
Geostrophic wind 69, 70
Giant nuclei 101, 164
Glaze 143
Gradient wind 71
Gram atom 430
Gram molecule 430
Gray body 23

Grazing 483
Green flash 146
Green house effect 433
Green lungs of the earth 29, 432, 443
Gustiness 114

H

Hadley cell 230
Hail 142
Hailstorm 216, 427
Halo 145
Hardness of water 468
　　temporary 468
　　permanent 468
Harmattan 128
Hatcheries 514
Haze 144
Heat waves 423
Heat-transport 26
Herbivories 482, 483
High volume sampler 459
Horizontal wind shear 521
Humidity 104, 109, 353, 458
Hurricane 208, 334
Hydrological cycle 464
Hydrometeors 141
Hydrostatic equation 160

I

ICAO Standard atmosphere 533
Igneous rocks 315
Incineration 476
Index of thermal efficiency 396
Infrasonic 478
INSAT 356, 359, 350, 382
Instability 162, 194, 354

9 789352 300389